GCSE

# chemistry

second edition

GCSE

# chemistry

second edition

B Earl and LDR Wilford

JOHN MURRAY

## International hazard warning symbols

You will need to be familiar with these symbols when undertaking practical experiments in the laboratory.

Corrosive.
These substances attack or destroy living tissues, including eyes and skin.

Harmful.
These substances are similar to toxic substances but less dangerous.

Irritant.
These substances are not corrosive but can cause reddening or blistering of the skin.

Oxidising.
These substances provide oxygen which allows other materials to burn more fiercely.

Toxic.
These substances can cause death.

Highly flammable.
These substances can easily catch fire.

First published in 1995
by John Murray (Publishers) Ltd, a member of the Hodder Headline Group
338 Euston Road
London NW1 3BH

Reprinted 1996, 1997, 1998, 1999, 2000 (twice)

Second edition 2001

Reprinted 2002, 2003 (twice)



Layouts by Fiona Webb
Artwork by Wearset
Cover design by John Townson/Creation
Typeset in 11½/13pt Bembo by Wearset, Boldon, Tyne and Wear
Printed in Great Britain by Butler and Tanner, Frome and London

A catalogue entry for this title can be obtained from the British Library

ISBN 0 7195 8616 X

# Contents

## 6 Electrolysis and its uses



## 7 Acids, bases and salts



## 8 Inorganic carbon chemistry

## 11 Rates of reaction



## 12 The petroleum industry



## 13 Energy sources

## 14 The (wider) organic manufacturing industry



## 15 Nitrogen

# Preface to the reader

This textbook has been written to help you in your study of chemistry to GCSE. Although you will be following a GCSE specification for only one particular examination group, this book contains the material needed by all the groups. For this reason it is not expected that you will need to study or learn everything in this textbook.

The different chapters in this book are split up into short topics. At the end of many of these topics are questions to test whether you have understood what you have read. At the end of each chapter there are larger study questions. Try to answer as many of the questions as you can as you come across them because asking and answering questions is at the heart of your study of chemistry.

A selection of examination questions, selected from examination papers published by the different examination groups, is included at the end of the book. In many cases they are designed to test your ability to apply your chemical knowledge. The questions may provide certain facts and ask you to make an interpretation of them. In such cases, the factual information may not be covered in the text.

To help draw attention to the more important words, scientific terms are printed in bold the first time they are used. There are also checklists at the end of each chapter summarising the important points covered.

This textbook will provide you with the information you need for your particular specification. We hope you enjoy using this book.

B Earl & LDR Wilford

We use coloured strips at the edges of pages to define different areas of chemistry:

- 'starter' chapters – basic principles
- physical chemistry
- inorganic chemistry
- organic chemistry and the living world.

# All about matter



Chemistry is about what **matter** is like and how it behaves, and our explanations and predictions of its behaviour. What is matter? This word is used to cover all the substances and materials from which the physical universe is composed. There are many millions of different substances known, and all of them can be categorised as solids, liquids or gases (Figure 1.1). These are what we call the **three states of matter**.

a solid

b liquid

c gas

**Figure 1.1** Water in three different states.

## *Solids, liquids and gases*

A **solid**, at a given temperature, has a definite volume and shape which may be affected by changes in temperature. Solids usually increase slightly in size when heated (**expansion**) (Figure 1.2) and usually decrease in size if cooled (**contraction**).

A **liquid**, at a given temperature, has a fixed volume and will take up the shape of any container into which it is poured. Like a solid, a liquid's volume is slightly affected by changes in temperature.

A **gas**, at a given temperature, has neither a definite shape nor a definite volume. It will take up the shape of any container into which it is placed and will spread out evenly within it. Unlike those of solids and liquids, the volumes of gases are affected quite markedly by changes in temperature.

Liquids and gases, unlike solids, are relatively **compressible**. This means that their volume can be reduced by the application of pressure. Gases are much more compressible than liquids.

**Figure 1.2** Without expansion gaps between the rails, the track would buckle in hot weather.

## *The kinetic theory of matter*

The **kinetic theory** helps to explain the way in which matter behaves. The evidence is consistent with the idea that all matter is made up of tiny **particles**. This theory explains the physical properties of matter in terms of the movement of its constituent particles.

The main points of the theory are:

- all matter is made up of tiny, moving particles, invisible to the naked eye. Different substances have different types of particles (atoms, molecules or ions) which have different sizes
- the particles move all the time. The higher the temperature, the faster they move on average
- heavier particles move more slowly than lighter ones at a given temperature.

The kinetic theory can be used as a scientific model to explain how the arrangement of particles relates to the properties of the three states of matter.

### Explaining the states of matter

In a solid the particles attract one another. There are attractive forces between the particles which hold them close together. The particles have little freedom of movement and can only vibrate about a fixed position. They are arranged in a regular manner, which explains why many solids form crystals.

**a** A model of a chrome alum crystal.

**b** An actual chrome alum crystal.

**Figure 1.3**

It is possible to model such crystals by using spheres to represent the particles (Figure 1.3a). If the spheres are built up in a regular way then the shape compares very closely with that of a part of a chrome alum crystal (Figure 1.3b).

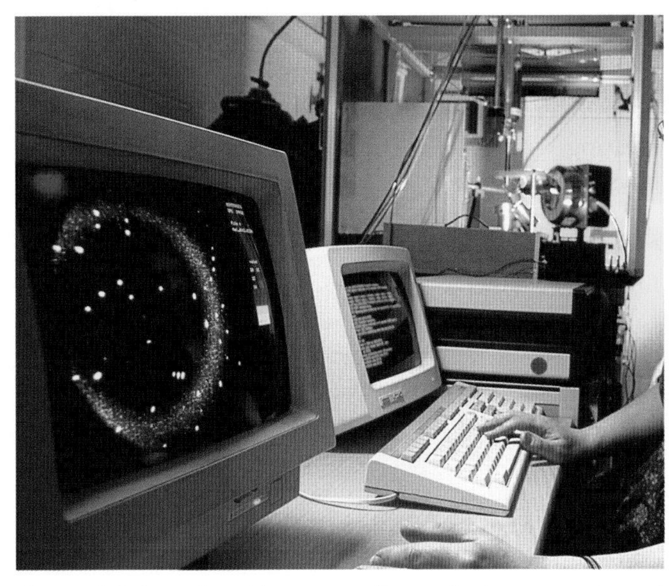

**Figure 1.4** A modern X-ray crystallography instrument, used for studying crystal structure.

Studies using X-ray crystallography (Figure 1.4) have confirmed how the particles are arranged in crystal structures. When crystals of a pure substance form under a given set of conditions, the particles present are always packed in the same way. However, the particles may be packed in different ways in crystals of different substances. For example, common salt (sodium chloride) has its particles arranged to give cubic crystals as shown in Figure 1.5.

**Figure 1.5** Sodium chloride crystals.

In a liquid the particles are still close together but they move around in a random way and often collide with one another. The forces of attraction between the particles in a liquid are weaker than those in a solid. Particles in the liquid form of a substance have more energy on average than the particles in the solid form of the same substance.

In a gas the particles are relatively far apart. They are free to move anywhere within the container in which they are held. They move randomly at very high velocities, much more rapidly than those in a liquid. They collide with each other, but less often than in a liquid, and they also collide with the walls of the container. They exert virtually no forces of attraction on each other because they are relatively far apart. Such forces, however, are very significant. If they did not exist we could not have solids or liquids (see Changes of state, p. 4).

The arrangement of particles in solids, liquids and gases is shown in Figure 1.6.

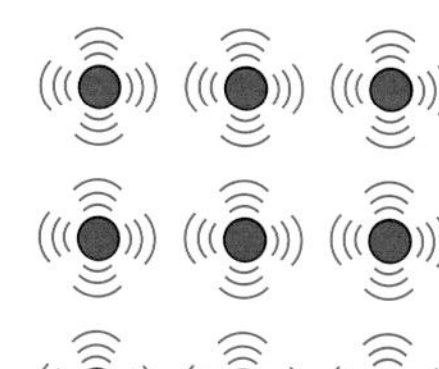

**solid**
Particles only vibrate about fixed positions. Regular structure.

**liquid**
Particles have some freedom and can move around each other. Collide often.

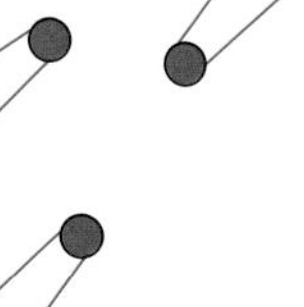

**gas**
Particles move freely and at random in all the space available. Collide less often than in liquid.

**Figure 1.6** The arrangement of particles in solids, liquids and gases.

## Questions

1. When a metal such as copper is heated it expands. Explain what happens to the metal particles as the solid metal expands.
2. Use your research skills on the Internet to find out about the technique of X-ray crystallography and how this technique can be used to determine the crystalline structure of solid substances such as sodium chloride.

## Changes of state

The kinetic theory model can be used to explain how a substance changes from one state to another. If a solid is heated the particles vibrate faster as they gain energy. This makes them 'push' their neighbouring particles further away from themselves. This causes an increase in the volume of the solid and the solid expands. Expansion has taken place.

Eventually, the heat energy causes the forces of attraction to weaken. The regular pattern of the structure breaks down. The particles can now move around each other. The solid has melted. The temperature at which this takes place is called the **melting point** of the substance. The temperature of a pure melting solid will not rise until it has all melted. When the substance has become a liquid there are still very significant forces of attraction between the particles, which is why it is a liquid and not a gas.

Solids which have high melting points have stronger forces of attraction between their particles than those which have low melting points. A list of some substances with their corresponding melting and boiling points is shown in Table 1.1.

**Table 1.1**

| Substance | Melting point/°C | Boiling point/°C |
|---|---|---|
| Aluminium | 661 | 2467 |
| Ethanol | −117 | 79 |
| Magnesium oxide | 2827 | 3627 |
| Mercury | −30 | 357 |
| Methane | −182 | −164 |
| Oxygen | −218 | −183 |
| Sodium chloride | 801 | 1413 |
| Sulphur | 113 | 445 |
| Water | 0 | 100 |

If the liquid is heated the particles will move around even faster as their average energy increases. Some particles at the surface of the liquid have enough energy to overcome the forces of attraction between themselves and the other particles in the liquid and they escape to form a gas. The liquid begins to **evaporate** as a gas is formed.

Eventually, a temperature is reached at which the particles are trying to escape from the liquid so quickly that bubbles of gas actually start to form inside the bulk of the liquid. This temperature is called the **boiling point** of the substance. At the boiling point the pressure of the gas created above the liquid equals that in the air – **atmospheric pressure**.

Liquids with high boiling points have stronger forces between their particles than liquids with low boiling points.

When a gas is cooled the average energy of the particles decreases and the particles move closer together. The forces of attraction between the particles now become significant and cause the gas to **condense** into a liquid. When a liquid is cooled it **freezes** to form a solid. In each of these changes energy is given out.

Changes of state are examples of **physical changes**. Whenever a physical change of state occurs, the temperature remains constant during the change (see Heating and cooling curves, opposite). During a physical change no new substance is formed.

## An unusual state of matter

**Liquid crystals** are an unusual state of matter (Figure 1.7). These substances look like liquids, flow like liquids but have some order in the arrangement of the particles, and so in some ways they behave like crystals.

Liquid crystals are now part of our everyday life. They are widely used in displays for digital watches, calculators and lap-top computer displays (Figure 1.8), and in portable televisions. They are also useful in thermometers because liquid crystals change colour as the temperature rises and falls.

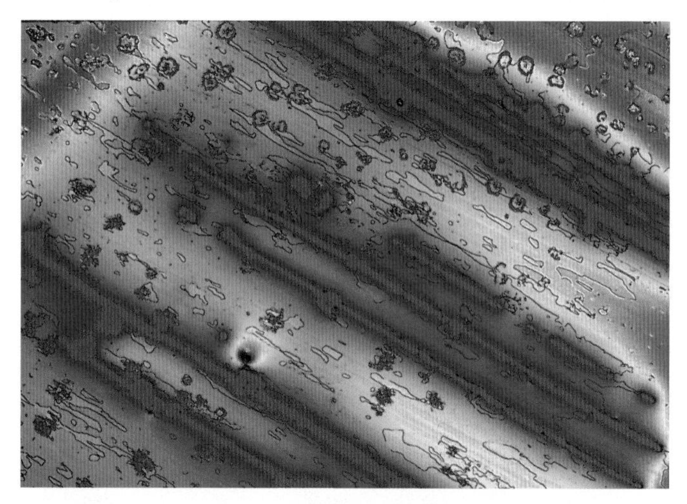

**Figure 1.7** A polarised light micrograph of liquid crystals.

**Figure 1.8** Liquid crystals are used in this computer display.

## An unusual change of state

There are a few substances that when they are heated change directly from a solid to a gas without ever becoming a liquid. This rapid spreading out of the particles is called **sublimation**. Cooling causes a change from a gas directly back to a solid. Examples of substances that behave in this way are carbon dioxide (Figure 1.9) and iodine.

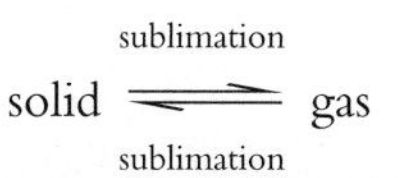

**Figure 1.9** Dry ice (solid carbon dioxide) sublimes on heating and can be used to create special effects on stage.

Carbon dioxide is a white solid called dry ice at temperatures below −78°C. When heated to just above −78°C it changes into carbon dioxide gas. The changes of state are summarised in Figure 1.10.

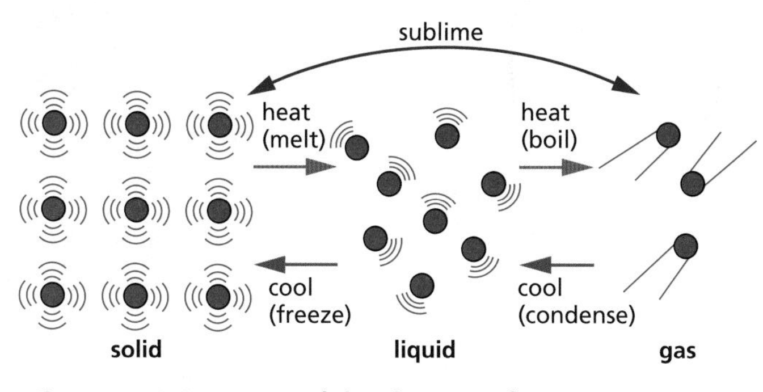


**Figure 1.10** Summary of the changes of state.

## Heating and cooling curves

The graph shown in Figure 1.11 was drawn by plotting the temperature of water as it was heated steadily from −15°C to 110°C. You can see from the curve that changes of state have taken place. When the temperature was first measured only ice was present. After a short space of time the curve flattens, showing that even though heat energy is being put in, the temperature remains constant.

In ice the particles of water are close together and are attracted to one another. For ice to melt the particles must obtain sufficient energy to overcome the forces of attraction between the water particles to allow relative movement to take place. This is where the heat energy is going.

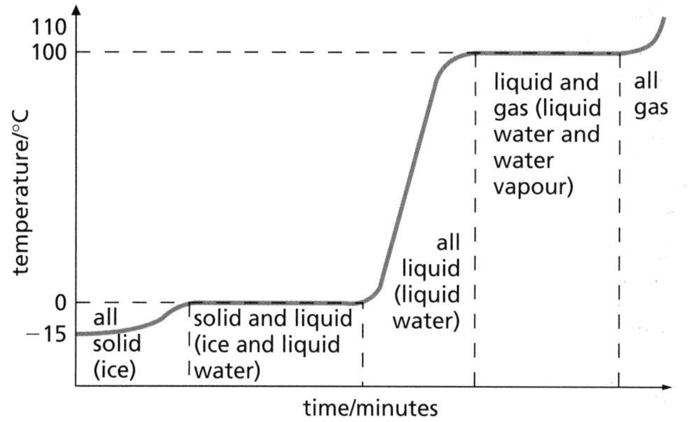


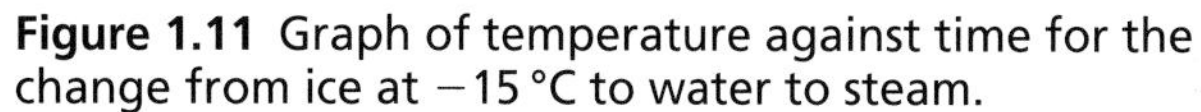

**Figure 1.11** Graph of temperature against time for the change from ice at −15 °C to water to steam.

The temperature will begin to rise again only after all the ice has melted. Generally, the heating curve for a pure solid always stops rising at its melting point and gives rise to a sharp melting point. The addition or presence of impurities lowers the melting point. You can try to find the melting point of a substance using the apparatus shown in Figure 1.12.

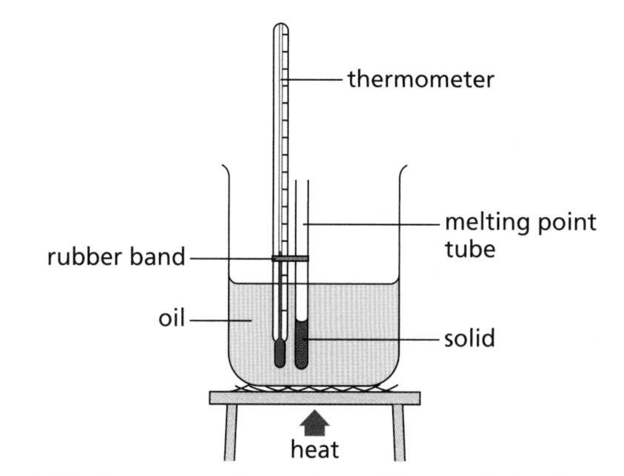


**Figure 1.12** Apparatus shown here if heated slowly can be used to find the melting point of a substance such as the solid in the melting point tube.

In the same way, if you want to boil a liquid such as water you have to give it some extra energy. This can be seen on the graph (Figure 1.11) where the curve levels out at 100 °C – the boiling point of water.

The reverse processes of condensing and freezing occur on cooling. This time, however, energy is given out when the gas condenses to the liquid and the liquid freezes to give the solid.

### *Questions*

1. Write down as many uses as you can for liquid crystals.
2. Why do gases expand more than solids for the same increase in temperature?
3. Ice on a car windscreen will disappear as you drive along, even without the heater on. Explain why this happens.
4. When salt is placed on ice the ice melts. Explain why this happens.
5. Draw and label the graph you would expect to produce if water at 100 °C was allowed to cool to a temperature of −5 °C.

## Diffusion – evidence for moving particles

When you walk past a cosmetics counter in a department store you can usually smell the perfumes. For this to happen gas particles must be leaving open perfume bottles and be spreading out through the air in the store. This spreading out of a gas is called **diffusion** and it takes place in a haphazard and random way.

All gases diffuse to fill the space available to them. As you can see from Figure 1.13, after a day the brown–red fumes of gaseous bromine have spread evenly throughout both gas jars from the liquid present in the lower gas jar.

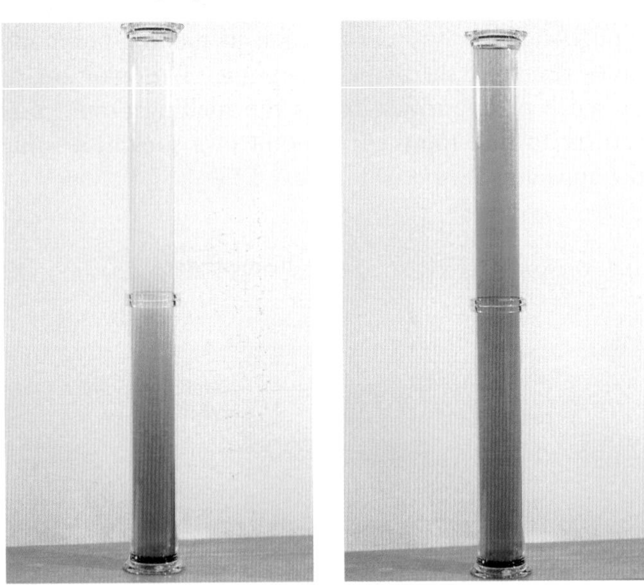

**Figure 1.13** After 24 hours the bromine fumes have diffused throughout both gas jars.

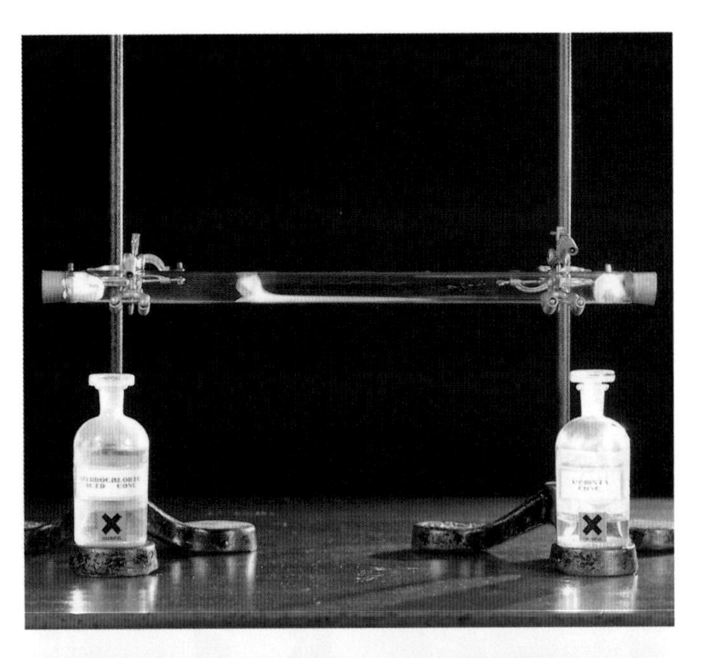

**Figure 1.14** Hydrochloric acid (left) and ammonia (right) diffuse at different rates.

Gases diffuse at different rates. If one piece of cotton wool is soaked in concentrated ammonia solution and another is soaked in concentrated hydrochloric acid and these are put at opposite ends of a dry glass tube, then after a few minutes a white cloud of ammonium chloride appears (Figure 1.14). This shows the position at which the two gases meet and react. The white cloud forms in the position shown because the ammonia particles are lighter than the hydrogen chloride particles (released from the hydrochloric acid) and so move faster. Generally, light particles move faster than heavier ones at a given temperature.

Diffusion also takes place in liquids (Figure 1.15) but it is a much slower process than in gases. This is because the particles of a liquid move much more slowly.

When diffusion takes place between a liquid and a gas it is known as **intimate mixing**. The kinetic theory can be used to explain this process. It states that collisions are taking place between particles in a liquid or a gas and that there is sufficient space between the particles of one substance for the particles of the other substance to move into.

a b

**Figure 1.15** Diffusion within nickel(II) sulphate solution can take days to reach the stage shown on the right.

## Brownian motion

Evidence for the movement of particles in liquids came to light in 1827 when a botanist, Robert Brown, observed that fine pollen grains on the surface of water were not stationary. Through his microscope he noticed that the grains were moving about in a random way. It was 96 years later, in 1923, that another scientist called Norbert Wiener explained what Brown had observed. He said that the pollen grains were moving because the much smaller and faster-moving water particles were constantly colliding with them (Figure 1.16a).

This random motion of visible particles (pollen grains) caused by much smaller, invisible ones (water particles) is called **Brownian motion** (Figure 1.16b), after the scientist who first observed this phenomenon.

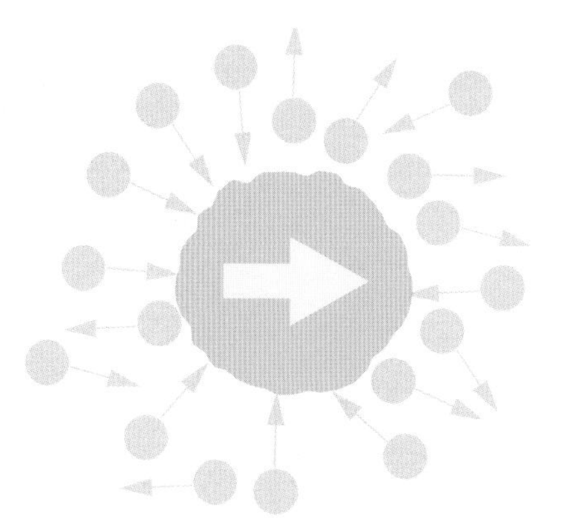

**a** Pollen particles being bombarded by water molecules.

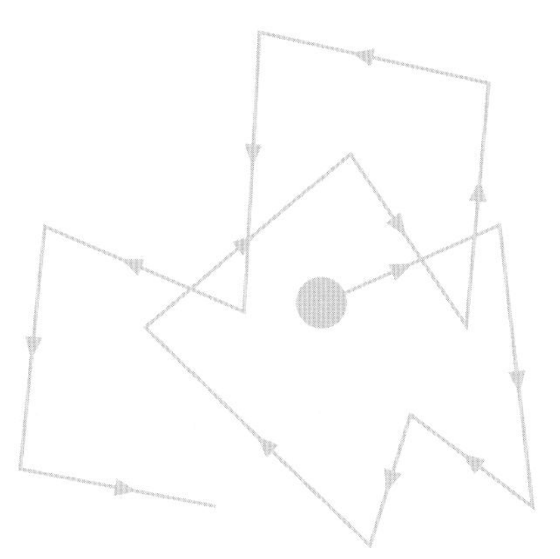

**b** Brownian motion causes the random motion of the visible particles.

**Figure 1.16**

## Questions

1. When a jar of coffee is opened, people in all parts of the room soon notice the smell. Use the kinetic theory to explain how this happens.
2. Describe, with the aid of diagrams, the diffusion of nickel(II) sulphate solution.
3. Explain why diffusion in gases is faster than in liquids.

# *Gas laws*

What do you think has caused the difference between the balloons in Figure 1.17?

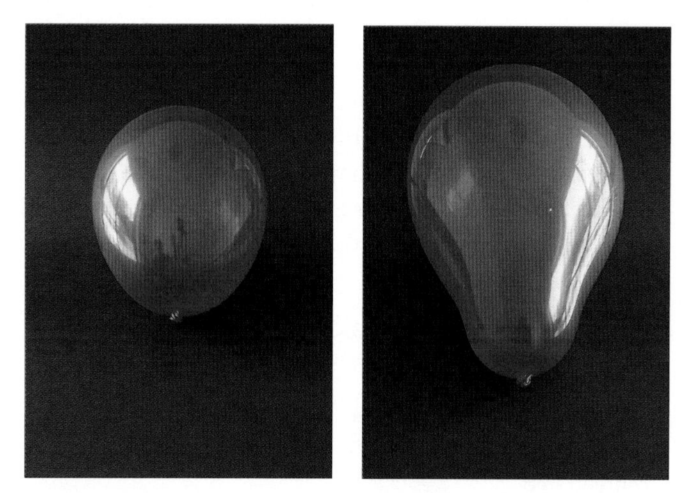

**Figure 1.17** The balloon is partly inflated at room temperature (left) and becomes fully expanded at a higher temperature.

The **pressure** inside a balloon is caused by the gas particles striking the inside surface of the balloon (Figure 1.18). There is an increased pressure inside the balloon at a higher temperature due to the gas particles having more energy, moving around faster and so striking the inside surface of the balloon more frequently; this in turn leads to an increase in pressure. Since the balloon is an elastic envelope, the increased pressure causes the skin to stretch and the volume to increase. This increase in volume with increased temperature is a property of all gases. In 1781, from this sort of observation, a French scientist called Jacques Alexandre César Charles concluded that when the temperature of a gas increased the volume also increased, at a constant pressure. This law is known as Charles' Law.

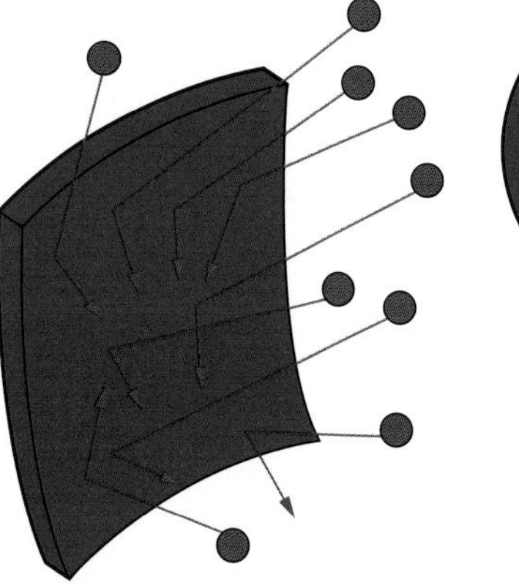

**Figure 1.18** The gas particles striking the surface create the pressure.

## Investigating Charles' Law

Charles' Law applies to a fixed mass of gas. The air column in the glass tube shown in Figure 1.19a has been trapped by a drop of concentrated acid, which moves up the tube as the gas expands. The length of the air column can be taken as a measure of the volume of air that is trapped. Readings are taken of the temperature and the length of the air column as the water bath is heated. If these data are plotted, then a volume against temperature graph like the one shown in Figure 1.19b is produced.

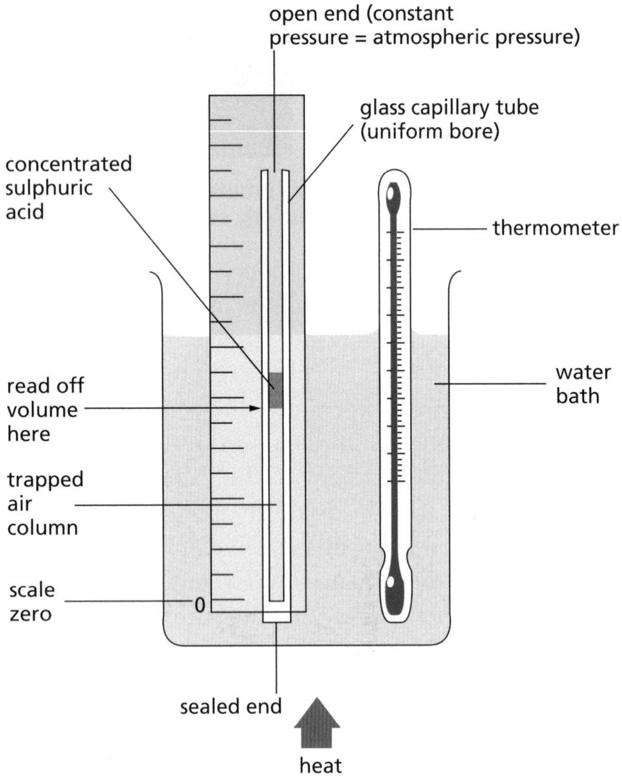

**a** Charles' Law apparatus.

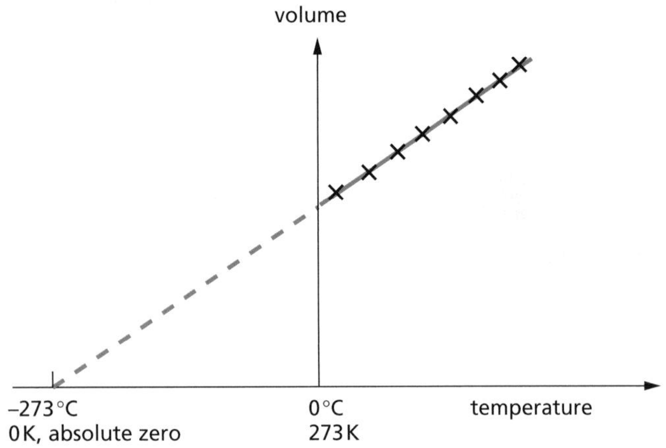

**b** Graph of volume against temperature.

**Figure 1.19**

The graph shows that at −273 °C the volume of the gas should contract to zero! This temperature is called **absolute zero** and it is the lowest possible temperature. At this temperature, theoretically, particles have no motion and therefore possess no energy.

We can define a new temperature scale, which was proposed by Lord Kelvin in 1854, called the Kelvin scale of temperature, which has 0 K at absolute zero. Kelvins are the same size as degrees on the Celsius scale. On the Kelvin scale, water freezes at 273 K and boils at 373 K. Note that we write 273 K without a ° (degree) sign. In general, to convert a Celsius temperature to a Kelvin temperature add 273.

$$K = °C + 273$$

Charles' Law states the volume, $V$, of a fixed mass of gas is directly proportional to its absolute temperature, $T$, if the pressure is kept constant. Later, it was found that a more accurate representation of Charles' Law required the temperature to be measured on the absolute temperature scale. Mathematically,

$$V \propto T$$

or

$$V = constant \times T$$

or

$$\frac{V}{T} = \text{constant}$$

## Boyle's Law

You can feel the increased pressure of a gas on your finger by pushing in the piston of a bicycle pump. As you push, you squash the same number of particles into a smaller volume (Figure 1.20). This squashing means they hit the walls of the pump more often so increasing the pressure.

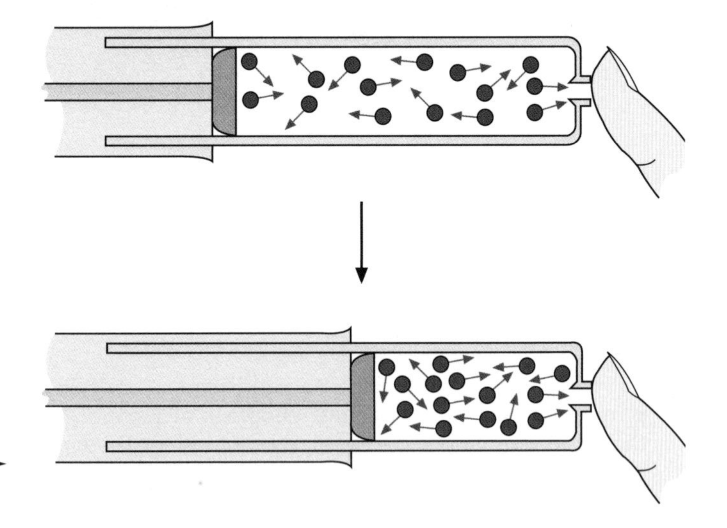

**Figure 1.20** A higher pressure is created by pushing in the piston.

Figure 1.21 Robert Boyle (1627–1691).

In 1662 a scientist called Robert Boyle (Figure 1.21) deduced from experiments he carried out on various gases that when the pressure was increased, the volume of the gas was reduced.

Why is it important to learn about how the volume of a gas changes with pressure? One popular use of this knowledge is in scuba diving (scuba stands for self-contained underwater breathing apparatus). To breathe under water, divers carry tanks of compressed gas (Figure 1.22). As they dive deeper, the water exerts more and more pressure on their bodies, and on the gas in their tanks. So that they can still breathe, the 'air' in the tanks has to be regulated and the pressure reduced so that it is about the same as that in the surrounding water.

Divers can descend fairly quickly as the volume of air in their bodies decreases as the pressure increases. When they want to return to the surface, though, they have to take it slowly because from 10 m below the surface to the surface itself the pressure doubles. This means that if divers come up too quickly or while holding their breath they can burst their lungs.

Figure 1.22 Scuba diving can be dangerous even at shallow depths, as nitrogen in the compressed air can dissolve in body tissues and make the diver feel 'drunk'.

## Investigating Boyle's Law

A car foot pump is used to increase the pressure on a fixed mass of air trapped by oil in a strong glass tube (Figure 1.23). The length of the air column is measured at different pressures.

A graph plotted of volume against $1/p$ gives a straight line through the origin. This demonstrates Boyle's Law, which states that the volume, $V$, of a fixed mass of gas is inversely proportional to its pressure, $p$, if the temperature is constant. Mathematically,

$$V \propto \frac{1}{p}$$

or

$$V = \frac{\text{constant}}{p}$$

or

$$pV = \text{constant}$$

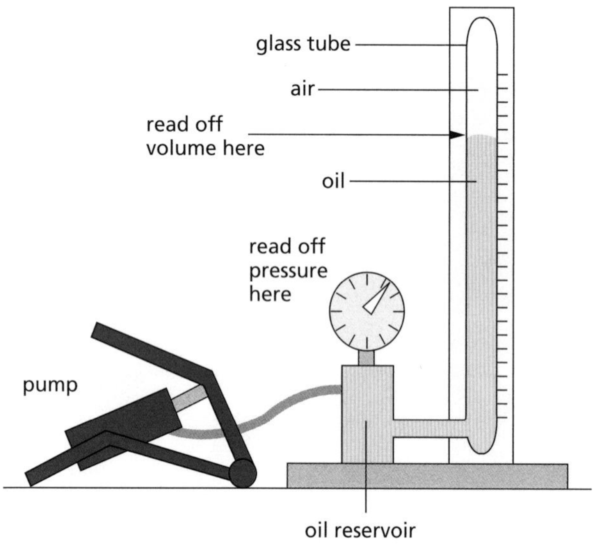

**Figure 1.23** Boyle's Law apparatus.

## Combining the gas laws

By combining Boyle's and Charles' Laws it is possible to show the relationship between the pressure, $p$, the volume, $V$, and the temperature, $T$, of a fixed mass of gas as:

$$\frac{pV}{T} = \text{constant}$$

or

$$\frac{p_1V_1}{T_1} = \frac{p_2V_2}{T_2}$$

where $p_1$, $V_1$ and $T_1$ and $p_2$, $V_2$ and $T_2$ are the pressure (in pascals), volume and temperature (in kelvin) in two different situations.

An example of the use of this relationship is given next. In the decomposition of hydrogen peroxide, if the volume of oxygen gas collected at 40 °C and a pressure of $1 \times 10^5$ Pa was 100 cm$^3$, what would be the volume of the gas at a temperature of 10 °C and a pressure of $2 \times 10^5$ Pa?

Using

$$\frac{p_1V_1}{T_1} = \frac{p_2V_2}{T_2}$$

$p_1 = 1 \times 10^5$ Pa $\qquad p_2 = 2 \times 10^5$ Pa

$V_1 = 100\,\text{cm}^3 \qquad V_2 = ?$

$T_1 = (40 + 273) = 313\,\text{K} \qquad T_2 = (10 + 273)\,\text{K} = 283\,\text{K}$

$$\frac{(1 \times 10^5) \times 100}{313} = \frac{(2 \times 10^5) \times V_2}{283}$$

$$V_2 = \frac{(1 \times 10^5) \times 100 \times 283}{313 \times (2 \times 10^5)}$$

$$V_2 = 45.21\,\text{cm}^3$$

## Questions

1. When a gas is heated the particles move more quickly. Explain what will happen to the volume of the heated gas if the pressure is kept constant.
2. A bubble of methane gas rises from the bottom of the North Sea.
   a. What will happen to the size of the bubble as it rises to the surface?
   b. Explain your answer to **a**.
3. A gas syringe contains 50 cm$^3$ of oxygen gas at 20 °C. If the temperature was increased to 45 °C, what would be the volume occupied by this gas, assuming constant pressure throughout?
4. A bicycle pump contains 50 cm$^3$ of air at a pressure of $1 \times 10^5$ Pa. What would be the volume of the air if the pressure was increased to $2.1 \times 10^5$ Pa at constant temperature?
5. If the volume of a gas collected at 60 °C and $1 \times 10^5$ Pa pressure was 70 cm$^3$, what would be the volume at a temperature of 0 °C and a pressure of $4 \times 10^5$ Pa?

## Checklist

**After studying Chapter 1 you should know and understand the following terms.**

- **Absolute temperature** A temperature measured with respect to absolute zero on the Kelvin scale. Absolute zero is the lowest possible temperature for all substances. The Kelvin scale is usually denoted by *T*.

$$T\text{K} = {}^\circ\text{C} + 273$$

- **Atmospheric pressure** The pressure exerted by the atmosphere on the surface of the Earth due to the weight of the air.
- **Boiling point** The temperature at which the pressure of the gas created above the liquid equals atmospheric pressure.
- **Boyle's Law** At a constant temperature the volume of a given mass of gas is inversely proportional to the pressure.

$$V \propto \frac{1}{p}$$

- **Charles' Law** At constant pressure the volume of a given mass of gas is directly proportional to the absolute temperature.

$$V \propto T$$

- **Condensation** The change of a vapour or a gas into a liquid. This process is accompanied by the evolution of heat.
- **Diffusion** The process by which different substances mix as a result of the random motions of their particles.
- **Evaporation** A process occurring at the surface of a liquid involving the change of state of a liquid into a vapour at a temperature below the boiling point.
- **Kinetic theory** A theory which accounts for the bulk properties of matter in terms of the constituent particles.
- **Matter** Anything which occupies space and has a mass.
- **Melting point** The temperature at which a solid begins to liquefy. Pure substances have a sharp melting point.
- **Solids, liquids and gases** The three states of matter to which all substances belong.
- **Sublimation** The direct change of state from solid to gas and the reverse process.

# All about matter
# *Additional questions*

**1 a** Draw diagrams to show the arrangement of particles in:
(i) solid lead
(ii) molten lead
(iii) gaseous lead.
**b** Explain how the particles move in these three states of matter.
**c** Explain, using the kinetic theory, what happens to the particles in oxygen as it is cooled down.

**2** Explain the meaning of each of the following terms. In your answer include an example to help with your explanation.
**a** Expansion.
**b** Contraction.
**c** Physical change.
**d** Sublimation.
**e** Diffusion.
**f** Random motion.
**g** Brownian motion.

**3 a** Why do solids not diffuse?
**b** Give two examples of diffusion of gases and liquids found in the house.

**4** Use the kinetic theory to explain the following:
**a** When you take a block of butter out of the fridge, it is quite hard. However, after 15 minutes it is soft enough to spread.
**b** When you come home from school and open the door you can smell your tea being cooked.
**c** A football is blown up until it is hard on a hot summer's day. In the evening the football feels softer.
**d** When a person wearing perfume enters a room it takes several minutes for the smell to reach the back of the room.
**e** A windy day is a good drying day.

**5** The apparatus shown below was set up. Give explanations for the following observations.

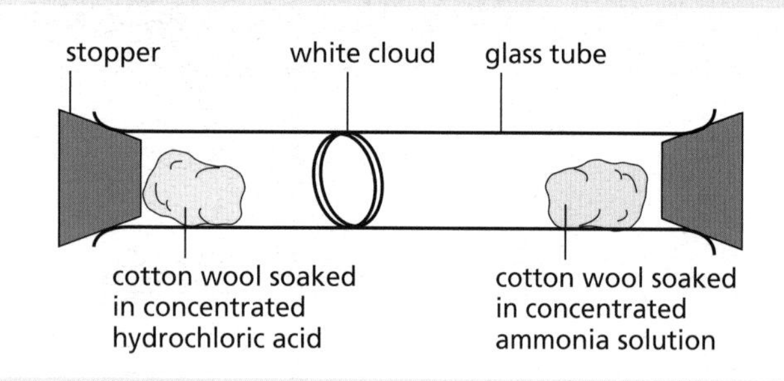


**a** The formation of a white cloud.
**b** It took a few minutes before the white cloud formed.
**c** The white cloud formed further from the cotton wool soaked in ammonia.
**d** Cooling the concentrated ammonia and hydrochloric acid before carrying out the experiment increased the time taken for the white cloud to form.

**6** The following diagram shows the three states of matter and how they can be interchanged.

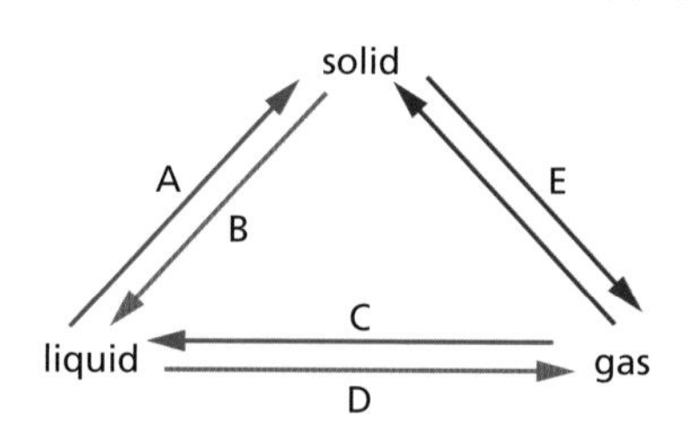


**a** Name the changes **A** to **E**.
**b** Name a substance which will undergo change **E**.
**c** Name a substance which will undergo changes from solid to liquid to gas between 0 °C and 100 °C.
**d** Describe what happens to the particles of the solid during change **E**.
**e** Which of the changes **A** to **E** will involve:
(i) an input of heat energy?
(ii) an output of heat energy?

**7** Some nickel(II) sulphate solution was carefully placed in the bottom of a beaker of water. The beaker was then covered and left for several days.

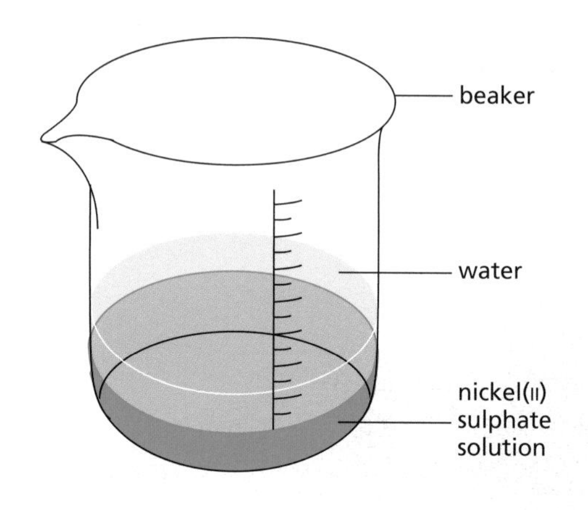


**a** Describe what you would see after:
(i) a few hours
(ii) several days.
**b** Explain your answer to **a** using your ideas of the kinetic theory of particles.
**c** What is the name of the physical process that takes place in this experiment?

**8** An electric light bulb has a volume of 200 $cm^3$. It contains argon gas at a pressure of $1.1 \times 10^5$ Pa and a temperature of 25 °C. When the light is switched on, the pressure increases to a steady $1.8 \times 10^5$ Pa. What is the temperature of the argon in the electric light bulb that creates this new steady pressure?

# 2 Elements, compounds and mixtures



The universe is made up of a very large number of substances (Figure 2.1), and our own world is no exception. If this vast array of substances is examined more closely, it is found that they are made up of some basic substances which were given the name **elements** in 1661 by Robert Boyle, who we met in Chapter 1.

In 1803, John Dalton (Figure 2.2) suggested that each element was composed of its own kind of particles, which he called **atoms**. Atoms are much too small to be seen. We now know that about $20 \times 10^6$ of them would stretch over a length of only 1 cm.

**Figure 2.1** The planets in the universe are made of millions of substances. These are made up mainly from just 91 elements which occur naturally on the Earth.

**Figure 2.2** John Dalton (1766–1844).

## Elements

Robert Boyle used the name element for any substance that cannot be broken down further, into a simpler substance. This definition can be extended to include the fact that each element is made up of only one kind of atom. The word atom comes from the Greek word *atomos* meaning 'unsplittable'.

For example, aluminium is an element which is made up of only aluminium atoms. It is not possible to obtain a simpler substance chemically from the aluminium atoms. You can only make more complicated substances from it, such as aluminium oxide, aluminium nitrate or aluminium sulphate.

There are 115 elements which have now been identified. Twenty-four of these do not occur in nature and have been made artificially by scientists. They include elements such as plutonium, curium and unnilpentium. Ninety-one of the elements occur naturally and range from some very reactive gases, such as fluorine and chlorine, to gold and platinum, which are unreactive elements.

All elements can be classified according to their various properties. A simple way to do this is to classify them as **metals** or **non-metals** (Figures 2.3 and 2.4). Table 2.1 shows the physical data for some common metallic and non-metallic elements.

You will notice that many metals have high densities, high melting points and high boiling points, and that most non-metals have low densities, low melting points and low boiling points. Table 2.2 summarises the different properties of metals and non-metals.

A discussion of the chemical properties of metals is given in Chapters 3, 4 and 9. The chemical properties of certain non-metals are discussed in Chapters 3, 4, 15 and 16.

**a** Gold is very decorative.

**b** Titanium has many uses in the aerospace industry.

**c** These coins contain nickel.

**Figure 2.3** Some metals.

**Table 2.1** Physical data for some metallic and non-metallic elements at room temperature and pressure.

| Element | Metal or non-metal | Density/g cm$^{-3}$ | Melting point/°C | Boiling point/°C |
|---|---|---|---|---|
| Aluminium | Metal | 2.70 | 660 | 2580 |
| Copper | Metal | 8.92 | 1083 | 2567 |
| Gold | Metal | 19.29 | 1065 | 2807 |
| Iron | Metal | 7.87 | 1535 | 2750 |
| Lead | Metal | 11.34 | 328 | 1740 |
| Magnesium | Metal | 1.74 | 649 | 1107 |
| Nickel | Metal | 8.90 | 1453 | 2732 |
| Silver | Metal | 10.50 | 962 | 2212 |
| Zinc | Metal | 7.14 | 420 | 907 |
| Carbon | Non-metal | 2.25 | 2652 | Sublimes |
| Hydrogen | Non-metal | 0.07[a] | −259 | −253 |
| Nitrogen | Non-metal | 0.88[b] | −210 | −196 |
| Oxygen | Non-metal | 1.15[c] | −218 | −183 |
| Sulphur | Non-metal | 2.07 | 113 | 445 |

Source: Earl B., Willford L.D.R. Chemistry data book. Nelson Blackie, 1991 [a] At −254 °C [b] At −197 °C [c] At −184 °C.

**Table 2.2** How the properties of metals and non-metals compare.

| Property | Metal | Non-metal |
|---|---|---|
| Physical state at room temperature | Usually solid (occasionally liquid) | Solid, liquid or gas |
| Malleability | Good | No – usually soft or brittle when solid |
| Ductility | Good | |
| Appearance (solids) | Shiny (lustrous) | Dull |
| Melting point | Usually high | Usually low |
| Boiling point | Usually high | Usually low |
| Density | Usually high | Usually low |
| Conductivity (thermal and electrical) | Good | Very poor |

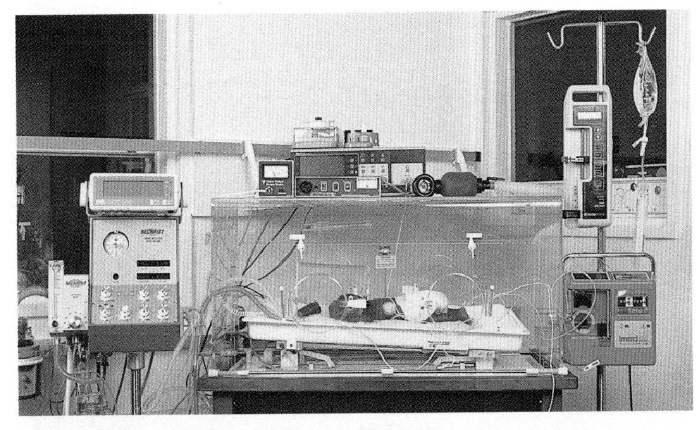

**a** A premature baby needs oxygen.

**b** Artists often use charcoal (carbon) to produce an initial sketch.

**c** Neon is used in advertising signs.

**Figure 2.4** Some non-metals.

## Atoms – the smallest particles

Everything is made up of billions of atoms. The atoms of all elements are extremely small; in fact they are too small to be seen. The smallest atom known is hydrogen, with each atom being represented as a sphere having a diameter of 0.000 000 07 mm (or $7 \times 10^{-8}$ mm) (Table 2.3). Atoms of different elements have different diameters as well as different masses. How many atoms of hydrogen would have to be placed side by side along the edge of your ruler to fill just one of the 1 mm division?

**Table 2.3** Sizes of atoms.

| Atom | Diameter of atom/mm |
|---|---|
| Hydrogen | $7 \times 10^{-8}$ |
| Oxygen | $12 \times 10^{-8}$ |
| Sulphur | $20.8 \times 10^{-8}$ |

Chemists use shorthand symbols to label the elements and their atoms. The symbol consists of one, two or three letters, the first of which must be a capital. Where several elements have the same initial letter, a second letter of the name or subsequent letter is added. For example, **C** is used for **carbon**, **Ca** for **calcium** and **Cl** for **chlorine**. Some symbols seem to have no relationship to the name of the element, for example **Na** for **sodium** and **Pb** for **lead**. These symbols come from their Latin names, natrium for sodium and plumbum for lead. A list of some common elements and their symbols is given in Table 2.4 (overleaf).

Table 2.4 Some common elements and their symbols.

| Element | Symbol | Physical state at room temperature and pressure |
|---|---|---|
| Aluminium | Al | Solid |
| Argon | Ar | Gas |
| Barium | Ba | Solid |
| Boron | B | Solid |
| Bromine | Br | Liquid |
| Calcium | Ca | Solid |
| Carbon | C | Solid |
| Chlorine | Cl | Gas |
| Chromium | Cr | Solid |
| Copper (Cuprum) | Cu | Solid |
| Fluorine | F | Gas |
| Germanium | Ge | Solid |
| Gold (Aurum) | Au | Solid |
| Helium | He | Gas |
| Hydrogen | H | Gas |
| Iodine | I | Solid |
| Iron (Ferrum) | Fe | Solid |
| Lead (Plumbum) | Pb | Solid |
| Magnesium | Mg | Solid |
| Mercury (Hydragyrum) | Hg | Liquid |
| Neon | Ne | Gas |
| Nitrogen | N | Gas |
| Oxygen | O | Gas |
| Phosphorus | P | Solid |
| Potassium (Kalium) | K | Solid |
| Silicon | Si | Solid |
| Silver (Argentum) | Ag | Solid |
| Sodium (Natrium) | Na | Solid |
| Sulphur | S | Solid |
| Tin (Stannum) | Sn | Solid |
| Zinc | Zn | Solid |

The complete list of the elements with their corresponding symbols is shown in the periodic table on page 40.

## Molecules

The atoms of some elements are joined together in small groups. These small groups of atoms are called **molecules**. For example, the atoms of the elements hydrogen, oxygen, nitrogen, fluorine, chlorine, bromine and iodine are each joined in pairs and they are known as **diatomic** molecules. In the case of phosphorus and sulphur the atoms are joined in larger numbers, four and eight respectively ($P_4$, $S_8$). In chemical shorthand the molecule of chlorine shown in Figure 2.5 is written as $Cl_2$.

Cl — Cl

a As a letter-and-stick model.

b As a space-filling model.

Figure 2.5 A chlorine molecule.

The gaseous elements helium, neon, argon, krypton, xenon and radon are composed of separate and individual atoms. When an element exists as separate atoms, then the molecules are said to be **monatomic**. In chemical shorthand these monatomic molecules are written as He, Ne, Ar, Kr, Xe and Rn respectively.

Molecules are not always formed by atoms of the same type joining together. For example, water exists as molecules containing oxygen and hydrogen atoms.

## Questions

1 How would you use a similar chemical shorthand to write a representation of the molecules of iodine and fluorine?

2 Using the periodic table on page 40 write down the symbols for the following elements and give their physical states at room temperature.
  a chromium
  b krypton
  c osmium.

## Compounds

Compounds are pure substances which are formed when two or more elements chemically combine together. Water is a simple compound formed from the elements hydrogen and oxygen (Figure 2.6). This combining of the elements can be represented by a word equation:

hydrogen + oxygen → water

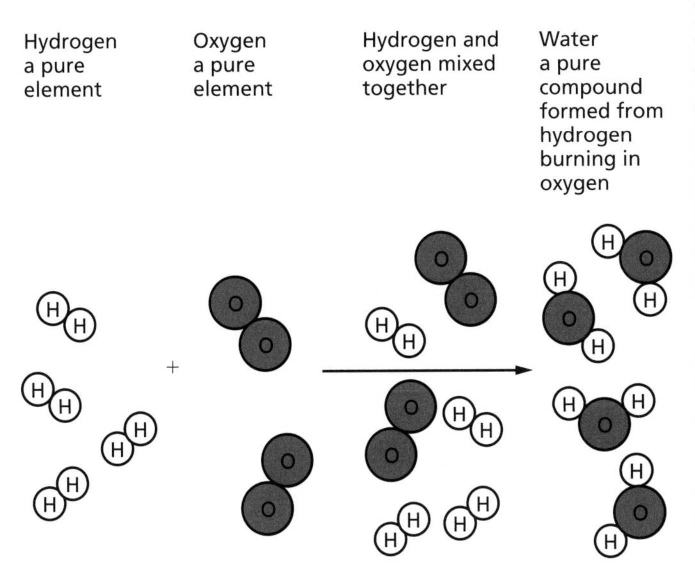


**Figure 2.6** The element hydrogen reacts with the element oxygen to produce the compound water.

Water molecules contain two atoms of hydrogen and one atom of oxygen, and hence water has the **chemical formula** $H_2O$. Elements other than hydrogen will also react with oxygen to form compounds called oxides. For example, magnesium reacts violently with oxygen gas to form the white powder magnesium oxide (Figure 2.7). This reaction is accompanied by a release of energy as new chemical bonds are formed.

When a new substance is formed during a chemical reaction, a **chemical change** has taken place.

magnesium + oxygen → magnesium oxide

When substances such as hydrogen and magnesium combine with oxygen in this way they are said to have been **oxidised**. The process is known as **oxidation**.

**Reduction** is the opposite of oxidation. In this process oxygen is removed instead of being added. For example, the oxygen has to be removed in the extraction of iron from iron(III) oxide. This can be done in a blast furnace with carbon monoxide. The iron(III) oxide loses oxygen to the carbon monoxide and is reduced to iron while carbon monoxide is oxidised to carbon dioxide. You will deal in more detail with this extraction process in Chapter 9.

iron(III) oxide + carbon monoxide → iron + carbon dioxide

Both **red**uction and **ox**idation have taken place in this chemical process, and so this is known as a **redox** reaction. A further discussion of oxidation and reduction takes place in Chapter 6.

**Figure 2.7** Magnesium burns brightly in oxygen to produce magnesium oxide.

### More about formulae

The formula of a compound is made up from the symbols of the elements present and numbers to show the ratio in which the different atoms are present. Carbon dioxide has the formula $CO_2$. This tells you that it contains one carbon atom for every two oxygen atoms. The 2 in the formula tells you that there are two oxygen atoms present in each molecule of carbon dioxide.

Table 2.5 (overleaf) shows the names and formulae of some common compounds which you will meet in your study of chemistry.

The ratio of atoms within a chemical compound is usually constant. Compounds are made up of fixed proportions of elements: they have a fixed composition. Chemists call this the **Law of constant composition**.

**Table 2.5** Names and formulae of some common compounds.

| Compound | Formula |
|---|---|
| Ammonia | $NH_3$ |
| Calcium hydroxide | $Ca(OH)_2$ |
| Carbon dioxide | $CO_2$ |
| Copper sulphate | $CuSO_4$ |
| Ethanol (alcohol) | $C_2H_5OH$ |
| Glucose | $C_6H_{12}O_6$ |
| Hydrochloric acid | $HCl$ |
| Nitric acid | $HNO_3$ |
| Sodium carbonate | $Na_2CO_3$ |
| Sodium hydroxide | $NaOH$ |
| Sulphuric acid | $H_2SO_4$ |

## Balancing chemical equations

Word equations are a useful way of representing chemical reactions but a better and more useful method is to produce a **balanced chemical equation**. This type of equation gives the formulae of the reactants and the products as well as showing the relative numbers of each particle involved. Balanced equations often include the physical state symbols:

(s) = solid, (l) = liquid, (g) = gas, (aq) = aqueous solution

The word equation to represent the reaction between iron and sulphur is:

$$\text{iron} + \text{sulphur} \xrightarrow{\text{heat}} \text{iron(II) sulphide}$$

When we replace the words with symbols for the reactants and the products and include their physical state symbols, we obtain:

$$Fe(s) + S(s) \xrightarrow{\text{heat}} FeS(s)$$

Since there is the same number of each type of atom on both sides of the equation this is a **balanced** chemical equation.

In the case of magnesium reacting with oxygen, the word equation was:

$$\text{magnesium} + \text{oxygen} \xrightarrow{\text{heat}} \text{magnesium oxide}$$

When we replace the words with symbols for the reactants and the products and include their physical state symbols, it is important to remember that oxygen is a diatomic molecule:

$$Mg(s) + O_2(g) \xrightarrow{\text{heat}} MgO(s)$$

In the equation there are two oxygen atoms on the left-hand side ($O_2$) but only one on the right (MgO). We cannot change the formula of magnesium oxide, so to produce the necessary two oxygen atoms on the right-hand side we will need 2MgO – this means $2 \times MgO$. The equation now becomes:

$$Mg(s) + O_2(g) \xrightarrow{\text{heat}} 2MgO(s)$$

There are now two atoms of magnesium on the right-hand side and only one on the left. By placing a 2 in front of the magnesium, we obtain the following balanced chemical equation:

$$2Mg(s) + O_2(g) \xrightarrow{\text{heat}} 2MgO(s)$$

This balanced chemical equation now shows us that two atoms of magnesium react with one molecule of oxygen gas when heated to produce two units of magnesium oxide.

## Instrumental techniques

Elements and compounds can be detected and identified by a variety of instrumental methods. Scientists have developed instrumental techniques that allow us to probe and discover which elements are present in the substance as well as how the atoms are arranged within the substance.

Many of the instrumental methods that have been developed are quite sophisticated. Some methods are suited to identifying elements. For example, atomic absorption spectroscopy allows the element to be identified and also allows the quantity of the element that is present to be found (Figure 2.8).

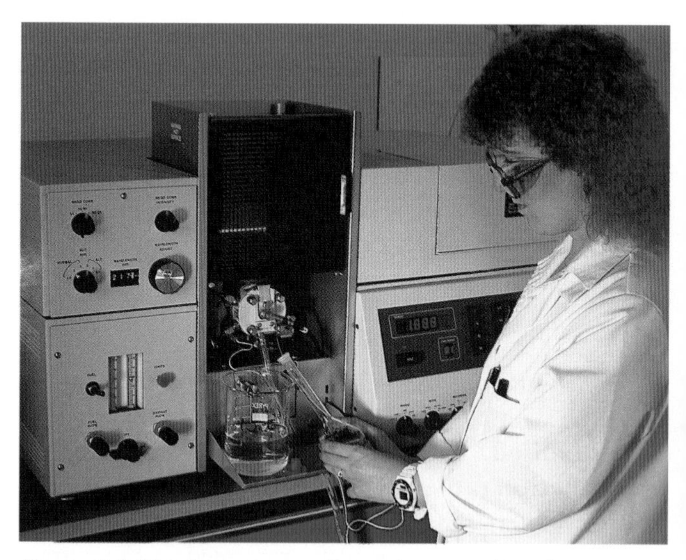

**Figure 2.8** This instrument allows the quantity of a particular element to be found. It is used extensively throughout industry for this purpose. It will allow even tiny amounts of a particular element to be found.

Some methods are particularly suited to the identification of compounds. For example, infrared spectroscopy is used to identify compounds by showing the presence of particular groupings of atoms (Figure 2.9).

Infrared spectroscopy is used by the pharmaceutical industry to identify and discriminate between drugs that are similar in structure, for example penicillin-type drugs. Used both with organic and inorganic molecules, this method assumes that each compound has a unique infrared spectrum. Samples can be solid, liquid or gas and are tiny. However, Ne, He, $O_2$, $N_2$ or $H_2$ cannot be used.

This method is also used to monitor environmental pollution, and has biological uses in monitoring tissue physiology including oxygenation, respiratory status and blood flow damage.

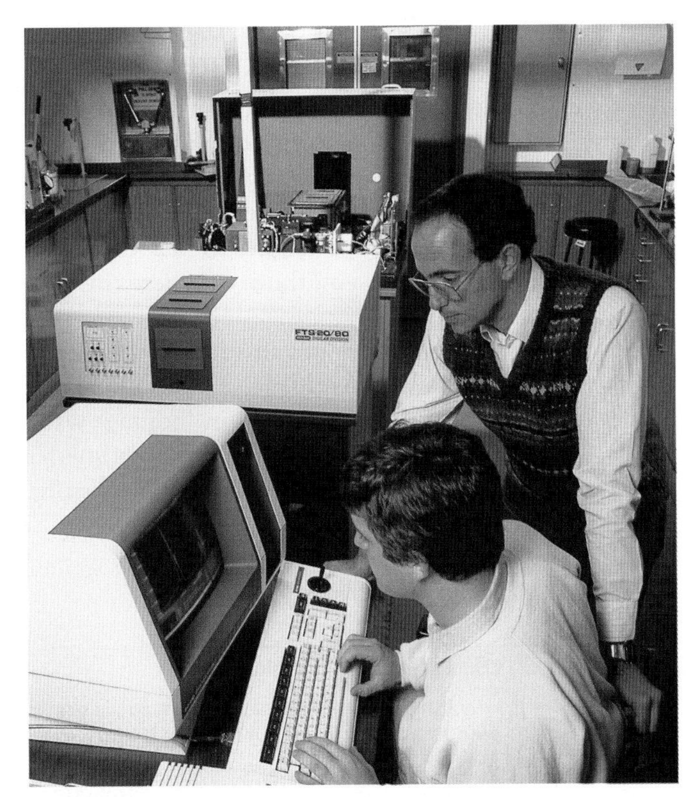

**Figure 2.9** This is a modern infrared spectrometer. It is used in analysis to obtain a so-called fingerprint spectrum of a substance that will allow the substance to be identified.

Forensic scientists make use of both these techniques because they are very accurate but they only require tiny amounts of sample – often only small amounts of sample are found at crime scenes. Other techniques utilised are nuclear magnetic resonance spectroscopy and ultraviolet/visible spectroscopy.

## Questions

1 Write the word and balanced chemical equations for the reactions which take place between:
 a calcium and oxygen
 b copper and oxygen.

2 Write down the ratio of the atoms present in each formula for each of the compounds shown in Table 2.5.

3 Iron is extracted from iron(III) oxide in a blast furnace by a redox reaction. What does the term 'redox reaction' mean?

# *Mixtures*

Many everyday things are not pure substances, they are mixtures. A mixture contains more than one substance (elements and/or compounds). An example of a common mixture is sea water (Figure 2.10).

**Figure 2.10** Sea water is a common mixture.

## What is the difference between mixtures and compounds?

There are differences between compounds and mixtures. This can be shown by considering the reaction between iron filings and sulphur. A mixture of iron filings and sulphur looks different from the individual elements (Figure 2.11). This mixture has the properties of both iron and sulphur; for example, a magnet can be used to separate the iron filings from the sulphur (Figure 2.12).

**Figure 2.11** The elements (left to right) iron and sulphur, and below a mixture of iron and sulphur and black iron(II) sulphide.

**Figure 2.12** A magnet will separate the iron from the mixture.

Substances in a mixture have not undergone a chemical reaction and it is possible to separate them provided that there is a suitable difference in their physical properties (p. 25). If the mixture of iron and sulphur is heated a chemical reaction occurs and a new substance is formed called iron(II) sulphide (Figure 2.11). The word equation for this reaction is:

$$\text{iron} + \text{sulphur} \xrightarrow{\text{heat}} \text{iron(II) sulphide}$$

During the reaction heat energy is given out as new chemical bonds are formed. This is called an **exothermic** reaction and accompanies a chemical change (Chapter 13, p. 190). The iron(II) sulphide formed has totally different properties to the mixture of iron and sulphur (Table 2.6). Iron(II) sulphide, for example, would not be attracted towards a magnet.

In iron(II) sulphide, FeS, one atom of iron has combined with one atom of sulphur. No such ratio exists in a mixture of iron and sulphur, because the atoms have not chemically combined. Table 2.7 summarises how mixtures and compounds compare.

# *Separating mixtures*

Many mixtures contain useful substances mixed with unwanted material. In order to obtain these useful substances, chemists often have to separate them from the impurities. Chemists have developed many different methods of separation, particularly for separating compounds from complex mixtures. Which separation method they use depends on what is in the mixture and the properties of the substances present. It also depends on whether the substances to be separated are solids, liquids or gases.

## Separating solid/liquid mixtures

If a solid substance is added to a liquid it may **dissolve** to form a **solution**. In this case the solid is said to be **soluble** and is called the **solute**. The liquid it has dissolved in is called the **solvent**. An example of this type of process is when sugar is added to tea or coffee. What other examples can you think of where this type of process takes place?

Sometimes the solid does not dissolve in the liquid. This solid is said to be **insoluble**. For example, tea leaves themselves do not dissolve in boiling water when tea is made from them, although the soluble materials from which tea is made are seen to dissolve from them.

**Table 2.6** Different properties of iron, sulphur, an iron/sulphur mixture and iron(II) sulphide.

| Substance | Appearance | Effect of a magnet | Effect of dilute hydrochloric acid |
|---|---|---|---|
| Iron | Dark grey powder | Attracted to it | Very little action when cold. When warm, a gas is produced with a lot of bubbling (effervescence) |
| Sulphur | Yellow powder | None | No effect when hot or cold |
| Iron/sulphur mixture | Dirty yellow powder | Iron powder attracted to it | Iron powder reacts as above |
| Iron(II) sulphide | Dark grey solid | No effect | A foul-smelling gas is produced with some effervescence |

**Table 2.7** The major differences between mixtures and compounds.

| Mixture | Compound |
|---|---|
| It contains two or more substances | It is a single substance |
| The composition can vary | The composition is always the same |
| No chemical change takes place when a mixture is formed | When the new substance is formed it involves chemical change |
| The properties are those of the individual elements | The properties are very different to those of the component elements |
| The components may be separated quite easily by physical means | The components can only be separated by one or more chemical reactions |

## *Question*

**1** Make a list of some other common mixtures, stating what they are mixtures of.

## Filtration

When a cup of tea is poured through a tea strainer you are carrying out a **filtering** process. **Filtration** is a common separation technique used in chemistry laboratories throughout the world. It is used when a solid needs to be separated from a liquid. For example, sand can be separated from a mixture with water by filtering through filter paper as shown in Figure 2.13.

The filter paper contains holes that, although too small to be seen, are large enough to allow the molecules of water through but not the sand particles. It acts like a sieve. The sand gets trapped in the filter paper and the water passes through it. The sand is called the **residue** and the water is called the **filtrate**.

**Figure 2.13** It is important when filtering not to overfill the filter paper.

## Decanting

Carrots do not dissolve in water. When you have boiled some carrots it is easy to separate them from the water by pouring it off. This process is called **decanting**. This technique is used quite often to separate an insoluble solid, which has settled at the bottom of a flask, from a liquid. A further example is the decanting of old red wine or port.

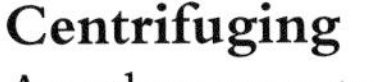

## Centrifuging

Another way to separate a solid from a liquid is to use a **centrifuge**. This technique is sometimes used instead of filtration. It is usually used when the solid particles are so small that they spread out (disperse) throughout the liquid and remain in **suspension**. They do not settle to the bottom of the container, as heavier particles would do, under the force of gravity. The technique of **centrifuging** or **centrifugation** involves the suspension being spun round very fast in a centrifuge so that the solid gets flung to the bottom of the tube (Figure 2.14a and b).

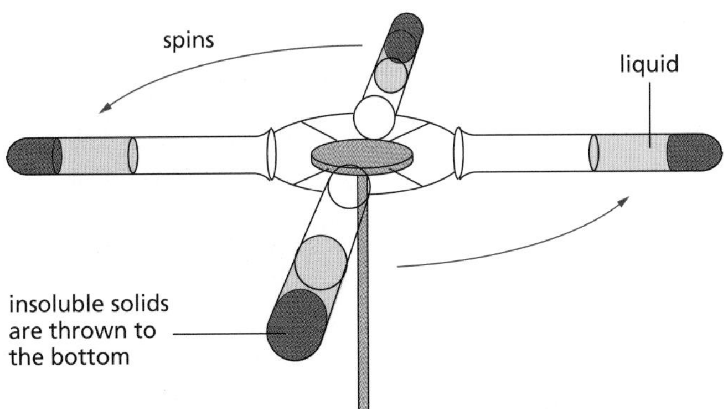


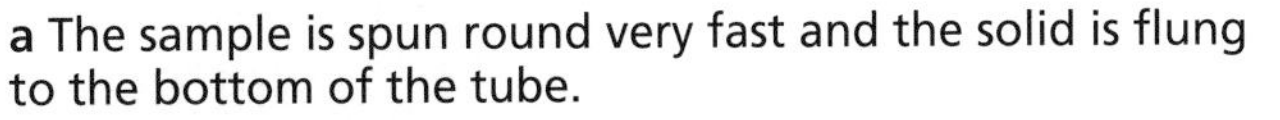

**a** The sample is spun round very fast and the solid is flung to the bottom of the tube.

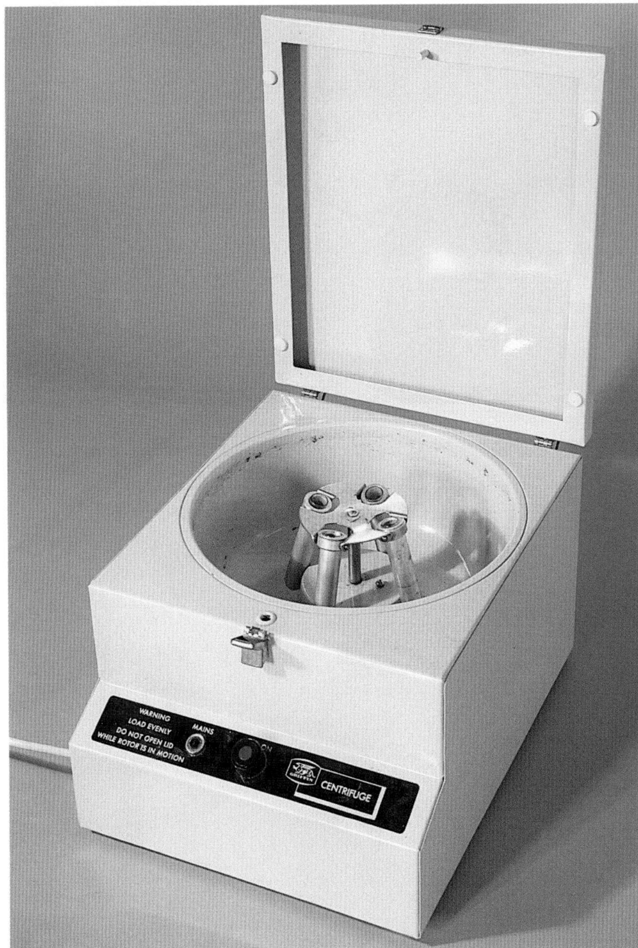


**b** An open centrifuge.

**Figure 2.14**

The pure liquid can be decanted after the solid has been forced to the bottom of the tube. This method of separation is used extensively to separate blood cells from blood plasma (Figure 2.15). In this case, the solid particles (the blood cells) are flung to the bottom of the tube, allowing the liquid plasma to be decanted.

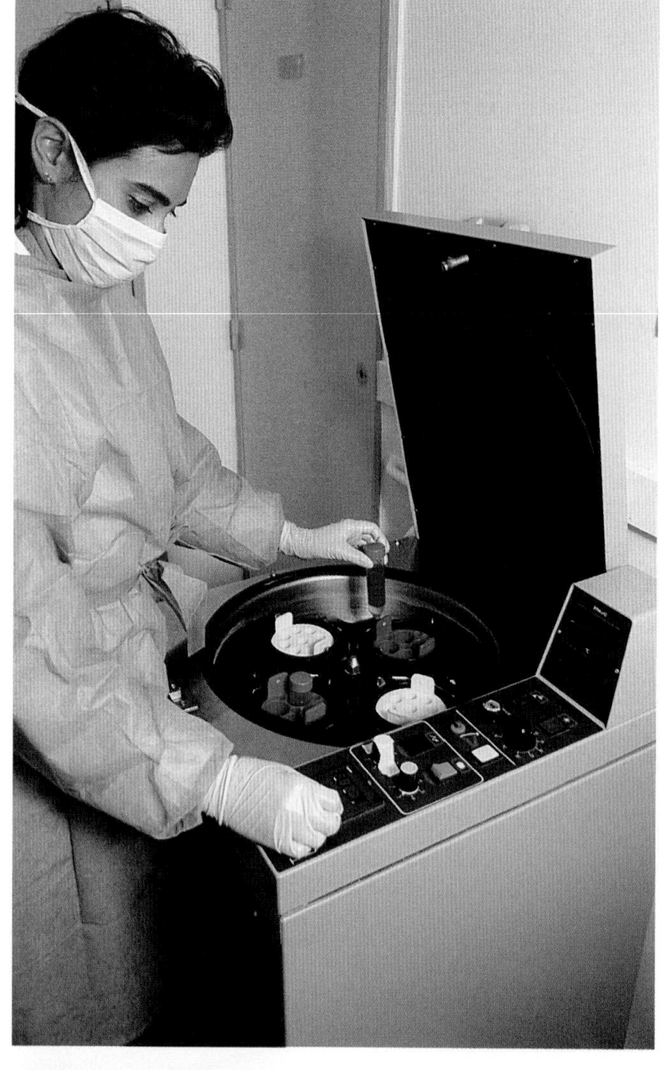

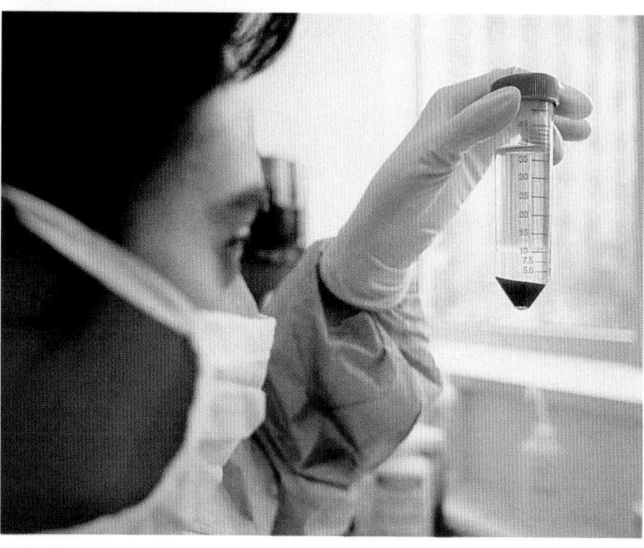

Figure 2.15 Whole blood (top) is separated by centrifuging into blood cells and plasma (bottom).

## Evaporation

If the solid has dissolved in the liquid it cannot be separated by filtering or centrifuging. Instead, the solution can be heated so that the liquid evaporates completely and leaves the solid behind. The simplest way to obtain salt from its solution is by slow evaporation as shown in Figure 2.16.

Figure 2.16 Apparatus used to slowly evaporate a solvent.

## Crystallisation

In many parts of the world salt is obtained from sea water on a vast scale. This is done by using the heat of the sun to evaporate the water to leave a saturated solution of salt known as brine. A **saturated solution** is defined as one that contains as much solute as can be dissolved at a particular temperature. When the solution is saturated the salt begins to **crystallise**, and it is removed using large scoops (Figure 2.17).

Figure 2.17 Salt is obtained in Rio de Janeiro, Brazil, by evaporation of sea water.

### Simple distillation

If we want to obtain the solvent from a solution, then the process of **distillation** can be carried out. The apparatus used in this process is shown in Figure 2.18.

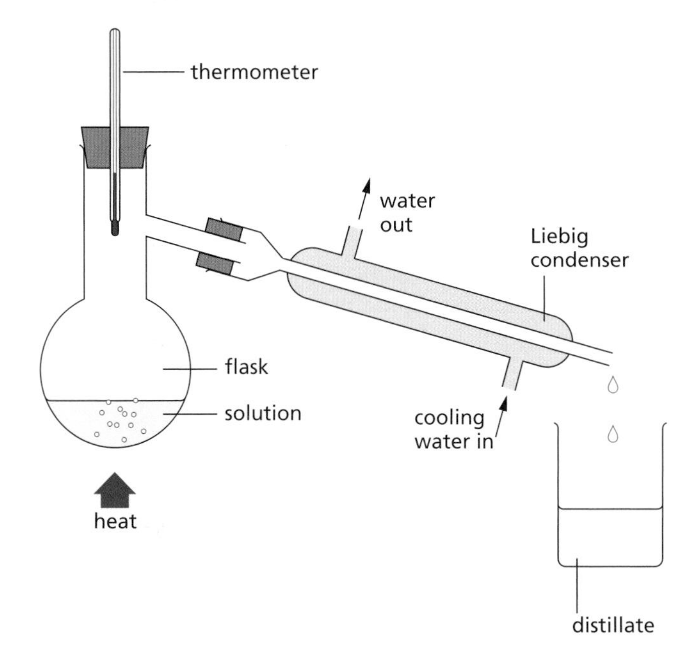


**Figure 2.18** Water can be obtained from salt water by distillation.

Water can be obtained from salt water using this method. The solution is heated in the flask until it boils. The steam rises into the Liebig condenser, where it condenses back into water. The salt is left behind in the flask. In hot and arid countries such as Saudi Arabia this sort of technique is used on a much larger scale to obtain pure water for drinking (Figure 2.19). This process is carried out in a desalination plant.

**Figure 2.19** This plant produces large quantities of drinking water in Bahrain.

## Separating liquid/liquid mixtures

In recent years there have been many oil tanker disasters, just like the one shown in Figure 2.20. These have resulted in millions of litres of oil being washed into the sea. Oil and water do not mix easily. They are said to be **immiscible**. When cleaning up disasters of this type, a range of chemicals can be added to the oil to make it more soluble. This results in the oil and water mixing with each other. They are now said to be **miscible**. The following techniques can be used to separate mixtures of liquids.

**Figure 2.20** Millions of litres of oil are spilt in such disasters and the cleaning-up operation is a slow and costly process.

### Liquids which are immiscible

If two liquids are immiscible they can be separated using a **separating funnel**. The mixture is poured into the funnel and the layers allowed to separate. The lower layer can then be run off by opening the tap as shown in Figure 2.21.

**Figure 2.21** The water is more dense than the oil and so sinks to the bottom of the separating funnel. When the tap is opened the water can be run off.

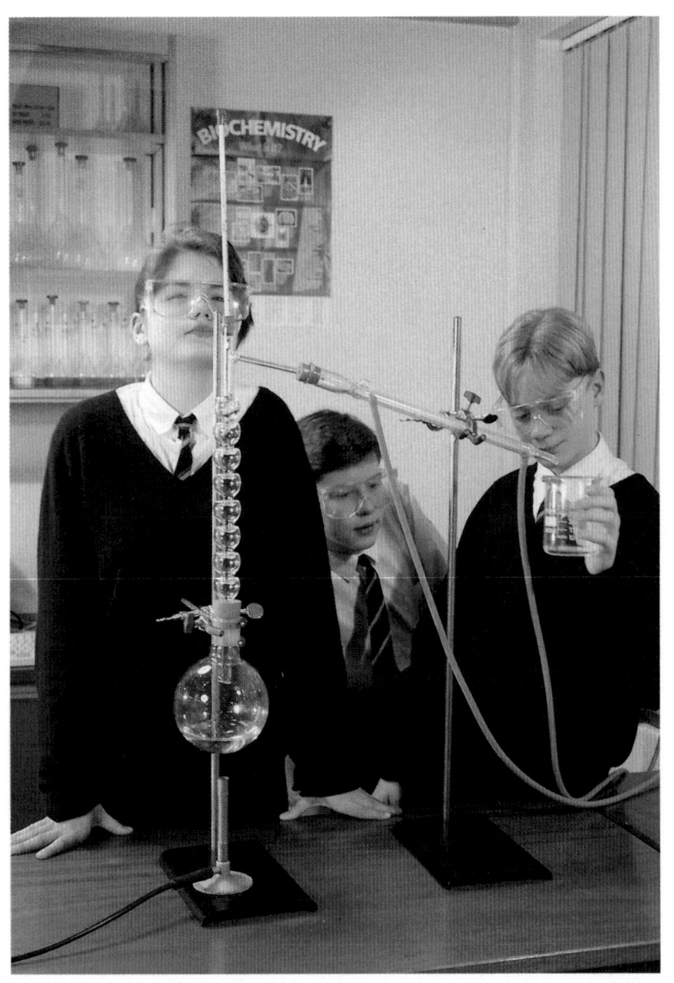


### Liquids which are miscible

If miscible liquids are to be separated, then this can be done by **fractional distillation**. The apparatus used for this process is shown in Figure 2.22 and could be used to separate a mixture of ethanol and water.

Fractional distillation relies upon the liquids having different boiling points. When an ethanol and water mixture is heated the vapours of ethanol and water boil off at different temperatures and can be condensed and collected separately.

Ethanol boils at 78 °C whereas water boils at 100 °C. When the mixture is heated the vapour produced is mainly ethanol with some steam. Because water has the higher boiling point of the two, it condenses out from the mixture with ethanol. This is what takes place in the fractionating column. The water condenses and drips back into the flask while the ethanol vapour moves up the column and into the condenser, where it condenses into liquid ethanol and is collected in the receiving flask. When all the ethanol has distilled over, the temperature reading on the thermometer rises steadily to 100 °C, showing that the steam is now entering the condenser. At this point the receiver can be changed and the condensing water can now be collected.

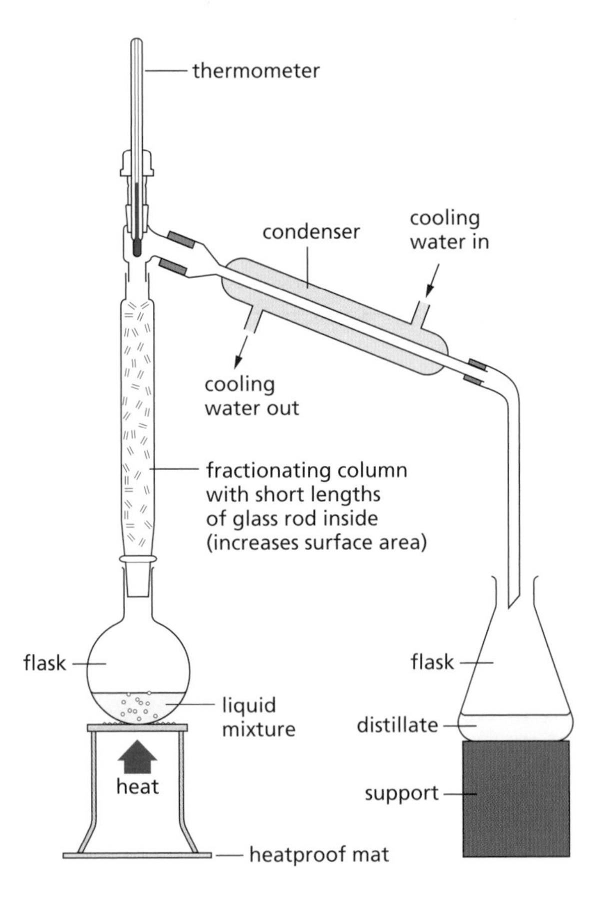


**Figure 2.22** Typical fractional distillation apparatus.

As well as separating miscible liquids such as crude oil (Figure 2.23a), fractional distillation can also separate individual gases, such as nitrogen, from the mixture we call air (Figure 2.23b).

**a** Fractional distillation unit for crude oil.

**b** Gases from the air are extracted in this fractional distillation plant.

**Figure 2.23**

## Separating solid/solid mixtures

You saw earlier in this chapter (p. 19) that it was possible to separate iron from sulphur using a magnet. In that case we were using one of the physical properties of iron, that is, the fact that it is magnetic. In a similar way, it is possible to separate scrap iron from other metals by using a large electromagnet like the one shown in Figure 2.24.

It is essential that when separating solid/solid mixtures you pay particular attention to the individual physical properties of the components. If, for example, you wish to separate two solids, one of which sublimes, then this property should dictate the method you employ.

**Figure 2.24** Magnetic separation of iron-containing materials.

In the case of an iodine/salt mixture the iodine sublimes but salt does not. Iodine can be separated by heating the mixture in a fume cupboard as shown in Figure 2.25. The iodine sublimes and re-forms on the cool inverted funnel.

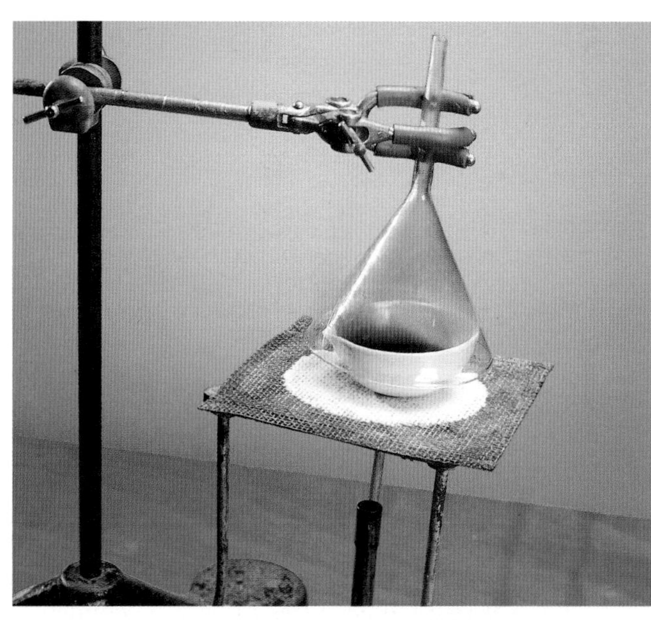

**Figure 2.25** Apparatus used to separate an iodine/salt mixture. The iodine sublimes on heating.

### Chromatography

What happens if you have to separate two or more solids that are soluble? This type of problem is encountered when you have mixtures of coloured materials such as inks and dyes. A technique called **chromatography** is widely used to separate these materials so that they can be identified.

There are several types of chromatography; however, they all follow the same basic principles. The simplest kind is paper chromatography. To separate the different-coloured dyes in a sample of black ink, a spot of the ink is put on to a piece of chromatography paper. This paper is then set in a suitable solvent as shown in Figure 2.26 on the next page.

As the solvent moves up the paper, the dyes are carried with it and begin to separate. They separate because the substances have different solubilities in the solvent and are absorbed to different degrees by the chromatography paper. As a result, they are separated gradually as the solvent moves up the paper. The **chromatogram** in Figure 2.26b shows how the ink contains three dyes, P, Q and R. Numerical measurements known as **$R_f$ values** can be obtained from chromatograms. An $R_f$ value is defined as the ratio of the distance travelled by the solute (for example P, Q or R) to the distance travelled by the solvent.

a Chromatographic separation of black ink.

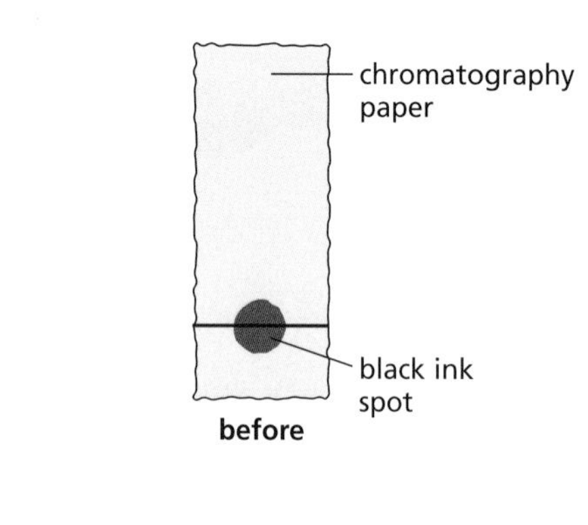


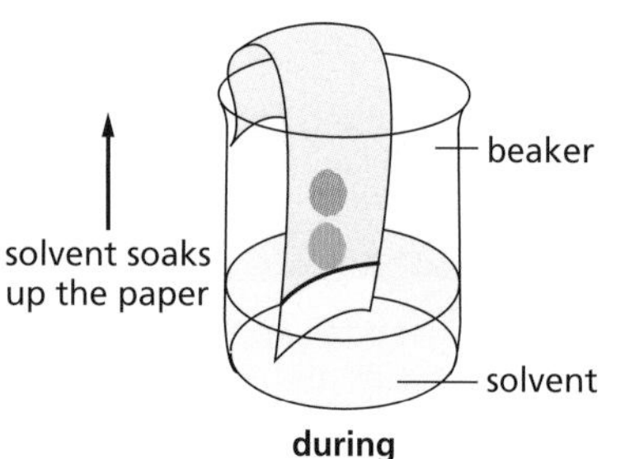


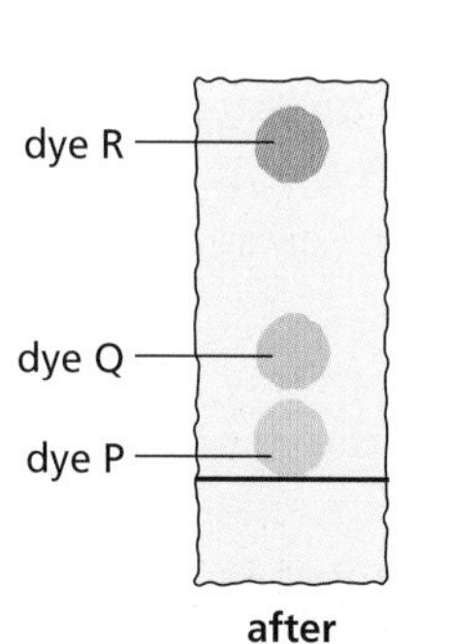


b The black ink separates into three dyes: P, Q and R.

**Figure 2.26**

Chromatography and electrophoresis (separation according to charge) are used extensively in medical research and forensic science laboratories to separate a variety of mixtures (Figure 2.27).

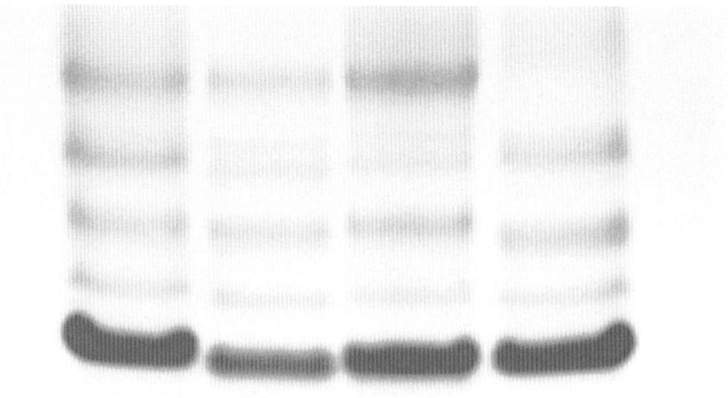

**Figure 2.27** Protein samples are separated by electrophoresis in medical research laboratories.

The substances to be separated do not have to be coloured. Colourless substances can be made visible by spraying the chromatogram with a **locating agent**. The locating agent will react with the colourless substances to form a coloured product. In other situations the position of the substances on the chromatogram may be located using ultraviolet light.

## Solvent extraction

Sugar can be obtained from crushed sugar cane by adding water. The water dissolves the sugar from the sugar cane (Figure 2.28). This is an example of **solvent extraction**. In a similar way some of the green substances can be removed from ground-up grass using ethanol. The substances are extracted from a mixture by using a solvent which dissolves only those substances required.

**Figure 2.28** Sugar can be extracted from sugar cane by using a suitable solvent.

Drugs have to be manufactured to a very high degree of purity (Figure 2.29). To ensure that the highest possible purity is obtained, the drugs are dissolved in a suitable solvent and subjected to fractional crystallisation.

**Figure 2.29** Drugs are manufactured to a high degree of purity by fractional crystallisation.

## Questions

1. Use your research techniques (including the Internet) to obtain as many examples as you can in which a centrifuge is used.
2. What is the difference between simple distillation and fractional distillation?
3. Describe how you would use chromatography to show whether blue ink contains a single pure dye or a mixture of dyes.
4. Explain the following terms, with the aid of examples:
   **a** miscible
   **b** immiscible
   **c** evaporation
   **d** condensation
   **e** solvent extraction.
5. Devise a method for obtaining salt (sodium chloride) from sea water in the school laboratory.

## *Gels, sols, foams and emulsions*

Gels, sols, foams and emulsions are all examples of mixtures which are formed by mixing two substances (or phases) which cannot mix. These mixtures are often referred to as **colloids**. Colloids are formed if the suspended particles are between 1 nm and 1000 nm in size ($1\,\text{nm} = 1 \times 10^{-9}\,\text{m}$).

Generally colloids cannot be separated by filtration since the size of the dispersed particles is smaller than that of the pores found in the filter paper. Look closely at the food substances shown in Figure 2.30 to see examples of these mixtures.

**a** These jelly-like mixtures of solid and liquid in fruit jelly and cold custard are examples of 'gels'.

**b** Emulsion paint is an example of a 'sol'.

**c** These foams have been formed by trapping bubbles of gas in liquids or solids.

**d** Emulsions are formed by mixing immiscible liquids.

**Figure 2.30**

When you mix a solid with a liquid you sometimes get a gel. A gel is a semi-solid which can move around but not as freely as a liquid. Within a gel the solid makes a kind of network which traps the liquid and makes it unable to flow freely (Figure 2.31).

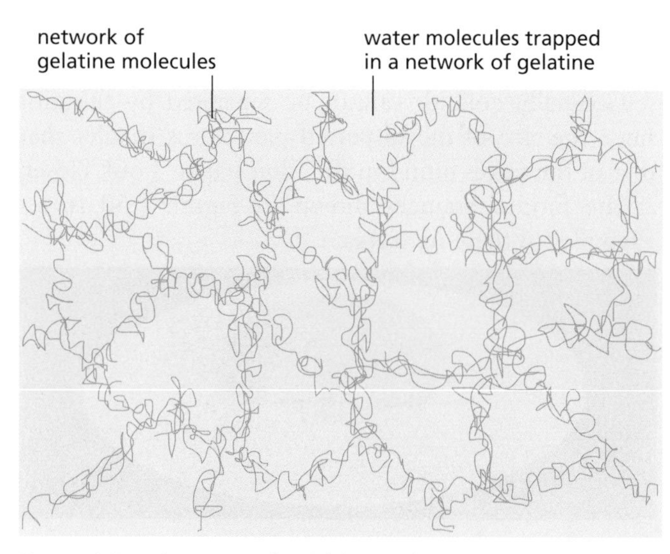


**Figure 2.31** The network within a gel.

A gelatine gel is made with warm water. Gelatine is a protein. Proteins are natural polymers (Chapter 14, p. 211) and the molecules of protein are very large. The large molecules disperse in water to form a gel. As the gelatine in water mixture cools, the gelatine molecules are attracted to each other and form a continuous network. In this way, the jelly you eat as a pudding is formed. The kind of gel which you put into your hair is made from water and an oil (Figure 2.32).

**Figure 2.32** Hair gel is a mixture of water and an oil plus a perfume.

A **sol** is similar to a gel; however, the mixture will flow, for example emulsion paint, or PVA glue.

When you pour out a glass of fizzy drink, the frothy part at the top of the drink is a gas/liquid mixture called a **foam**. The gas, carbon dioxide, has formed tiny bubbles in the liquid but has not dissolved in it. If left to stand, foams like this one collapse as the tiny bubbles join together to form bigger bubbles which then escape. It is possible to form solid foams where the gases are trapped in a solid structure. This happens in foam rubber and bread (Figure 2.33).

**Emulsions** are mixtures of liquids which are immiscible. Earlier in this chapter you found out that when two liquids are immiscible they do not mix but form two different layers. Oil and water are like this but if you shake the mixture it becomes cloudy.

**Figure 2.33** Examples of solid foams.

The apparent mixing that you see is due to the fact that one of the liquids has been broken into tiny droplets which float suspended in the other liquid. If the mixture of oil and water is now left to stand the two layers will re-form. To make emulsions, such as mayonnaise, an **emulsifier** is used to stop the droplets joining back together again to form a separate layer. The emulsifier used when making mayonnaise is egg yolk. If you examine the ingredients on the side of many packets found in your kitchen cupboard you will find that emulsifiers have 'E-numbers' in the range E322 to E494. For example, ammonium phosphatide E442 is used as the emulsifier in cocoa and chocolate. Other food additives such as colourings and preservatives are also given E-numbers but in different ranges to that of the emulsifiers.

It is worth noting that gels, foams and emulsions are all examples of different kinds of solutions. In true solutions the two phases completely mix together but in these systems the two phases are separate.

To produce a stable colloid, the particles dispersed must not only be of the right size (1–1000 nm) but also be prevented from joining back together (coagulating). One way of doing this is to ensure that all the particles possess the same electrical charge. This causes the particles to repel one another.

A colloidal suspension can be destroyed by bringing the dispersed particles together. This process is known as **flocculation**. A method of doing this involves adding ionic substances such as aluminium chloride or aluminium sulphate to the particular colloid. The dispersed particles interact with the added highly charged ions and form particles which are large enough either to settle out under the force of gravity or simply be filtered out. During the treatment of water, aluminium sulphate is added to water prior to filtering to remove suspended solids (Figure 2.34).

**Figure 2.34** Water is treated to remove suspended solids by the addition of aluminium sulphate.

## *Questions*

**1** Explain the following terms:
- **a** colloid
- **b** emulsifier
- **c** foam
- **d** 'E' number
- **e** sol.

**2** Use your research skills (including the Internet) to obtain information about as many common gels, sols, foams and emulsions as you can, other than those given in the text.

# *Mixtures for strength*

## Composite materials

Composite materials are those that combine the properties of two constituents in order to get the exact properties needed for a particular job.

Glass-reinforced plastic (GRP) is an example of a composite material combining the properties of two different materials. It is made by embedding short fibres of glass in a matrix of plastic. The glass fibres give the plastic extra strength so that it does not break when it is bent or moulded into shape. The finished product has the lightness of plastic as well as the strength and flexibility of the glass fibres (Figure 2.35).

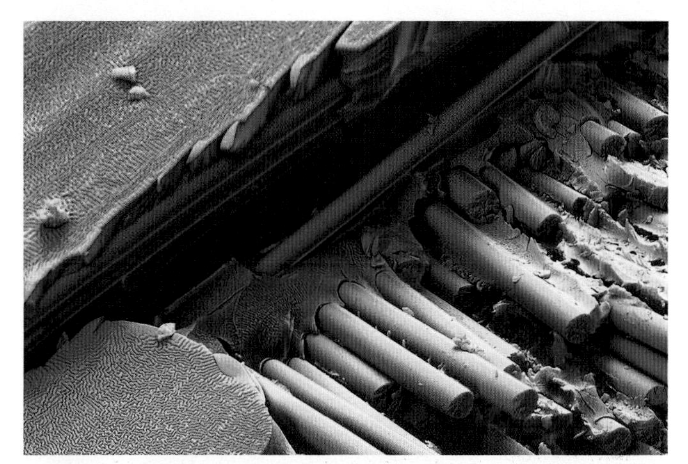

**Figure 2.35** GRP consists of glass fibres (rod shapes) embedded in plastic, in this case polyester.

**Figure 2.36** The glass-reinforced plastic used to make boats like this is a composite material.

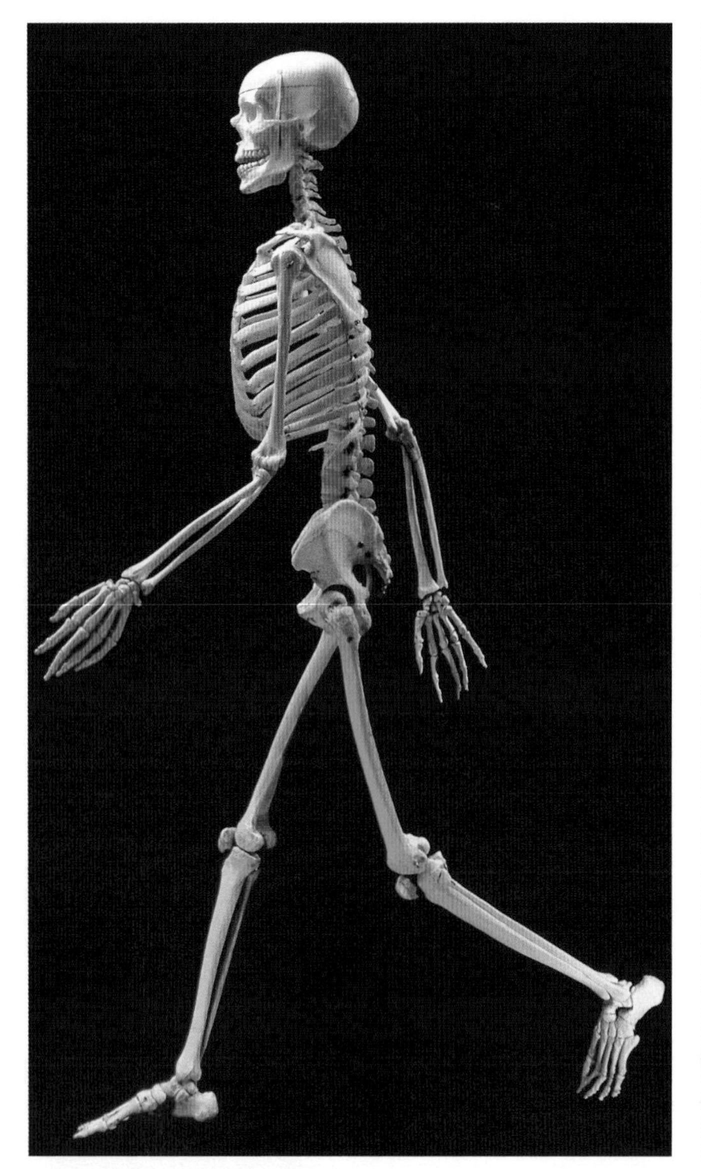

**Figure 2.37** Bone is a composite material.

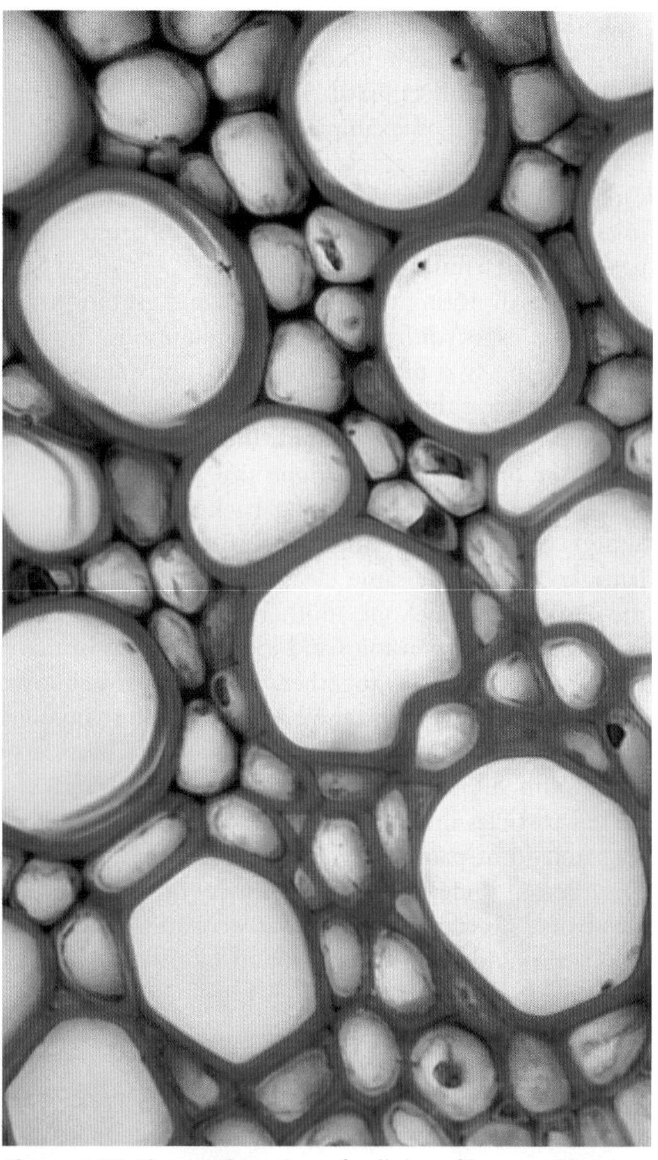

**Figure 2.38** The combination of cellulose fibres and lignin makes the cell walls hard, thick and very strong. These properties reinforce the cells against collapse.

With a little investigation you will find that many composite materials are found in the natural world. Our bones, for example, are a composite material formed from strands of the protein collagen and the mineral calcium phosphate (Figure 2.37). The calcium phosphate is hard and therefore gives strength to the bone. Another example is wood. Wood consists of cellulose fibres mixed with lignin (Figure 2.38), which is largely responsible for the strength of the wood.

## *Questions*

**1** Why are composite materials often used instead of single materials?

**2** Using the information in the text and any other information available to you, give a use other than those already mentioned for each of the following composite materials:

**a** reinforced concrete
**b** glass-reinforced plastic
**c** laminate
**d** glass fibre.

## Checklist

**After studying Chapter 2 you should know and understand the following terms.**

- **Atom** The smallest part of an element that can exist as a stable entity.
- **Centrifuging** The separation of the components of a mixture by rapid spinning. The denser particles are flung to the bottom of the containing tubes. The liquid can then be decanted off.
- **Chemical change** A permanent change in which a new substance is formed.
- **Chemical formula** A shorthand method of representing chemical elements and compounds.
- **Chromatography** A technique employed for the separation of mixtures of dissolved substances.
- **Colloid** Systems in which there are two or more phases, with one (the dispersed phase) distributed in the other (the continuous phase). One of the phases has particles in the range 1 to 1000 nm ($1\,nm = 1 \times 10^{-9}\,m$).
- **Composite materials** Materials which combine the properties of two substances in order to get the exact properties required for a particular job.
- **Compound** A substance formed by the combination of two or more elements in fixed proportions.
- **Crystallisation** The process of forming crystals from a liquid.
- **Decanting** The process of removing a liquid from a solid which has settled or from an immiscible heavier liquid by pouring.
- **Diatomic molecule** A molecule containing two atoms, for example hydrogen, $H_2$, and oxygen, $O_2$.
- **Distillation** The process of boiling a liquid and then condensing the vapour produced back into a liquid. It is used to purify liquids and to separate mixtures of liquids.
- **Element** A substance which cannot be further divided into simpler substances by chemical methods.
- **Emulsion** The apparent mixing of two immiscible liquids by the use of an emulsifier which breaks down one of the liquids into tiny droplets. The droplets of this liquid float suspended in the other liquid so that they do not separate out into different layers.
- **Exothermic reaction** A chemical reaction in which heat energy is produced.
- **Filtrate** The liquid which passes through the filter paper during filtration.
- **Filtration** The process of separating a solid from a liquid, using a fine filter paper which does not allow the solid to pass through.
- **Flocculation** The destruction of a colloidal suspension by bringing the dispersed particles together.
- **Foam** A mixture formed between a gas and a liquid. The gas forms tiny bubbles in the liquid but has not dissolved in it.
- **Gel** A mixture formed between a solid and a liquid in which the solid forms a network which traps the liquid so that it cannot flow freely.
- **Immiscible** When two liquids form two layers when mixed together, they are said to be immiscible.
- **Insoluble** If the solute does not dissolve in the solvent it is said to be insoluble.
- **Instrumental techniques** Instrumental methods of analysis that are particularly useful when the amount of sample is very small. Examples are atomic absorption spectroscopy and infrared spectroscopy.
- **Law of constant composition** Compounds always have the same elements joined together in the same proportions.
- **Metals** A class of chemical elements which have a characteristic lustrous appearance and which are good conductors of heat and electricity.
- **Miscible** When two liquids form a homogeneous layer when mixed together, they are said to be miscible.
- **Mixture** A system of two or more substances that can be separated by physical means.
- **Molecule** A group of atoms chemically bonded together.

*continued*

- **Monatomic molecule** A molecule which consists of only one atom, for example neon and argon.
- **Non-metals** A class of chemical elements that are typically poor conductors of heat and electricity.
- **Oxidation** The process of combining with oxygen.
- **Redox reaction** A reaction which involves the two processes of reduction and oxidation.
- **Reduction** The process of removing oxygen.
- **Residue** The solid left behind in the filter paper after filtration has taken place.
- ***R*$_f$ value** This is the ratio of the distance travelled by the solute to the distance travelled by the solvent in chromatography.
- **Saturated solution** This is a solution which contains as much dissolved solute as it can at a particular temperature.
- **Sol** A mixture formed between a solid and a liquid, which then forms a network that can flow.
- **Soluble** If the solute dissolves in the solvent it is said to be soluble.
- **Solution** This is formed when a substance (solute) disappears (dissolves) into another substance (solvent).

$$\text{solute} + \text{solvent} \xrightarrow{\text{dissolves}} \text{solution}$$

# Elements, compounds and mixtures
# *Additional questions*

1 Define the following terms using specific examples to help with your explanation:
 a element
 b metal
 c non-metal
 d compound
 e molecule
 f mixture
 g flocculation
 h gel
 i foam
 j emulsion
 k sol.

2 Which of the substances listed below are:
 a metallic elements?
 b non-metallic elements?
 c compounds?
 d mixtures?
 Silicon, sea water, calcium, argon, water, air, carbon monoxide, iron, sodium chloride, diamond, brass, copper, dilute sulphuric acid, sulphur, oil, nitrogen, ammonia.

3 At room temperature and pressure (rtp), which of the substances listed below is:
 a a solid element?
 b a liquid element?
 c a gaseous mixture?
 d a solid mixture?
 e a liquid compound?
 f a solid compound?
 Bromine, carbon dioxide, helium, steel, air, oil, marble, copper, water, sand, tin, bronze, mercury, salt.

4 A student heated a mixture of iron filings and sulphur strongly. He saw a red glow spread through the mixture as the reaction continued. At the end of the experiment a black solid had been formed.
 a Explain what the red glow indicates.
 b Give the chemical name of the black solid.
 c Write a word equation and a balanced chemical equation to represent the reaction which has taken place.
 d The black solid is a compound. Explain the difference between the mixture of sulphur and iron and the compound formed by the chemical reaction between them.

5 Name the method which is most suitable for separating the following:
 a the sediment formed at the bottom of a sherry bottle
 b oxygen from liquid air
 c red blood cells from plasma
 d petrol and kerosene from crude oil
 e coffee grains from coffee solution
 f pieces of steel from engine oil
 g amino acids from fruit juice solution
 h ethanol and water.

6 The table below shows the melting points, boiling points and densities of substances **A** to **D**.

| Substance | Melting point/ °C | Boiling point/ °C | Density/ g cm$^{-3}$ |
|---|---|---|---|
| A | 1110 | 2606 | 9.1 |
| B | −266 | −252 | 0.07 |
| C | 40 | 94 | 1.6 |
| D | −14 | 60 | 0.9 |

 a Which substance is a gas at room temperature?
 b Which substance is a liquid at room temperature?
 c Which substances are solids at room temperature?
 d Which substance is most likely to be a metal?
 e Which substance will be a liquid at −260 °C?
 f What is the melting point of the least dense non-metal?
 g Which substance is a gas at 72 °C?

7 a How many atoms of the different elements are there in the formulae of the compounds given below?
 (i) nitric acid, $HNO_3$
 (ii) methane, $CH_4$
 (iii) copper nitrate, $Cu(NO_3)_2$
 (iv) ethanoic acid, $CH_3COOH$
 (v) sugar, $C_{12}H_{22}O_{11}$
 (vi) phenol, $C_6H_5OH$
 (vii) ammonium sulphate, $(NH_4)_2SO_4$
 b Balance the following equations:
 (i) $Zn(s) + O_2(g) \rightarrow ZnO(s)$
 (ii) $Fe(s) + Cl_2(g) \rightarrow FeCl_3(s)$
 (iii) $Li(s) + O_2(g) \rightarrow Li_2O(s)$
 (iv) $H_2(g) + O_2(g) \rightarrow H_2O(g)$
 (v) $Mg(s) + CO_2(g) \rightarrow MgO(s) + C(s)$

8 Carbon-fibre-reinforced plastic (CRP) is used in the manufacture of golf clubs and tennis rackets.
 a What are composite materials?
 b Which two substances are used to manufacture this composite material?

 Consider the data below.

| Material | Strength/ GPa | Stiffness/ GPa | Density/ g cm$^{-3}$ | Relative cost |
|---|---|---|---|---|
| Aluminium | 0.2 | 75 | 2.7 | low |
| Steel | 1.1 | 200 | 7.8 | low |
| CRP | 1.8 | 195 | 1.6 | high |

 c Discuss the advantages and disadvantages of using the three materials above in the manufacture of golf clubs.

# 3 Atomic structure and the periodic table



We have already seen in Chapter 2 that everything you see around you is made out of tiny particles called atoms (Figure 3.1). When John Dalton developed his atomic theory, about 200 years ago (1807/1808), he stated that the atoms of any one element were identical and that each atom was 'indivisible'. Scientists in those days believed that atoms were solid particles like marbles.

However, in the last hundred years or so it has been proved by great scientists, such as Niels Bohr, Albert Einstein, Henry Moseley, Joseph Thomson, Ernest Rutherford and James Chadwick, that atoms are in fact made up of even smaller 'sub-atomic' particles. The most important of these are **electrons**, **protons** and **neutrons**, although 70 sub-atomic particles have now been discovered.

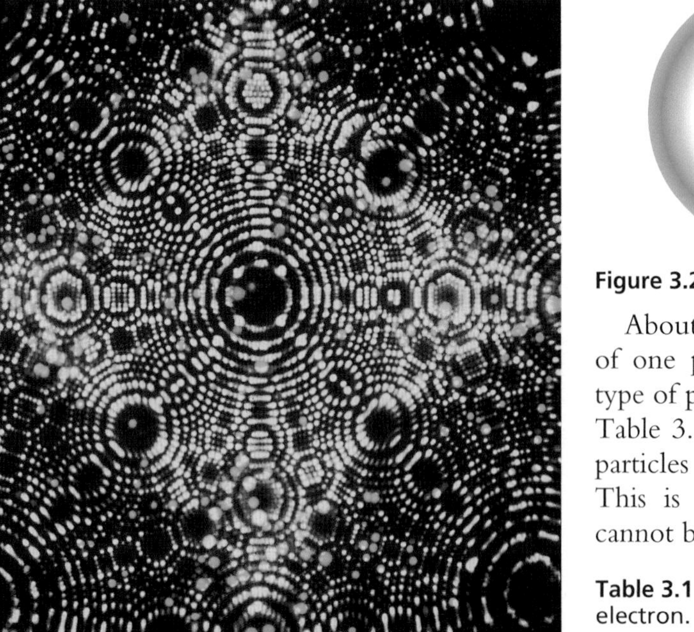

**Figure 3.1** Atoms. A field ion micrograph of atoms of the metal element iridium. The tiny dots are the locations of individual atoms, and the ring-like patterns are facets of a single crystal.

## Inside atoms

The three sub-atomic particles are found in distinct and separate regions. The protons and neutrons are found in the centre of the atom, which is called the **nucleus**. The neutrons have no charge and protons are positively charged. The nucleus occupies only a very small volume of the atom but is very dense.

The rest of the atom surrounding the nucleus is where electrons are most likely to be found. The electrons are negatively charged and move around very quickly in **electron shells** or **energy levels**. The electrons are held within the atom by an **electrostatic force of attraction** between themselves and the positive charge of the protons in the nucleus (Figure 3.2).

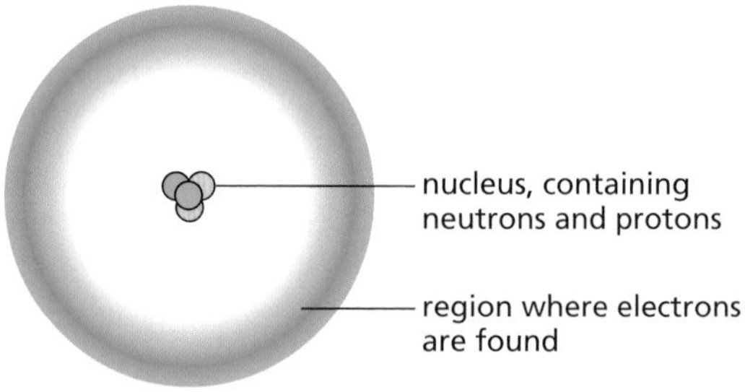

nucleus, containing neutrons and protons
region where electrons are found

**Figure 3.2** Diagram of an atom.

About 1837 electrons are equal in mass to the mass of one proton or one neutron. A summary of each type of particle, its mass and relative charge is shown in Table 3.1. You will notice that the masses of all these particles are measured in **atomic mass units (amu)**. This is because they are so light that their masses cannot be measured usefully in grams.

**Table 3.1** Characteristics of a proton, a neutron and an electron.

| Particle | Symbol | Relative mass/amu | Relative charge |
|---|---|---|---|
| Proton | p | 1 | +1 |
| Neutron | n | 1 | 0 |
| Electron | e | 1/1837 | −1 |

Although atoms contain electrically charged particles, the atoms themselves are electrically neutral (they have no overall electric charge). This is because atoms contain equal numbers of electrons and protons. For example, the diagram in Figure 3.3 represents the atom of the non-metallic element helium. The atom of helium possesses two protons, two neutrons and two electrons. The electrical charge of the protons in the nucleus is, therefore, balanced by the opposite charge of the two electrons.

## Atomic number and mass number

The number of protons in the nucleus of an atom is called the **atomic number** (or proton number) and is given the symbol $\boldsymbol{Z}$. Hence in the diagram shown in Figure 3.3, the helium atom has an atomic number of 2, since it has two protons in its nucleus. Each element has its own atomic number and no two different elements have the same atomic number. For example, a different element, lithium, has an atomic number of 3, since it has three protons in its nucleus.

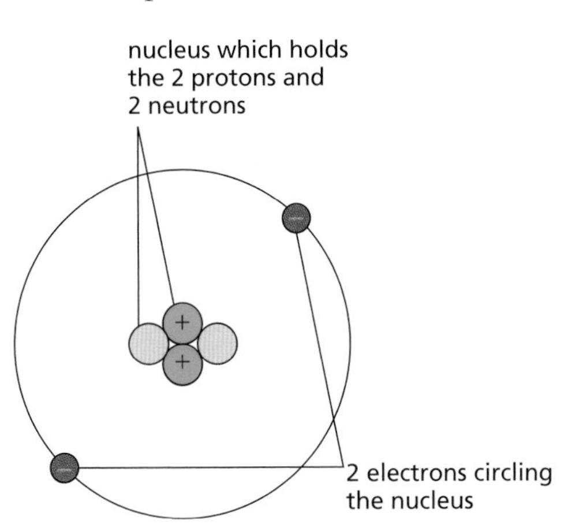


**Figure 3.3** An atom of helium has two protons, two electrons and two neutrons.

Neutrons and protons have a similar mass. Electrons possess very little mass. So the mass of any atom depends on the number of neutrons and protons in its nucleus. The total number of protons and neutrons found in the nucleus of an atom is called the **mass number** (or nucleon number) and is given the symbol $\boldsymbol{A}$.

$$\underset{(A)}{\text{mass number}} = \underset{(Z)}{\text{atomic number}} + \underset{\text{neutrons}}{\text{number of}}$$

Hence, in the example shown in Figure 3.3 the helium atom has a mass number of 4, since it has two protons and two neutrons in its nucleus. If we consider the metallic element lithium, it has three protons and four neutrons in its nucleus. It therefore has a mass number of 7.

The atomic number and mass number of an element are usually written in the following shorthand way:

mass number ($A$) ↘

${}^{4}_{2}\text{He}$←symbol of the element

atomic number ($Z$) ↗

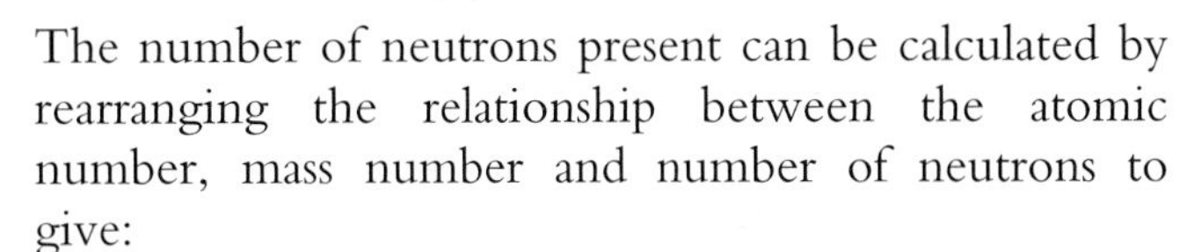

The number of neutrons present can be calculated by rearranging the relationship between the atomic number, mass number and number of neutrons to give:

$$\underset{\text{neutrons}}{\text{number of}} = \underset{(A)}{\text{mass number}} - \underset{(Z)}{\text{atomic number}}$$

For example, the number of neutrons in one atom of ${}^{24}_{12}\text{Mg}$ is:

$$\underset{(A)}{24} - \underset{(Z)}{12} = 12$$

and the number of neutrons in one atom of ${}^{207}_{82}\text{Pb}$ is:

$$\underset{(A)}{207} - \underset{(Z)}{82} = 125$$

Table 3.2 shows the number of protons, neutrons and electrons in the atoms of some common elements.

**Table 3.2** Number of protons, neutrons and electrons in some elements.

| Element | Symbol | Atomic number | Number of electrons | Number of protons | Number of neutrons | Mass number |
|---|---|---|---|---|---|---|
| Hydrogen | H | 1 | 1 | 1 | 0 | 1 |
| Helium | He | 2 | 2 | 2 | 2 | 4 |
| Carbon | C | 6 | 6 | 6 | 6 | 12 |
| Nitrogen | N | 7 | 7 | 7 | 7 | 14 |
| Oxygen | O | 8 | 8 | 8 | 8 | 16 |
| Fluorine | F | 9 | 9 | 9 | 10 | 19 |
| Neon | Ne | 10 | 10 | 10 | 10 | 20 |
| Sodium | Na | 11 | 11 | 11 | 12 | 23 |
| Magnesium | Mg | 12 | 12 | 12 | 12 | 24 |
| Sulphur | S | 16 | 16 | 16 | 16 | 32 |
| Potassium | K | 19 | 19 | 19 | 20 | 39 |
| Calcium | Ca | 20 | 20 | 20 | 20 | 40 |
| Iron | Fe | 26 | 26 | 26 | 30 | 56 |
| Zinc | Zn | 30 | 30 | 30 | 35 | 65 |

## Ions

An ion is an electrically charged particle. When an atom loses one or more electrons it becomes a positively charged ion. For example, during the chemical reactions of potassium, each atom loses an electron to form a positive ion, $K^+$.

$$
{}_{19}K^+ \quad \begin{array}{ll} 19 \text{ protons} & = 19+ \\ 18 \text{ electrons} & = 18- \\ \hline \text{Overall charge} & = 1+ \end{array}
$$

When an atom gains one or more electrons it becomes a negatively charged ion. For example, during some of the chemical reactions of chlorine it gains an electron to form a negative ion, $Cl^-$.

$$
{}_{17}Cl^- \quad \begin{array}{ll} 17 \text{ protons} & = 17+ \\ 18 \text{ electrons} & = 18- \\ \hline \text{Overall charge} & = 1- \end{array}
$$

Table 3.3 shows some common ions. You will notice from Table 3.3 that:

- some ions contain more than one type of atom, for example $NO_3^-$
- an ion may possess more than one unit of charge (either negative or positive), for example $Al^{3+}$, $O^{2-}$ or $SO_4^{2-}$.

**Table 3.3** Some common ions.

| Name | Formula |
|---|---|
| Lithium ion | $Li^+$ |
| Sodium ion | $Na^+$ |
| Potassium ion | $K^+$ |
| Magnesium ion | $Mg^{2+}$ |
| Calcium ion | $Ca^{2+}$ |
| Aluminium ion | $Al^{3+}$ |
| Zinc ion | $Zn^{2+}$ |
| Ammonium ion | $NH_4^+$ |
| Fluoride ion | $F^-$ |
| Chloride ion | $Cl^-$ |
| Bromide ion | $Br^-$ |
| Hydroxide ion | $OH^-$ |
| Oxide ion | $O^{2-}$ |
| Sulphide ion | $S^{2-}$ |
| Carbonate ion | $CO_3^{2-}$ |
| Nitrate ion | $NO_3^-$ |
| Sulphate ion | $SO_4^{2-}$ |

## Isotopes

Not all of the atoms in a sample of chlorine, for example, will be identical. Some atoms of the same element can contain different numbers of neutrons. Atoms of the same element which have different numbers of neutrons are called **isotopes**.

The two isotopes of chlorine are shown in Figure 3.4. Generally, isotopes behave in the same way during chemical reactions. The only effect of the extra neutrons is to alter the mass of the atom and properties which depend on it, such as density. Some other examples of atoms with isotopes are shown in Table 3.4.

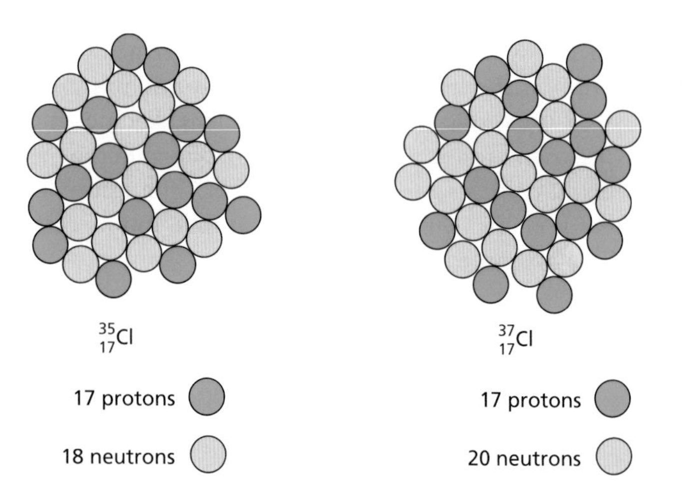


**Figure 3.4** The two isotopes of chlorine.

Some of the atoms of certain isotopes are unstable because of the extra number of neutrons, and they are said to be radioactive. The best known elements which have radioactive isotopes are uranium and carbon. The major isotopes of these elements are shown in Table 3.4.

**Table 3.4** Some atoms and their isotopes.

| Element | Symbol | Particles present |
|---|---|---|
| Hydrogen | $^{1}_{1}H$ | 1 e, 1 p, 0 n |
| (Deuterium) | $^{2}_{1}H$ | 1 e, 1 p, 1 n |
| (Tritium) | $^{3}_{1}H$ | 1 e, 1 p, 2 n |
| Carbon | $^{12}_{6}C$ | 6 e, 6 p, 6 n |
| | $^{13}_{6}C$ | 6 e, 6 p, 7 n |
| | $^{14}_{6}C$ | 6 e, 6 p, 8 n |
| Oxygen | $^{16}_{8}O$ | 8 e, 8 p, 8 n |
| | $^{17}_{8}O$ | 8 e, 8 p, 9 n |
| | $^{18}_{8}O$ | 8 e, 8 p, 10 n |
| Strontium | $^{86}_{38}Sr$ | 38 e, 38 p, 48 n |
| | $^{88}_{38}Sr$ | 38 e, 38 p, 50 n |
| | $^{90}_{38}Sr$ | 38 e, 38 p, 52 n |
| Uranium | $^{235}_{92}U$ | 92 e, 92 p, 143 n |
| | $^{238}_{92}U$ | 92 e, 92 p, 146 n |

## The mass spectrometer

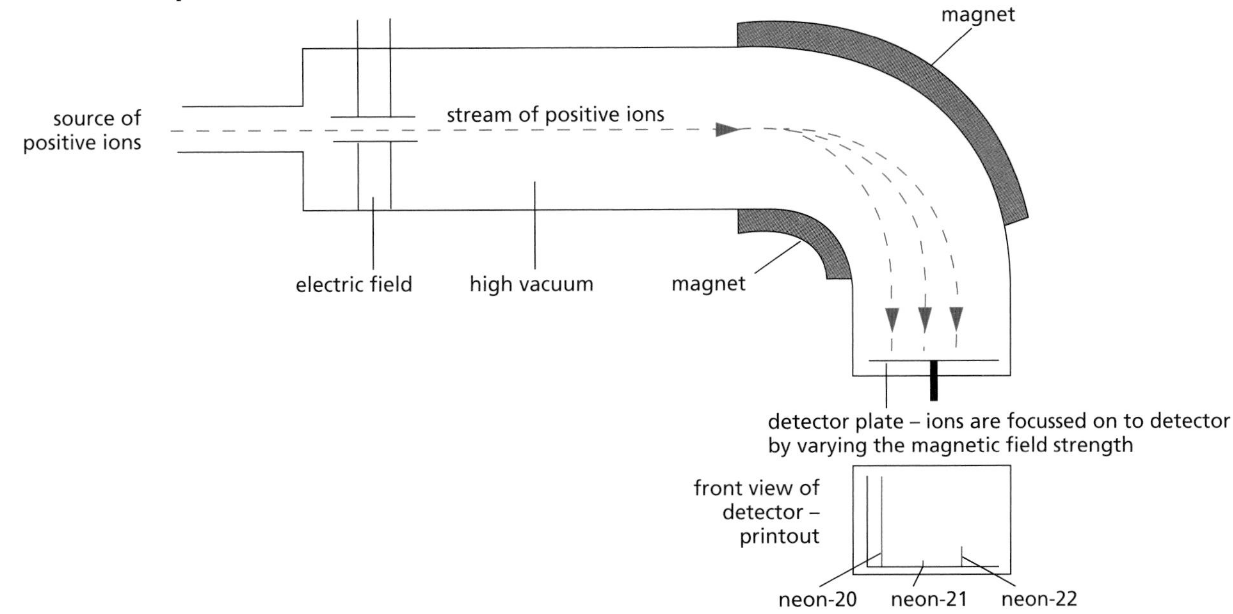


**Figure 3.5** A diagram of a mass spectrometer.

How do we know isotopes exist? They were first discovered by scientists using apparatus called a mass spectrometer (Figure 3.5). The first mass spectrometer was built by the British scientist Francis Aston in 1919 and enabled scientists to compare the relative masses of atoms accurately for the first time.

A vacuum exists inside a mass spectrometer. A sample of the vapour of the element is injected into a chamber where it is bombarded by electrons. The subsequent collisions that take place cause the atoms in the vapour to lose one of their electrons and so form positive ions. The beam of positive ions is accelerated by an electric field and then deflected by a magnetic field. The amount of deflection depends on the different masses of the positive ions. The lighter ions, which were formed from the lighter isotopes, are deflected more than the heavier ones. In this way particles with different masses can be separated and identified. A detector counts the number of each of the ions that fall upon it and so a measure of the percentage abundance of each isotope is obtained. A typical mass spectrum for chlorine is shown in Figure 3.6.

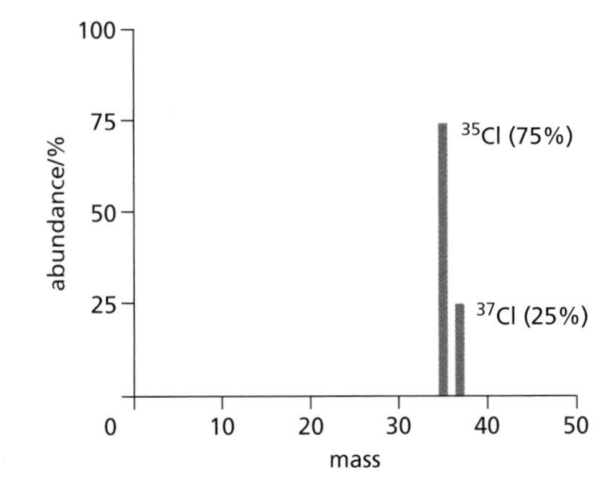


**Figure 3.6** The mass spectrum for chlorine.

## Relative atomic mass

The average mass of a large number of atoms of an element is called its **relative atomic mass** (symbol $\boldsymbol{A_r}$). This quantity takes into account the percentage abundance of all the isotopes of an element which exist.

In 1961 the International Union of Pure and Applied Chemistry (IUPAC) recommended that the standard used for the $A_r$ scale was carbon-12. An atom of carbon-12 was taken to have a mass of 12 amu. The $A_r$ of an element is now defined as the average mass of its isotopes compared with one-twelfth the mass of one atom of carbon-12:

$$A_r = \frac{\text{average mass of isotopes of the element}}{\frac{1}{12} \times \text{mass of 1 atom of carbon-12}}$$

Note: $\frac{1}{12}$ of the mass of one carbon-12 atom = 1 amu.

For example, chlorine has two isotopes:

| | $^{35}_{17}$Cl | $^{37}_{17}$Cl |
|---|---|---|
| % Abundance | 75 | 25 |

Hence the 'average mass' of a chlorine atom is:

$$\frac{(75 \times 35) + (25 \times 37)}{100} = 35.5$$

$$A_r = \frac{35.5}{1}$$

$$= 35.5 \text{ amu}$$

### Questions

**1** Calculate the number of neutrons in the following atoms:
**a** $^{27}_{13}$Al **b** $^{31}_{15}$P **c** $^{262}_{107}$Uns **d** $^{190}_{76}$Os

**2** Given that the percentage abundance of $^{20}_{10}$Ne is 90% and that of $^{22}_{10}$Ne is 10%, calculate the $A_r$ of neon.

## The arrangement of electrons in atoms

The nucleus of an atom contains the heavier sub-atomic particles – the protons and the neutrons. The electrons, the lightest of the sub-atomic particles, move around the nucleus at great distances from the nucleus relative to their size. They move very fast in electron energy levels very much as the planets orbit the Sun.

It is not possible to give the exact position of an electron in an energy level. However, we can state that electrons can only occupy certain, definite energy levels and that they cannot exist between them. Each of the electron energy levels can hold only a certain number of electrons.

- First energy level holds up to 2 electrons.
- Second energy level holds up to 8 electrons.
- Third energy level holds up to 18 electrons.

There are further energy levels which contain increasing numbers of electrons.

The third energy level can be occupied by a maximum of 18 electrons. However, when eight electrons have occupied this level a certain stability is given to the atom and the next two electrons go into the fourth energy level, and then the remaining ten electrons complete the third energy level.

The electrons fill the energy levels starting from the energy level nearest to the nucleus, which has the lowest energy. When this is full (with two electrons) the next electron goes into the second energy level. When this energy level is full with eight electrons, then the electrons begin to fill the third and fourth energy levels as stated above.

For example, a $^{16}_{8}O$ atom has an atomic number of 8 and therefore has eight electrons. Two of the eight electrons enter the first energy level, leaving six to occupy the second energy level, as shown in Figure 3.7. The electron configuration for oxygen can be written in a shorthand way as 2,6.

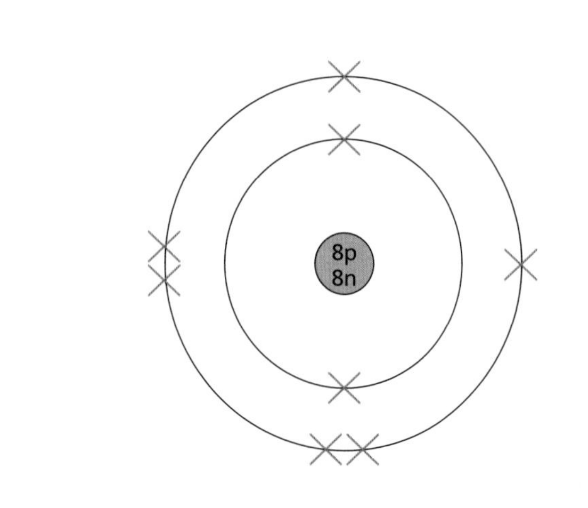


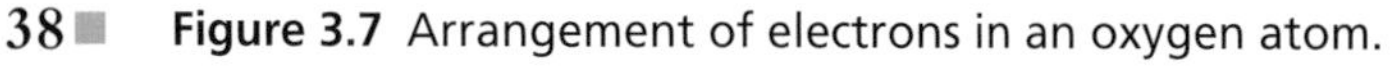
**Figure 3.7** Arrangement of electrons in an oxygen atom.

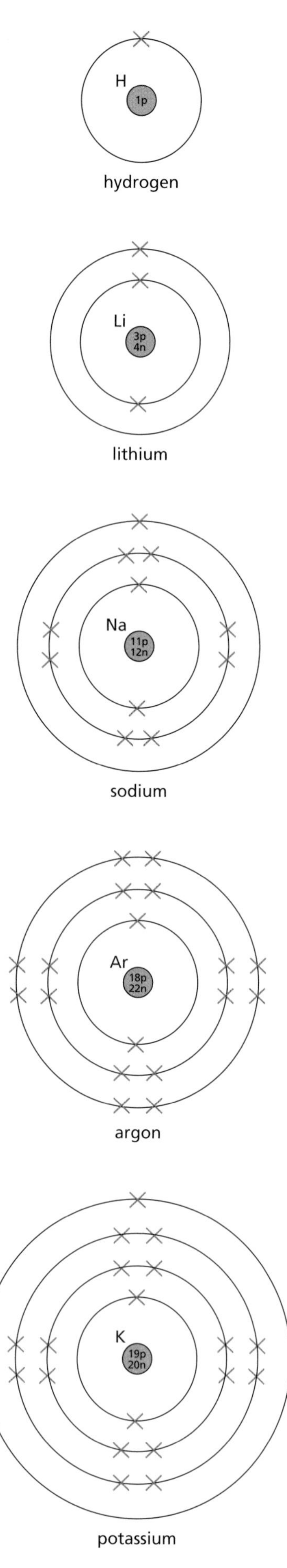


**Figure 3.8** Electron arrangements of hydrogen, lithium, sodium, argon and potassium.

There are 115 elements, and Table 3.5 shows the way in which the electrons are arranged in the first 20 of these elements. The way in which the electrons are distributed is called the **electron structure** or **electron configuration**. Figure 3.8 shows the electron configuration of a selection of atoms.

**Table 3.5** Electron arrangement in the first 20 elements.

| Element | Symbol | Atomic number | Number of electrons | Electron structure |
|---|---|---|---|---|
| Hydrogen | H | 1 | 1 | 1 |
| Helium | He | 2 | 2 | 2 |
| Lithium | Li | 3 | 3 | 2,1 |
| Beryllium | Be | 4 | 4 | 2,2 |
| Boron | B | 5 | 5 | 2,3 |
| Carbon | C | 6 | 6 | 2,4 |
| Nitrogen | N | 7 | 7 | 2,5 |
| Oxygen | O | 8 | 8 | 2,6 |
| Fluorine | F | 9 | 9 | 2,7 |
| Neon | Ne | 10 | 10 | 2,8 |
| Sodium | Na | 11 | 11 | 2,8,1 |
| Magnesium | Mg | 12 | 12 | 2,8,2 |
| Aluminium | Al | 13 | 13 | 2,8,3 |
| Silicon | Si | 14 | 14 | 2,8,4 |
| Phosphorus | P | 15 | 15 | 2,8,5 |
| Sulphur | S | 16 | 16 | 2,8,6 |
| Chlorine | Cl | 17 | 17 | 2,8,7 |
| Argon | Ar | 18 | 18 | 2,8,8 |
| Potassium | K | 19 | 19 | 2,8,8,1 |
| Calcium | Ca | 20 | 20 | 2,8,8,2 |

## Questions

1. How many electrons may be accommodated in the first three energy levels?
2. What is the same about the electron structures of:
   a. lithium, sodium and potassium?
   b. beryllium, magnesium and calcium?

## *The periodic table*

The periodic table was devised in 1869 by the Russian Dmitri Mendeleev, who was the Professor of Chemistry at St Petersburg University (Figure 3.9). His periodic table was based on the chemical and physical properties of the 63 elements that had been discovered at that time.

**Figure 3.9** Dmitri Mendeleev (1834–1907).

However, other scientists had also attempted to categorise the known elements. In 1817, Johann Döbereiner noticed that the atomic weight (now called atomic mass) of strontium fell midway between the weights of calcium and barium. These were elements which possessed similar chemical properties. They formed a **triad** of elements. Other triads were also discovered, composed of:

chlorine, bromine, iodine

lithium, sodium, potassium

He called this the '**Law of Triads**'. This encouraged other scientists to search for patterns.

In 1865, John Newlands, an English chemist, arranged the 56 known elements in order of increasing atomic weight. He realised when he did this that every eighth element in the series was similar.

H **Li** Be B C N O F **Na** Mg Al Si P S Cl **K**

He likened this to music and called it the '**Law of Octaves**'. It fell down, however, because some of the weights were inaccurate and there were elements that had not been discovered then.

| Period | Group | | | | | | | |
|---|---|---|---|---|---|---|---|---|
| | 1 | 2 | 3 | 4 | 5 | 6 | 7 | 8 |
| 1 | H | | | | | | | |
| 2 | Li | Be | B | C | N | O | F | |
| 3 | Na | Mg | Al | Si | P | S | Cl | |
| 4 | K | Ca | * | Ti | V | Cr | Mn | Fe Co Ni |
| | Cu | Zn | * | † | As | Se | Br | |

**Figure 3.10** Mendeleev's periodic table. He left gaps for undiscovered elements.

Mendeleev's classification proved to be the most successful. Mendeleev arranged all the 63 known elements in order of increasing atomic weight but in such a way that elements with similar properties were in the same vertical column. He called the vertical columns **groups** and the horizontal rows **periods** (Figure 3.10). If necessary he left gaps in the table.

As a scientific idea, Mendeleev's periodic table was tested by making predictions about elements that were unknown at that time but could possibly fill the gaps. Three of these gaps are shown by the symbols * and † in Figure 3.10. As new elements were discovered, they were found to fit easily into the classification. For example, Mendeleev predicted the properties of the missing element 'eka-silicon' (†). He predicted the colour, density and melting point as well as its atomic weight.

In 1886 the element we now know as germanium was discovered in Germany by Clemens Winkler; its properties were almost exactly those Mendeleev had predicted. In all, Mendeleev predicted the atomic weight of ten new elements, of which seven were eventually discovered – the other three, atomic weights 45, 146 and 175 do not exist!

The success of Mendeleev's predictions showed that his ideas were probably correct. His periodic table was quickly accepted by scientists as an important summary of the properties of the elements.

Mendeleev's periodic table has been modified in the light of work carried out by Rutherford and Moseley. Discoveries about sub-atomic particles led them to realise that the elements should be arranged by atomic number. In the modern periodic table the 115 known elements are arranged in order of increasing atomic number (Figure 3.11).

| Period | Group 1 | 2 | | | | | | | | | | | 3 | 4 | 5 | 6 | 7 | 0 |
|---|---|---|---|---|---|---|---|---|---|---|---|---|---|---|---|---|---|---|
| 1 | | | | | $^{1}_{1}$H Hydrogen | | | | | | | | | | | | | $^{4}_{2}$He Helium |
| 2 | $^{7}_{3}$Li Lithium | $^{9}_{4}$Be Beryllium | | | | | | | | | | | $^{11}_{5}$B Boron | $^{12}_{6}$C Carbon | $^{14}_{7}$N Nitrogen | $^{16}_{8}$O Oxygen | $^{19}_{9}$F Fluorine | $^{20}_{10}$Ne Neon |
| 3 | $^{23}_{11}$Na Sodium | $^{24}_{12}$Mg Magnesium | | | | | | | | | | | $^{27}_{13}$Al Aluminium | $^{28}_{14}$Si Silicon | $^{31}_{15}$P Phosphorus | $^{32}_{16}$S Sulphur | $^{35.5}_{17}$Cl Chlorine | $^{40}_{18}$Ar Argon |
| 4 | $^{39}_{19}$K Potassium | $^{40}_{20}$Ca Calcium | $^{45}_{21}$Sc Scandium | $^{48}_{22}$Ti Titanium | $^{51}_{23}$V Vanadium | $^{52}_{24}$Cr Chromium | $^{55}_{25}$Mn Manganese | $^{56}_{26}$Fe Iron | $^{59}_{27}$Co Cobalt | $^{59}_{28}$Ni Nickel | $^{63.5}_{29}$Cu Copper | $^{65}_{30}$Zn Zinc | $^{70}_{31}$Ga Gallium | $^{73}_{32}$Ge Germanium | $^{75}_{33}$As Arsenic | $^{79}_{34}$Se Selenium | $^{80}_{35}$Br Bromine | $^{84}_{36}$Kr Krypton |
| 5 | $^{85}_{37}$Rb Rubidium | $^{88}_{38}$Sr Strontium | $^{89}_{39}$Y Yttrium | $^{91}_{40}$Zr Zirconium | $^{93}_{41}$Nb Niobium | $^{96}_{42}$Mo Molybdenum | $^{99}_{43}$Tc Technetium | $^{101}_{44}$Ru Ruthenium | $^{103}_{45}$Rh Rhodium | $^{106}_{46}$Pd Palladium | $^{108}_{47}$Ag Silver | $^{112}_{48}$Cd Cadmium | $^{115}_{49}$In Indium | $^{119}_{50}$Sn Tin | $^{122}_{51}$Sb Antimony | $^{128}_{52}$Te Tellurium | $^{127}_{53}$I Iodine | $^{131}_{54}$Xe Xenon |
| 6 | $^{133}_{55}$Cs Caesium | $^{137}_{56}$Ba Barium | • | $^{178.5}_{72}$Hf Hafnium | $^{181}_{73}$Ta Tantalum | $^{184}_{74}$W Tungsten | $^{186}_{75}$Re Rhenium | $^{190}_{76}$Os Osmium | $^{192}_{77}$Ir Iridium | $^{195}_{78}$Pt Platinum | $^{197}_{79}$Au Gold | $^{201}_{80}$Hg Mercury | $^{204}_{81}$Tl Thallium | $^{207}_{82}$Pb Lead | $^{209}_{83}$Bi Bismuth | $^{209}_{84}$Po Polonium | $^{210}_{85}$At Astatine | $^{222}_{86}$Rn Radon |
| 7 | $^{223}_{87}$Fr Francium | $^{226}_{88}$Ra Radium | • | $^{261}_{104}$Rf Rutherfordium | $^{262}_{105}$Db Dubnium | $^{263}_{106}$Sg Seaborgium | $^{262}_{107}$Bh Bohrium | $^{269}_{108}$Hs Hassium | $^{268}_{109}$Mt Meitnerium | $^{269}_{110}$Uun Ununnilium | $^{272}_{111}$Uuu Unununium | $^{277}_{112}$Uub Ununbium | | $^{285}_{114}$Uuq Ununquadium | | $^{289}_{116}$Uuh Ununhexium | | $^{293}_{118}$Uno Ununoctium |

| | | | | | | | | | | | | | | |
|---|---|---|---|---|---|---|---|---|---|---|---|---|---|---|
| $^{139}_{57}$La Lanthanum | $^{140}_{58}$Ce Cerium | $^{141}_{59}$Pr Praseodymium | $^{144}_{60}$Nd Neodymium | $^{147}_{61}$Pm Promethium | $^{150}_{62}$Sm Samarium | $^{152}_{63}$Eu Europium | $^{157}_{64}$Gd Gadolinium | $^{159}_{65}$Tb Terbium | $^{162}_{66}$Dy Dysprosium | $^{165}_{67}$Ho Holmium | $^{167}_{68}$Er Erbium | $^{169}_{69}$Tm Thulium | $^{173}_{70}$Yb Ytterbium | $^{175}_{71}$Lu Lutetium |
| $^{227}_{89}$Ac Actinium | $^{232}_{90}$Th Thorium | $^{231}_{91}$Pa Protactinium | $^{238}_{92}$U Uranium | $^{237}_{93}$Np Neptunium | $^{244}_{94}$Pu Plutonium | $^{243}_{95}$Am Americium | $^{247}_{96}$Cm Curium | $^{247}_{97}$Bk Berkelium | $^{251}_{98}$Cf Californium | $^{252}_{99}$Es Einsteinium | $^{257}_{100}$Fm Fermium | $^{258}_{101}$Md Mendelevium | $^{259}_{102}$No Nobelium | $^{260}_{103}$Lr Lawrencium |

**Key**

- reactive metals
- transition metals
- poor metals
- metalloids
- non-metals
- noble gases

**Figure 3.11** The modern periodic table.

**Figure 3.12** Transition elements have a wide range of uses, both as elements and as alloys.

Those elements with similar chemical properties are found in the same columns or **groups**. There are eight groups of elements. The first column is called group 1; the second group 2; and so on up to group 7. The final column in the periodic table is called group 0. Some of the groups have been given names.

Group 1: The alkali metals
Group 2: The alkaline earth metals
Group 7: The halogens
Group 0: Inert gases or noble gases

The horizontal rows are called **periods** and these are numbered 1–7 going down the periodic table.

Between groups 2 and 3 is the block of elements known as the transition elements (Figure 3.12).

The periodic table can be divided into two as shown by the bold line that starts beneath boron, opposite. The elements to the left of this line are metals (fewer than three-quarters) and those on the right are non-metals (fewer then one-quarter). The elements which lie on this dividing line are known as metalloids (Figure 3.13). These elements behave in some ways as metals and in others as non-metals.

**Figure 3.13** The metalloid silicon is used to make silicon 'chips'.

## Electron configuration and the periodic table

Now that the number of electrons in the outer energy level has been established, it can be seen that it corresponds with the number of the group in the periodic table in which the element is found. For example, the elements shown in Table 3.6 have one electron in their outer energy level and they are all found in group 1. The elements in group 0, however, are an exception to this rule, as they have two or eight electrons in their outer energy level. The outer electrons are mainly responsible for the chemical properties of any element, and, therefore, elements in the same group have similar chemical properties (Tables 3.7 and 3.8).

**Table 3.6** Electron configuration of the first three elements of group 1.

| Element | Symbol | Atomic number | Electron configuration |
|---|---|---|---|
| Lithium | Li | 3 | 2,1 |
| Sodium | Na | 11 | 2,8,1 |
| Potassium | K | 19 | 2,8,8,1 |

**Table 3.7** Electron configuration of the first three elements of group 2.

| Element | Symbol | Atomic number | Electron configuration |
|---|---|---|---|
| Beryllium | Be | 4 | 2,2 |
| Magnesium | Mg | 12 | 2,8,2 |
| Calcium | Ca | 20 | 2,8,8,2 |

**Table 3.8** Electron configuration of the first three elements in group 7.

| Element | Symbol | Atomic number | Electron configuration |
|---|---|---|---|
| Fluorine | F | 9 | 2,7 |
| Chlorine | Cl | 17 | 2,8,7 |
| Bromine | Br | 35 | 2,8,18,7 |

## Group 1 – the alkali metals

Group 1 consists of the five metals lithium, sodium, potassium, rubidium and caesium, and the radioactive element francium. Lithium, sodium and potassium are commonly available for use in school. They are all very reactive metals and they are stored under oil to prevent them coming into contact with water or air. These three metals have the following properties.

- They are good conductors of electricity and heat.
- They are soft metals.
- They are metals with low densities.
- They have shiny surfaces when freshly cut with a knife (Figure 3.14).

**Figure 3.14** Freshly cut sodium.

- They burn in oxygen or air, with characteristic flame colours, to form white solid oxides. For example, lithium reacts with the oxygen in the air to form white lithium oxide, according to the following equation:

  lithium + oxygen → lithium oxide

  $$4Li(s) + O_2(g) \rightarrow 2Li_2O(s)$$

  These group 1 oxides all dissolve in water to form alkaline solutions of the metal hydroxide.

  lithium oxide + water → lithium hydroxide

  $$Li_2O(s) + H_2O(l) \rightarrow 2LiOH(aq)$$

- They react vigorously with water to give an alkaline solution of the metal hydroxide as well as producing hydrogen gas.

  For example:

  potassium + water → potassium hydroxide + hydrogen gas

  $$2K(s) + 2H_2O(l) \rightarrow 2KOH(aq) + H_2(g)$$

  Of these three metals, potassium is the most reactive towards water (Figure 3.15), followed by sodium and then lithium. Such gradual changes we call **trends**. Trends are useful to chemists as they allow predictions to be made about elements we have not observed in action.
- They react vigorously with halogens, such as chlorine, to form metal halides, for example sodium chloride (Figure 3.16).

  sodium + chorine → sodium chloride

  $$2Na(s) + Cl_2(g) \rightarrow 2NaCl(s)$$

Considering the group as a whole, the further down the group you go the more reactive the metals become. Francium is, therefore, the most reactive of the group 1 metals.

a Potassium reacts very vigorously with cold water.

**b** An alkaline solution is produced when potassium reacts with water.

**Figure 3.15**

**Figure 3.16** A very vigorous reaction takes place when sodium burns in chlorine gas. Sodium chloride is produced.

Table 3.6 shows the electron configuration of the first three elements of group 1. You will notice in each case that the outer energy level contains only one electron. When these elements react they lose this outer electron, and in doing so become more stable, because they obtain the electron configuration of a noble gas. You will learn more about the stable nature of these gases later in this chapter (p. 48).

When, for example, the element sodium reacts it loses its outer electron. This requires energy to overcome the electrostatic attractive forces between the outer electron and the positive nucleus (Figure 3.17).

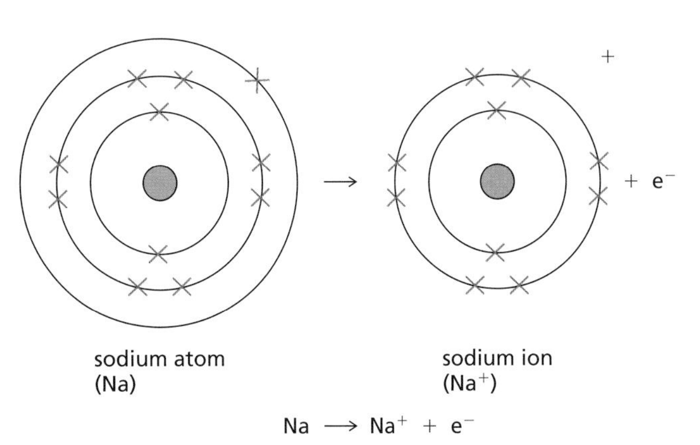


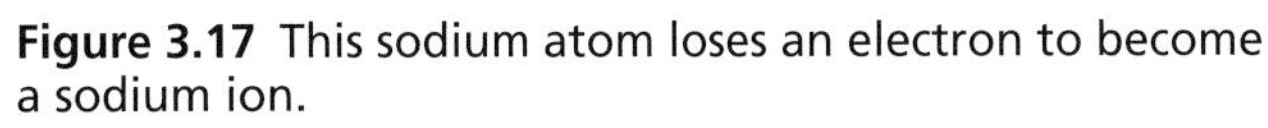


**Figure 3.17** This sodium atom loses an electron to become a sodium ion.

Look at Figure 3.18. Why do you think potassium is more reactive than lithium or sodium?

Potassium is more reactive because less energy is required to remove the outer electron from its atom than for lithium or sodium. This is because as you go down the group the size of the atoms increases and the outer electron gets further away from the nucleus and becomes easier to remove.

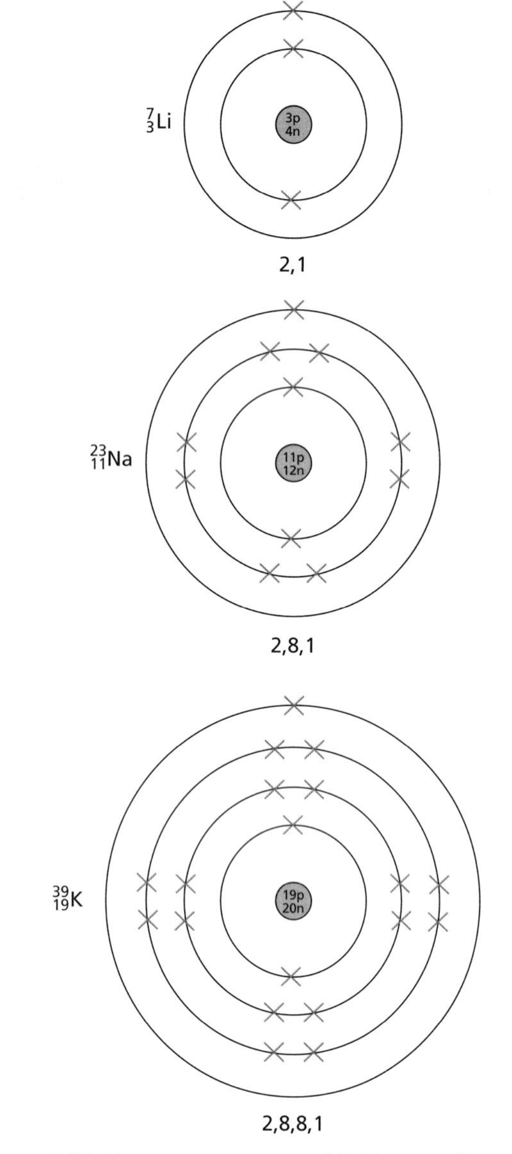

**Figure 3.18** Electron structures of lithium, sodium and potassium.

## Questions

**1** Write word and balanced chemical equations for the reactions between:
**a** sodium and oxygen
**b** sodium and water.

**2** Write word and balanced chemical equations for the reactions between:
**a** magnesium and water
**b** calcium and oxygen.

**3** Account for the fact that calcium is more reactive than magnesium.

## Group 2 – the alkaline earth metals

Group 2 consists of the five metals beryllium, magnesium, calcium, strontium and barium, and the radioactive element radium. Magnesium and calcium are generally available for use in school. These metals have the following properties.

- They are harder than those in group 1.
- They are silvery-grey in colour when pure and clean. They tarnish quickly, however, when left in air due to the formation of a metal oxide on their surfaces (Figure 3.19).
- They are good conductors of heat and electricity.
- They burn in oxygen or air with characteristic flame colours to form solid white oxides. For example:

$$\text{magnesium} + \text{oxygen} \rightarrow \text{magnesium oxide}$$
$$2Mg(s) + O_2(g) \rightarrow 2MgO(s)$$

- They react with water, but they do so much less vigorously than the elements in group 1. For example:

$$\text{calcium} + \text{water} \rightarrow \text{calcium hydroxide} + \text{hydrogen gas}$$
$$Ca(s) + 2H_2O(l) \rightarrow Ca(OH)_2(aq) + H_2(g)$$

Considering the group as a whole, the further down the group you go, the more reactive the elements become.

**Figure 3.19** Tarnished (left) and cleaned-up magnesium.

## Flame colours

If a clean nichrome wire is dipped into a metal compound and then held in the hot part of a Bunsen flame, the flame can become coloured (Figure 3.20). Certain metal ions may be detected in their compounds by observing their flame colours (Table 3.9).

**Table 3.9** Characteristic flame colours of some metal ions.

| | **Metal** | **Flame colour** |
|---|---|---|
| Group 1 | Lithium | Crimson |
| | Sodium | Golden yellow |
| | Potassium | Lilac |
| | Rubidium | Red |
| | Caesium | Blue |
| Group 2 | Calcium | Brick red |
| | Strontium | Crimson |
| | Barium | Apple green |
| Others | Lead | Blue–white |
| | Copper | Green |

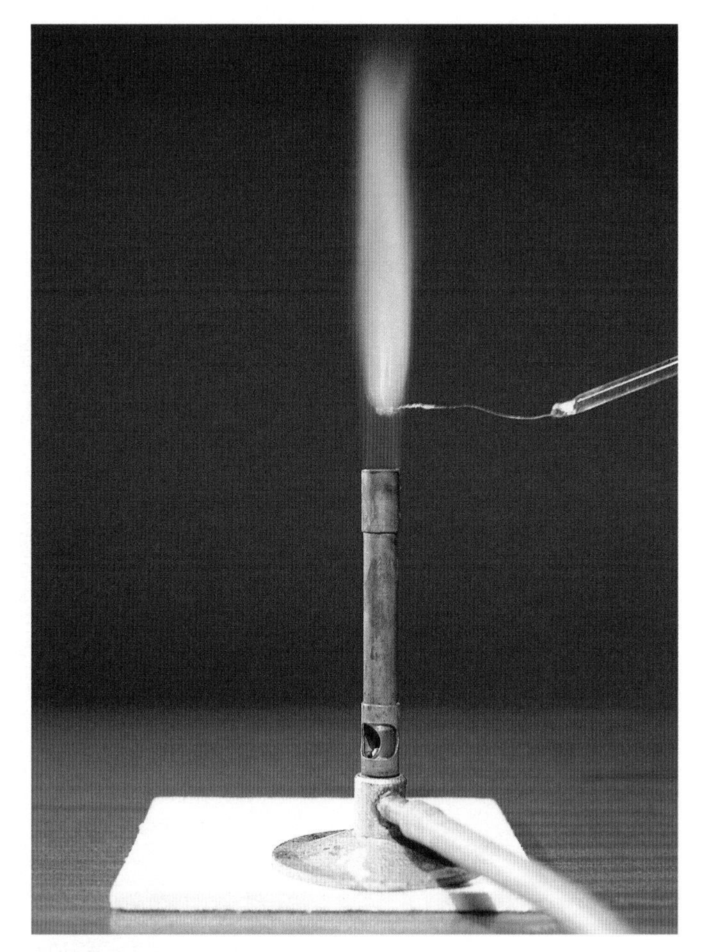

**Figure 3.20** The green colour is characteristic of copper.

A flame colour is obtained as a result of the electrons in the particular ions being excited when they absorb energy from the flame which is then emitted as visible light. The different electron configurations of the different ions, therefore, give rise to the different colours (Figure 3.21).

**Figure 3.21** Different metal ions are used to produce the colours seen in fireworks.

## Group 7 – the halogens

Group 7 consists of the four elements fluorine, chlorine, bromine and iodine, and the radioactive element astatine. Of these five elements, chlorine, bromine and iodine are generally available for use in school.

- These elements are coloured and darken going down the group (Table 3.10).
- They exist as diatomic molecules, for example $Cl_2$, $Br_2$ and $I_2$.
- They show a gradual change from a gas ($Cl_2$), through a liquid ($Br_2$), to a solid ($I_2$) (Figure 3.22).
- They form molecular compounds with other non-metallic elements, for example HCl.
- They react with hydrogen to produce the hydrogen halides, which dissolve in water to form acidic solutions (pH < 7).

$$\text{hydrogen} + \text{chlorine} \rightarrow \text{hydrogen chloride}$$
$$H_2(g) + Cl_2(g) \rightarrow 2HCl(g)$$

$$\text{hydrogen chloride} + \text{water} \rightarrow \text{hydrochloric acid}$$
$$HCl(g) + H_2O \rightarrow HCl(aq) \rightarrow H^+(aq) + Cl^-(aq)$$

- They react with metals to produce ionic metal halides, for example chlorine and iron produce iron chloride.

$$\text{iron} + \text{chlorine} \rightarrow \text{iron(III) chloride}$$
$$2Fe(s) + 3Cl_2(g) \rightarrow 2FeCl_3(s)$$

To find out about the extraction of bromine, see Chapter 10, p. 152.

**Table 3.10** Colours of some halogens.

| **Halogen** | **Colour** |
|---|---|
| Chlorine | Pale green |
| Bromine | Red–brown |
| Iodine | Purple–black |

**a** Chlorine, bromine and iodine.

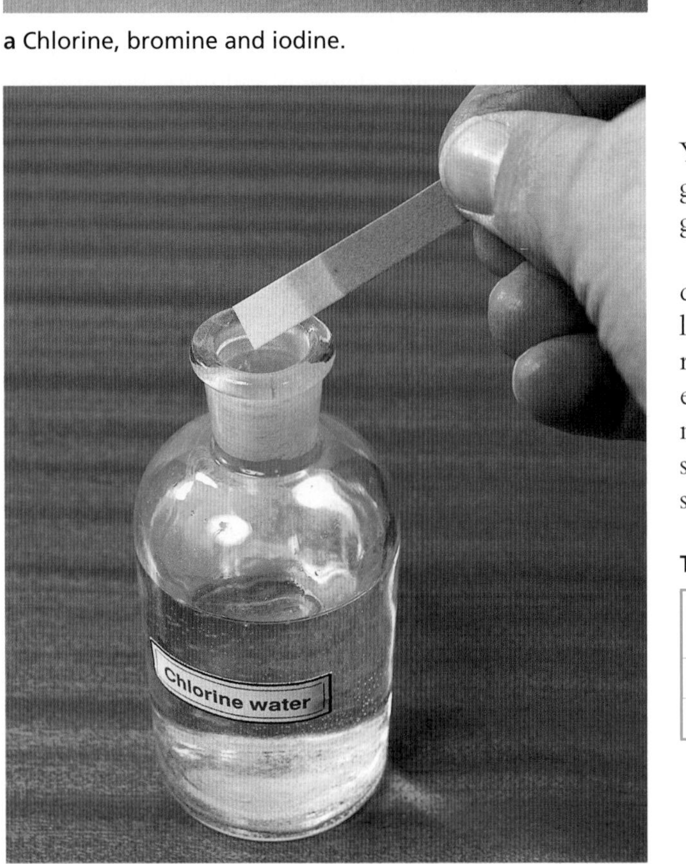


**b** Chlorine gas bleaches moist indicator paper.

**Figure 3.22**

## Displacement reactions

If chlorine is bubbled into a solution of potassium bromide the less reactive halogen, bromine, is **displaced** by the more reactive halogen, chlorine, as you can see from Figure 3.23:

potassium bromide + chlorine → potassium chloride + bromine

$$2KBr(aq) + Cl_2(g) \rightarrow 2KCl(aq) + Br_2(aq)$$

**Figure 3.23** Bromine being displaced.

The observed order of reactivity of the halogens, confirmed by similar displacement reactions, is:

**Decreasing reactivity**

⟶

chlorine bromine iodine

You will notice that, in contrast to the elements of groups 1 and 2, the order of reactivity decreases on going down the group.

Table 3.11 shows the electron configuration for chlorine and bromine. In each case the outer energy level contains seven electrons. When these elements react they gain one electron per atom to gain the stable electron configuration of a noble gas. You will learn more about the stable nature of these gases in the next section. For example, when chlorine reacts it gains a single electron and forms a negative ion (Figure 3.24).

**Table 3.11** Electron configuration of chlorine and bromine.

| Element | Symbol | Atomic number | Electron configuration |
|---|---|---|---|
| Chlorine | Cl | 17 | 2,8,7 |
| Bromine | Br | 35 | 2,8,18,7 |

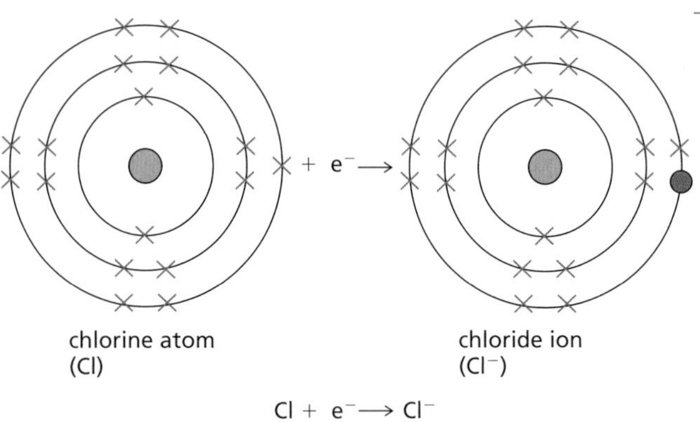


**Figure 3.24** A chlorine atom gains an electron to form a chloride ion.

Chlorine is more reactive than bromine because the incoming electron is being more strongly attracted into the outer energy level of the smaller atom. The attractive force on it will be greater than in the case of bromine, since the outer energy level of chlorine is closer to the nucleus. As you go down the group this outermost extra electron is further from the nucleus. It will, therefore, be held less securely, and the resulting reactivity of the elements in group 7 will decrease down the group.

The halogens and halogenic compounds are used in a multitude of different ways (Figure 3.25).

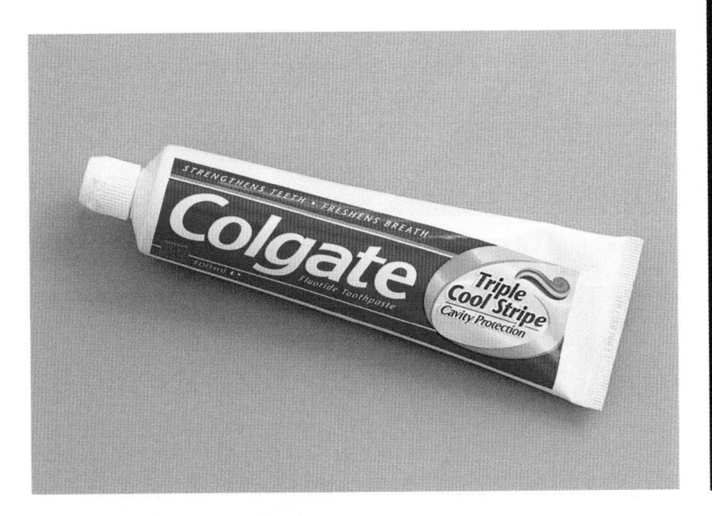


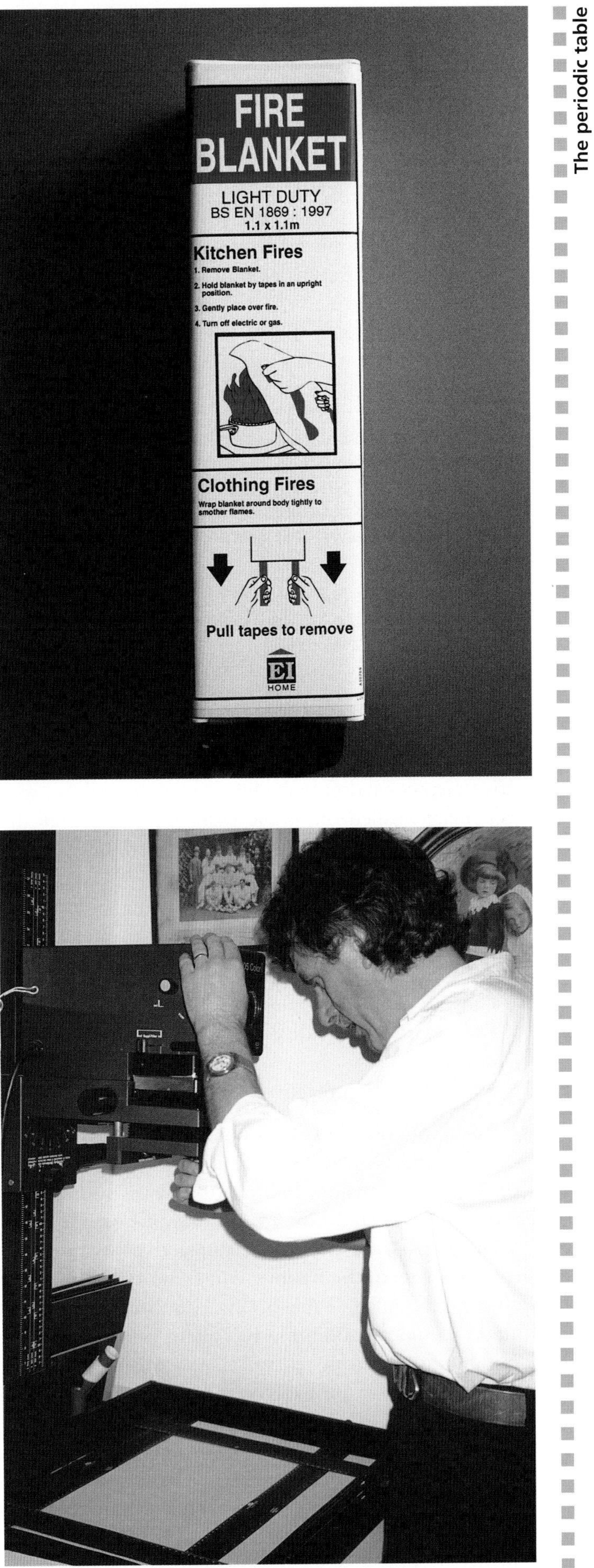


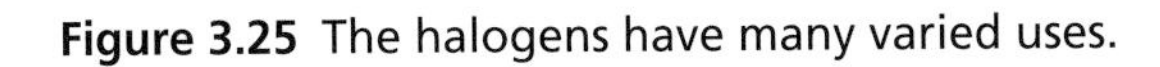
**Figure 3.25** The halogens have many varied uses.

- Fluorine is used in the form of fluorides in drinking water and toothpaste because it reduces tooth decay by hardening the enamel on teeth.
- Chlorine is used to make PVC plastic as well as household bleaches. It is also used to kill bacteria and viruses in drinking water (Chapter 10, p. 155).
- Bromine (Chapter 10, p. 152) is used to make disinfectants, medicines and fire retardants.
- Iodine is used in medicines and disinfectants and also as a photographic chemical.

## Questions

1 Write word and balanced chemical equations for the reactions between:
   a bromine and potassium iodide solution
   b bromine and potassium chloride solution.
   If no reaction will take place, write 'no reaction' and explain why.

2 Write down the names and symbols for the noble gases not given in Table 3.12 and use your research skills to find a use for each.

## Group 0 – the noble gases

Helium, neon, argon, krypton, xenon and the radioactive element radon make up a most unusual group of non-metals, called the noble gases. They were all discovered after Mendeleev had published his periodic table. They were discovered between 1894 and 1900, mainly through the work of the British scientists Sir William Ramsay (Figure 3.26a) and Lord John William Strutt Rayleigh (Figure 3.26b).

- They are colourless gases.
- They exist as individual atoms, for example He, Ne and Ar.
- They are very unreactive.

No compounds of helium, neon or argon have ever been found. However, more recently a number of compounds of xenon and krypton with fluorine and oxygen have been produced, for example $XeF_6$.

These gases are chemically unreactive because they have electron configurations which are stable and very difficult to change (Table 3.12). They are so stable that other elements attempt to attain these electron configurations during chemical reactions (Chapter 4, p. 54). You have probably seen this in your study of the elements of groups 1, 2 and 7.

Although unreactive, they have many uses. Argon, for example, is the gas used to fill light bulbs to prevent the tungsten filament reacting with air. Neon is used extensively in advertising signs and in lasers. Further uses of these gases are discussed in Chapter 10, p. 151.

a Sir William Ramsay (1852–1916).

b Lord Rayleigh (1842–1919).

**Figure 3.26** Both helped to discover the noble gases and won the Nobel Prize in Chemistry in 1904 for their work.

**Table 3.12** Electron configuration of helium, neon and argon.

| Element | Symbol | Atomic number | Electron configuration |
|---|---|---|---|
| Helium | He | 2 | 2 |
| Neon | Ne | 10 | 2,8 |
| Argon | Ar | 18 | 2,8,8 |

Helium is separated from natural gas by the liquefaction of the other gases. The other noble gases are obtained in large quantities by the fractional distillation of liquid air (Chapter 10, p. 150).

## Transition elements

This block of metals includes many you will be familiar with, for example copper, iron, nickel, zinc and chromium (Figure 3.27).

- They are harder and stronger than the metals in groups 1 and 2.
- They have much higher densities than the metals in groups 1 and 2.
- They have high melting points (except for mercury, which is a liquid at room temperature).
- They are less reactive metals.
- They form a range of brightly coloured compounds (Figure 3.28).
- They are good conductors of heat and electricity.
- They show catalytic activity (Chapter 11, p. 165) as elements and compounds. For example, iron is used in the industrial production of ammonia gas (Haber process, Chapter 15, p. 220).
- They do not react (corrode) so quickly with oxygen and/or water.
- They form more than one simple ion. For example, copper forms $Cu^+$ and $Cu^{2+}$ in compounds such as $Cu_2O$ and $CuSO_4$; iron forms $Fe^{2+}$ and $Fe^{3+}$ in compounds such as $FeSO_4$ and $FeCl_3$.

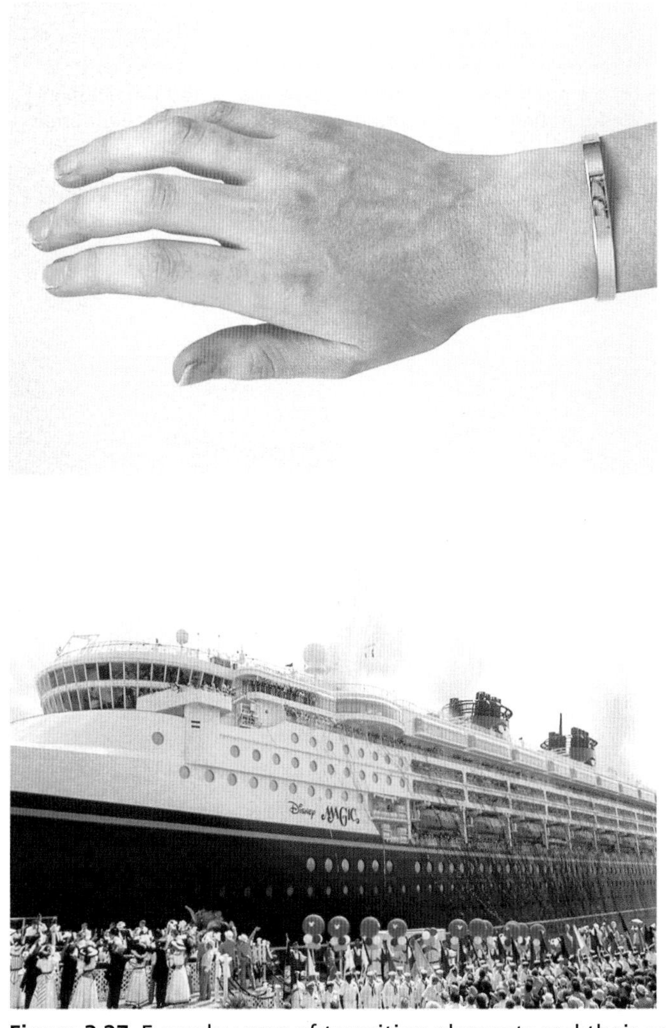

**Figure 3.27** Everyday uses of transition elements and their compounds. They are often known as the 'everyday metals'.

a Some solutions of coloured transition element compounds.

b The coloured compounds of transition elements can be seen in these pottery glazes.

**Figure 3.28**

## *Questions*

1 Look at the photographs in Figure 3.27 (p. 49) and decide which properties are important when considering the particular use the metal is being put to.

2 Which groups in the periodic table contain:
 a only metals?
 b only non-metals?
 c both metals and non-metals?

## The position of hydrogen

Hydrogen is often placed by itself in the periodic table. This is because the properties of hydrogen are unique. However, profitable comparisons can be made with the other elements. It is often shown at the top of either group 1 or group 7, but it cannot fit easily into the trends shown by either group, see Table 3.13.

**Table 3.13** Comparison of hydrogen with lithium and fluorine.

| Lithium | Hydrogen | Fluorine |
|---|---|---|
| Solid | Gas | Gas |
| Forms a positive ion | Forms positive or negative ions | Forms a negative ion |
| 1 electron in outer energy level | 1 electron in outer energy level | 1 electron short of a full outer energy level |
| Loses 1 electron to form a noble gas configuration | Needs 1 electron to form a noble gas configuration | Needs 1 electron to form a noble gas configuration |

## Checklist

**After studying Chapter 3 you should know and understand the following terms.**

- **Atomic mass unit** Exactly $\frac{1}{12}$ of the mass of one atom of the most abundant isotope of carbon-12.
- **Atomic number (proton number)** Symbol *Z*. The number of protons in the nucleus of an atom. The number of electrons present in an atom. The order of the element within the periodic table.
- **Displacement reaction** A reaction in which a more reactive element displaces a less reactive element from solution.
- **Electron** A fundamental sub-atomic particle with a negative charge present in all atoms within energy levels around the nucleus.
- **Electron configuration** A shorthand method of describing the arrangement of electrons within the energy levels of an atom.
- **Electron energy levels** The allowed energies of electrons in atoms.
- **Electrostatic force of attraction** A strong force of attraction between opposite charges.
- **Group** A vertical column of the periodic table containing elements with similar properties with the same number of electrons in their outer energy levels. They have an increasing number of inner energy levels as you descend the group.
- **Ion** An atom or group of atoms which has either lost one or more electrons, making it positively charged, or gained one or more electrons, making it negatively charged.
- **Isotopes** Atoms of the same element which possess different numbers of neutrons. They differ in mass number (nucleon number).
- **Mass number (nucleon number)** Symbol *A*. The total number of protons and neutrons found in the nucleus of an atom.
- **Mass spectrometer** A device in which atoms or molecules are ionised and then accelerated. Ions are separated according to their mass.
- **Metalloid (semi-metal)** Any of the class of chemical elements intermediate in properties between metals and non-metals, for example boron and silicon.
- **Neutron** A fundamental, uncharged sub-atomic particle present in the nuclei of atoms.
- **Periodic table** A table of elements arranged in order of increasing atomic number to show the similarities of the chemical elements with related electron configurations.
- **Periods** Horizontal rows of the periodic table. Within a period the atoms of all the elements have the same number of occupied energy levels but have an increasing number of electrons in the outer energy level.
- **Proton** A fundamental sub-atomic particle which has a positive charge equal in magnitude to that of an electron. Protons occur in all nuclei.
- **Relative atomic mass** Symbol $A_r$.

$$A_r = \frac{\text{average mass of isotopes of the element}}{\frac{1}{12} \times \text{mass of a carbon-12 atom}}$$

# Atomic structure and the periodic table

# *Additional questions*

**1** An atom **X** has an atomic number of 19 and relative atomic mass of 39.

**a** How many electrons, protons and neutrons are there in an atom of **X**?

**b** How many electrons will there be in the outer energy level (shell) of an atom of **X**?

**c** Write down the symbol for the ion **X** will form.

**d** Which group of the periodic table would **X** be in?

**e** (i) How would you expect **X** to react with water?
(ii) Write a word and balanced chemical equation for this reaction.

**2** The atomic number of barium (Ba) is 56. It is in group 2 of the periodic table.

**a** How many electrons would you expect a barium atom to contain in its outer energy level?

**b** How would you expect barium to react with chlorine? Write a word and balanced chemical equation for this reaction.

**c** How would you expect barium to react with water? Write a word and balanced chemical equation for this reaction.

**d** Write down the formulae of the bromide and sulphate of barium.

**3** Find the element germanium (Ge) in the periodic table.

**a** Which group of the periodic table is this element in?

**b** How many electrons will it have in its outer energy level (shell)?

**c** Is germanium a metal or a non-metal?

**d** What is the formula of the chloride of germanium?

**e** Name and give the symbols of the other elements in this group.

**4** Three members of the halogens are: $^{35.5}_{17}Cl$, $^{80}_{35}Br$ and $^{127}_{53}I$.

**a** (i) Write down the electron structure of an atom of chlorine.
(ii) Why is the relative atomic mass of chlorine not a whole number?
(iii) How many protons are there in an atom of bromine?
(iv) How many neutrons are there in an atom of iodine?
(v) State and account for the order of reactivity of these elements.

**b** When potassium is allowed to burn in a gas jar of chlorine, in a fume cupboard, clouds of white smoke are produced.
(i) Why is this reaction carried out in a fume cupboard?
(ii) What does the white smoke consist of?
(iii) Write a word and balanced chemical equation for this reaction.
(iv) Describe what you would expect to see when potassium is allowed to burn safely in a gas jar of bromine vapour. Write a word and balanced chemical equation for this reaction.

**5** 'By using displacement reactions it is possible to deduce the order of reactivity of the halogens.' Discuss this statement with reference to the elements bromine, iodine and chlorine only.

**6** Use the information given in the table below to answer the questions below concerning the elements **Q**, **R**, **S**, **T** and **X**.

| Element | Atomic number | Mass number | Electron structure |
|---|---|---|---|
| Q | 3 | 7 | 2,1 |
| R | 20 | 40 | 2,8,8,2 |
| S | 18 | 40 | 2,8,8 |
| T | 8 | 18 | 2,6 |
| X | 19 | 39 | 2,8,8,1 |

**a** Which element has 22 neutrons in each atom?

**b** Which element is a noble gas?

**c** Which two elements form ions with the same electron structure as neon?

**d** Which two elements are in the same group of the periodic table and which group is this?

**e** Place the elements in the table above into the periods in which they belong.

**f** Which is the most reactive metal element shown in the table?

**g** (i) Which of the above elements is calcium?
(ii) What colour flame would compounds containing calcium produce?

**7** **a** $^{69}_{31}Ga$ and $^{71}_{31}Ga$ are isotopes of gallium. With reference to this example, explain what you understand by the term isotope.

**b** A sample of gallium contains 60% of atoms of $^{69}_{31}Ga$ and 40% of atoms of $^{71}_{31}Ga$. Calculate the relative atomic mass of this sample of gallium.

**8** Copy and complete the following table with reference to the periodic table on p. 40.

| Element name | Symbol | Atomic number | Mass number | Number of neutrons | $^{A}_{Z}X$ |
|---|---|---|---|---|---|
| | | 5 | 11 | | |
| | | | 40 | 22 | |
| | | 14 | 28 | | |
| | | | 20 | 10 | |
| | | 26 | | 30 | |
| | | | 84 | 48 | |
| | | 52 | | 76 | |

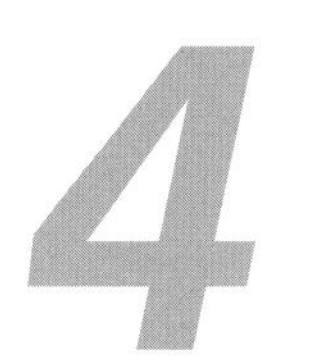

# Bonding and structure



In Chapter 3 we have seen that the noble gases are stable or unreactive because they have full electron energy levels (Figure 4.1). It would seem that when elements react to form compounds they do so to achieve full electron energy levels. This idea forms the basis of the electronic theory of chemical bonding.

**Figure 4.1** The unreactive noble gas neon is used in lasers.

## Ionic bonding

Ionic bonds are usually found in compounds that contain metals combined with non-metals. When this type of bond is formed, electrons are transferred from the metal atoms to the non-metal atoms during the chemical reaction. In doing this, the atoms become more stable by getting full outer energy levels. For example, consider what happens when sodium and chlorine combine to make sodium chloride (Figure 4.2).

sodium + chlorine → sodium chloride

**Figure 4.2** The properties of salt are very different from those of the sodium and chlorine it was made from. To get your salt you would not eat sodium or inhale chlorine.

Sodium has just one electron in its outer energy level ($^{11}$Na 2,8,1). Chlorine has seven electrons in its outer energy level ($^{17}$Cl 2,8,7). When these two elements react, the outer electron of each sodium atom is transferred to the outer energy level of a chlorine atom (Figure 4.3). In this way both the atoms obtain full outer energy levels and become 'like' the nearest noble gas. The sodium atom has become a sodium ion with an electron configuration like neon, while the chlorine atom has become a chloride ion with an electron configuration like argon. Only the outer electrons are important in bonding, so we can simplify the diagrams by missing out the inner energy levels (Figure 4.4).

The charges on the sodium and chloride ions are equal but opposite. They balance each other and the resulting formula for sodium chloride is NaCl. These oppositely charged ions attract each other and are pulled, or **bonded**, to one another by strong electrostatic forces. This type of bonding is called **ionic bonding**. The alternative name, **electrovalent bonding**, is derived from the fact that there are electrical charges on the atoms involved in the bonding.

Figure 4.5 shows the electron transfers that take place between a magnesium ion and an oxide ion during the formation of magnesium oxide.

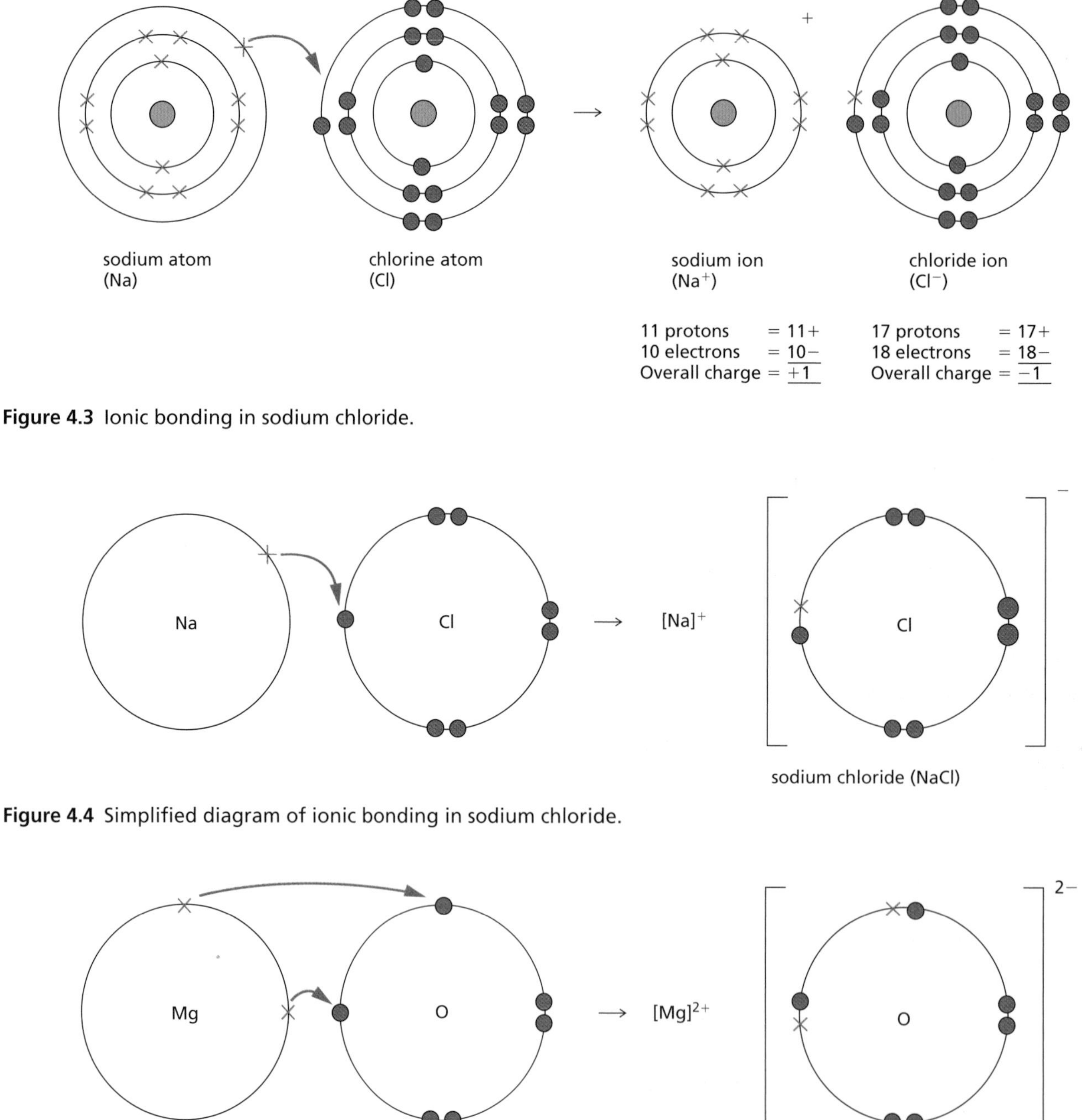


**Figure 4.3** Ionic bonding in sodium chloride.

**Figure 4.4** Simplified diagram of ionic bonding in sodium chloride.

**Figure 4.5** Simplified diagram of ionic bonding in magnesium oxide.

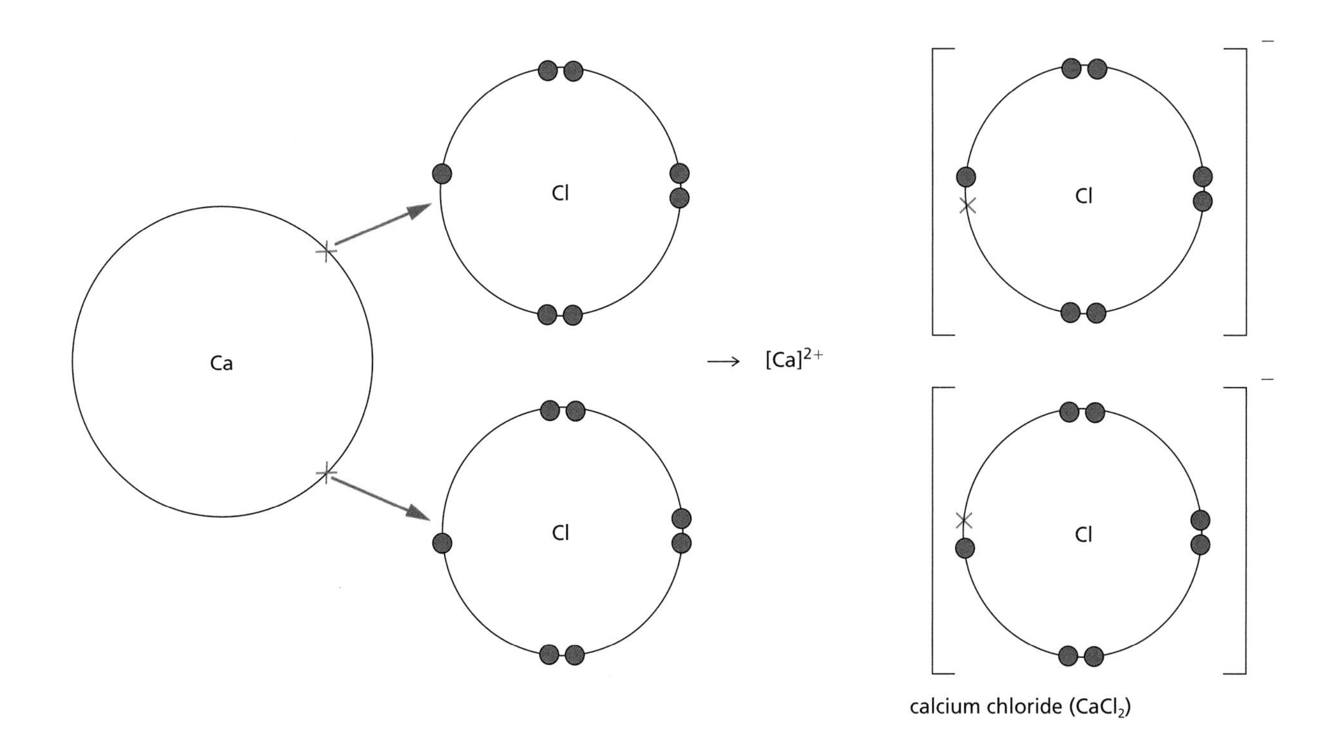


**Figure 4.6** The transfer of electrons that takes place during the formation of calcium chloride.

Magnesium obtains a full outer energy level by losing two electrons. These are transferred to the oxygen atom. In magnesium oxide, the $Mg^{2+}$ and $O^{2-}$ are oppositely charged and are attracted to one another. The formula for magnesium oxide is MgO.

Figure 4.6 shows the electron transfers that take place during the formation of calcium chloride. When these two elements react, the calcium atom gives each of the two chlorine atoms one electron. In this case, a compound is formed containing two chloride ions ($Cl^-$) for each calcium ion ($Ca^{2+}$). The chemical formula is $CaCl_2$.

## Question

1 Draw diagrams to represent the bonding in each of the following ionic compounds:
  a magnesium fluoride ($MgF_2$)
  b potassium fluoride (KF)
  c lithium chloride (LiCl)
  d calcium oxide (CaO).

## Ionic structures

Ionic structures are solids at room temperature and have high melting and boiling points. The ions are packed together in a regular arrangement called a **lattice**. Within the lattice, oppositely charged ions attract one another strongly. Scientists, using **X-ray diffraction** (Figure 4.7a, overleaf), have obtained photographs that indicate the way in which the ions are arranged (Figure 4.7b). The electron density map of sodium chloride is shown in Figure 4.7c.

Figure 4.7d shows the structure of sodium chloride as determined by the X-ray diffraction technique. The study of crystals using X-ray diffraction was pioneered by Sir William Bragg and his son Sir Lawrence Bragg in 1912. X-rays are a form of electromagnetic radiation. They have a much shorter wavelength than light therefore it is possible to use them to investigate extremely small structures.

When X-rays are passed through a crystal of sodium chloride, for example, you get a pattern of spots called a diffraction pattern (Figure 4.7b). This pattern can be recorded on photographic film and used to work out how the ions or atoms are arranged in the crystal. Crystals give particular diffraction patterns depending on their structure, and this makes X-ray diffraction a particularly powerful technique in the investigation of crystal structures.

Figure 4.7d shows only a tiny part of a small crystal of sodium chloride. Many millions of sodium ions and chloride ions would be arranged in this way in a crystal of sodium chloride to make up the giant ionic structure. Each sodium ion in the lattice is surrounded by six chloride ions, and each chloride ion is surrounded by six sodium ions.

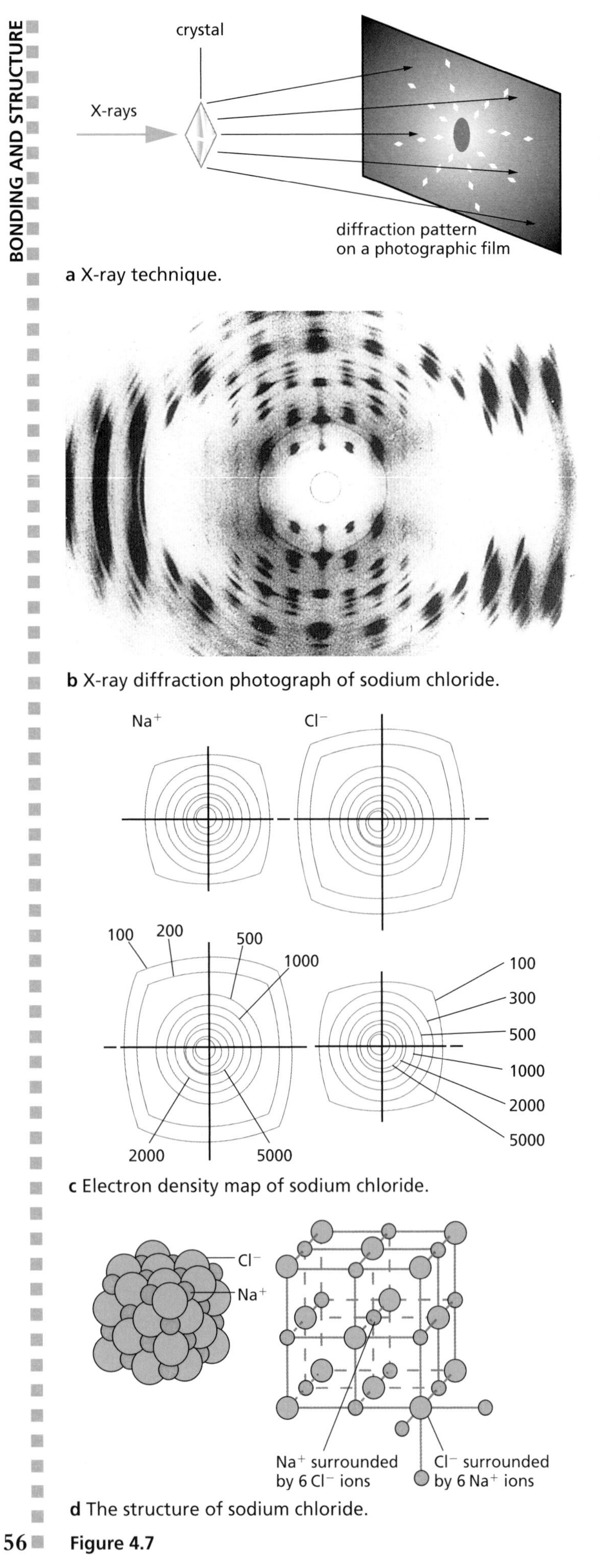


**a** X-ray technique.

**b** X-ray diffraction photograph of sodium chloride.

**c** Electron density map of sodium chloride.

**d** The structure of sodium chloride.

**Figure 4.7**

Not all ionic substances form the same structures. Caesium chloride (CsCl), for example, forms a different structure due to the larger size of the caesium ion compared with that of the sodium ion. This gives rise to the structure shown in Figure 4.8, which is called a body-centred cubic structure. Each caesium ion is surrounded by eight chloride ions and, in turn, each chloride ion is surrounded by eight caesium ions.

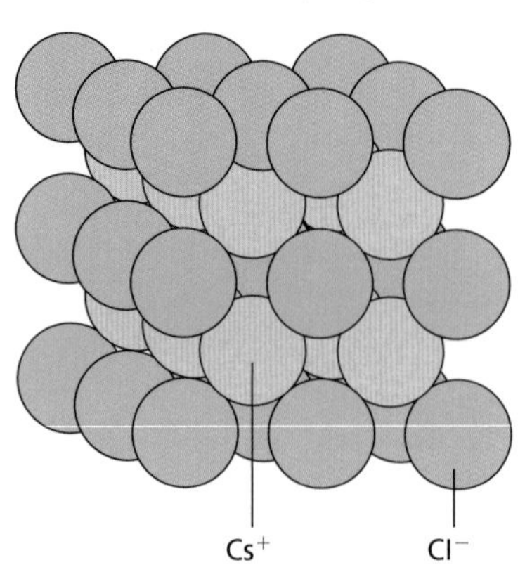


**a** The arrangement of ions.

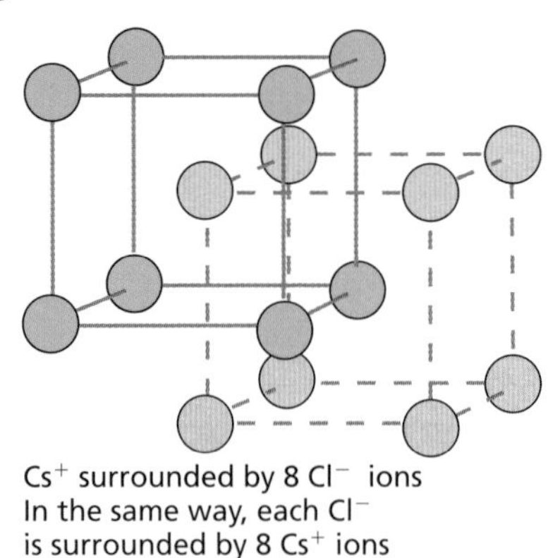

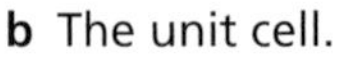


**b** The unit cell.

**Figure 4.8** Structure of caesium chloride.

## Properties of ionic compounds

Ionic compounds have the following properties.

- They are usually solids at room temperature, with high melting points. This is due to the strong electrostatic forces holding the crystal lattice together. A lot of energy is therefore needed to separate the ions and melt the substance.
- They are usually hard substances.
- They usually cannot conduct electricity when solid, because the ions are not free to move.
- They mainly dissolve in water. This is because water molecules are able to bond with both the positive and the negative ions, which breaks up the lattice and keeps the ions apart.
- They usually conduct electricity when in the molten state or in aqueous solution. The forces of attraction between the ions are weakened and the ions are free to move. This allows an electric current to be passed through the molten sodium chloride. For a further discussion of this process see Chapter 6 (p. 84).

## Formulae of ionic substances

On p. 54 we saw that ionic compounds contain positive and negative ions, whose charges balance. For example, sodium chloride contains one $Na^+$ ion for every $Cl^-$ ion, giving rise to the formula NaCl. This method can be used to write down formulae which show the ratio of the number of ions present in any ionic compound.

The formula of magnesium chloride is $MgCl_2$. This formula is arrived at by each $Mg^{2+}$ ion combining with two $Cl^-$ ions, and once again the charges balance. The size of the charge on an ion is a measure of its **valency** or **combining power**. $Na^+$ has a valency of 1, but $Mg^{2+}$ has a valency of 2. $Na^+$ can bond (combine) with only one $Cl^-$ ion, whereas $Mg^{2+}$ can bond with two $Cl^-$ ions.

Some elements, such as copper and iron, possess two ions with different valencies. Copper can form the $Cu^+$ ion and the $Cu^{2+}$ ion, and therefore it can form two different compounds with chlorine, CuCl (copper(I) chloride) and $CuCl_2$ (copper(II) chloride). Iron forms the $Fe^{2+}$ and $Fe^{3+}$ ions.

Table 4.1 shows the valencies of a series of ions you will normally meet in your study of chemistry.

You will notice that Table 4.1 includes groups of atoms which have net charges. For example, the nitrate ion is a single unit composed of one nitrogen atom and three oxygen atoms and has one single negative charge. The formula, therefore, of magnesium nitrate would be $Mg(NO_3)_2$. You will notice that the $NO_3$ has been placed in brackets with a 2 outside the bracket. This indicates that there are two nitrate ions present for every magnesium ion. The ratio of the atoms present is therefore:

$$Mg\ (N\ O_3)_2$$

$$1Mg:2N:6O$$

### Questions

1 Using the information in Table 4.1, write the formulae for:

a copper(I) oxide

b zinc phosphate

c iron(III) chloride

d lead bromide.

2 Using the formulae in your answer to question **1**, write down the ratio of atoms present for each of the compounds.

**Table 4.1** Valencies of some common substances.

| | Valency | | | | | |
|---|---|---|---|---|---|---|
| | 1 | | 2 | | 3 | |
| Metals | Lithium | ($Li^+$) | Magnesium | ($Mg^{2+}$) | Aluminium | ($Al^{3+}$) |
| | Sodium | ($Na^+$) | Calcium | ($Ca^{2+}$) | Iron | ($Fe^{3+}$) |
| | Potassium | ($K^+$) | Copper | ($Cu^{2+}$) | | |
| | Silver | ($Ag^+$) | Zinc | ($Zn^{2+}$) | | |
| | Copper | ($Cu^+$) | Iron | ($Fe^{2+}$) | | |
| | | | Lead | ($Pb^{2+}$) | | |
| | | | Barium | ($Ba^{2+}$) | | |
| Non-metals | Fluoride | ($F^-$) | Oxide | ($O^{2-}$) | | |
| | Chloride | ($Cl^-$) | Sulphide | ($S^{2-}$) | | |
| | Bromide | ($Br^-$) | | | | |
| | Hydrogen | ($H^+$) | | | | |
| Groups of atoms | Hydroxide | ($OH^-$) | Carbonate | ($CO_3^{2-}$) | Phosphate | ($PO_4^{3-}$) |
| | Nitrate | ($NO_3^-$) | Sulphate | ($SO_4^{2-}$) | | |
| | Ammonium | ($NH_4^+$) | | | | |
| | Hydrogencarbonate | ($HCO_3^-$) | | | | |

## Covalent bonding

Another way in which atoms can gain the stability of the noble gas electron configuration is by sharing the electrons in their outer energy levels. This occurs between non-metal atoms, and the bond formed is called a **covalent bond**. The simplest example of this type of bonding can be seen by considering the hydrogen molecule, $H_2$.

Each hydrogen atom in the molecule has one electron. In order to obtain a full outer energy level and gain the electron configuration of the noble gas helium each of the hydrogen atoms must have two electrons. To do this, the two hydrogen atoms allow their outer energy levels to overlap (Figure 4.9). A molecule of hydrogen is formed, with two hydrogen atoms sharing a pair of electrons (Figure 4.10). This shared pair of electrons is known as a single covalent bond and is represented by a single line as in hydrogen:

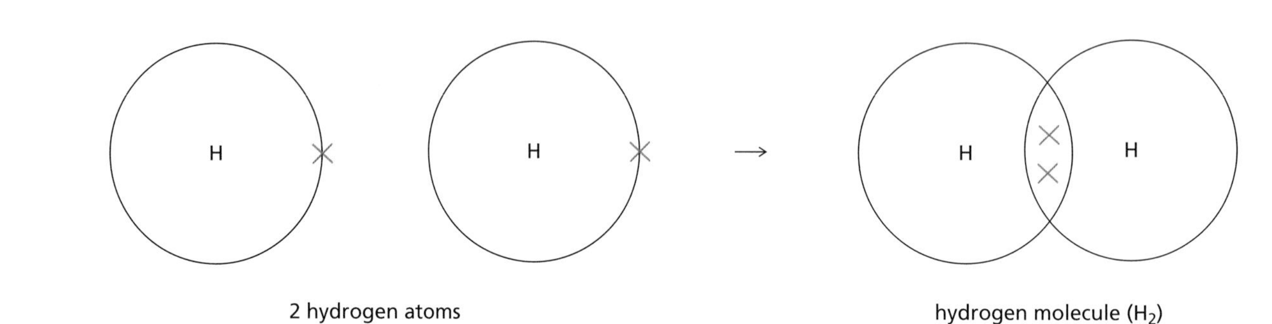


**Figure 4.9** Electron sharing to form the single covalent bond in $H_2$ molecules.

0.175 0.175 0.200 0.200 H H 0.150 0.125 0.100 0.075 0.050 0.025 0.074 nm

electron densities in electrons per cubic atomic unit of length

**a** Model of hydrogen molecule.

**b** The electron density map of a hydrogen molecule.

**Figure 4.10**

## Other covalent compounds

**Methane** (natural gas) is a gas whose molecules contain atoms of carbon and hydrogen. The electron structures are:

$_{6}C$ 2,4 $\quad$ $_{1}H$ 1

The carbon atom needs four more electrons to attain the electron configuration of the noble gas neon. Each hydrogen atom needs only one electron to form the electron configuration of helium. Figure 4.11 shows how the atoms gain these electron configurations by the sharing of electrons. You will note that only the outer electron energy levels are shown. Figure 4.12 shows the shape of the methane molecule.

**Ammonia** is a gas containing the elements nitrogen and hydrogen. It is used in large amounts to make fertilisers. The electron configurations of the two elements are:

$_{7}N$ 2,5 $\quad$ $_{1}H$ 1

The nitrogen atom needs three more electrons to obtain the noble gas structure of neon. Each hydrogen requires only one electron to form the noble gas structure of helium. The nitrogen and hydrogen atoms share electrons, forming three single covalent bonds (Figure 4.13, overleaf). Unlike methane the shape of an ammonia molecule is pyramidal (Figure 4.14, overleaf).

**Water** is a liquid containing the elements hydrogen and oxygen. The electron configurations of the two elements are:

$_{8}O$ 2,6 $\quad$ $_{1}H$ 1

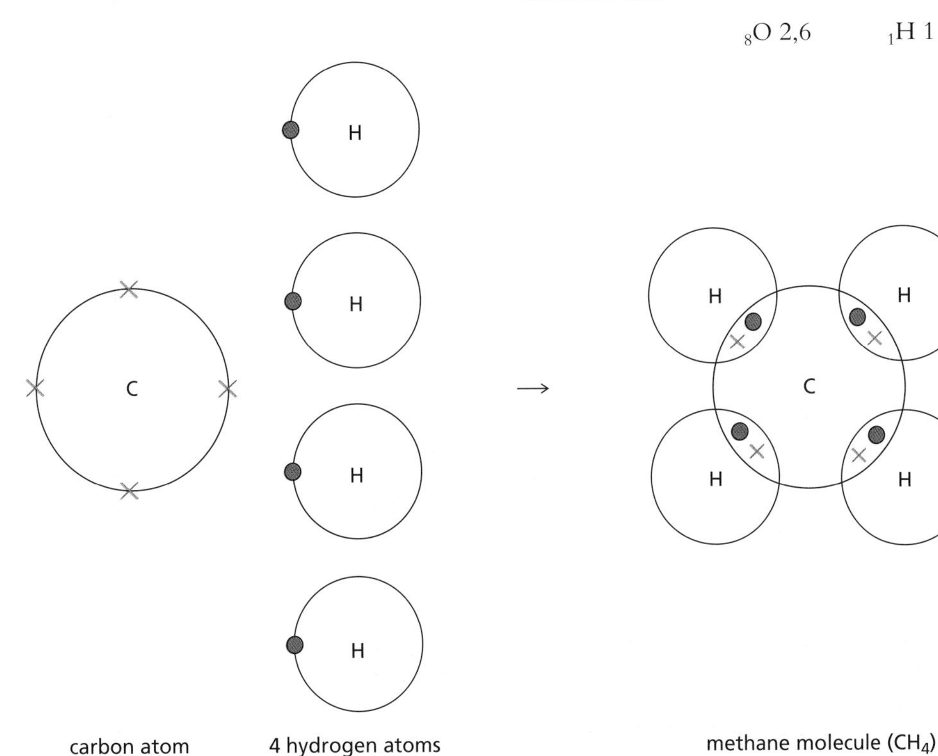


**Figure 4.11** Formation of methane.

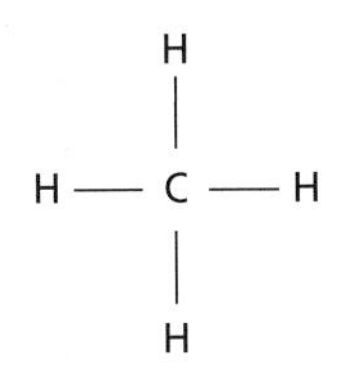


the methane molecule is tetrahedral

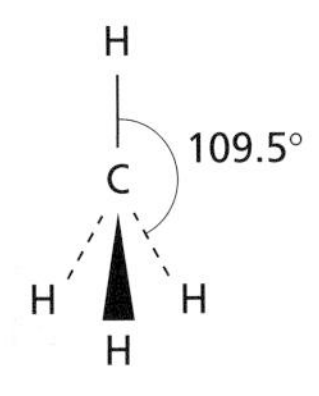


**a** Methane molecule.

**b** Model of the methane molecule. The tetrahedral shape is caused by the repulsion of the C–H bonding pairs of the electrons.

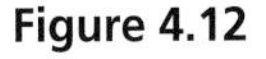

**Figure 4.12**

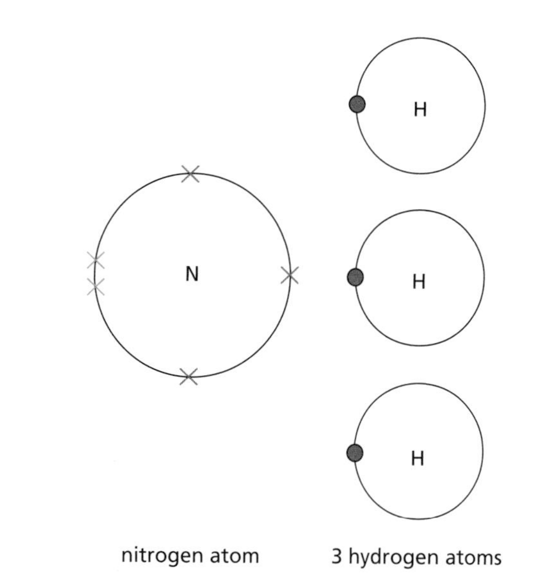


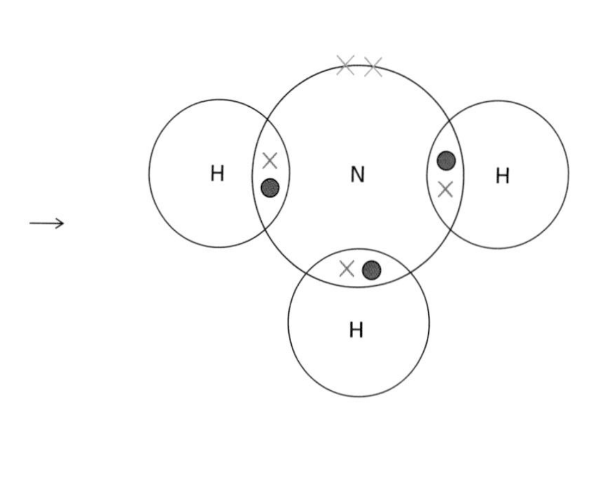

**Figure 4.13** The bonding in ammonia.

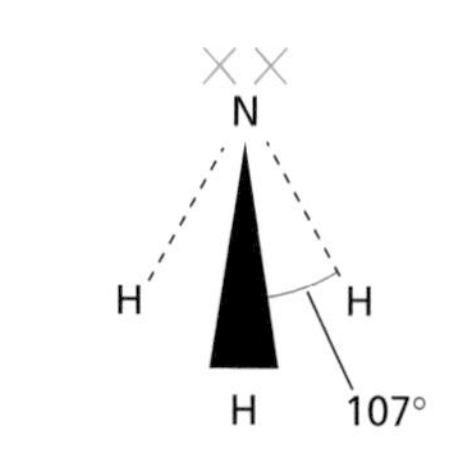


**a** Ammonia molecule.

**b** Model of the ammonia molecule. The pyramidal shape is caused by the repulsion between the bonding pairs of electrons as well as the lone pair (or non-bonding pair) of electrons.

**Figure 4.14**

The oxygen atom needs two electrons to gain the electron configuration of neon. Each hydrogen requires one more electron to gain the electron configuration of helium. Again, the oxygen and hydrogen atoms share electrons, forming a water molecule with two single covalent bonds as shown in Figure 4.15. A water molecule is V-shaped (Figure 4.16).

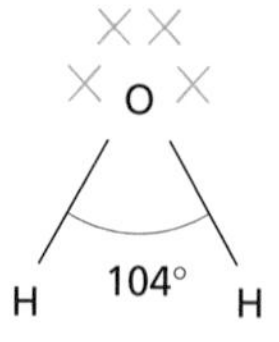


**a** Water molecule.

**b** Model of a water molecule. This is a V-shaped molecule.

**Figure 4.16**

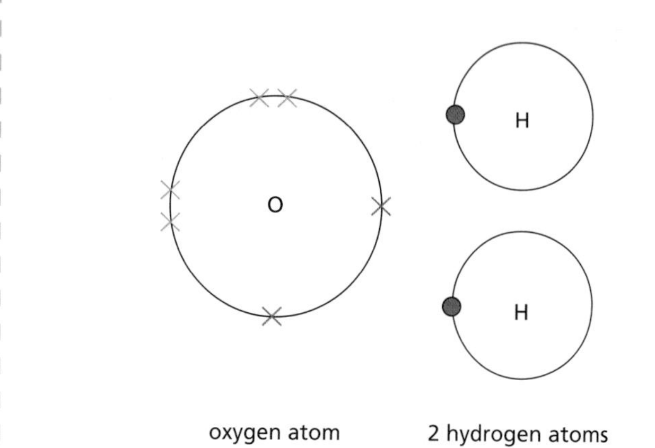


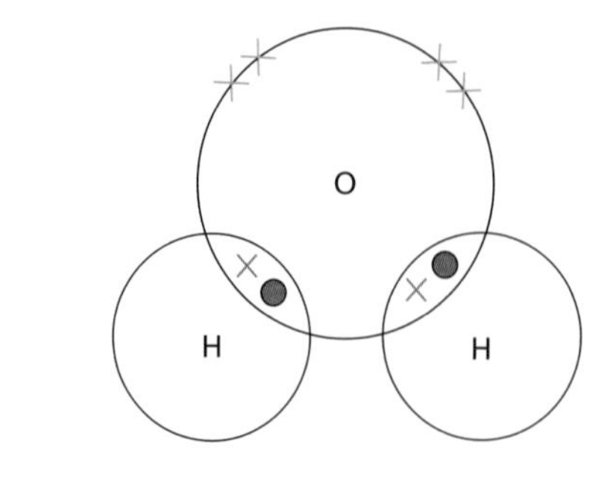


**Figure 4.15** Formation of water.

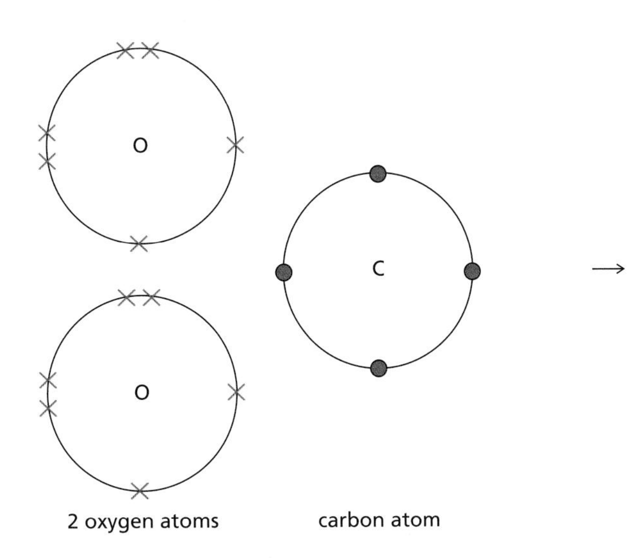


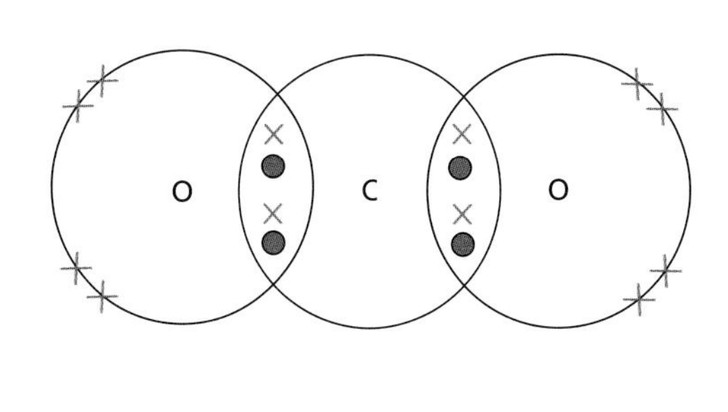

**Figure 4.17** Formation of carbon dioxide.

**Carbon dioxide** is a gas containing the elements carbon and oxygen. The electron configurations of the two elements are:

$_{6}C$ 2,4 $\quad$ $_{8}O$ 2,6

In this case each carbon atom needs to share four electrons to gain the electron configuration of neon. Each oxygen needs to share two electrons to gain the electron configuration of neon. This is achieved by forming two **double covalent bonds** in which two pairs of electrons are shared in each case, as shown in Figure 4.17. Carbon dioxide is a linear molecule (Figure 4.18).

O=C=O

**a** Carbon dioxide molecule. Note the double covalent bond is represented by a double line.

**b** Model of the linear carbon dioxide molecule.

**Figure 4.18**

## Questions

1 Draw diagrams to represent the bonding in each of the following covalent compounds:
   - **a** tetrachloromethane ($CCl_4$)
   - **b** oxygen gas ($O_2$)
   - **c** hydrogen sulphide ($H_2S$)
   - **d** hydrogen chloride (HCl)
   - **e** ethene ($C_2H_4$)
   - **f** methanol ($CH_3OH$)
   - **g** nitrogen ($N_2$).

2 Explain why the water molecule in Figure 4.16 is V-shaped.

## Covalent structures

Compounds containing covalent bonds have molecules whose structures can be classified as either **simple molecular** or **giant molecular**.

Simple molecular structures are simple, formed from only a few atoms. They have strong covalent bonds between the atoms within a molecule (**intra-molecular bonds**) but have weak bonds between the molecules (intermolecular bonds) (Figure 4.19). One type of weak bond between molecules is known as the **van der Waals' bond** (or force), and these forces increase steadily with the increasing size of the molecule. Examples of simple molecules are iodine, methane, water and ethanol.

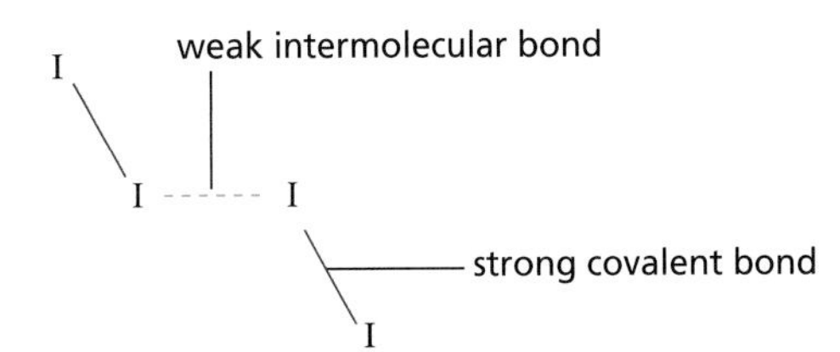


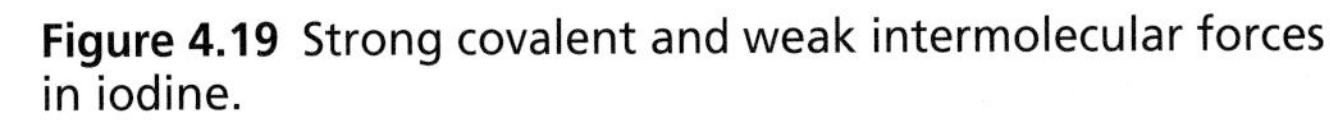

**Figure 4.19** Strong covalent and weak intermolecular forces in iodine.

**Giant molecular** structures contain many hundreds of thousands of atoms joined by strong covalent bonds. Examples of substances showing this type of structure are diamond, graphite, silicon(IV) oxide (Figure 4.20) and plastics such as polythene (polymers, Chapter 14, p. 197). Plastics are a tangled mass of very long molecules in which the atoms are joined together by strong covalent bonds to form long chains. Molten plastics can be made into fibres by being forced through hundreds of tiny holes in a 'spinneret' (Figure 4.21). This process aligns the long chains of atoms along the length of the fibre. For a further discussion of plastics, see Chapter 14, p. 197.

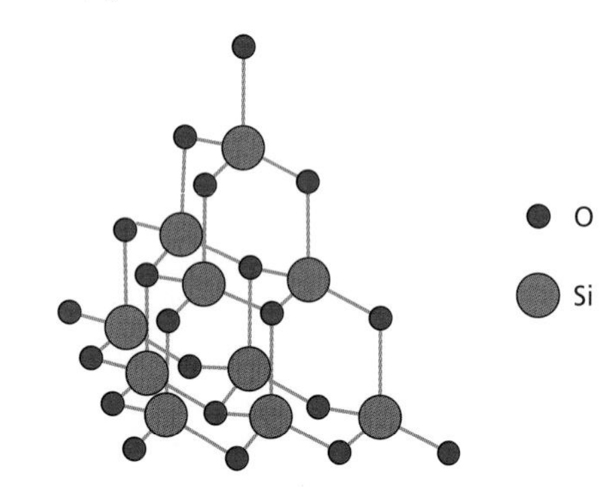


**a** The silicon(IV) oxide structure in quartz.

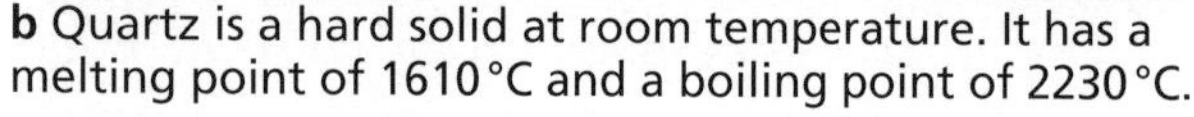

**b** Quartz is a hard solid at room temperature. It has a melting point of 1610 °C and a boiling point of 2230 °C.

**Figure 4.20**

**Figure 4.21** These are magnified nylon fibres formed by forcing molten plastic through hundreds of tiny holes.

## Properties of covalent compounds

Covalent compounds have the following properties.

- As simple molecular substances, they are usually gases, liquids or solids with low melting and boiling points. The melting points are low because of the weak intermolecular forces of attraction which exist between simple molecules. Giant molecular substances have higher melting points, because the whole structure is held together by strong covalent bonds within the giant molecule.
- Generally, they do not conduct electricity when molten or dissolved in water. This is because they do not contain ions. However, some molecules actually react with water to form ions. For example, hydrogen chloride gas produces aqueous hydrogen ions and chloride ions when it dissolves in water:

$$HCl(g) \xrightarrow{\text{water}} H^{+}(aq) + Cl^{-}(aq)$$

- Generally, they do not dissolve in water. However, water is an excellent solvent and can interact with and dissolve some covalent molecules better than others.

## Allotropy

When an element can exist in more than one physical form in the same state it is said to exhibit **allotropy** (or polymorphism). Each of the different physical forms is called an **allotrope**. Allotropy is actually quite a common feature of the elements in the periodic table. Some examples of elements which show allotropy are sulphur, tin, iron and carbon.

## Allotropes of carbon

Carbon is a non-metallic element which exists in more than one solid structural form. Its allotropes are called **graphite** and **diamond**. Each of the allotropes has a different structure (Figures 4.22 and 4.23) and so the allotropes exhibit different physical properties (Table 4.2). The different physical properties that they exhibit lead to the allotropes being used in different ways (Table 4.3).

**Table 4.2** Physical properties of graphite and diamond.

| Property | Graphite | Diamond |
|---|---|---|
| Appearance | A dark grey, shiny solid | A colourless transparent crystal which sparkles in light |
| Electrical conductivity | Conducts electricity | Does not conduct electricity |
| Hardness | A soft material with a slippery feel | A very hard substance |
| Density/g $cm^{-3}$ | 2.25 | 3.51 |

Table 4.3 Uses of graphite and diamond.

| Graphite | Diamond |
|---|---|
| Pencils | Jewellery |
| Electrodes | Glass cutters |
| Lubricant | Diamond-studded saws |
| | Drill bits |
| | Polishers |

## Graphite

Figure 4.22a shows the structure of graphite. This is a layer structure. Within each layer each carbon atom is bonded to three others by strong covalent bonds. Each layer is therefore a giant molecule. Between these layers there are weak forces of attraction (van der Waals' forces) and so the layers will pass over each other easily.

With only three covalent bonds formed between carbon atoms within the layers, an unbonded electron is present on each carbon atom. These 'spare' (or **delocalised**) electrons form electron clouds between the layers and it is because of these spare electrons that graphite conducts electricity.

In recent years a set of interesting compounds known as **graphitic compounds** have been developed. In these compounds different atoms have been fitted in between the layers of carbon atoms to produce a substance with a greater electrical conductivity than pure graphite. Graphite is also used as a component in certain sports equipment, such as tennis and squash rackets.

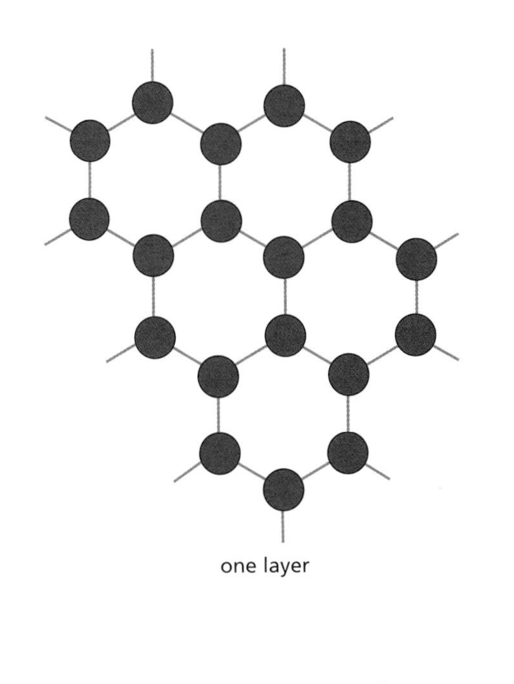


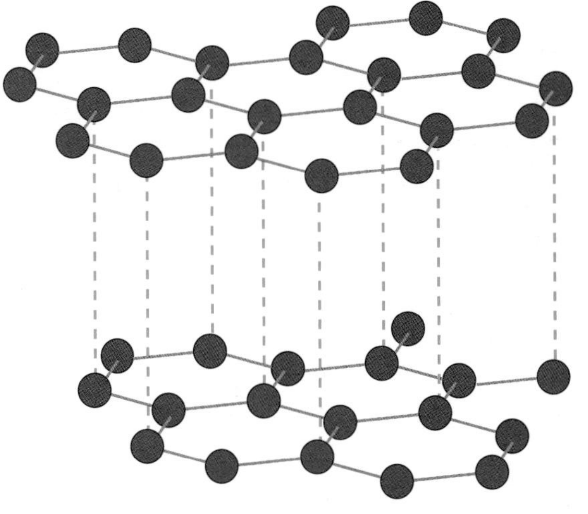


**a** A portion of the graphite structure.

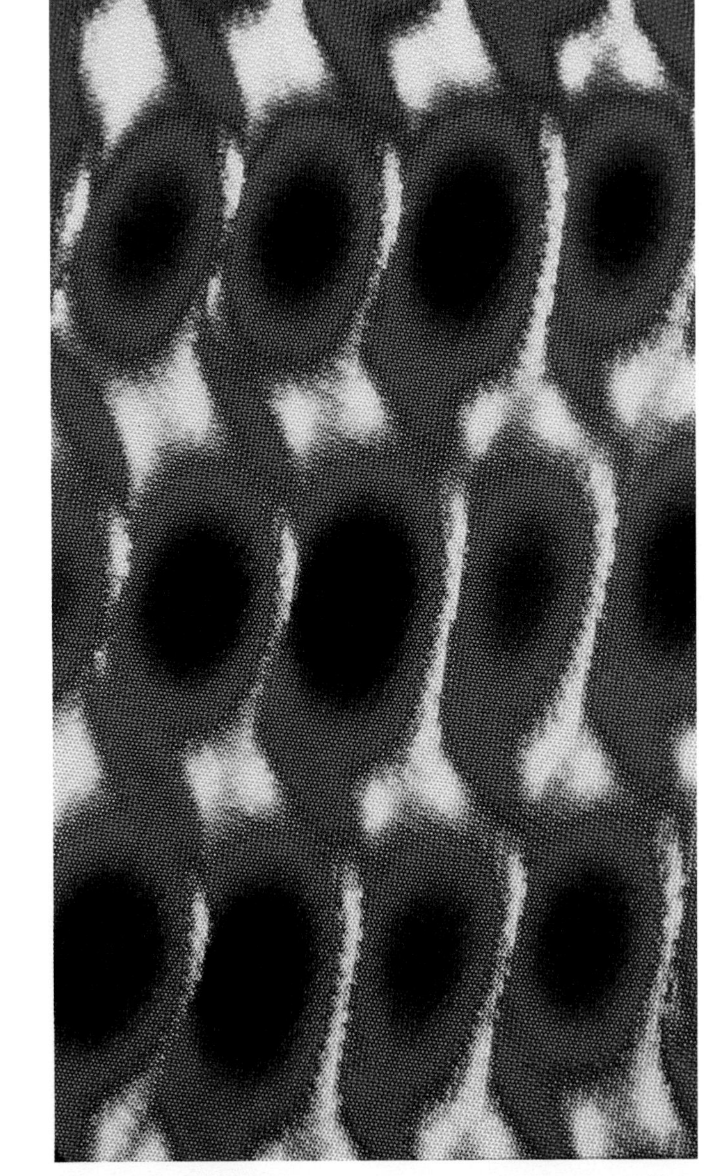

**b** A piece of graphite as seen through a scanning tunnelling microscope.

**Figure 4.22**

### Diamond

Figure 4.23 shows the diamond structure. Each of the carbon atoms in the giant structure is covalently bonded to four others. They form a tetrahedral arrangement. This bonding scheme gives rise to a very rigid, three-dimensional structure and accounts for the extreme hardness of the substance. All the outer energy level electrons of the carbon atoms are used to form covalent bonds, so there are no electrons available to enable diamond to conduct electricity.

It is possible to manufacture both allotropes of carbon. Diamond is made by heating graphite to about 300 °C at very high pressures. Diamond made by this method is known as industrial diamond. Graphite can be made by heating a mixture of coke and sand at a very high temperature in an electric arc furnace for about 24 hours.

The various uses of graphite and diamond result from their differing properties (Figure 4.24).

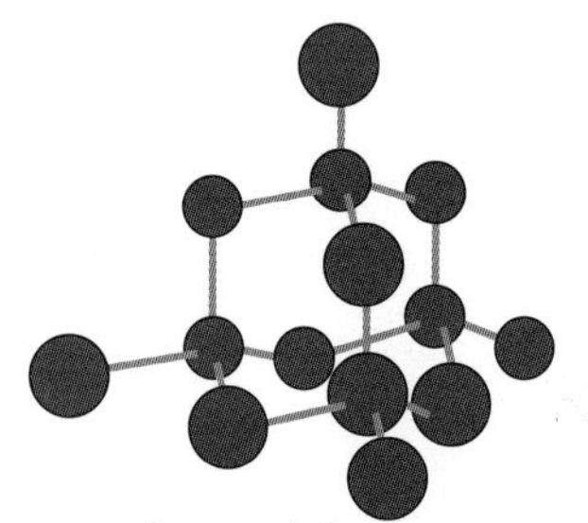

a small part of the structure

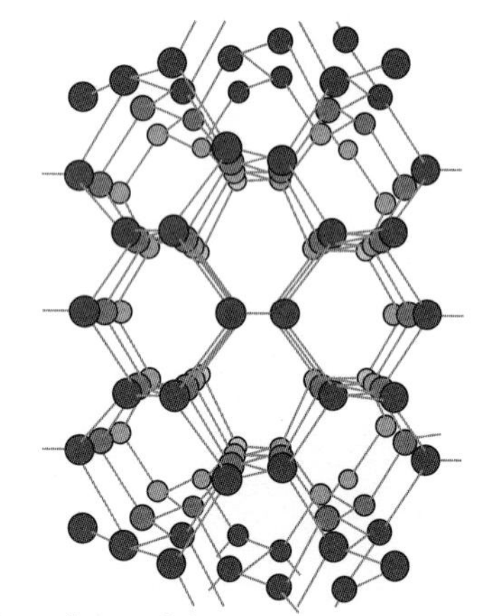

a view of a much larger part of the structure

**a** The structure of diamond.

**b** The Regent Diamond has been worn by Queen Elizabeth II.

**Figure 4.23**

**Figure 4.24** Uses of graphite (as a pencil and in a squash racket) and diamond (as a toothed saw to cut marble and on a dentist's drill).

## Buckminsterfullerene – an unusual form of carbon

In 1985 a new form of carbon was obtained by Richard Smalley and Robert Curl of Rice University, Texas. It was formed by the action of a laser beam on a sample of graphite. The structure of buckminsterfullerene can be seen in Figure 4.25.

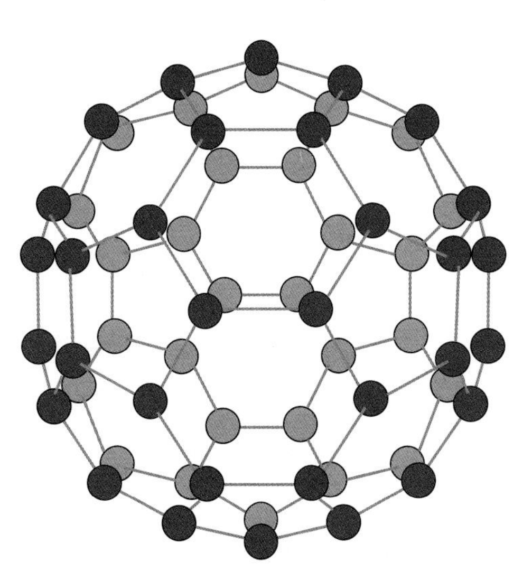

**Figure 4.25** Buckminsterfullerene – a 'bucky ball' ($C_{60}$).

This spherical structure is composed of 60 carbon atoms covalently bonded together. Further spherical forms of carbon, 'bucky balls', containing 70, 72 and 84 carbon atoms have been identified and it has led to a whole new branch of inorganic carbon chemistry. It is thought that this type of molecule exists in chimney soot. Chemists have suggested that due to the large surface area of the bucky balls they may have uses as catalysts (Chapter 11, p. 165). Also they may have uses as superconductors.

Buckminsterfullerene is named after an American architect, Buckminster Fuller, who built complex geometrical structures (Figure 4.26).

### Question

1 Explain the difference between ionic and covalent bonding. Discuss in what ways the electron structure of the noble gases is important in both of these theories of bonding.

**Figure 4.26** $C_{60}$ has a structure similar to a football and to the structure of the dome at the Expo in Montreal, Canada.

# Glasses and ceramics

## Glasses

**Glasses** are irregular giant molecular structures held together by strong covalent bonds. Glass can be made by heating silicon(IV) oxide with other substances until a thick viscous liquid is formed. As this liquid cools, the atoms present cannot move freely enough to return to their arrangement within the pure silicon(IV) oxide structure. Instead they are forced to form a disordered arrangement as shown in Figure 4.27. Glass is called a **supercooled liquid**.

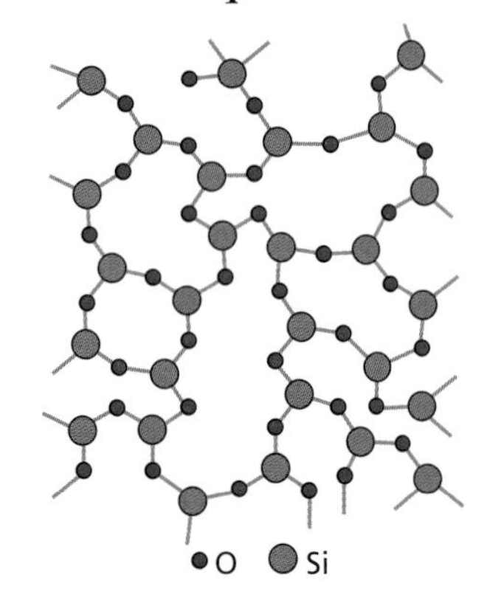


**Figure 4.27** Two-dimensional structure of silicon(IV) oxide.

The glass used in bottles and windows is **soda glass**. This type of glass is made by heating a mixture of sand (silicon(IV) oxide), soda (sodium carbonate) and lime (calcium oxide). Pyrex is a borosilicate glass (Figure 4.28). It is made by incorporating some boron oxide into the silicon(IV) oxide structure so that silicon atoms are replaced by boron atoms. This type of glass is tougher than soda glass and more resistant to temperature changes. It is, therefore, used in the manufacture of cooking utensils and laboratory glassware.

**Figure 4.28** This glassware is made from Pyrex.

## Ceramics

The word **ceramic** comes from the Greek word *keramos* meaning pottery or 'burnt stuff'. Clay dug from the ground contains a mixture of several materials. The main one is a mineral called kaolinite, $Al_2Si_2O_5(OH)_4$, in which the atoms are arranged in layers in a giant structure. While wet, the clay can be moulded because the kaolinite crystals move over one another. However, when it is dry the clay becomes rigid because the crystals stick together.

During firing in a furnace, the clay is heated to a temperature of 1000 °C. A complicated series of chemical changes take place, new minerals are formed and some of the substances in the clay react to form a type of glass. The material produced at the end of the firing, the ceramic, consists of many minute mineral crystals bonded together with glass.

Modern ceramic materials now include zirconium oxide ($ZrO_2$), titanium carbide (TiC), and silicon nitride (SiN). There are now many more uses of these new ceramic materials. For example, vehicle components such as ceramic bearings do not need lubrication – even at high speeds. In space technology, ceramic tiles protect the Space Shuttle from intense heat during its re-entry into the Earth's atmosphere (Figure 4.29).

**Figure 4.29** The ceramic tiles protect the Space Shuttle from re-entry temperatures of 1500 °C. Each tile has fibres coated with silica.

## Questions

1 Draw up a table to summarise the properties of the different types of substances you have met in this chapter. Your table should include examples from ionic substances, covalent substances (simple and giant), ceramics and glasses.

2 Use your research skills, including suitable websites, to discover details of recently developed bioceramics as well as ceramics used as superconductors.

## *Metallic bonding*

Another way in which atoms obtain a more stable electron structure is found in metals. The electrons in the outer energy level of the atom of a metal move freely throughout the structure (they are **delocalised** forming a mobile 'sea' of electrons (Figure 4.30)). When the metal atoms lose these electrons, they become positive ions. Therefore, metals consist of positive ions embedded in moving clouds electrons. The negatively charged electrons attract all the positive metal ions and bond them together with strong electrostatic forces of attraction as a single unit. This is the **metallic bond**.

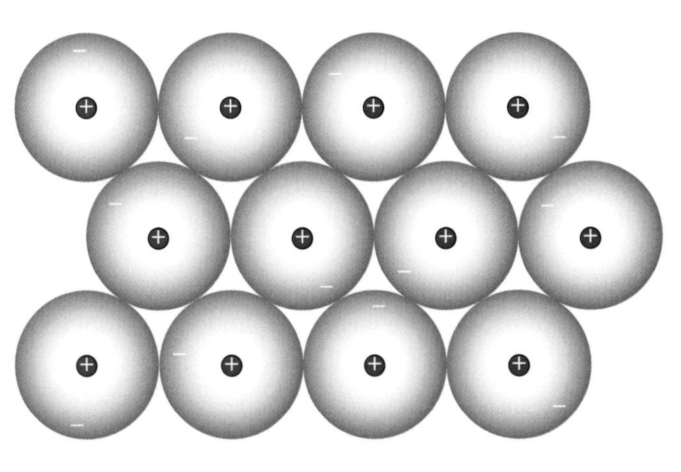

**Figure 4.30** Metals consist of positive ions surrounded by a 'sea' of electrons.

### Properties of metals

Metals have the following properties.

- They usually have high melting and boiling points due to the strong attraction between the positive metal ions and the mobile 'sea' of electrons.
- They conduct electricity due to the mobile electrons within the metal structure. When a metal is connected in a circuit, the electrons move towards the positive terminal while at the same time electrons are fed into the other end of the metal from the negative terminal.
- They are malleable and ductile. Unlike the fixed bonds in diamond, metallic bonds are not rigid but are still strong. If a force is applied to a metal, rows of ions can slide over one another. They reposition themselves and the strong bonds re-form as shown in Figure 4.31. Malleable means that metals can be hammered into different shapes. Ductile means that the metals can be pulled out into thin wires.
- They have high densities because the atoms are very closely packed in a regular manner as can be seen in Figure 4.32. Different metals show different types of packing and in doing so they produce the arrangements of ions shown in Figure 4.33.

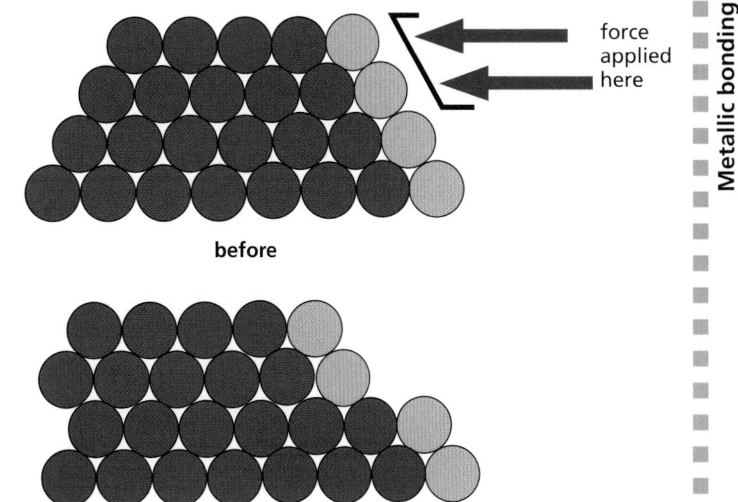


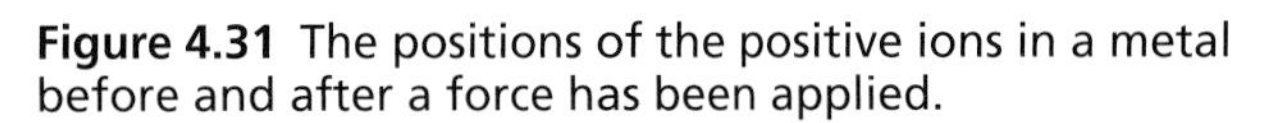

**Figure 4.31** The positions of the positive ions in a metal before and after a force has been applied.

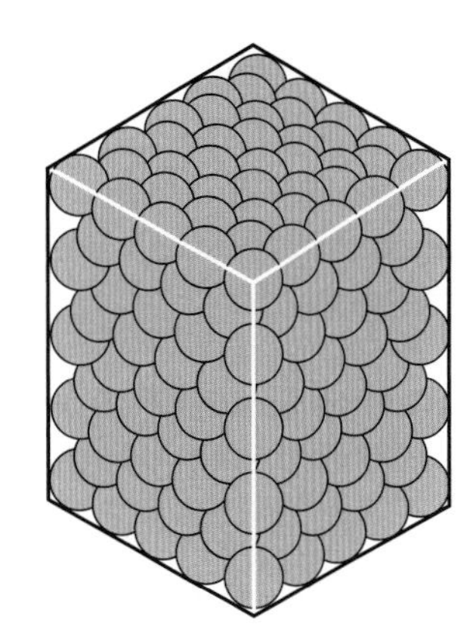

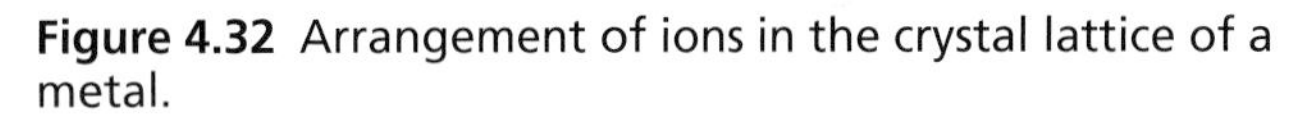

**Figure 4.32** Arrangement of ions in the crystal lattice of a metal.

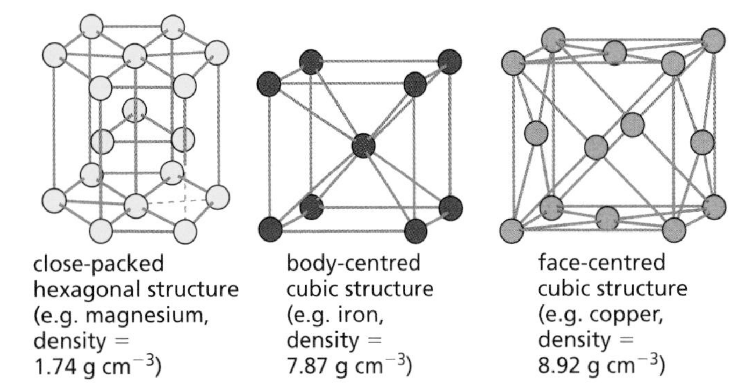


**Figure 4.33** Relating different structures to the density of metal.

## *Questions*

1. Explain the terms:
   a malleable
   b ductile.
2. Explain why metals are able to conduct heat and electricity.
3. Explain why the melting point of magnesium (649 °C) is much higher than the melting point of sodium (97.9 °C).

## Checklist

**After studying Chapter 4 you should know and understand the following terms.**

- **Allotropy** The existence of an element in two or more different forms in the same physical state.
- **Ceramics** Materials such as pottery made from inorganic chemicals by high-temperature processing. Other modern ceramics include zirconium oxide and silicon nitride.
- **Covalent bond** A chemical bond formed by the sharing of one or more pairs of electrons between two atoms.
- **Giant ionic structure** A lattice held together by the electrostatic forces of attraction between ions.
- **Giant molecular substance** A substance containing thousands of atoms per molecule.
- **Glass** A supercooled liquid which forms a hard, brittle substance that is usually transparent and resistant to chemical attack.
- **Ionic (electrovalent) bond** A strong electrostatic force of attraction between oppositely charged ions.
- **Intermolecular bonds** Attractive forces which act between molecules, for example van der Waals' forces.
- **Intramolecular bonds** Forces which act within a molecule, for example covalent bonds.
- **Lattice** A regular three-dimensional arrangement of atoms/ions in a crystalline solid.
- **Metallic bond** An electrostatic force of attraction between the mobile 'sea' of electrons and the regular array of positive metal ions within the solid metal.
- **Simple molecular substance** These substances possess between one and a few hundred atoms per molecule.
- **Supercooled liquid** One which has cooled below its freezing point without solidification.
- **Valency** The combining power of an atom or group of atoms. The valency of an ion is equal to its charge.
- **X-ray diffraction** A technique often used to study crystal structures.

# Bonding and structure

# *Additional questions*

**1** Draw diagrams to show the bonding in each of the following compounds:

**a** calcium fluoride ($CaF_2$)
**b** oxygen ($O_2$)
**c** magnesium chloride ($MgCl_2$)
**d** tetrachloromethane ($CCl_4$).

**2** Use the information given in Table 4.1 to work out the formula for:

**a** silver oxide
**b** zinc chloride
**c** potassium sulphate
**d** calcium nitrate
**e** iron(II) nitrate
**f** copper(II) carbonate
**g** iron(III) hydroxide
**h** aluminium fluoride.

**3** Atoms of elements **X**, **Y** and **Z** have 16, 17 and 19 electrons, respectively. Atoms of argon have 18 electrons.

**a** Determine the formulae of the compounds formed by the combination of the atoms of the elements:
(i) **X** and **Z**
(ii) **Y** and **Z**
(iii) **X** with itself.

**b** In each of the cases shown in **a**(i)–(iii) above, name the type of chemical bond formed.

**c** Give two properties you would expect to be shown by the compounds formed in **a**(ii) and **a**(iii).

**4** The diagram shows the arrangement of the outer electrons only in a molecule of ethanoic acid.

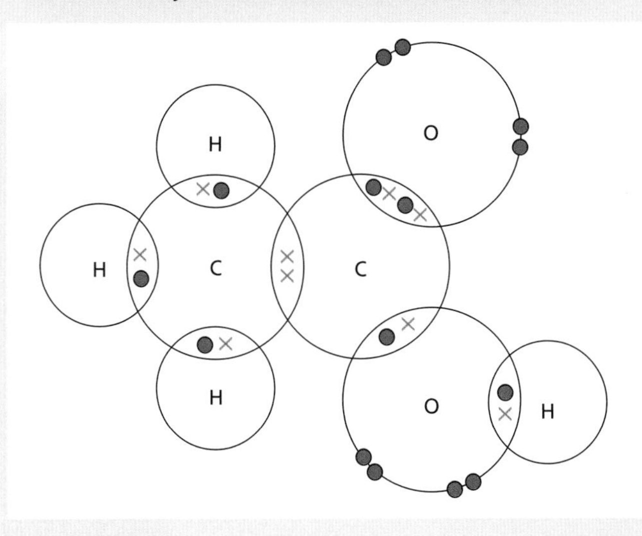


**a** Name the different elements found in this compound.
**b** What is the total number of atoms present in this molecule?
**c** Between which two atoms is there a double covalent bond?
**d** How many single covalent bonds does each carbon atom have?
**e** Write a paragraph explaining the sorts of properties you would expect this sort of substance to have.

**5** Make a summary table of the properties of substances with covalent structures. Your table should include examples of both simple molecular and giant molecular substances.

**6** The elements sodium and chlorine react together to form the compound sodium chloride, which has a giant ionic lattice structure.

**a** What type of structure do the elements (i) sodium and (ii) chlorine have?
**b** Draw a diagram to represent how the ions are arranged in the crystal lattice of sodium chloride.
**c** Explain how the ions are held together in this crystal lattice.
**d** Draw diagrams to show how the electrons are arranged in a sodium ion and a chloride ion. (The atomic numbers of sodium and chlorine are 11 and 17, respectively.)
**e** Make a table showing the properties of the three substances sodium, chlorine and sodium chloride. Include in your table:
(i) the physical state at room temperature
(ii) solubility in (or reaction with) water
(iii) colour
(iv) electrical conductivity.

**7** Explain the following.

**a** Ammonia is a gas at room temperature.
**b** The melting points of sodium chloride and iodine are so different.
**c** Metals generally are good conductors of electricity.
**d** Buckminsterfullerene is an allotrope of carbon.
**e** Metals usually have high melting and boiling points.

**8** Discuss the following with reference to diamond and graphite.

**a** Diamond is one of the hardest substances known.
**b** Graphite is a good lubricating agent.
**c** Graphite conducts electricity but diamond does not.

# 5 Chemical calculations



Chemists often need to know how much of a substance has been formed or used up during a chemical reaction (Figure 5.1). This is particularly important in the chemical industry, where the substances being reacted (the **reactants**) and the substances being produced (the **products**) are worth thousands of pounds. Waste costs money!

To solve this problem they need a way of counting atoms, ions or molecules. Atoms, ions and molecules are very tiny particles and it is impossible to measure out a dozen or even a hundred of them. Instead, chemists weigh out a very large number of particles. This number is $6 \times 10^{23}$ atoms, ions or molecules and is called **Avogadro's constant** after the famous Italian scientist Amedeo Avogadro (1776–1856). An amount of substance containing $6 \times 10^{23}$ particles is called a **mole** (often abbreviated to mol).

So, a mole of the element magnesium is $6 \times 10^{23}$ atoms of magnesium and a mole of the element carbon is $6 \times 10^{23}$ atoms of carbon (Figure 5.2).

**Figure 5.1** The chemists at paint manufacturers need to know how much pigment is going to be produced.

**a** A mole of magnesium.

**b** A mole of carbon.

**Figure 5.2**

## Calculating moles

You have already seen in Chapter 3 how we can compare the masses of all the other atoms with the mass of carbon atoms. This is the basis of the **relative atomic mass scale**. Chemists have found by experiment that if you take the relative atomic mass of an element in grams, it always contains $6 \times 10^{23}$ or one mole of its atoms.

### Moles and elements

For example, the relative atomic mass ($A_r$) of iron is 56, so one mole of iron is 56 g. Therefore, 56 g of iron contains $6 \times 10^{23}$ atoms.

The $A_r$ for aluminium is 27. In 27 g of aluminium it is found that there are $6 \times 10^{23}$ atoms. Therefore, 27 g of aluminium is one mole of aluminium atoms.

The mass of a substance present in any number of moles can be calculated using the relationship:

$$\text{mass (in grams)} = \text{number of moles} \times \text{mass of 1 mole of the element}$$

**Example 1**

Calculate the mass of (a) 2 moles and (b) 0.25 mole of iron ($A_r$: Fe = 56).

**a** mass of 2 moles of iron
= number of moles × relative atomic mass ($A_r$)
= 2 × 56
= 112 g

**b** mass of 0.25 mole of iron
= number of moles × relative atomic mass ($A_r$)
= 0.25 × 56
= 14 g

If we know the mass of the element then it is possible to calculate the number of moles of that element using:

$$\text{number of moles} = \frac{\text{mass of the element}}{\text{mass of 1 mole of that element}}$$

**Example 2**

Calculate the number of moles of aluminium present in (a) 108 g and (b) 13.5 g of the element ($A_r$: Al = 27).

**a** number of moles of aluminium

$$= \frac{\text{mass of aluminium}}{\text{mass of 1 mole of aluminium}}$$

$$= \frac{108}{27} = 4 \text{ moles}$$

**b** number of moles of aluminium

$$= \frac{\text{mass of aluminium}}{\text{mass of 1 mole of aluminium}}$$

$$= \frac{13.5}{27} = 0.5 \text{ mole}$$

### Moles and compounds

The idea of the mole has been used so far only with elements and atoms. However, it can also be used with compounds (Figure 5.3).

We cannot discuss the atomic mass of a molecule or of a compound because more than one type of atom is involved. Instead, we have to discuss the **relative formula mass (RFM)**. This is the sum of the relative atomic masses of all those elements shown in the formula of the substance.

What is the mass of 1 mole of water ($H_2O$) molecules? ($A_r$: H = 1; O = 16)

From the formula of water, $H_2O$, you will see that 1 mole of water molecules contains 2 moles of hydrogen (H) atoms and 1 mole of oxygen (O) atoms. The mass of 1 mole of water molecules is therefore:

$$(2 \times 1) + (1 \times 16) = 18 \text{ g}$$

The mass of 1 mole of a compound is called its molar mass. If you write the molar mass of a compound without any units then it is the relative formula mass, often called the **relative molecular mass ($M_r$)**. So the relative formula mass of water is 18.

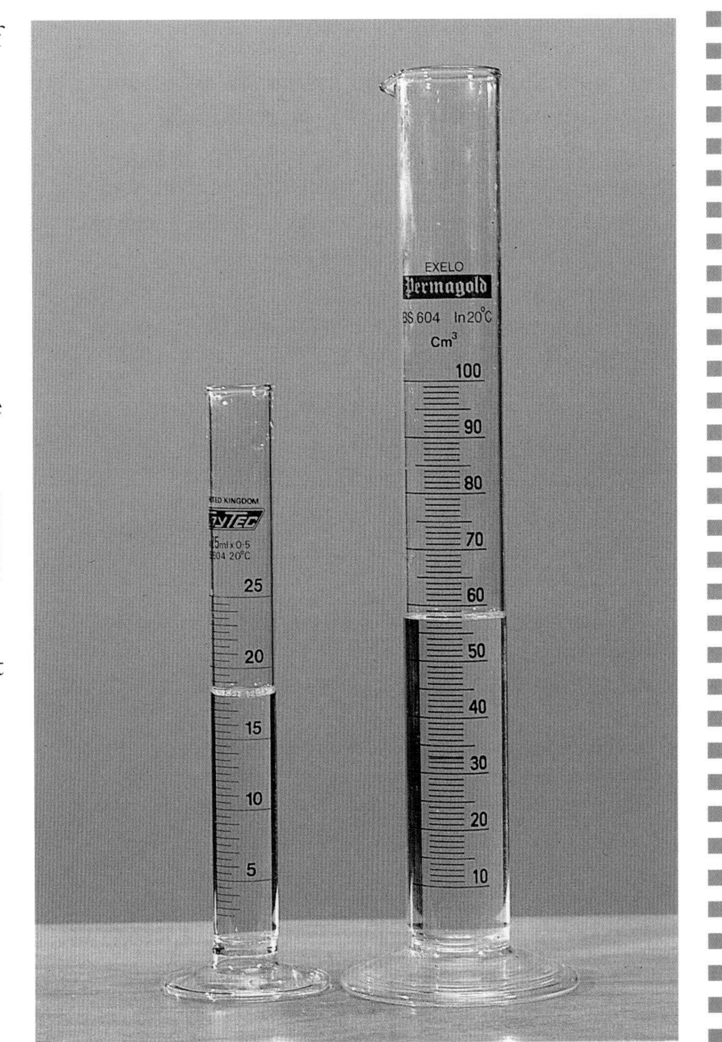


**Figure 5.3** One mole of water ($H_2O$) (left) and one mole of ethanol ($C_2H_5OH$) (right) in separate measuring cylinders.

Now follow these examples to help you learn and understand more about moles and compounds.

**Example 1**
What is (a) the mass of 1 mole and (b) the relative formula mass (RFM) of ethanol, $C_2H_5OH$? ($A_r$: H = 1; C = 12; O = 16)

**a** One mole of $C_2H_5OH$ contains 2 moles of carbon atoms, 6 moles of hydrogen atoms and 1 mole of oxygen atoms. Therefore:
mass of 1 mole of ethanol
$= (2 \times 12) + (6 \times 1) + (1 \times 16)$
$= 46\,g$

**b** The RFM of ethanol is 46.

**Example 2**
What is (a) the mass of 1 mole and (b) the RFM of nitrogen gas, $N_2$? ($A_r$: N = 14)

**a** Nitrogen is a diatomic gas. Each nitrogen molecule contains two atoms of nitrogen. Therefore:
mass of 1 mole of $N_2$
$= 2 \times 14$
$= 28\,g$

**b** The RFM of $N_2$ is 28.

The mass of a compound found in any number of moles can be calculated using the relationship:

$$\text{mass of compound} = \text{number of moles of the compound} \times \text{mass of 1 mole of the compound}$$

**Example 3**
Calculate the mass of (a) 3 moles and (b) 0.2 moles of carbon dioxide gas, $CO_2$. ($A_r$: C = 12; O = 16)

**a** One mole of $CO_2$ contains 1 mole of carbon atoms and 2 moles of oxygen atoms. Therefore:
mass of 1 mole of $CO_2$
$= (1 \times 12) + (2 \times 16)$
$= 44\,g$
mass of 3 moles of $CO_2$
$=$ number of moles $\times$ mass of 1 mole of $CO_2$
$= 3 \times 44$
$= 132\,g$

**b** mass of 0.2 mole of $CO_2$
$=$ number of moles $\times$ mass of 1 mole of $CO_2$
$= 0.2 \times 44$
$= 8.8\,g$

If we know the mass of the compound then we can calculate the number of moles of the compound using the relationship:

$$\text{number of moles of compound} = \frac{\text{mass of compound}}{\text{mass of 1 mole of the compound}}$$

**Example 4**
Calculate the number of moles of magnesium oxide, MgO, in (a) 80 g and (b) 10 g of the compound. ($A_r$: O = 16; Mg = 24)

**a** One mole of MgO contains 1 mole of magnesium atoms and 1 mole of oxygen atoms. Therefore:
mass of 1 mole of MgO
$= (1 \times 24) + (1 \times 16)$
$= 40\,g$

number of moles of MgO in 80 g

$$= \frac{\text{mass of MgO}}{\text{mass of 1 mole of MgO}}$$

$$= \frac{80}{40}$$

$= 2$ moles

**b** number of moles of MgO in 10 g

$$= \frac{\text{mass of MgO}}{\text{mass of 1 mole of MgO}}$$

$$= \frac{10}{40}$$

$= 0.25$ mole

## Moles and gases

Many substances exist as gases. If we want to find the number of moles of a gas we can do this by measuring the volume rather than the mass.

Chemists have shown by experiment that:

One mole of any gas occupies a volume of approximately $24\,dm^3$ (24 l) at room temperature and pressure (rtp).

Therefore, it is relatively easy to convert volumes of gases into moles and moles of gases into volumes using the following relationship:

$$\text{number of moles of a gas} = \frac{\text{volume of the gas (in } dm^3 \text{ at rtp)}}{24\,dm^3}$$

or

$$\text{volume of a gas (in } dm^3 \text{ at rtp)} = \text{number of moles of gas} \times 24\,dm^3$$

**Example 1**
Calculate the number of moles of ammonia gas, $NH_3$, in a volume of $72\,dm^3$ of the gas measured at rtp.

$$\text{number of moles of ammonia} = \frac{\text{volume of ammonia in dm}^3}{24\,dm^3} = \frac{72}{24} = 3$$

**Example 2**
Calculate the volume of carbon dioxide gas, $CO_2$, occupied by (a) 5 moles and (b) 0.5 mole of the gas measured at rtp.

**a** volume of $CO_2$
$= \text{number of moles of } CO_2 \times 24\,dm^3$
$= 5 \times 24$
$= 120\,dm^3$

**b** volume of $CO_2$
$= \text{number of moles of } CO_2 \times 24\,dm^3$
$= 0.5 \times 24$
$= 12\,dm^3$

The volume occupied by one mole of any gas must contain $6 \times 10^{23}$ molecules. Therefore, it follows that equal volumes of all gases measured at the same temperature and pressure must contain the same number of molecules. This idea was also first put forward by Amedeo Avogadro and is called **Avogadro's Law**.

## Moles and solutions

Chemists often need to know the concentration of a solution. Sometimes it is measured in grams per cubic decimetre ($g\,dm^{-3}$) but more often concentration is measured in **moles per cubic decimetre** (**mol $dm^{-3}$**). When 1 mole of a substance is dissolved in water and the solution is made up to $1\,dm^3$ ($1000\,cm^3$), a **1 molar** (**1 M**) solution is produced. Chemists do not always need to make up such large volumes of solution. A simple method of calculating the concentration is by using the relationship:

$$\text{concentration (in mol dm}^{-3}) = \frac{\text{number of moles}}{\text{volume (in dm}^3)}$$

**Example 1**
Calculate the concentration (in $mol\,dm^{-3}$) of a solution of sodium hydroxide, NaOH, which was made by dissolving 10 g of solid sodium hydroxide in water and making up to $250\,cm^3$. ($A_r$: Na = 23; O = 16; H = 1)

1 mole of NaOH contains 1 mole of sodium, 1 mole of oxygen and 1 mole of hydrogen. Therefore:

$$\text{mass of 1 mole of NaOH} = (1 \times 23) + (1 \times 16) + (1 \times 1) = 40\,g$$

number of moles of NaOH in 10 g

$$= \frac{\text{mass of NaOH}}{\text{mass of 1 mole of NaOH}} = \frac{10}{40} = 0.25$$

$$(250\,cm^3 = \frac{250}{1000}\,dm^3 = 0.25\,dm^3)$$

concentration of the NaOH solution

$$= \frac{\text{number of moles of NaOH}}{\text{volume of solution (dm}^3)} = \frac{0.25}{0.25} = 1\,mol\,dm^{-3} \text{ (or 1 M)}$$

Sometimes chemists need to know the mass of a substance that has to be dissolved to prepare a known volume of solution at a given concentration. A simple method of calculating the number of moles and so the mass of substance needed is by using the relationship:

$$\text{number of moles} = \text{concentration (in mol dm}^{-3}) \times \text{volume of solution (in dm}^3)$$

**Example 2**
Calculate the mass of potassium hydroxide, KOH, that needs to be used to prepare $500\,cm^3$ of a $2\,mol\,dm^{-3}$ (2 M) solution in water. ($A_r$: H = 1; O = 16; K = 39)

number of moles of KOH

$$= \text{concentration of solution (mol dm}^{-3}) \times \text{volume of solution (dm}^3)$$

$$= 2 \times \frac{500}{1000} = 1$$

1 mole of KOH contains 1 mole of potassium, 1 mole of oxygen and 1 mole of hydrogen. Therefore:

$$\text{mass of 1 mole of KOH} = (1 \times 39) + (1 \times 16) + (1 \times 1) = 56\,g$$

Therefore:

mass of KOH in 1 mole
$= \text{number of moles} \times \text{mass of 1 mole}$
$= 1 \times 56$
$= 56\,g$

## Questions

Use the values of $A_r$ which follow to answer the questions below:

H = 1; C = 12; N = 14; O = 16; Ne = 20; Na = 23; Mg = 24; S = 32; K = 39; Fe = 56; Cu = 63.5; Zn = 65.

One mole of any gas at rtp occupies $24\,dm^3$.

1 Calculate the number of moles in:
 a 2g of neon atoms
 b 4g of magnesium atoms
 c 24g of carbon atoms.

2 Calculate the mass of:
 a 0.1 mole of oxygen molecules
 b 5 moles of sulphur atoms
 c 0.25 mole of sodium atoms.

3 Calculate the number of moles in:
 a 9.8g of sulphuric acid ($H_2SO_4$)
 b 40g of sodium hydroxide (NaOH)
 c 720g of iron(II) oxide (FeO).

4 Calculate the mass of:
 a 2 moles of zinc oxide (ZnO)
 b 0.25 mole of hydrogen sulphide ($H_2S$)
 c 0.35 mole of copper(II) sulphate ($CuSO_4$).

5 Calculate the number of moles at rtp in:
 a $2\,dm^3$ of carbon dioxide ($CO_2$)
 b $240\,dm^3$ of sulphur dioxide ($SO_2$)
 c $20\,cm^3$ of carbon monoxide (CO).

6 Calculate the volume of:
 a 0.3 mole of hydrogen chloride (HCl)
 b 4.4g of carbon dioxide
 c 34g of ammonia ($NH_3$)

7 Calculate the concentration of solutions containing:
 a 0.2 mole of sodium hydroxide dissolved in water and made up to $100\,cm^3$
 b 9.8g of sulphuric acid dissolved in water and made up to $500\,cm^3$.

8 Calculate the mass of:
 a copper(II) sulphate ($CuSO_4$) which needs to be used to prepare $500\,cm^3$ of a $0.1\,mol\,dm^{-3}$ solution
 b potassium nitrate ($KNO_3$) which needs to be used to prepare $200\,cm^3$ of a $2\,mol\,dm^{-3}$ solution.

# *Calculating formulae*

If we have 1 mole of a compound, then the formula shows the number of moles of each element in that compound. For example, the formula for lead(II) bromide is $PbBr_2$. This means that 1 mole of lead(II) bromide contains 1 mole of lead ions and 2 moles of bromide ions. If we do not know the formula of a compound, we can find the masses of the elements present experimentally and these masses can be used to work out the formula of that compound.

## Finding the formula

### Magnesium oxide

When magnesium ribbon is heated strongly, it burns very brightly to form the white powder called magnesium oxide.

magnesium + oxygen → magnesium oxide

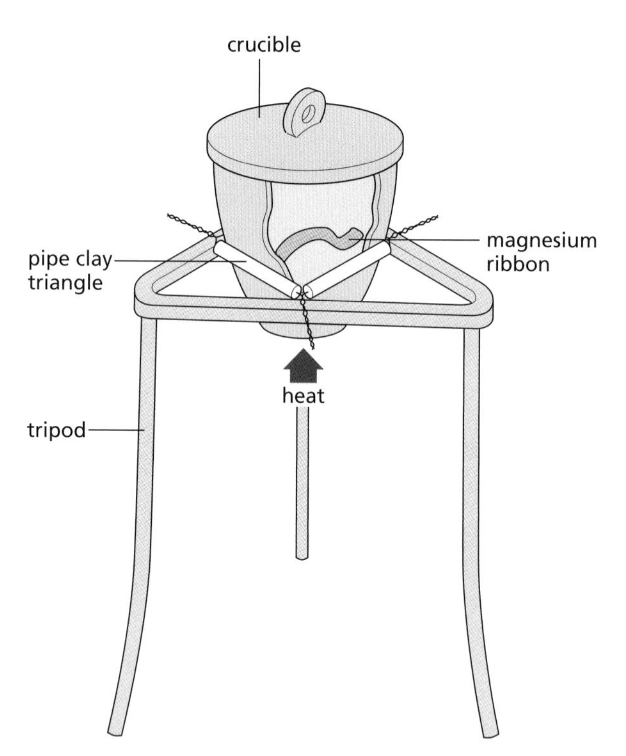


**Figure 5.4** Apparatus used to determine magnesium oxide's formula.

**Table 5.1** Data from experiment shown in Figure 5.4.

| | |
|---|---|
| Mass of crucible | 14.63 g |
| Mass of crucible and magnesium | 14.87 g |
| Mass of crucible and magnesium oxide | 15.03 g |
| Mass of magnesium used | 0.24 g |
| Mass of oxygen which has reacted with the magnesium | 0.16 g |

The data shown in Table 5.1 were obtained from an experiment using the apparatus shown in Figure 5.4 to find the formula for this white powder, magnesium oxide. From these data we can calculate the number of moles of each of the reacting elements. ($A_r$: O = 16; Mg = 24 ).

| | **Mg** | **O** |
|---|---|---|
| Masses reacting (g) | 0.24 | 0.16 |
| Number of moles | $\frac{0.24}{24}$ | $\frac{0.16}{16}$ |
| | = 0.01 | = 0.01 |
| Ratio of moles | 1 | 1 |
| Formula | MgO | |

This formula is the **empirical formula** of the compound. It shows the simplest ratio of the atoms present.

### Unknown compound 1

In another experiment an unknown organic compound was found to contain 0.12 g of carbon and 0.02 g of hydrogen. Calculate the empirical formula of the compound. ($A_r$: H = 1; C = 12)

| | **C** | **H** |
|---|---|---|
| Masses (g) | 0.12 | 0.02 |
| Number of moles | $\frac{0.12}{12}$ | $\frac{0.02}{1}$ |
| | = 0.01 | = 0.02 |
| Ratio of moles | 1 | 2 |
| Empirical formula | $CH_2$ | |

From our knowledge of bonding (Chapter 4, p. 59) we know that a molecule of this formula cannot exist. However, molecules with the following formulae do exist: $C_2H_4$, $C_3H_6$, $C_4H_8$ and $C_5H_{10}$. All of these formulae show the same ratio of carbon atoms to hydrogen atoms, $CH_2$, as our unknown. To find out which of these formulae is the actual formula for the unknown organic compound, we need to know the mass of one mole of the compound.

Using a mass spectrometer, the relative molecular mass ($M_r$) of this organic compound was found to be 56. We need to find out the number of empirical formulae units present:

$M_r$ of the empirical formula unit
$= (1 \times 12) + (2 \times 1)$
$= 14$

Number of empirical formula units present

$$= \frac{M_r \text{ of compound}}{M_r \text{ of empirical formula unit}}$$

$$= \frac{56}{14}$$

$$= 4$$

Therefore, the actual formula of the unknown organic compound is $4 \times CH_2 = C_4H_8$.

This substance is called butene. $C_4H_8$ is the **molecular formula** for this substance and shows the **actual** numbers of atoms of each element present in one molecule of the substance.

Sometimes the composition of a compound is given as a percentage by mass of the elements present. In cases such as this the procedure shown in the next example is followed.

### Unknown compound 2

Calculate the empirical formula of an organic compound containing 92.3% carbon and 7.7% hydrogen by mass. The $M_r$ of the organic compound is 78. What is its molecular formula? ($A_r$: H = 1; C = 12)

| | **C** | **H** |
|---|---|---|
| % by mass | 92.3 | 7.7 |
| in 100 g | 92.3 g | 7.7 g |
| moles | $\frac{92.3}{12}$ | $\frac{7.7}{1}$ |
| | = 7.7 | = 7.7 |
| Ratios of moles | 1 | 1 |
| Empirical formula | CH | |

$M_r$ of the empirical formula unit CH
$= 12 + 1$
$= 13$

Number of empirical formula units present

$$= \frac{M_r \text{ of compound}}{M_r \text{ of empirical formula unit}}$$

$$= \frac{78}{13}$$

$$= 6$$

The molecular formula of the organic compound is $6 \times CH = C_6H_6$. This is a substance called benzene.

## Questions

Use the following values of $A_r$ to answer the questions below: H = 1; C = 12; O = 16; Ca = 40.

1 Determine the empirical formula of an oxide of calcium formed when 0.4 g of calcium reacts with 0.16 g of oxygen.

2 Determine the empirical formula of an organic hydrocarbon compound which contains 80% by mass of carbon and 20% by mass of hydrogen. If the $M_r$ of the compound is 30, what is its molecular formula?

## Moles and chemical equations

When we write a balanced chemical equation we are indicating the numbers of moles of reactants and products involved in the chemical reaction. Consider the reaction between magnesium and oxygen.

$$\text{magnesium} + \text{oxygen} \rightarrow \text{magnesium oxide}$$
$$2Mg(s) + O_2(g) \rightarrow 2MgO(s)$$

This shows that 2 moles of magnesium react with 1 mole of oxygen to give 2 moles of magnesium oxide.

Using the ideas of moles and masses we can use this information to calculate the quantities of the different chemicals involved.

| $2Mg(s)$ + | $O_2(g)$ | → $2MgO(s)$ |
|---|---|---|
| 2 moles | 1 mole | 2 moles |
| $2 \times 24$ | $1 \times (16 \times 2)$ | $2 \times (24 + 16)$ |
| = 48 g | = 32 g | = 80 g |

You will notice that the total mass of reactants is equal to the total mass of product. This is true for any chemical reaction and it is known as the **Law of conservation of mass**. This law was understood by the Greeks but was first clearly formulated by Antoine Lavoisier in 1774. Chemists can use this idea to calculate masses of products formed and reactants used in chemical processes before they are carried out.

### Example using a solid

Lime (calcium oxide, CaO) is used in the manufacture of mortar. It is manufactured in large quantities by a company called Tilcon at their Swindon quarry in North Yorkshire (see Figure 5.5) by heating limestone (calcium carbonate, $CaCO_3$).

**Figure 5.5** Calcium oxide production at Swindon quarry.

The equation for the process is:

| $CaCO_3(s)$ | → $CaO(s)$ + | $CO_2(g)$ |
|---|---|---|
| 1 mole | 1 mole | 1 mole |
| $(40 + 12 + (3 \times 16))$ | $40 + 16$ | $12 + (2 \times 16)$ |
| = 100 g | = 56 g | = 44 g |

Calculate the amount of lime produced when 10 tonnes of limestone are heated (Figure 5.6). ($A_r$: C = 12; O = 16; Ca = 40)

$$1 \text{ tonne (t)} = 1000\,\text{kg}$$
$$1\,\text{kg} = 1000\,\text{g}$$

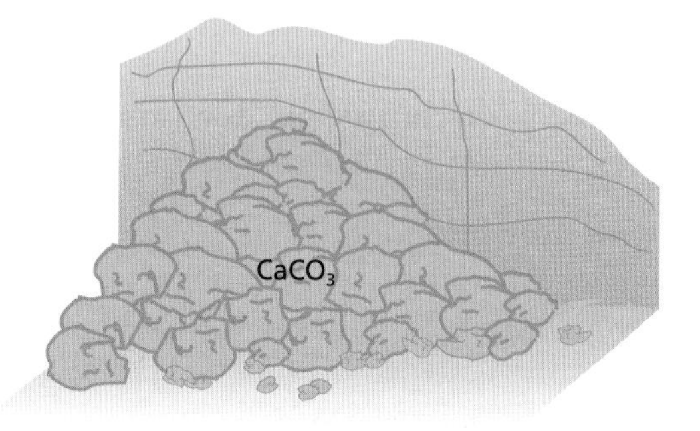


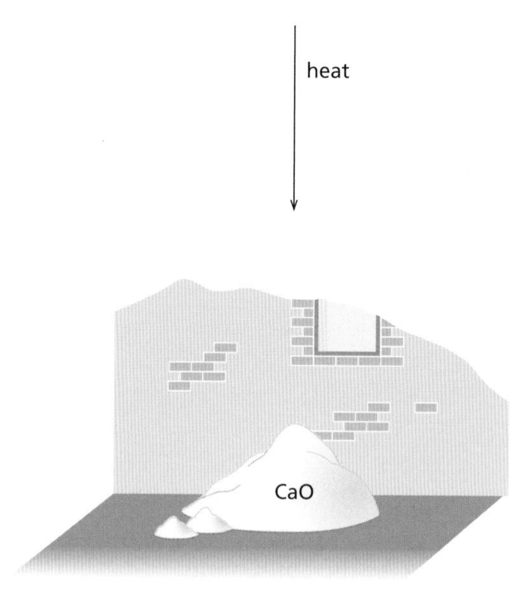


**Figure 5.6** How much lime is produced?

From this relationship between grams and tonnes we can replace the masses in grams by masses in tonnes.

| | $CaCO_3(s)$ → | $CaO(s)$ + | $CO_2(g)$ |
|---|---|---|---|
| | 100 t | 56 t | 44 t |
| Hence | 10 t | 5.6 t | 4.4 t |

The equation now shows that 100 t of limestone will produce 56 t of lime. Therefore, 10 t of limestone will produce 5.6 t of lime.

### Example using a gas

Many chemical processes involve gases. The volume of a gas is measured more easily than its mass. This example shows how chemists work out the volumes of gaseous reactants and products needed using Avogadro's Law and the idea of moles.

Some rockets use hydrogen gas as a fuel. When hydrogen burns in oxygen it forms water vapour. Calculate the volumes of (a) $O_2(g)$ used and (b) water, $H_2O(g)$, produced if 960 $dm^3$ of hydrogen gas, $H_2(g)$, were burned in oxygen. ($A_r$: H = 1; O = 16) Assume 1 mole of any gas occupies a volume of 24 $dm^3$.

| $2H_2(g)$ | $+ O_2(g)$ | $\rightarrow 2H_2O(g)$ |
|---|---|---|
| 2 moles | 1 mole | 2 moles |
| $2 \times 24$ | $1 \times 24$ | $2 \times 24$ |
| $= 48\ dm^3$ | $= 24\ dm^3$ | $= 48\ dm^3$ |

Therefore:

| | | | |
|---|---|---|---|
| $(\times 2)$ | $96\ dm^3$ | $48\ dm^3$ | $96\ dm^3$ |
| $(\times 10)$ | $960\ dm^3$ | $480\ dm^3$ | $960\ dm^3$ |

When 960 $dm^3$ of hydrogen are burned in oxygen:

**a** 480 $dm^3$ of oxygen are required and
**b** 960 $dm^3$ of $H_2O(g)$ are produced.

### Example 1 using a solution

Chemists usually carry out reactions using solutions. If they know the concentration of the solution(s) they are using they can find out the quantities reacting.

Calculate the volume of 1 mol $dm^{-3}$ solution of $H_2SO_4$ required to react completely with 6 g of magnesium. ($A_r$: Mg = 24)

number of moles of magnesium

$$= \frac{\text{mass of magnesium}}{\text{mass of 1 mole of magnesium}}$$

$$= \frac{6}{24}$$

$$= 0.25$$

| $Mg(s)$ | $+ H_2SO_4(aq)$ | $\rightarrow MgSO_4(aq)$ | $+ H_2(g)$ |
|---|---|---|---|
| 1 mole | 1 mole | 1 mole | 1 mole |
| 0.25 mol | 0.25 mol | 0.25 mol | 0.25 mol |

We can see that 0.25 mol of $H_2SO_4(aq)$ is required. Using:

volume of $H_2SO_4(aq)$ ($dm^3$)

$$= \frac{\text{moles of } H_2SO_4}{\text{concentration of } H_2SO_4\ (\text{mol dm}^{-3})}$$

$$= \frac{0.25}{1}$$

$$= 0.25\ dm^3 \text{ or } 250\ cm^3$$

### Example 2 using a solution

What is the concentration of sodium hydroxide solution used in the following neutralisation reaction: 40 $cm^3$ of 0.2 mol $dm^{-3}$ solution of hydrochloric acid just neutralised 20 $cm^3$ of sodium hydroxide solution.

number of moles of HCl used
$= \text{concentration (mol dm}^{-3}) \times \text{volume (dm}^3)$
$= 0.2 \times 0.04$
$= 0.08$

| $HCl(aq)$ | + | $NaOH(aq)$ | $\rightarrow$ | $NaCl(aq)$ | + | $H_2O(l)$ |
|---|---|---|---|---|---|---|
| 1 mole | | 1 mole | | 1 mole | | 1 mole |
| 0.08 mol | | 0.08 mol | | 0.08 mol | | 0.08 mol |

You will see that 0.08 mole of NaOH would have been present. The concentration of the NaOH(aq) will be given by:

concentration of NaOH (mol $dm^{-3}$)

$$= \frac{\text{number of moles of NaOH}}{\text{volume of NaOH (dm}^3)}$$

$$\left(\text{volume of NaOH in dm}^3 = \frac{20}{1000} = 0.02\right)$$

$$= \frac{0.08}{0.02}$$

$= 4$ mol $dm^{-3}$ or 4 M

## Questions

Use the following $A_r$ values to answer the questions below: O = 16; Mg = 24; S = 32; K = 39; Cu = 63.5.

1. Calculate the mass of sulphur dioxide produced by burning 16 g of sulphur in an excess of oxygen in the Contact process (see p. 231).
2. Calculate the mass of sulphur which, when burned in excess oxygen, produces 640 g of sulphur dioxide in the Contact process.
3. Calculate the mass of copper required to produce 159 g of copper(II) oxide when heated in excess oxygen.
4. In the rocket mentioned previously in which hydrogen is used as a fuel, calculate the volume of hydrogen used to produce 24 $dm^3$ of water ($H_2O(g)$).
5. Calculate the volume of 2 mol $dm^{-3}$ solution of sulphuric acid required to react with 24 g of magnesium.
6. What is the concentration of potassium hydroxide solution used in the following neutralisation reaction? 20 $cm^3$ of 0.2 mol $dm^{-3}$ solution of hydrochloric acid just neutralised 15 $cm^3$ of potassium hydroxide solution.

## Checklist

**After studying Chapter 5 you should know and understand the following terms.**

- **Avogadro's Law** Equal volumes of all gases measured under the same conditions of temperature and pressure contain equal numbers of molecules.

- **Calculating moles**

— **Compounds:**

mass of compound (in grams) = number of moles × mass of 1 mole of compound

$$\text{number of moles} = \frac{\text{mass of compound}}{\text{mass of 1 mole of compound}}$$

— **Elements:**

mass of element (in grams) = number of moles × mass of 1 mole of the element

$$\text{number of moles} = \frac{\text{mass of the element}}{\text{mass of 1 mole of that element}}$$

— **Gases:**

1 mole of any gas occupies 24 $dm^3$ (litres) at room temperature and pressure (rtp).

$$\text{number of moles of gas} = \frac{\text{volume of the gas (in dm}^3\text{ at rtp)}}{24\,\text{dm}^3}$$

— **Solutions:**

$$\text{concentration of a solution (in mol dm}^{-3}) = \frac{\text{number of moles of solute}}{\text{volume (in dm}^3)}$$

number of moles = concentration (in mol $dm^{-3}$) × volume of solution (in $dm^3$)

- **Empirical formula** A formula showing the simplest ratio of atoms present.

- **Mole** The amount of substance which contains $6 \times 10^{23}$ atoms, ions or molecules. This number is called Avogadro's constant.

  Atoms – 1 mole of atoms has a mass equal to the relative atomic mass ($A_r$) in grams.

  Molecules – 1 mole of molecules has a mass equal to its relative molecular mass ($M_r$) in grams.

- **Molecular formula** A formula showing the actual number of atoms of each element present in one molecule.

- **Relative formula mass (RFM)** The sum of the relative atomic masses of all those elements shown in the formula of the substance. This is often referred to as the **relative molecular mass** ($M_r$).

# Chemical calculations

# *Additional questions*

Use the data in the table below to answer the questions which follow:

| Element | $A_r$ |
|---|---|
| H | 1 |
| C | 12 |
| N | 14 |
| O | 16 |
| Na | 23 |
| Mg | 24 |
| Si | 28 |
| S | 32 |
| Cl | 35.5 |
| Fe | 56 |

**1** Calculate the mass of:
- **a** 1 mole of:
  - (i) chlorine molecules
  - (ii) iron(III) oxide.
- **b** 0.5 mole of:
  - (i) magnesium nitrate
  - (ii) ammonia.

**2** Calculate the volume occupied, at rtp, by the following gases. (One mole of any gas occupies a volume of 24 $dm^3$ at rtp.)
- **a** 12.5 moles of sulphur dioxide gas.
- **b** 0.15 mole of nitrogen gas.

**3** Calculate the number of moles of gas present in the following:
- **a** 36 $cm^3$ of sulphur dioxide
- **b** 144 $dm^3$ of hydrogen sulphide.

**4** Using the following experimental information to determine the empirical formula of an oxide of silicon.

| | |
|---|---|
| Mass of crucible | 18.20 g |
| Mass of crucible + silicon | 18.48 g |
| Mass of crucible + oxide of silicon | 18.80 g |

**5 a** Calculate the empirical formula of an organic liquid containing 26.67% of carbon, 2.22% of hydrogen with the rest being oxygen.
- **b** The $M_r$ of the liquid is 90. What is its molecular formula?

**6** Iron is extracted from its ore, haematite, in the blast furnace. The main extraction reaction is:

$$Fe_2O_3(s) + 3CO(g) \rightarrow 2Fe(s) + 3CO_2(g)$$

- **a** Name the reducing agent in this process.
- **b** Name the oxide of iron shown in the equation.
- **c** Explain why this is a **redox** reaction.
- **d** Calculate the mass of iron which will be produced from 640 tonnes of haematite.

**7** Consider the following information about the newly discovered element, vulcium, whose symbol is Vu 'Vulcium is a solid at room temperature. It is easily cut by a penknife to reveal a shiny surface which tarnishes quite rapidly. It reacts violently with water, liberating a flammable gas and forms a solution with a pH of 13. When vulcium reacts with chlorine, it forms a white crystalline solid containing 29.5% chlorine.'
($A_r$: Vu = 85)
- **a** Calculate the empirical formula of vulcium chloride.
- **b** To which group of the periodic table should vulcium be assigned?
- **c** Write a word and balanced chemical equation for the reaction between vulcium and chlorine.
- **d** What other information in the description supports the assignment of group you have given to vulcium?
- **e** What type of bonding is present in vulcium chloride?
- **f** Write a word and balanced chemical equation for the reaction between vulcium and water.
- **g** Write the formulae for:
  - (i) vulcium sulphate
  - (ii) vulcium carbonate
  - (iii) vulcium hydroxide.

  Look at the periodic table (p. 40) to find out the real name of vulcium.

**8** 0.048 g of magnesium was reacted with excess dilute hydrochloric acid at room temperature and pressure. The hydrogen gas given off was collected.
- **a** Write a word and balanced symbol equation for the reaction taking place.
- **b** Draw a diagram of an apparatus which could be used to carry out this experiment and collect the hydrogen gas.
- **c** How many moles of magnesium were used?
- **d** Using the equation you have written in your answer to **a**, calculate the number of moles of hydrogen and hence the volume of this gas produced.
- **e** Calculate the volume of a solution containing 0.1 mol $dm^{-3}$ hydrochloric acid which would be needed to react exactly with 0.048 g of magnesium.

# 6 Electrolysis and its uses



What do all the items in the photographs shown in Figure 6.1 have in common? They all involve electricity through a process known as **electrolysis**. Electrolysis is the process of splitting up (decomposing) substances by passing an electric current through them. The substance which is decomposed is called the **electrolyte** (Figure 6.2, opposite). An electrolyte is a substance that conducts electricity when in the molten state or in solution. The electricity is carried through the electrolyte by **ions**. The electric currrent enters and leaves the electrolyte through **electrodes**, which are usually made of unreactive metals such as platinum or of the non-metal carbon (**inert** electrodes). The names given to the two electrodes are **cathode**, the negative electrode which attracts **cations** (positively charged ions), and **anode**, the positive electrode which attracts **anions** (negatively charged ions).

Electrolysis is very important in industry. To help you to understand what is happening in the process shown in the photographs, we will first consider the electrolysis of aluminium oxide.

**a** This picture frame has been silver plated using an electroplating process involving electrolysis.

**b** Aluminium is produced by electrolysis.

**c** This watch has a thin coating of gold over steel; the thin coating is produced by electrolysis.

**Figure 6.1**

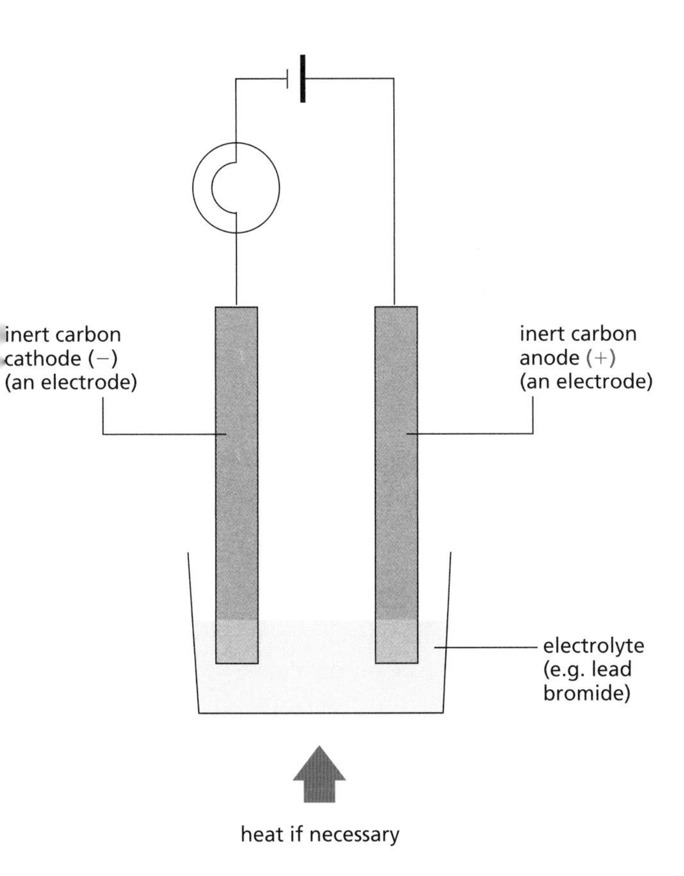


**Figure 6.2** The important terms used in electrolysis.

## Electrolysis of aluminium oxide

Aluminium is the most abundant metallic element in the Earth's crust. It was discovered in 1825 by Hans Christian Oersted in Copenhagen, Denmark, but was first isolated by Friedrich Wöhler in 1827. It makes up 8% of the crust and is found in the minerals bauxite (Figure 6.3), cryolite and mica, as well as in clay.

In the nineteenth century Napoleon III owned a very precious dinner service. It was said to be made of a metal more precious than gold. That metal was aluminium. The reason it was precious was that it was very rarely found as the pure metal. Aluminium is a reactive metal and as such was very difficult to extract from its ore. Reactive metals hold on tightly to the element(s) they have combined with and many are extracted from their ores by electrolysis.

Today we use aluminium in very large quantities. The annual production in the world is 19.5 million tonnes, of which 248 000 tonnes are produced in the UK. The commercial extraction of aluminium has been made possible by two scientists, working independently of each other, who discovered a method using electrolysis. The two scientists were Charles Martin Hall (USA), who discovered the process in 1886, and the French chemist Paul Héroult, who discovered the process independently in the same year.

**Figure 6.3** Bauxite mining in Australia.

The process they developed, often called the Hall–Héroult process, involves the electrolysis of aluminium oxide (alumina). The process involves the following stages.

- Bauxite, an impure form of aluminium oxide, is first treated with sodium hydroxide to obtain pure aluminium oxide, removing impurities such as iron(III) oxide and sand.
- The purified aluminium oxide is then dissolved in molten cryolite ($Na_3AlF_6$). Cryolite, a mineral found naturally in Greenland, is used to reduce the working temperature of the Hall–Héroult cell from 2017 °C (the melting point of pure aluminium oxide) to between 800 and 1000 °C. Therefore, the cryolite provides a considerable saving in the energy requirements of the process. In recent years it has become necessary to manufacture the cryolite.
- The molten mixture is then electrolysed in a cell similar to that shown in Figure 6.4.

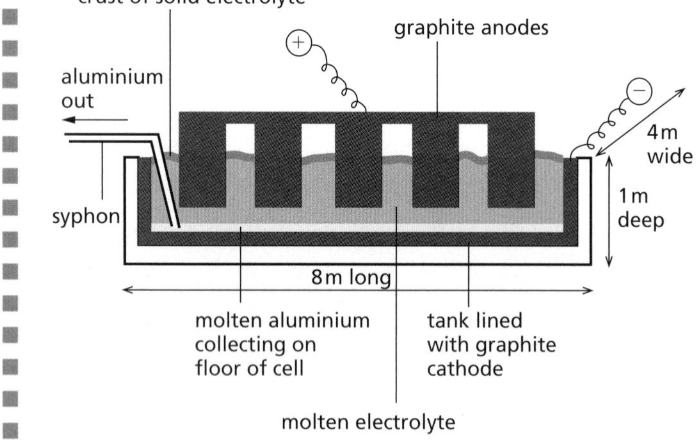


**Figure 6.4** The Hall–Héroult cell is used in industry to extract aluminium by electrolysis.

The anodes of this process are blocks of graphite which are lowered into the molten mixture from above. The cathode is the graphite lining of the steel vessel containing the cell.

Aluminium oxide is an ionic compound. When it is melted the ions become mobile, as the strong electrostatic forces of attraction between them are broken by the input of heat energy. During electrolysis the negatively charged oxide ions are attracted to the anode (the positive electrode), where they lose electrons to form oxygen gas.

$$\text{oxide ions} \rightarrow \text{oxygen molecules} + \text{electrons}$$
$$2O^{2-}(l) \rightarrow O_2(g) + 4e^-$$

The positive aluminium ions are attracted to the cathode (the negative electrode). They gain electrons to form molten aluminium metal.

$$\text{aluminium ions} + \text{electrons} \rightarrow \text{aluminium metal}$$
$$Al^{3+}(l) + 3e^- \rightarrow Al(l)$$

The gain of electrons which takes place at the cathode is called **reduction**. The loss of electrons which takes place at the anode is called **oxidation**. This is a further way of considering oxidation and reduction (Chapter 2, p. 17). A handy way of remembering it is **oil rig** (**o**xidation **i**s **l**oss, **r**eduction **i**s **g**ain of electrons).

The overall reaction which takes place in the cell is:

$$\text{aluminium oxide} \xrightarrow{\text{electrolysis}} \text{aluminium} + \text{oxygen}$$
$$2Al_2O_3(l) \longrightarrow 4Al(l) + 3O_2(g)$$

The molten aluminium collects at the bottom of the cell and it is syphoned out at regular intervals. No problems arise with other metals being deposited, since the cryolite is largely 'unaffected' by the flow of electricity. Problems do arise, however, with the graphite anodes. At the working temperature of the cell, the oxygen liberated reacts with the graphite anodes, producing carbon dioxide.

$$\text{carbon} + \text{oxygen} \rightarrow \text{carbon dioxide}$$
$$C(s) + O_2(g) \rightarrow CO_2(g)$$

The anodes burn away and have to be replaced on a regular basis.

The electrolysis of aluminium oxide is a continuous process in which vast amounts of electricity are used. Approximately 15 kWh of electricity are used to produce 1 kg of aluminium. In order to make the process an economic one, a cheap form of electricity is required. Hydroelectric power (HEP) is usually used for this process. The plant shown in Figure 6.5 uses an HEP scheme to provide some of the electrical energy required for this process. Further details about HEP are given in Chapter 13, p. 187.

**Figure 6.5** An aluminium smelting plant.

Using cheap electrical energy has allowed aluminium to be produced in such large quantities that it is the second most widely used metal after iron. It is used in the manufacture of electrical cables, cars, bikes and cooking foil as well as in alloys (Chapter 9, p. 140) such as duralumin, which is used in the manufacture of aeroplane bodies (Figure 6.6).

**Figure 6.6** Aluminium is used in the manufacture of aeroplane bodies.

**Figure 6.7** Bauxite pollution of the Amazon in Brazil.

Environmental problems associated with the location of aluminium plants are concerned with:

- the effects of the extracted impurities, which form a red mud (Figure 6.7)
- the fine cryolite dust, which is emitted through very tall chimneys so as not to affect the surrounding area
- the claimed link between environmental aluminium and a degenerative brain disorder called Alzheimer's disease – it is thought that aluminium is a major influence on the early onset of this disease. However, the evidence is still inconclusive

## Anodising

This is a process in which the surface coating of oxide on aluminium ($Al_2O_3$) is made thicker. In this process the aluminium object is made the anode in a cell in which the electrolyte is dilute sulphuric acid. During the electrolysis process, oxygen is produced at the anode and combines with the aluminium. The oxide layer on the surface of the aluminium therefore increases. Dyes can be mixed with the electrolyte and so the new thicker coating of oxide is colourful and also decorative (Figure 6.8).

**Figure 6.8** The oxide layer on the surface of these rolls of aluminium has been thickened, and dyes added to obtain the vibrant colours.

## Question

1 Produce a flow chart to summarise the processes involved in the extraction of aluminium metal from bauxite.

## Extraction of sodium

Sodium, a very reactive metal, was first discovered by Sir Humphry Davy in 1807. It is now extracted by electrolysing molten sodium chloride in the Down's cell. Sodium chloride may be obtained from the evaporation of sea water or mined as rock salt (Figure 6.9). The largest salt mines in England are those in Cheshire. The rock salt is mined from large underground caverns. Rock salt for the chemical industry is obtained by dissolving the salt and pumping the brine to the surface.

**Figure 6.9** Rock salt mining in Argentina.

The first economic extraction of sodium was developed by Down in 1921 in North America using hydroelectric power generated by the Niagara Falls. This ensured the availability of cheap electricity.

The process involves the following.

- Sodium chloride, obtained from rock salt or sea water, is mixed with calcium chloride and melted. The calcium chloride is added to the sodium chloride electrolyte to reduce the working temperature of the cell from 801 °C (the melting point of sodium chloride) to 600 °C (the melting point of the mixture). This saves electrical energy and, therefore, makes the process more economical.
- The mixture is then electrolysed in a cell similar to the one in Figure 6.10.

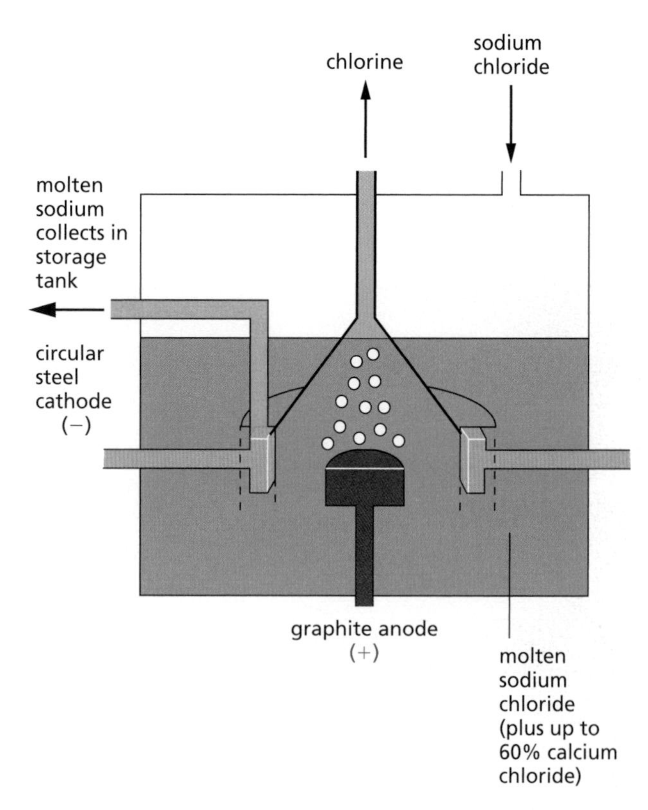


**Figure 6.10** The Down's cell used for the electrolysis of sodium chloride.

The chloride ions are attracted to the anode, where they lose electrons and form chlorine gas.

$$\text{chloride ions} \rightarrow \text{chlorine molecules} + \text{electrons}$$
$$2Cl^-(l) \rightarrow Cl_2(g) + 2e^-$$

The positive sodium ions are attracted to the cathode. They gain electrons to form molten sodium metal.

$$\text{sodium ions} + \text{electrons} \rightarrow \text{sodium metal}$$
$$Na^+(l) + e^- \rightarrow Na(l)$$

The overall reaction which takes place in the cell is:

$$\text{sodium chloride} \xrightarrow{\text{electrolysis}} \text{sodium} + \text{chlorine}$$
$$2NaCl(l) \longrightarrow 2Na(l) + Cl_2(g)$$

The cathode is a circle of steel around the graphite anode. At 600 °C sodium and chlorine would react violently together to re-form sodium chloride. To prevent this from happening, the Down's cell contains a steel gauze around the graphite anode to keep it and the cathode apart. The molten sodium floats on the electrolyte and is run off for storage.

A problem arises, however, in that calcium ions are also attracted to the cathode, where they form calcium metal. Therefore, the sodium which is run off contains a significant proportion of calcium. Fortunately, the calcium crystallises out when the mixture cools and relatively pure sodium metal remains (Figure 6.11).

**Figure 6.11** Molten sodium – the result of electrolysis of sodium chloride.

The worldwide production of sodium each year is 60–80 000 tonnes. Of this, 16 000 tonnes are produced per year in the UK.

Sodium is used as a liquid coolant in nuclear power stations as well as in street lighting and in the production of metals such as titanium. Chlorine gas (the co-product) is sold to make the process more economical (see p. 48 and p. 86 for the uses of chlorine).

## *Questions*

1 Use your research skills, including the Internet, to find uses, other than those given in the text, for sodium.

2 A student carries out the electrolysis of molten lead(II) bromide in a fume cupboard.
   **a** Draw a diagram to show a suitable apparatus the student could use to carry out this experiment.
   **b** Write anode and cathode reactions to represent the processes taking place during the electrolysis.
   **c** Why does this experiment need to be carried out in a fume cupboard?
   **d** Find uses for the anode product of this cell.

## *Electrolysis of aqueous solutions*

Other industrial processes involve the electrolysis of aqueous solutions. To help you to understand what is happening in these processes, we will first consider the electrolysis of water.

### Electrolysis of water

Pure water is a very poor conductor of electricity because there are so few ions in it. However, it can be made to decompose if an electric current is passed through it in a Hofmann voltameter, as in Figure 6.12.

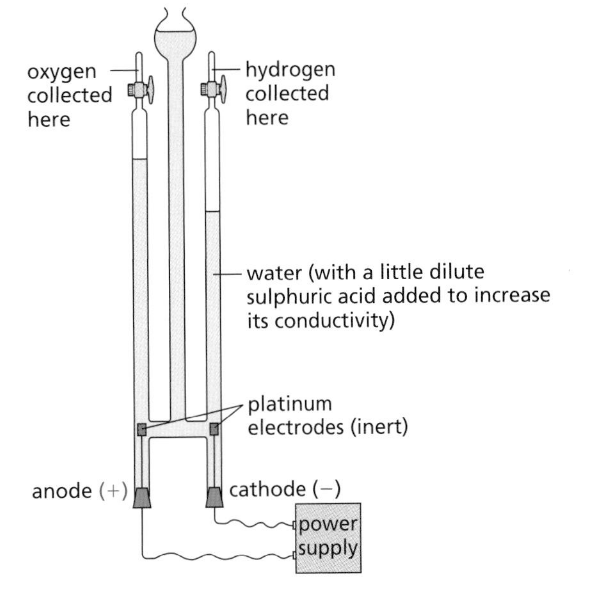


**Figure 6.12** A Hofmann voltameter used to electrolyse water.

To enable water to conduct electricity better, some dilute sulphuric acid (or sodium hydroxide solution) is added. When the power is turned on and an electric current flows through this solution, gases can be seen to be produced at the two electrodes and they are collected in the side arms of the apparatus. After about 20 minutes, roughly twice as much gas is produced at the cathode as at the anode.

The gas collected at the cathode burns with a squeaky pop, showing it to be hydrogen gas. For hydrogen to be collected in this way, the positively charged hydrogen ions must have moved to the cathode.

$$\text{hydrogen ions} + \text{electrons} \rightarrow \text{hydrogen molecules}$$
$$4H^+(aq) + 4e^- \rightarrow 2H_2(g)$$

If during this process the water molecules lose $H^+(aq)$, then the remaining portion must be hydroxide ions, $OH^-(aq)$. These ions are attracted to the anode. The gas collected at the anode relights a glowing splint, showing it to be oxygen.

This gas is produced in the following way.

$$\text{hydroxide ions} \rightarrow \text{water molecules} + \text{oxygen molecules} + \text{electrons}$$
$$4OH^-(aq) \rightarrow 2H_2O(l) + O_2(g) + 4e^-$$

This experiment was first carried out by Sir Humphry Davy, who confirmed by this experiment that the formula for water was $H_2O$.

## Chlor-alkali industry

The electrolysis of saturated sodium chloride solution (brine) is the basis of a major industry. Rock salt (sodium chloride) is mined in the United Kingdom in Cheshire, Lancashire, Staffordshire and Cleveland. Three very important substances are produced in this electrolysis process – chlorine, sodium hydroxide and hydrogen. The electrolytic process is a very expensive one, requiring vast amounts of electricity. The process is economical only because all three products have a large number of uses (Figure 6.13).

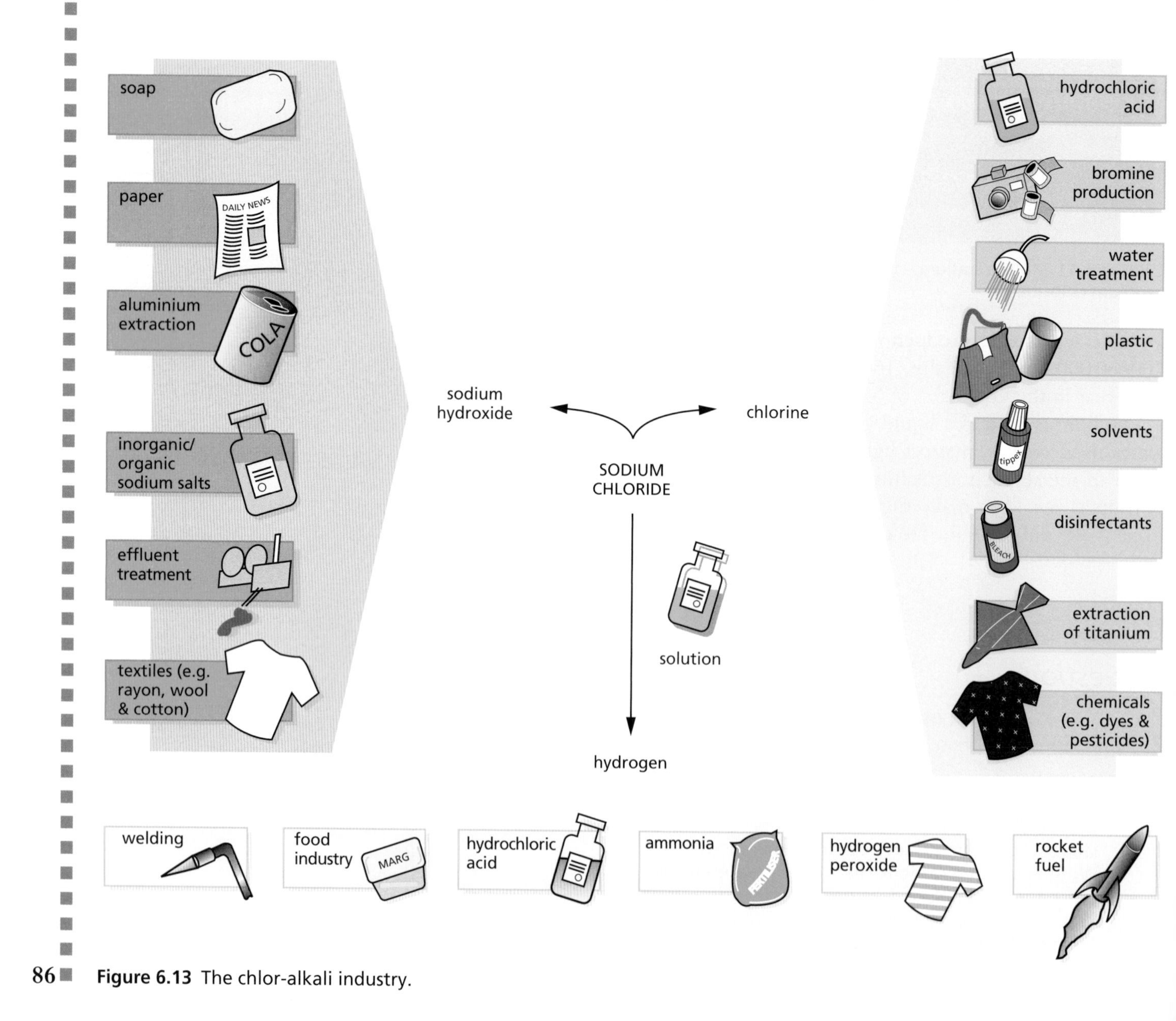


**Figure 6.13** The chlor-alkali industry.

There are two well-established methods for electrolysing brine, the **diaphragm cell** and the **mercury cell**. However, recent developments in electrolysis technology, by chemical engineers, have produced the **membrane cell** (Figure 6.14). This method is now preferred to the other two because it produces a purer product, it causes less pollution and it is cheaper to run.

The brine is first purified to remove calcium, strontium and magnesium compounds by a process of ion exchange (see Chapter 8, p. 120).

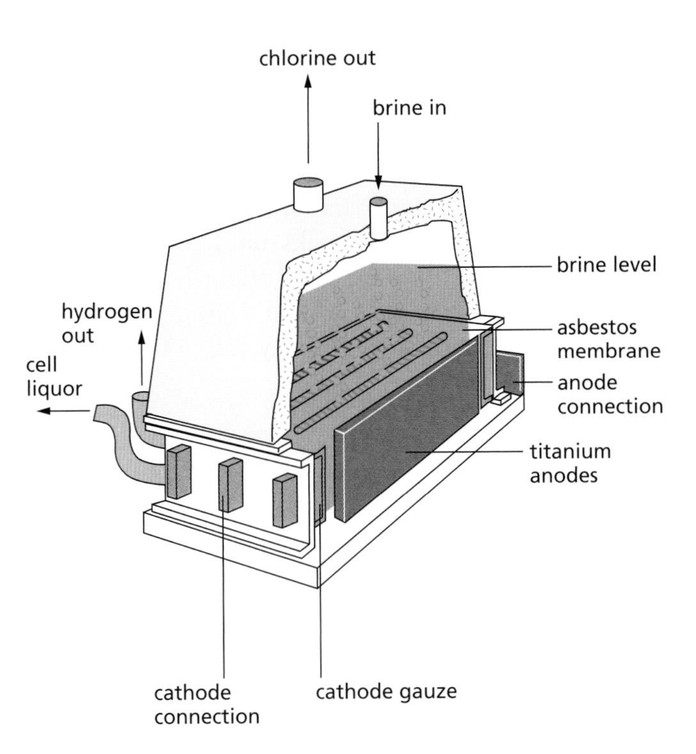


**a** A section through the membrane cell.

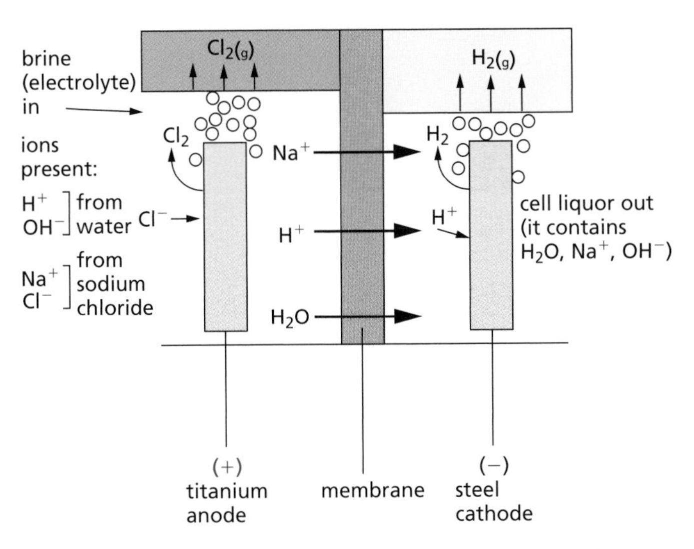


**b** A diagrammatic representation of the reactions going on inside the cell.

**Figure 6.14**

The membrane cell is used continuously, with fresh brine flowing into the cell as the process breaks up the brine. The cell has been designed to ensure that the products do not mix. The ions in this concentrated sodium chloride solution are:

| | | |
|---|---|---|
| from the water: | $H^+(aq)$ | $OH^-(aq)$ |
| from the sodium chloride: | $Na^+(aq)$ | $Cl^-(aq)$ |

When the current flows, the chloride ions, $Cl^-(aq)$, are attracted to the anode. Chlorine gas is produced by the electrode process.

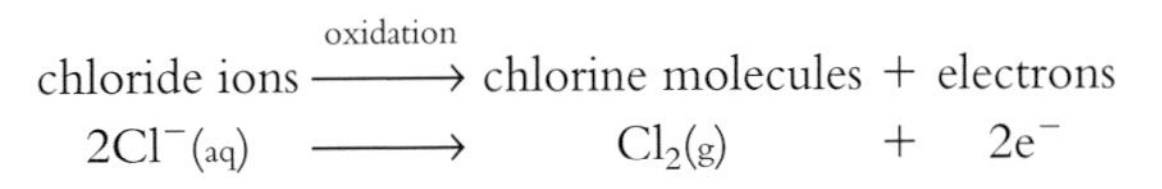

$$\text{chloride ions} \xrightarrow{\text{oxidation}} \text{chlorine molecules} + \text{electrons}$$
$$2Cl^-(aq) \longrightarrow Cl_2(g) + 2e^-$$

This leaves a high concentration of sodium ions, $Na^+(aq)$, around the anode.

The hydrogen ions, $H^+(aq)$, are attracted to the cathode and hydrogen gas is produced.

$$\text{hydrogen ions} + \text{electrons} \xrightarrow{\text{reduction}} \text{hydrogen molecules}$$
$$2H^+(aq) + 2e^- \longrightarrow H_2(g)$$

This leaves a high concentration of hydroxide ions, $OH^-(aq)$, around the cathode. The sodium ions, $Na^+(aq)$, are drawn through the membrane, where they combine with the $OH^-(aq)$ to form sodium hydroxide, NaOH, solution. The annual production in the UK is 1.2 million tonnes and worldwide is 41 million tonnes.

## *Questions*

**1** Suggest a reason for only 'roughly' twice as much hydrogen gas being produced at the cathode as oxygen gas at the anode in the electrolysis of water.

**2** Account for the following observations which were made when concentrated sodium chloride solution, to which a little universal indicator had been added, was electrolysed in the laboratory in a Hofmann voltameter.
- **a** The universal indicator initially turns red in the region of the anode, but as the electrolysis proceeds it loses its colour.
- **b** The universal indicator turns blue in the region of the cathode.

**3** Why is it important to remove compounds of calcium, strontium and magnesium before brine is electrolysed?

**4** The uses of sodium hydroxide can be separated on a percentage basis as follows:

| | |
|---|---|
| Neutralisation | 5% |
| Paper manufacture | 5% |
| Oil refining | 5% |
| Soap/detergents | 5% |
| Manufacture of rayon and acetate fibres | 16% |
| Manufacture of chemicals | 30% |
| Miscellaneous uses | 34% |

Use a graph plot program to create a 'pie' chart of this data.

## Purification of copper

Because copper is a very good conductor of electricity, it is used for electrical wiring and cables (Figure 6.15).

**Figure 6.15** The copper used in electrical wiring has to be very pure.

However, even small amounts of impurities cut down this conductivity quite noticeably. The metal must be 99.99% pure to be used in this way. To ensure this level of purity, the newly extracted copper has to be purified by electrolysis (Figure 6.16).

The impure copper is used as the anode and is typically 1 m square, 35–50 mm thick and 330 kg in weight. The cathode is a 1 mm thick sheet and weighs about 5 kg; it is made from very pure copper. The electrolyte is a solution of copper(II) sulphate (0.3 mol $dm^{-3}$) acidified with a 2 mol $dm^{-3}$ solution of sulphuric acid to help the solution conduct electricity.

When the current flows, the copper moves from the impure anode to the pure cathode. Any impurities fall to the bottom of the cell and collect below the anode in the form of a slime. This slime is rich in precious metals and the recovery of these metals is an important aspect of the economics of the process. The electrolysis proceeds for about three weeks until the anodes are reduced to about 10% of their original size and the cathodes weigh between 100 and 120 kg. A potential of 0.25 V and a current density of 200 A $m^{-2}$ are usually used.

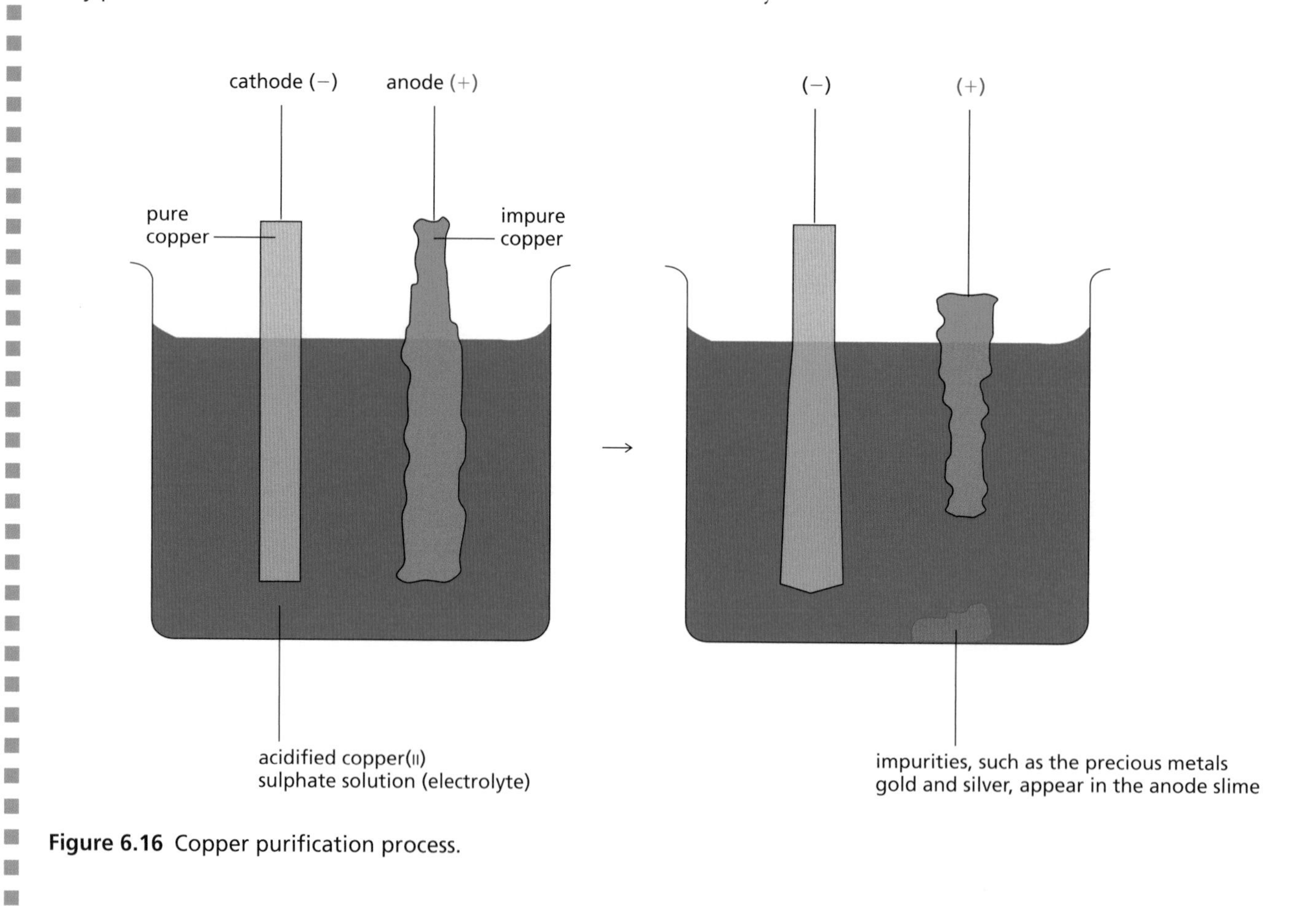


**Figure 6.16** Copper purification process.

The ions present in the solution are:

| | | |
|---|---|---|
| from the water: | $H^+(aq)$ | $OH^-(aq)$ |
| from the copper(II) sulphate: | $Cu^{2+}(aq)$ | $SO_4^{2-}(aq)$ |

During the process the impure anode loses mass because the copper atoms lose electrons and become copper ions, $Cu^{2+}(aq)$ (Figure 6.17).

copper atoms → copper ions + electrons

$$Cu(s) \rightarrow Cu^{2+}(aq) + 2e^-$$

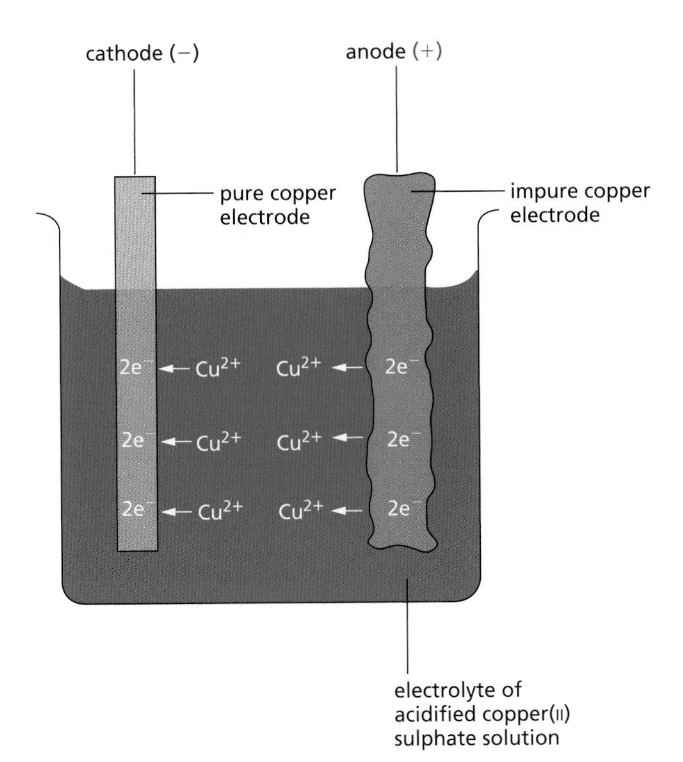


**Figure 6.17** The movement of ions in the purification of copper by electrolysis.

The electrons released at the anode travel around the external circuit to the cathode. There the electrons are passed on to the copper ions, $Cu^{2+}(aq)$, from the copper(II) sulphate solution and the copper is deposited or copper plated on to the cathode.

copper ions + electrons → copper atoms

$$Cu^{2+}(aq) + 2e^- \rightarrow Cu(s)$$

The annual production of copper worldwide is 12 million tonnes. We produce 57 000 tonnes in the UK, the vast majority of which is recycled.

## *Electrolysis guidelines*

The following points may help you work out the products of electrolysis in unfamiliar situations. They will also help you to remember what happens at each of the electrodes.

- Non-metals are produced at the anode whereas metals and hydrogen gas are produced at the cathode.
- At the anode, chlorine, bromine and iodine (the halogens) are produced in preference to oxygen.
- At the cathode, hydrogen is produced in preference to metals unless unreactive metals such as copper and nickel are present.

### *Questions*

1. Why do you think it is advantageous to use inert electrodes in the electrolysis processes?
2. Predict the products of electrolysis of a solution of copper(II) sulphate if carbon electrodes are used instead of those made from copper as referred to in the purification of copper section.
3. Predict the products of the electrolysis of concentrated hydrochloric acid using platinum electrodes.
4. Using your knowledge of electrolysis, predict the likely products of the electrolysis of copper(II) chloride solution, using platinum electrodes. Write electrode equations for the formation of these products.

## *Electroplating*

Electroplating is the process involving electrolysis to plate, or coat, one metal with another or a plastic with a metal. Often the purpose of electroplating is to give a protective coating to the metal beneath. For example, bath taps are chromium plated to prevent corrosion, and at the same time are given a shiny, more attractive finish (Figure 6.18).

**Figure 6.18** This tap has been chromium plated.

The electroplating process is carried out in a cell such as the one shown in Figure 6.19a. This is often known as the 'plating bath' and it contains a suitable electrolyte, usually a solution of a metal salt.

For silver plating the electrolyte is a solution of a silver salt. The article to be plated is made the cathode in the cell so that the metal ions move to it when the current is switched on. The cathode reaction in this process is:

$$\text{silver ions} + \text{electrons} \rightarrow \text{silver atoms}$$
$$Ag^+(aq) \quad + \quad e^- \quad \rightarrow \quad Ag(s)$$

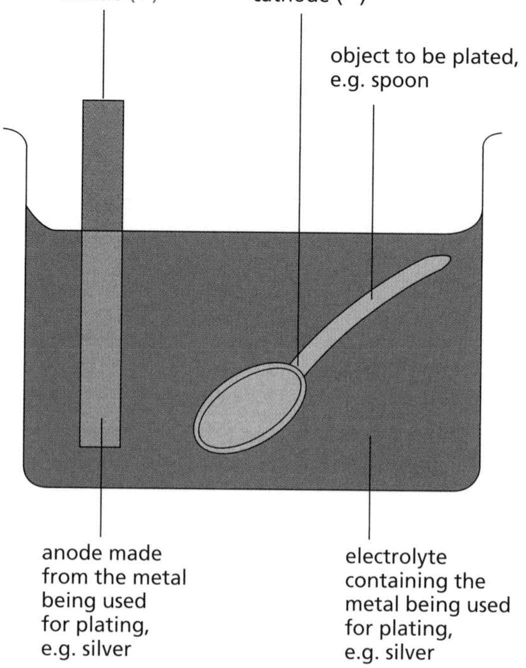

anode (+)
cathode (−)
object to be plated, e.g. spoon
anode made from the metal being used for plating, e.g. silver
electrolyte containing the metal being used for plating, e.g. silver

**a** Silver plating a spoon.

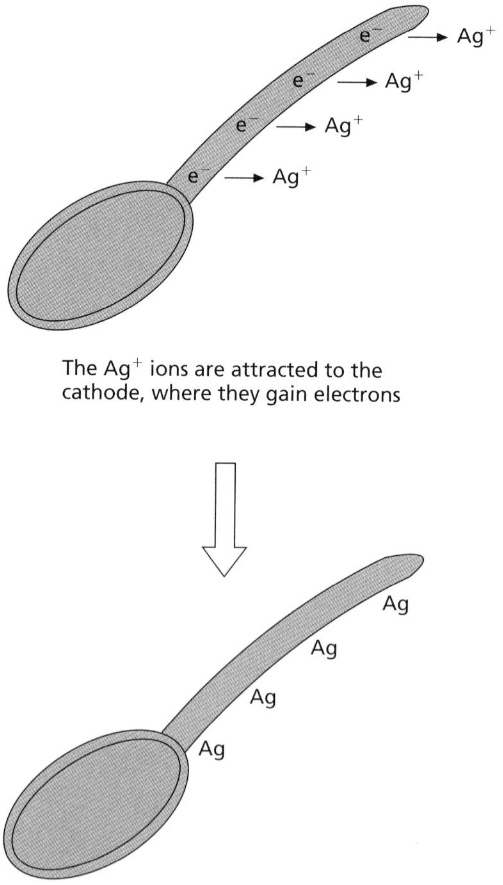

$e^-$ → $Ag^+$
$e^-$ → $Ag^+$
$e^-$ → $Ag^+$
$e^-$ → $Ag^+$

The $Ag^+$ ions are attracted to the cathode, where they gain electrons

Ag
Ag
Ag
Ag

A coating of silver forms on the spoon at the cathode

**b** Explaining silver plating.

**Figure 6.19**

## Plating plastics

Nowadays it is not only metals that are electroplated. Plastics have been developed that are able to conduct electricity. For example, the plastic poly(pyrrole) can be electroplated in the same way as the metals we have discussed above (Figure 6.20).

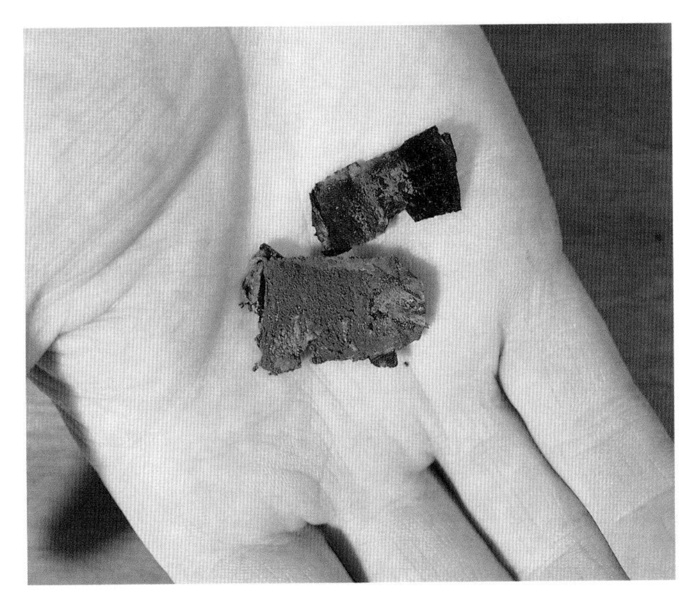

**Figure 6.20** This plastic has been coated with copper by electrolysis.

**Figure 6.21** This leaf has been electroplated.

## *Questions*

1 The leaf in Figure 6.21 has been copper plated. Suggest a suitable method for copper plating the leaf.

2 Explain why copper(II) chloride solution would not be used as an electrolyte in the electrolyte cell used for copper plating.

# *Calculations in electrolysis*

The quantity of electricity flowing through an electrolysis cell is measured in **coulombs** (C). If one **ampere** (A) is passed for one second, the quantity of electricity is said to be 1 coulomb.

$$\underset{\text{(coulombs)}}{\text{quantity of electricity}} = \underset{\text{(amperes)}}{\text{current}} \times \underset{\text{(seconds)}}{\text{time}}$$

Therefore, if 2 amps flow for 5 seconds then the quantity of electricity passed is 10 coulombs.

In the purification of copper you saw that copper was deposited at the cathode. The electrode equation is:

$$\underset{\text{1 mole}}{Cu^{2+}(aq)} + \underset{\text{2 moles}}{2e^-} \rightarrow \underset{\text{1 mole}}{Cu(s)}$$

This equation tells us that 1 mole of copper(II) ions combines with 2 moles of electrons to produce 1 mole of copper metal atoms (63.5 g).

An old name for a mole of electrons was the **faraday**. This unit is named after an English scientist, Michael Faraday (Figure 6.22), who carried out many significant experiments while investigating the nature of magnetism and electricity. So we can say that we need 2 moles of electrons of electricity to form 1 mole of copper atoms (63.5 g) at the cathode during this purification process.

**Figure 6.22** In 1883 Michael Faraday was the first scientist to measure the masses of elements produced during electrolysis.

From accurate electrolysis experiments it has been found that:

1 mole of electrons = 96 500 coulombs

Therefore, the quantity of electricity required to deposit 1 mole of copper atoms (63.5 g) is:

$2 \times 96\,500 = 193\,000$ coulombs (2 moles of electrons)

**Example 1**

Calculate the number of moles of electrons required to deposit 10 g of silver on the surface of a fork during an electroplating process. ($A_r$: Ag = 108)

Electrode equation: $Ag^+(aq) + e^- \rightarrow Ag(s)$

1 mole of silver is deposited by 1 mole of electrons.
Therefore, 108 g of silver is deposited by 1 mole of electrons.
Hence 1 g of silver is deposited by $\frac{1}{108}$ mole of electrons.
Therefore, 10 g of silver are deposited by

$$\frac{1}{108} \times 10 = 0.093 \text{ moles of electrons}$$

**Example 2**

Calculate the volume of oxygen gas, measured at room temperature and pressure (rtp), liberated at the anode in the electrolysis of acidified water by 2 moles of electrons. (1 mole of oxygen at rtp occupies a volume of 24 $dm^3$.)

Electrode equation: $4OH^-(aq) \rightarrow 2H_2O(l) + O_2(g) + 4e^-$

1 mole of oxygen gas is liberated by 4 moles of electrons.
Therefore, 24 $dm^3$ of oxygen is liberated by 4 moles of electrons.
Hence 12 $dm^3$ of oxygen would be liberated by 2 moles of electrons.

**Example 3**

The industrial production of aluminium uses a current of 25 000 amps. Calculate the time required to produce 10 kg of aluminium from the electrolysis of molten aluminium oxide.

Electrode equation: $Al^{3+}(l) + 3e^- \rightarrow Al(l)$

1 mole of aluminium is produced by 3 moles of electrons.
27 g of aluminium is produced by 3 moles of electrons.
Hence 1 g of aluminium would be produced by $\frac{3}{27}$ moles of electrons.

Therefore, 10 000 g of aluminium would be produced by

$$\frac{3}{27} \times 10\,000 \text{ faradays}$$

$= 1111.1$ faradays

So, the quantity of electricity

$= \text{number of faradays} \times 96\,500$
$= 1111.1 \times 96\,500$
$= 1.07 \times 10^8$ C

Time (seconds)

$$= \frac{\text{coulombs}}{\text{amps}}$$

$$= \frac{1.07 \times 10^8}{25\,000}$$

= 4280 seconds (71.3 minutes)

## Questions

1 Write equations which represent the discharge at the cathode of the following ions:
  a $K^+$
  b $Pb^{2+}$
  c $Al^{3+}$
  and at the anode of:
  d $Br^-$
  e $O^{2-}$
  f $F^-$

2 How many moles of electrons are required to discharge 1 mole of the following ions?
  a $Mg^{2+}$
  b $K^+$
  c $Al^{3+}$
  d $O^{2-}$

3 How many coulombs are required to discharge 1 mole of the following ions:
  (1 mole of electrons = 96 500 coulombs)
  a $Cu^{2+}$
  b $Na^+$
  c $Pb^{2+}$

4 Calculate the number of moles of electrons required to deposit 6.35 g of copper on a metal surface. ($A_r$: Cu = 63.5)

## Checklist

**After studying Chapter 6 you should know and understand the following terms.**

- **Anode** The positive electrode. It is positively charged because electrons are drawn away from it.
- **Cathode** The negative electrode. It is negatively charged because an excess of electrons move towards it.
- **Coulombs** The unit used to measure the quantity of electricity:

  1 coulomb = 1 amp × 1 second

- **Down's cell** The electrolysis cell in which sodium is extracted from molten sodium chloride to which 60% calcium chloride has been added to reduce the working temperature to 600 °C. This cell has a graphite anode surrounded by a cylindrical steel cathode.
- **Electrode** A point where the electric current enters and leaves the electrolytic cell. An inert electrode is usually made of platinum or carbon and does not react with the electrolyte or the substances produced at the electrodes themselves.
- **Electrolysis** A process in which a chemical reaction is caused by the passage of an electric current.
- **Electrolyte** A substance which will carry electric current only when it is molten or dissolved.
- **Electroplating** The process of depositing metals from solution in the form of a layer on other surfaces such as metal or plastic.
- **Hall–Héroult cell** The electrolysis cell in which aluminium is extracted from purified bauxite dissolved in molten cryolite at 900 °C. This cell has both a graphite anode and a graphite cathode.
- **Membrane cell** An electrolytic cell used for the production of sodium hydroxide, hydrogen and chlorine from brine in which the anode and cathode are separated by a membrane.
- **One mole of electrons** This is the same as one Faraday and is equivalent to 96 500 coulombs.
- **Oxidation** Takes place at the anode and involves a negative ion losing electrons.
- **Reduction** Takes place at the cathode and involves a positive ion gaining electrons.

# Electrolysis and its uses
# *Additional questions*

1 Explain the meaning of each of the following terms. Use a suitable example, in each case, to help with your explanation.
  a Anode.
  b Cathode.
  c Electrolysis.
  d Electrolyte.
  e Anion.
  f Cation.
  g Oxidation.
  h Reduction.

2 Copper is purified by electrolysis, as in the example shown below.

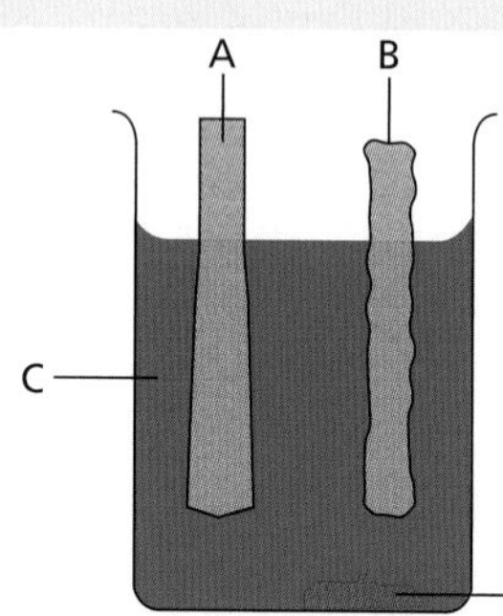


  a Name the materials used for the electrodes **A** and **B**.
  b Name the electrolyte **C** and substance **D**.
  c Why is substance **D** of economic importance in respect of this process.
  d Write equations for the reactions which take place at the cathode and anode during this process.
  e Draw a labelled diagram to show the cell after electrolysis has taken place.
  f Why has electrolyte **C** to be acidified with the dilute sulphuric acid?
  g Why does copper have to be 99.99% pure for use in electrical cables?

3 Copy and complete the table below, which shows the results of the electrolysis of four substances using inert electrodes.

| Electrolyte | Product at anode (positive electrode) | Product at cathode (negative electrode) |
|---|---|---|
| Molten aluminium oxide | | Aluminium |
| Concentrated sodium chloride solution | Chlorine | |
| Molten lithium chloride | | |
| Silver nitrate solution | | Silver |

  a State what you understand by 'inert electrodes'.
  b Explain why the lithium chloride solution becomes progressively more alkaline during electrolysis.
  c Explain why solid lithium chloride is a non-conductor of electricity, whereas molten lithium chloride and lithium chloride solution are good conductors of electricity.
  d During the electrolysis of molten aluminium chloride ($AlCl_3$) the carbon anodes are burned away. Explain why this should happen and write balanced chemical equations for the reactions that take place.

4 The industrial production of sodium uses a current of 15 000 amps. Calculate the time required to produce 100 kg of sodium from the electrolysis of molten sodium chloride. (1 mole of electrons = 96 500 coulombs. $A_r$: Na = 23)

5 Calculate the volume of hydrogen gas, measured at room temperature and pressure, liberated at the cathode in the electrolysis of acidified water by 18 moles of electrons of electricity. (One mole of any gas at rtp occupies a volume of 24 $dm^3$.)

6 Calculate the number of moles of electrons required to produce 10 g of gold deposited on the surface of some jewellery during an electroplating process. ($A_r$: Au = 197; note gold in solution is present as the $Au^{3+}$ ion)

7 A pupil carried out an experiment in a fume cupboard to find out how electricity affected different substances. Some of the substances were in aqueous solution, others were in the molten state. Carbon electrodes were used in each experiment and she wrote down her results in a table with these headings.

| Substance | What was formed at the cathode (−) | What was formed at the anode (+) |
|---|---|---|
| | | |

Make a table like the one shown and fill it in with what you think happened for each of the substances below.
  a Molten lead iodide.
  b Sugar solution.
  c Silver nitrate solution.
  d Copper(II) sulphate solution.
  e Molten sodium bromide.
  f Ethanol solution.

8 Sodium hydroxide is made by the electrolysis of brine.
  a Draw and label a simplified diagram of the cell used in this process. Make certain that you have labelled on the diagram:
    (i) the electrolyte
    (ii) the material of the electrodes
    (iii) the material of the membrane.
  b Write equations for the reactions which take place at the cathode and anode. State clearly whether a reaction is oxidation or reduction.
  c Give two large-scale uses of the products of this electrolytic process.
  d Comment on the following statement: 'This electrolytic process is a very expensive one'.
  e Both the membrane cell and the older mercury cell make sodium hydroxide of high purity. Explain why the membrane cell is now the preferred way of making sodium hydroxide.

# Acids, bases and salts



## Acids and alkalis

All the substances shown in Figure 7.1 contain an **acid** of one sort or another. Acids are certainly all around us. What properties do these substances have which make you think that they are acids or contain acids?

The word acid means 'sour' and all acids possess this property. They are also:

- soluble in water
- corrosive.

**Alkalis** are very different from acids. They are the chemical 'opposite' of acids.

- They will remove the sharp taste from an acid.
- They have a soapy feel.

Some common alkaline substances are shown in Figure 7.2.

It would be too dangerous to taste a liquid to find out if it was acidic. Chemists use substances called **indicators** which change colour when they are added to acids or alkalis. Many indicators are dyes which have been extracted from natural sources. For example, litmus is a purple dye which has been extracted from lichens. Litmus turns red when it is added to an acid and turns blue when added to an alkali. Some other indicators are shown in Table 7.1, along with the colours they turn in acids and alkalis.

**Table 7.1** Indicators and their colours in acid and alkaline solution.

| Indicator | Colour in acid solution | Colour in alkaline solution |
|---|---|---|
| Blue litmus | Red | Blue |
| Methyl orange | Pink | Yellow |
| Methyl red | Red | Yellow |
| Phenolphthalein | Colourless | Pink |
| Red litmus | Red | Blue |

**Figure 7.1** What do all these foods have in common?

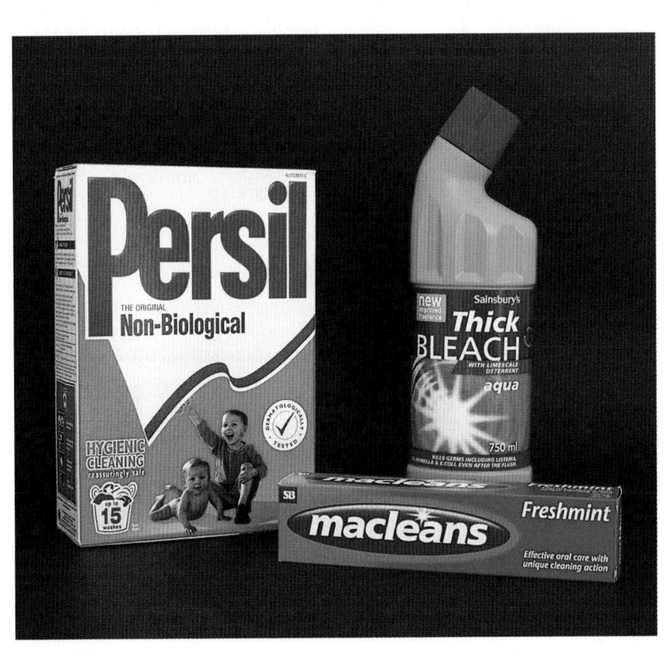


**Figure 7.2** Some common alkaline substances.

These indicators tell chemists whether a substance is acid or alkaline (Figure 7.3). To obtain an idea of how acidic or alkaline a substance is, we use another indicator known as a **universal indicator**. This indicator is a mixture of many other indicators. The colour shown by this indicator can be matched against a **pH scale**. The pH scale was developed by a Scandinavian chemist called Søren Sørenson, who was employed by the Carlsberg brewery. The pH scale runs from below 0 to 14. A substance with a pH of less than 7 is an acid. One with a pH of greater than 7 is alkaline. One with a pH of 7 is said to be neither acid nor alkaline, that is neutral. Water is the most common example of a neutral substance. Figure 7.4 shows the universal indicator colour range along with everyday substances with their particular pH values.

**Figure 7.3** Indicators tell you if a substance is acid or alkaline.

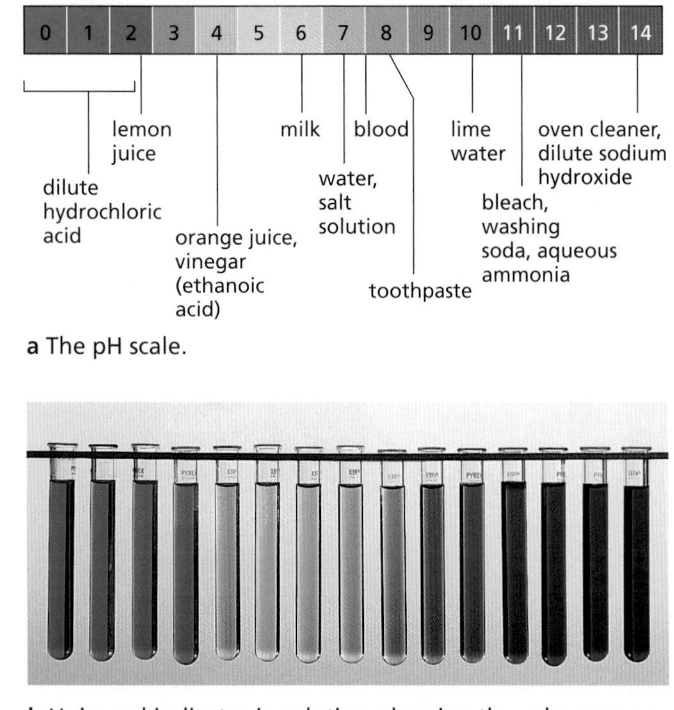


**a** The pH scale.

**b** Universal indicator in solution, showing the colour range.

**Figure 7.4**

Another way in which the pH of a substance can be measured is by using a pH meter (Figure 7.5). The pH electrode is placed into the solution and a pH reading is given on the digital display.

**Figure 7.5** A digital pH meter.

## Theories of acids and bases

There have been many attempts to define the difference between acids and bases. The first real attempt took place in 1777, when the Frenchman Antoine Lavoisier (Figure 7.6) suggested that acids were substances that contained oxygen.

**Figure 7.6** Antoine Lavoisier (1714–1794) at work in his laboratory.

It was not long after this that the 'hydro-halic' acids (HCl, HBr, and so on) were discovered and they had no oxygen present in them. This produced a modified theory in 1810 when the English chemist Sir Humphry Davy (1778–1829) suggested that all acids contain hydrogen as the important element; however, it was pointed out that there were many hydrogen-containing substances that were not acids. The German chemist Justus von Liebig (Figure 7.7) then made the next useful proposal about acids (1838) when he suggested that acids were substances that can react with metals to produce hydrogen gas.

**Figure 7.7** Justus von Liebig (1803–1873), who defined acids more closely.

This theory was followed in 1884 by the first really comprehensive theory of acids and bases, produced by the Swedish chemist Svante Arrhenius (1859–1927). He suggested that since these acid solutions were electrolytes (see Chapter 6) their solutions contained many ions. According to Arrhenius' theory, acids produce hydrogen ions ($H^+$) when they dissolve in water, whereas bases produce hydroxide ions ($OH^-$).

It was thus recognised that water plays an important part in the acidity of a substance. This led to the suggestion that the hydrogen ion cannot exist alone in aqueous solution. This was prompted by the fact that gaseous hydrogen chloride, HCl(g), is not acidic but when it dissolves in water an acidic solution is produced.

These ideas were rather limiting since they only applied to aqueous solutions. There were situations where acid–base reactions were taking place in solvents other than water, or even in no solvent at all. This problem was addressed in 1923 by the Danish chemist Johannes Brønsted (1879–1947) and the English chemist Thomas Lowry (1874–1936) when they independently proposed a more general definition of acids and bases, and the study of acids and bases took a great step forward. This theory became known as the **Brønsted–Lowry** theory of acids and bases.

**The Brønsted–Lowry theory**

This theory defined:

- an acid as an $H^+$ ion (or proton) donor
- a base as an $H^+$ ion (or proton) acceptor.

The theory explains why a pure acid behaves differently from its aqueous solution, since for an acid to behave as an $H^+$ ion donor it must have another substance present to accept the $H^+$ ion. So the water, in the aqueous acid solution, is behaving as a Brønsted–Lowry base and accepting an $H^+$ ion. Generally:

$$\underset{\text{acid}}{HA(aq)} + \underset{\text{base}}{H_2O(l)} \rightarrow H_3O^+(aq) + A^-(aq)$$

If a substance can behave both as a Brønsted–Lowry acid and as a Brønsted–Lowry base then it is called **amphoteric**. Water has this ability. As well as reacting with acids (above) it can also react with Brønsted–Lowry bases such as ammonia in the following way to form the base $OH^-$:

$$\underset{\text{base}}{NH_3(aq)} + \underset{\text{acid}}{H_2O(l)} \rightarrow NH_4^+(aq) + OH^-(aq)$$

The reaction between hydrogen chloride gas and ammonia can be described as an acid–base reaction under this theory. The hydrogen chloride molecule acts as a proton donor and the ammonia molecule acts as the proton acceptor (Figure 7.8).

$$\underset{\text{acid}}{HCl(g)} + \underset{\text{base}}{NH_3(g)} \rightarrow NH_4^+Cl^-(s)$$

**Figure 7.8** The hydrogen chloride molecule (from concentrated hydrochloric acid) acts as a hydrogen ion donor. The ammonia molecule (from concentrated ammonia) acts as a hydrogen ion acceptor.

## The relative strengths of acids and bases

The relative strength of an acid is found by comparing one acid with another. The strength of any acid depends upon how many molecules dissociate (split up) when the acid is dissolved in water. The relative strength of a base is found by comparing one base with another and is again dependent upon the dissociation of the base in aqueous solution.

### Strong and weak acids

A typical strong acid is hydrochloric acid. It is formed by dissolving hydrogen chloride gas in water. In hydrochloric acid the ions formed separate completely.

$$\text{hydrogen chloride} \xrightarrow{\text{water}} \text{hydrogen ions} + \text{chloride ions}$$

$$HCl(g) \xrightarrow{\text{water}} H^+(aq) + Cl^-(aq)$$

For hydrochloric acid *all* the hydrogen chloride molecules break up to form $H^+$ ions and $Cl^-$ ions. Any acid that behaves in this way is termed a **strong acid**. Both sulphuric acid and nitric acid also behave in this way and are therefore also termed strong acids. All these acids have a high concentration of hydrogen ions in solution ($H^+(aq)$).

When strong acids are neutralised by strong alkalis the following reaction takes place between hydrogen ions and hydroxide ions.

$$H^+(aq) + OH^-(aq) \rightarrow H_2O(l)$$

A **weak acid** such as ethanoic acid, which is found in vinegar, produces few hydrogen ions when it dissolves in water compared with a strong acid of the same concentration.

$$\text{ethanoic acid} \overset{\text{water}}{\rightleftharpoons} \text{hydrogen ions} + \text{ethanoate ions}$$

$$CH_3COOH(l) \overset{\text{water}}{\rightleftharpoons} H^+(aq) + CH_3COO^-(aq)$$

The $\rightleftharpoons$ sign means that the reaction is **reversible**. This means that if the ethanoic acid molecule breaks down to give hydrogen ions and ethanoate ions then they will react together to re-form the ethanoic acid molecule. The fact that fewer ethanoic acid molecules dissociate compared with a strong acid, and that the reaction is reversible, means that few oxonium ions are present in the solution. Other examples of weak acids are citric acid, found in oranges and lemons, carbonic acid, found in soft drinks, sulphurous acid (acid rain) (Figure 7.9) and ascorbic acid (vitamin C).

**Figure 7.9** Sulphurous acid is found in acid rain. It is a weak acid and is oxidised to sulphuric acid (a strong acid). Acid rain damages the environment quite badly.

All acids when in aqueous solution produce hydrogen ions, $H^+(aq)$. To say an acid is a strong acid does not mean it is concentrated. The *strength* of an acid tells you how easily it dissociates (ionises) to produce hydrogen ions. The *concentration* of an acid indicates the proportions of water and acid present in aqueous solution. It is important to emphasise that a strong acid is still a strong acid even when it is in dilute solution and a weak acid is still a weak acid even if it is concentrated.

### Strong and weak bases

An alkali is a base which produces hydroxide ions, $OH^-(aq)$, when dissolved in water. Sodium hydroxide is a **strong alkali** because when it dissolves in water its lattice breaks up completely to produce ions.

$$\text{sodium hydroxide} \xrightarrow{\text{water}} \text{sodium ions} + \text{hydroxide ions}$$

$$NaOH(s) \longrightarrow Na^+(aq) + OH^-(aq)$$

These substances which are strong alkalis produce large quantities of hydroxide ions. Other common, strong soluble bases include potassium hydroxide.

A **weak alkali**, such as ammonia, produces fewer hydroxide ions when it dissolves in water than a strong soluble base of the same concentration.

$$\text{ammonia} + \text{water} \rightleftharpoons \text{ammonium ions} + \text{hydroxide ions}$$

$$NH_3(g) + H_2O(l) \rightleftharpoons NH_4^+(aq) + OH^-(aq)$$

The ammonia molecules react with the water molecules to form ammonium ions and hydroxide ions. However, fewer ammonia molecules do this so only a low concentration of hydroxide ions is produced.

## Neutralising an acid

A common situation involving neutralisation of an acid is when you suffer from indigestion. This is caused by a build-up of acid in your stomach. Normally you treat it by taking an indigestion remedy containing a substance which will react with and neutralise the acid.

In the laboratory, if you wish to neutralise a common acid such as hydrochloric acid you can use an alkali such as sodium hydroxide. If the pH of the acid is measured as some sodium hydroxide solution is added to it, the pH increases. If equal volumes of the same concentration of hydrochloric acid and sodium hydroxide are added to one another, the resulting solution is found to have a pH of 7. The acid has been neutralised and a neutral solution has been formed.

| hydrochloric acid | + | sodium hydroxide | → | sodium chloride | + | water |
|---|---|---|---|---|---|---|
| $HCl(aq)$ | + | $NaOH(aq)$ | → | $NaCl(aq)$ | + | $H_2O(l)$ |

As we have shown, when both hydrochloric acid and sodium hydroxide dissolve in water the ions separate completely. We may therefore write:

$$H^+(aq)Cl^-(aq) + Na^+(aq)OH^-(aq) \rightarrow Na^+(aq) + Cl^-(aq) + H_2O(l)$$

You will notice that certain ions are unchanged on either side of the equation. They are called **spectator ions** and are usually taken out of the equation. The equation now becomes:

$$H^+(aq) + OH^-(aq) \rightarrow H_2O(l)$$

This type of equation is known as an **ionic equation**. The reaction between any acid and alkali in aqueous solution can be summarised by this ionic equation. It shows the ion which causes acidity ($H^+(aq)$) reacting with the ion which causes alkalinity ($OH^-(aq)$) to produce neutral water ($H_2O(l)$).

### Questions

**1** Complete the following equations:
**a** $CH_3COOH + NaOH$
**b** $H_2SO_4 + KOH$
**c** $NH_3 + HBr$.
In each case name the acid and the base. Also in parts **a** and **b** write the ionic equation for the reactions.

**2** Explain the terms 'concentration' and 'strength' as applied to acids.

**3** Explain what part water plays in the acidity of a solution.

## *Formation of salts*

In the example above, sodium chloride was produced as part of the neutralisation reaction. Compounds formed in this way are known as **normal salts**. A normal salt is a compound that has been formed when all the hydrogen ions of an acid have been replaced by metal ions or by the ammonium ion ($NH_4^+$).

Normal salts can be classified as those which are soluble in water or those which are insoluble in water. The following salts are soluble in cold water:

- all nitrates
- all common sodium, potassium and ammonium salts
- all chlorides except lead, silver and mercury
- all sulphates except lead, barium and calcium.

Salts are very useful substances, as you can see from Table 7.2 and Figure 7.10 (overleaf).

**Table 7.2** Useful salts.

| Salt | Use |
|---|---|
| Ammonium chloride | In torch batteries |
| Ammonium sulphate | In fertilisers |
| Calcium carbonate | Extraction of iron, making cement, glass making |
| Calcium chloride | In the extraction of sodium, drying agent (anhydrous) |
| Calcium sulphate | For making plaster boards, plaster casts for injured limbs |
| Iron(II) sulphate | In 'iron' tablets |
| Magnesium sulphate | In medicines |
| Potassium nitrate | In fertiliser and gunpowder manufacture |
| Silver bromide | In photography |
| Sodium carbonate | Glass making, softening water, making modern washing powders |
| Sodium chloride | Making hydrochloric acid, for food flavouring, hospital saline, in the Solvay process for the manufacture of sodium carbonate |
| Sodium stearate | In some soaps |
| Tin(II) fluoride | Additive to toothpaste |

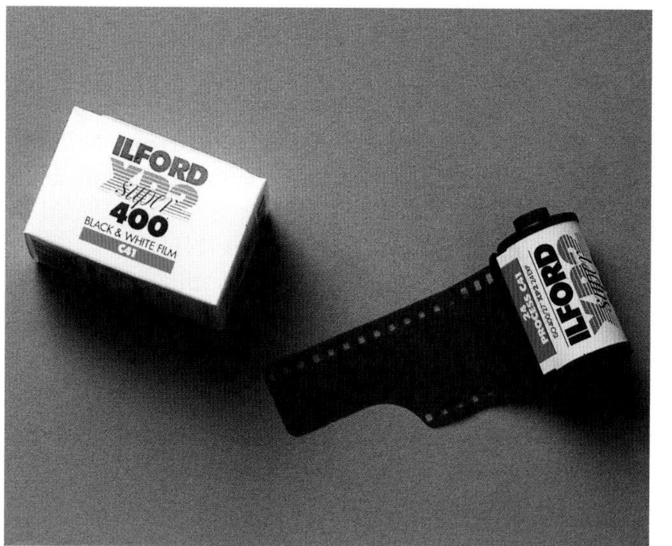


a Silver bromide is used in photography.

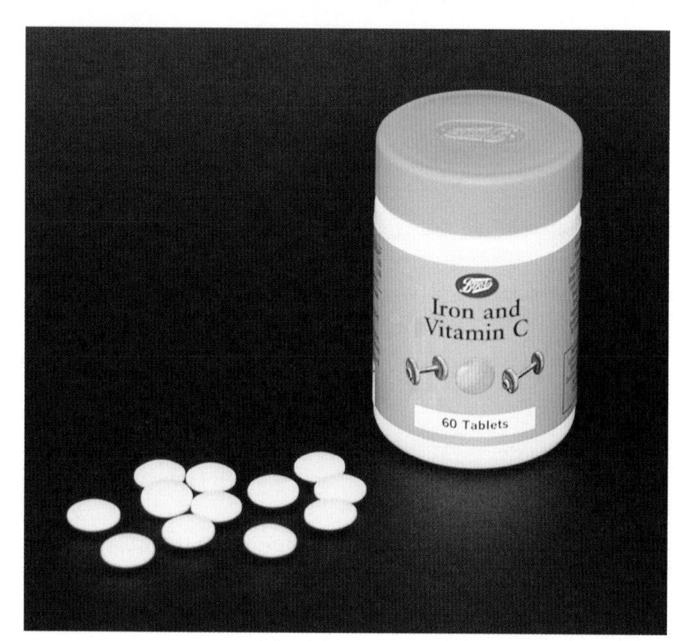


b Iron(II) sulphate is the main substance in these iron and vitamin C tablets.

**Figure 7.10** Some uses of salts.

If the acid being neutralised is hydrochloric acid, salts called **chlorides** are formed. Other types of salts can be formed with other acids. A summary of the different types of salt along with the acid they have been formed from is shown in Table 7.3.

**Table 7.3** Types of salt and the acids they are formed from.

| Acid | Type of salt | Example |
|---|---|---|
| Carbonic acid | Carbonates | Sodium carbonate ($Na_2CO_3$) |
| Ethanoic acid | Ethanoates | Sodium ethanoate ($CH_3COONa$) |
| Hydrochloric acid | Chlorides | Potassium chloride (KCl) |
| Nitric acid | Nitrates | Potassium nitrate ($KNO_3$) |
| Sulphuric acid | Sulphates | Sodium sulphate ($Na_2SO_4$) |

## Methods of preparing soluble salts

There are four general methods of preparing soluble salts:

### Acid + metal

This method can only be used with the less reactive metals. It would be very dangerous to use a reactive metal such as sodium in this type of reaction. The metals usually used in this method of salt preparation are the **MAZIT** metals, that is, **m**agnesium, **a**luminium, **z**inc, **i**ron and **t**in. A typical experimental method is given below.

Excess magnesium ribbon is added to dilute nitric acid. During this addition an effervescence is observed due to the production of hydrogen gas. In this reaction the hydrogen ions from the nitric acid gain electrons from the metal atoms as the reaction proceeds.

hydrogen ions + electrons (from metal) → hydrogen gas

$$2H^+ + 2e^- \rightarrow H_2(g)$$

How would you test the gas to show that it was hydrogen? What would be the name and formula of the compound produced during the test you suggested?

magnesium + nitric acid → magnesium nitrate + hydrogen

$$Mg(s) + 2HNO_3(aq) \rightarrow Mg(NO_3)_2(aq) + H_2(g)$$

The excess magnesium is removed by filtration (Figure 7.11).

**Figure 7.11** The excess magnesium is filtered in this way.

The magnesium nitrate solution is evaporated slowly to form a saturated solution of the salt (Figure 7.12).

The hot concentrated magnesium nitrate solution produced is tested by dipping a cold glass rod into it. If salt crystals form at the end of the rod the solution is ready to crystallise and is left to cool. Any crystals produced on cooling are filtered and dried between clean tissues.

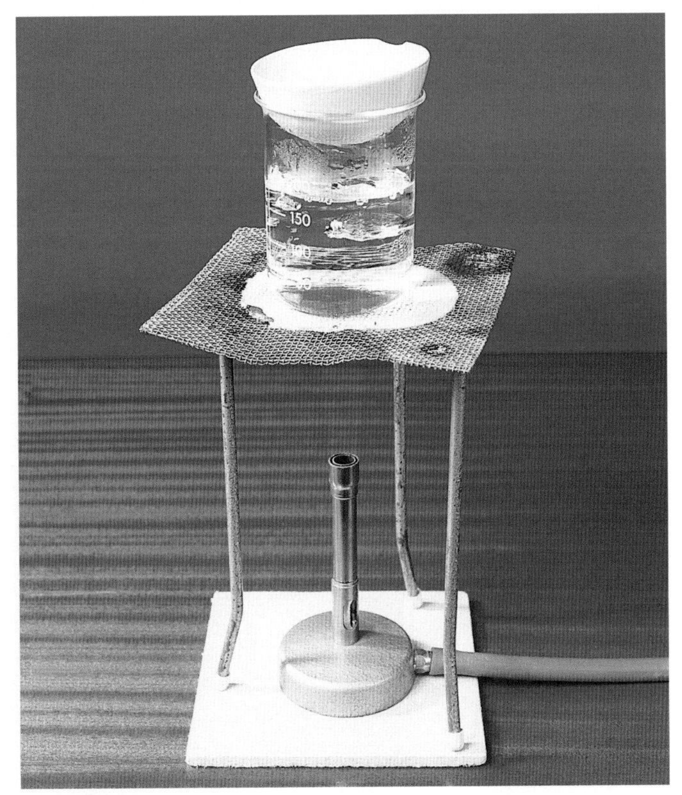

**Figure 7.12** The solution of magnesium nitrate is concentrated by slow evaporation.

## Acid + carbonate

This method can be used with any metal carbonate and any acid, providing the salt produced is soluble. The typical experimental procedure is similar to that carried out for an acid and a metal. For example, copper(II) carbonate would be added in excess to dilute nitric acid. Effervescence would be observed due to the production of carbon dioxide.

How would you test the gas to show it was carbon dioxide? Write an equation to help you explain what is happening during the test you have chosen.

| copper(II) carbonate | + | nitric acid | → | copper(II) nitrate | + | carbon dioxide | + | water |
|---|---|---|---|---|---|---|---|---|

$$CuCO_3(s) + 2HNO_3(aq) \rightarrow Cu(NO_3)_2(aq) + CO_2(g) + H_2O(l)$$

Metal carbonates contain carbonate ions, $CO_3^{2-}$. In this reaction the carbonate ions react with the hydrogen ions in the acid.

| carbonate ions | + | hydrogen ions | → | carbon dioxide | + | water |
|---|---|---|---|---|---|---|

$$CO_3^{2-}(aq) + 2H^+(aq) \rightarrow CO_2(g) + H_2O(l)$$

## Acid + alkali (soluble base)

This method is generally used for preparing the salts of very reactive metals, such as potassium or sodium. It would certainly be too dangerous to add the metal directly to the acid. In this case, we solve the problem indirectly and use an alkali which contains the particular reactive metal whose salt we wish to prepare.

A **base** is a substance which neutralises an acid, producing a salt and water as the only products. If the base is soluble the term alkali can be used, but there are several bases which are insoluble. In general, most metal oxides and hydroxides (as well as ammonia solution) are bases. Some examples of soluble and insoluble bases are shown in Table 7.4. Salts can be formed by this method only if the base is soluble.

**Table 7.4** Examples of soluble and insoluble bases.

| Soluble bases (alkalis) | Insoluble bases |
|---|---|
| Sodium hydroxide ($NaOH$) | Iron(III) oxide ($Fe_2O_3$) |
| Potassium hydroxide ($KOH$) | Copper(II) oxide ($CuO$) |
| Calcium hydroxide ($Ca(OH)_2$) | Lead(II) oxide ($PbO$) |
| Ammonia solution ($NH_3(aq)$) | Magnesium oxide ($MgO$) |

Because in this neutralisation reaction both reactants are in solution, a special technique called **titration** is required. Acid is slowly and carefully added to a measured volume of alkali using a burette (Figure 7.13) until the indicator, usually phenolphthalein, changes colour.

**Figure 7.13** The acid is added to the alkali until the indicator just changes colour.

An indicator is used to show when the alkali has been neutralised completely by the acid. This is called the **end-point**. Once you know where the end-point is, you can add the same volume of acid to the measured volume of alkali but this time without the indicator.

The solution which is produced can then be evaporated slowly to obtain the salt. For example,

| hydrochloric acid | + | sodium hydroxide | → | sodium chloride | + | water |
|---|---|---|---|---|---|---|
| HCl(aq) | + | NaOH(aq) | → | NaCl(aq) | + | $H_2O(l)$ |

As previously discussed on p. 98, this reaction can best be described by the ionic equation:

$$H^+(aq) + OH^-(aq) \rightarrow H_2O(l)$$

### Acid + insoluble base

This method can be used to prepare a salt of an unreactive metal, such as lead or copper. In these cases it is not possible to use a direct reaction of the metal with an acid so the acid is neutralised using the particular metal oxide (Figure 7.14).

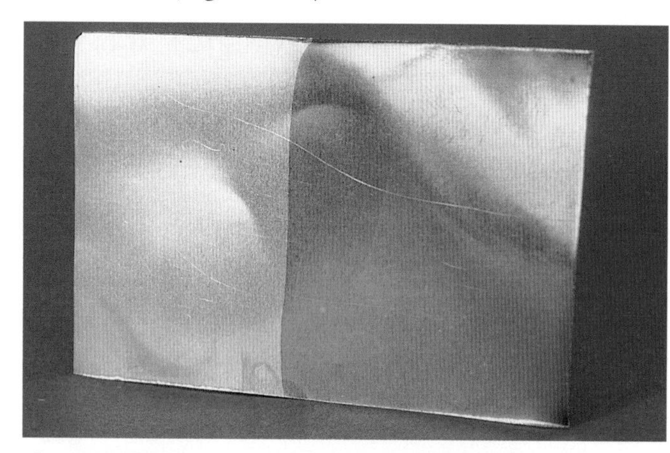

**Figure 7.14** The citric acid in the lemon juice reacts with oxide on the surface of the copper and the tarnish dissolves away.

The method is generally the same as that for a metal carbonate and an acid, though some warming of the reactants may be necessary. An example of such a reaction is the neutralisation of sulphuric acid by copper(II) oxide to produce copper(II) sulphate (Figure 7.15).

**Figure 7.15** After slow evaporation to concentrate the solution, the solution is left to crystallise. Crystals of copper(II) sulphate are produced.

| sulphuric acid | + | copper(II) oxide | → | copper(II) sulphate | + | water |
|---|---|---|---|---|---|---|
| $H_2SO_4(aq)$ | + | CuO(s) | → | $CuSO_4(aq)$ | + | $H_2O(l)$ |

Metal oxides contain the oxide ion, $O^{2-}$. The ionic equation for this reaction is therefore:

$$2H^+(aq) + O^{2-}(s) \rightarrow H_2O(l)$$

## Methods of preparing insoluble salts

An insoluble salt, such as barium sulphate, can be made by precipitation. In this case, solutions of the two chosen soluble salts are mixed (Figure 7.16). To produce barium sulphate, barium chloride and sodium sulphate can be used. The barium sulphate precipitate can be filtered off, washed with distilled water and dried. The reaction that has occurred is:

| barium chloride | + | sodium sulphate | → | barium sulphate | + | sodium chloride |
|---|---|---|---|---|---|---|
| $BaCl_2(aq)$ | + | $Na_2SO_4(aq)$ | → | $BaSO_4(s)$ | + | 2NaCl(aq) |

**Figure 7.16** When barium chloride solution is added to sodium sulphate a white precipitate of barium sulphate is produced.

The ionic equation for this reaction is:

$$Ba^{2+}(aq) + SO_4^{2-}(aq) \rightarrow BaSO_4(s)$$

This method is sometimes known as **double decomposition** and may be summarised as follows

| soluble salt | + | soluble salt | → | insoluble salt | + | soluble salt |
|---|---|---|---|---|---|---|
| (AX) | | (BY) | | (BX) | | (AY) |

It should be noted that even salts like barium sulphate dissolve to a very small extent. For example, 1 litre of water will dissolve $2.2 \times 10^{-3}$ g of barium sulphate at 25 °C. This substance and substances like this are said to be **sparingly soluble**.

## More about salts

You have already seen on p. 99 in Table 7.2 that salts are useful substances. Some of the salts shown in that table occur naturally and are mined, for example calcium sulphate (gypsum) and calcium carbonate (limestone). Many of the others must be made by the chemical industry, for example ammonium nitrate, iron(II) sulphate and silver bromide.

With acids such as sulphuric acid, which has two replaceable hydrogen ions per molecule, it is possible to replace only *one* of these with a metal ion. The salt produced is called an **acid salt**. An acid salt is one in which not all of the replaceable hydrogen ions of the acid have been replaced by metal ions or the ammonium ion. Some examples of acid salts are shown in Table 7.5.

**Table 7.5** Examples of acid salts.

| Acid | Type of acid salt | Example |
|---|---|---|
| Carbonic acid ($H_2CO_3$) | Hydrogencarbonate | Sodium hydrogencarbonate ($NaHCO_3$) |
| Sulphuric acid ($H_2SO_4$) | Hydrogensulphate | Potassium hydrogensulphate ($KHSO_4$) |

Sodium hydrogencarbonate is the acid salt used as the raising agent in the baking of cakes and bread, and is often called baking soda (Figure 7.17).

**Figure 7.17** Sodium hydrogencarbonate is used as a raising agent.

## Testing for different salts

Sometimes we want to analyse a salt and find out what is in it. There are simple chemical tests which allow us to identify the anion part of the salt. These are often called **spot tests**.

### Testing for a sulphate ($SO_4^{2-}$)

You have seen that barium sulphate is a useful insoluble salt (p. 102). Therefore, if you take a solution of a suspected sulphate and add it to a soluble barium salt (such as barium chloride) then a white precipitate of barium sulphate will be produced.

barium ion + sulphate ion → barium sulphate

$$Ba^{2+}(aq) + SO_4^{2-}(aq) \rightarrow BaSO_4(s)$$

A few drops of dilute hydrochloric acid are also added to this mixture. If the precipitate does not dissolve, then it is barium sulphate and the unknown salt was in fact a sulphate. If the precipitate does dissolve, then the unknown salt may have been a sulphite (containing the $SO_3^{2-}$ ion).

### Testing for a chloride ($Cl^-$), a bromide ($Br^-$) or an iodide ($I^-$)

Earlier in this chapter you saw that silver chloride is an insoluble salt (p. 99). Therefore, if you take a solution of a suspected chloride and add to it a small volume of dilute nitric acid followed by a small amount of a solution of a soluble silver salt (such as silver nitrate) a white precipitate of silver chloride will be produced.

chloride ion + silver ion → silver chloride

$$Cl^-(aq) + Ag^+(aq) \rightarrow AgCl(s)$$

If left to stand, the precipitate goes grey (Figure 7.18).

**Figure 7.18** If left to stand the white precipitate of silver chloride goes grey. This photochemical change plays an essential part in black and white photography.

In a similar way, a bromide and an iodide will react to produce either a cream precipitate of silver bromide (AgBr) or a yellow precipitate of silver iodide (AgI) (Figure 7.19).

**Figure 7.19** AgCl, a white precipitate, AgBr, a cream precipitate, and AgI, a yellow precipitate.

### Testing for a carbonate

If a small amount of an acid is added to some of the suspected carbonate (either solid or in solution) then effervescence occurs. If it is a carbonate then carbon dioxide gas is produced, which will turn limewater 'milky' (a cloudy white precipitate of calcium carbonate forms).

carbonate ions + hydrogen ions → carbon dioxide + water

$$CO_3^{2-}(aq) + 2H^+(aq) \rightarrow CO_2(g) + H_2O(l)$$

### Testing for a nitrate

By using Devarda's alloy (45% Al, 5% Zn, 50% Cu) in alkaline solution, nitrates are reduced to ammonia. The ammonia can be identified using damp indicator paper, which turns blue.

## Questions

1 Complete the word equations and write balanced chemical equations for the following soluble salt preparations:
magnesium + sulphuric acid →
calcium carbonate + hydrochloric acid →
zinc oxide + hydrochloric acid →
potassium hydroxide + nitric acid →
Also write ionic equations for each of the reactions.

2 Lead carbonate and lead iodide are insoluble. Which two soluble salts could you use in the preparation of each substance? Write
 a a word equation
 b a symbol equation
 c an ionic equation
to represent the reactions taking place.

3 An analytical chemist working for an environmental health organisation has been given a sample of water which is thought to have been contaminated by a sulphate, a carbonate and a chloride.
 a Describe how she could confirm the presence of these three types of salt by simple chemical tests.
 b Write ionic equations to help you explain what is happening during the testing process.

## Crystal hydrates

Some salts, such as sodium chloride, copper carbonate and sodium nitrate, crystallise in their anhydrous forms (without water). However, many salts produce **hydrates** when they crystallise from solution. A hydrate is a salt which incorporates water into its crystal structure. This water is referred to as **water of crystallisation**. The shape of the crystal hydrate is very much dependent on the presence of water of crystallisation. Some examples of crystal hydrates are given in Table 7.6 and shown in Figure 7.20.

**Table 7.6** Examples of crystal hydrates.

| Salt hydrate | Formula |
|---|---|
| Cobalt(II) chloride hexahydrate | $CoCl_2.6H_2O$ |
| Copper(II) sulphate pentahydrate | $CuSO_4.5H_2O$ |
| Iron(II) sulphate heptahydrate | $FeSO_4.7H_2O$ |
| Magnesium sulphate heptahydrate | $MgSO_4.7H_2O$ |
| Sodium carbonate decahydrate | $Na_2CO_3.10H_2O$ |
| Sodium hydrogensulphate monohydrate | $NaHSO_4.H_2O$ |
| Sodium sulphate decahydrate | $Na_2SO_4.10H_2O$ |

**Figure 7.20** Hydrate crystals (left to right): cobalt nitrate, calcium nitrate and nickel sulphate (top) and manganese sulphate, copper sulphate and chromium potassium sulphate (bottom).

When many hydrates are heated the water of crystallisation is driven away. For example, if crystals of copper(II) sulphate hydrate (blue) are heated strongly, they lose their water of crystallisation. Anhydrous copper(II) sulphate remains as a white powder:

copper(II) sulphate pentahydrate → anhydrous copper(II) sulphate + water

$$CuSO_4.5H_2O(s) \rightarrow CuSO_4(s) + 5H_2O(g)$$

When water is added to anhydrous copper(II) sulphate the reverse process occurs. It turns blue and the pentahydrate is produced (Figure 7.21). This is an extremely exothermic process.

$$CuSO_4(s) + 5H_2O(l) \rightarrow CuSO_4.5H_2O(s)$$

Because the colour change only takes place in the presence of water, the reaction is used to test for the presence of water.

**Figure 7.21** Anhydrous copper(II) sulphate is a white powder which turns blue when water is added to it.

Some crystal hydrates **effloresce**, that is they lose some or all of their water of crystallisation to the atmosphere. For example, when colourless sodium carbonate decahydrate crystals are left out in the air they become coated with a white powder, which is the monohydrate (Figure 7.22). The process is called **efflorescence**.

$$Na_2CO_3.10H_2O(s) \rightarrow Na_2CO_3.H_2O(s) + 9H_2O(g)$$

**Figure 7.22** A white powder forms on the surface of sodium carbonate decahydrate when it is left in the air.

With some substances, not necessarily salt hydrates, the reverse of efflorescence occurs. For example, if anhydrous calcium chloride is left in the air, it absorbs water vapour and eventually forms a very concentrated solution. This process is called **deliquescence**, and substances which behave like this are said to be **deliquescent**. Solid sodium hydroxide will deliquesce.

There are some substances which, if left out in the atmosphere, absorb moisture but do not change their state. For example, concentrated sulphuric acid, a colourless, viscous liquid, absorbs water vapour from the air and becomes a solution. Substances which do this are said to be **hygroscopic**.

## Calculation of water of crystallisation

Sometimes it is necessary to work out the percentage, by mass, of water of crystallisation in a hydrated salt. The method is the same as that used in Chapter 5, p. 75, but this time the '$H_2O$' is treated as an element in the calculation.

**Example**

Calculate the percentage by mass of water in the salt hydrate $MgSO_4.7H_2O$. ($A_r$: H = 1, O = 16, Mg = 24, S = 32)

$$M_r \text{ for } MgSO_4.7H_2O = 24 + 32 + (4 \times 16) + (7 \times 18) = 246$$

The mass of water as a fraction of the total mass of hydrate

$$= \frac{126}{246}$$

The percentage of water present

$$= \frac{126}{246} \times 100 = 51.2\%$$

## Questions

1 Calculate the percentage by mass of water in the following salt hydrates:
   a $CuSO_4.5H_2O$
   b $Na_2CO_3.10H_2O$
   c $Na_2S_2O_3.5H_2O$.
   ($A_r$: H = 1, O = 16, Na = 23, S = 32, Cu = 63.5)
2 Devise an experiment to determine the percentage of water of crystallisation present in a salt hydrate of your choice.

## Solubility of salts in water

Water is a very good solvent and will dissolve a whole range of solutes, including sodium chloride and copper(II) sulphate, as well as other substances such as sugar. You can dissolve more sugar than sodium chloride in 100 $cm^3$ of water at the same temperature. The sugar is said to be more **soluble** than the sodium chloride at the same temperature. We say that the sugar has a greater **solubility** than the sodium chloride. The solubility of a solute in water at a given temperature is the number of grams of that solute which can be dissolved in 100 g of water to produce a saturated solution at that temperature.

### Calculating solubility

**Example**

21.5 g of sodium chloride dissolve in 60 g of water at 25 °C. Calculate the solubility of sodium chloride in water at that temperature.

If 60 g of water dissolves 21.5 g of sodium chloride, then 1 g of water will dissolve:

$$\frac{21.5}{60} \text{ g of sodium chloride}$$

Therefore, 100 g of water will dissolve:

$$\frac{21.5}{60} \times 100 = 35.8\text{ g}$$

The solubility of sodium chloride at 25 °C = 35.8 g per 100 g of water.

### Solubility curves

It is a well-known fact that usually the amount of solute that a solvent will dissolve increases with temperature. We say that the solubility increases with increasing temperature. Figure 7.23 shows how the solubilities of copper(II) sulphate and potassium nitrate vary with temperature. These graphs of solubility against temperature are known as **solubility curves**. By using curves of this type you can find the solubility of the solute at any temperature.

From a solubility curve such as Figure 7.23, it can be seen that the solubility of potassium nitrate at 50 °C is 80 g per 100 g of water, whereas that of copper(II) sulphate is 35 g per 100 g of water.

When a saturated solution is cooled, some of the solute crystallises out of solution. Thus, by using solubility curves it is possible to determine the amount of solute that will crystallise out of solution at different temperatures. For example, if potassium nitrate is cooled from 70 °C to 40 °C then the difference between the two solubilities at these temperatures tells you the amount that crystallises out.

Solubility at 70 °C = 135 g
Solubility at 40 °C = 62 g
Difference = 73 g

Therefore the amount of potassium nitrate which crystallises out during cooling from 70 °C to 40 °C is 73 g.

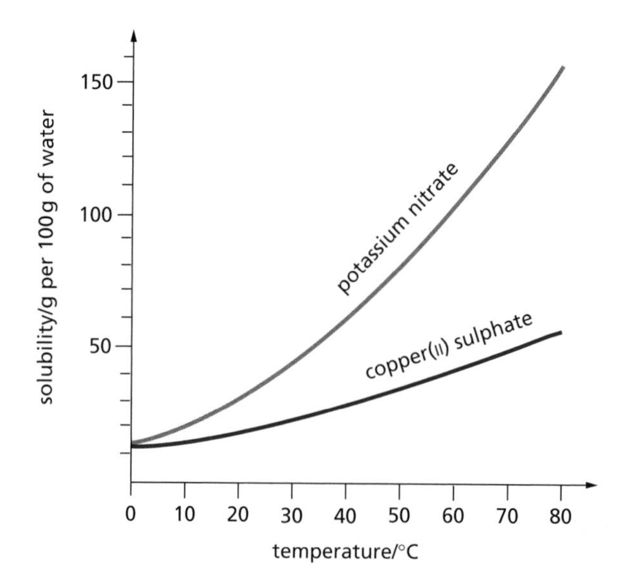


**Figure 7.23** Solubility curves for copper(II) sulphate and potassium nitrate.

## Questions

1. Calculate the solubility, at the temperature given, of the following salts in water:
   - **a** 12.1 g of copper(II) sulphate dissolved in 50 g of water at 30 °C
   - **b** 11.1 g of potassium chlorate(V) dissolved in 60 g of water at 50 °C
   - **c** 9.72 g of potassium chloride dissolved in 30g of water at 70 °C.
2. Use the data given below to plot a solubility curve for lead nitrate.

| Temperature/°C | 10 | 30 | 50 | 70 | 90 |
|---|---|---|---|---|---|
| Solubility/g per 100 g of water | 44.0 | 60.7 | 78.6 | 97.7 | 117.4 |

3. Using your solubility curve from question **2**, find the solubility of lead nitrate at the following temperatures:
   - **a** 25 °C
   - **b** 45 °C
   - **c** 75 °C.
4. Using your answers to question **3**, calculate the amount of lead nitrate that will crystallise out of solution when cooled from:
   - **a** 45 °C to 25 °C
   - **b** 75 °C to 25 °C
   - **c** 75 °C to 45 °C.
5. 25 g of a saturated solution containing lead nitrate is cooled from 54 °C to 27 °C. Calculate the amount of lead nitrate that would crystallise out of the solution.

## Titration

On p.101 you saw that it was possible to prepare a soluble salt by reacting an acid with a soluble base (alkali). The method used was that of **titration**. Titration can also be used to find the concentration of the alkali used. In the laboratory, the titration of hydrochloric acid with sodium hydroxide is carried out in the following way.

1 25 $cm^3$ of sodium hydroxide solution is pipetted into a conical flask to which a few drops of phenolphthalein indicator have been added (Figure 7.24). Phenolphthalein is pink in alkaline conditions but colourless in acid.

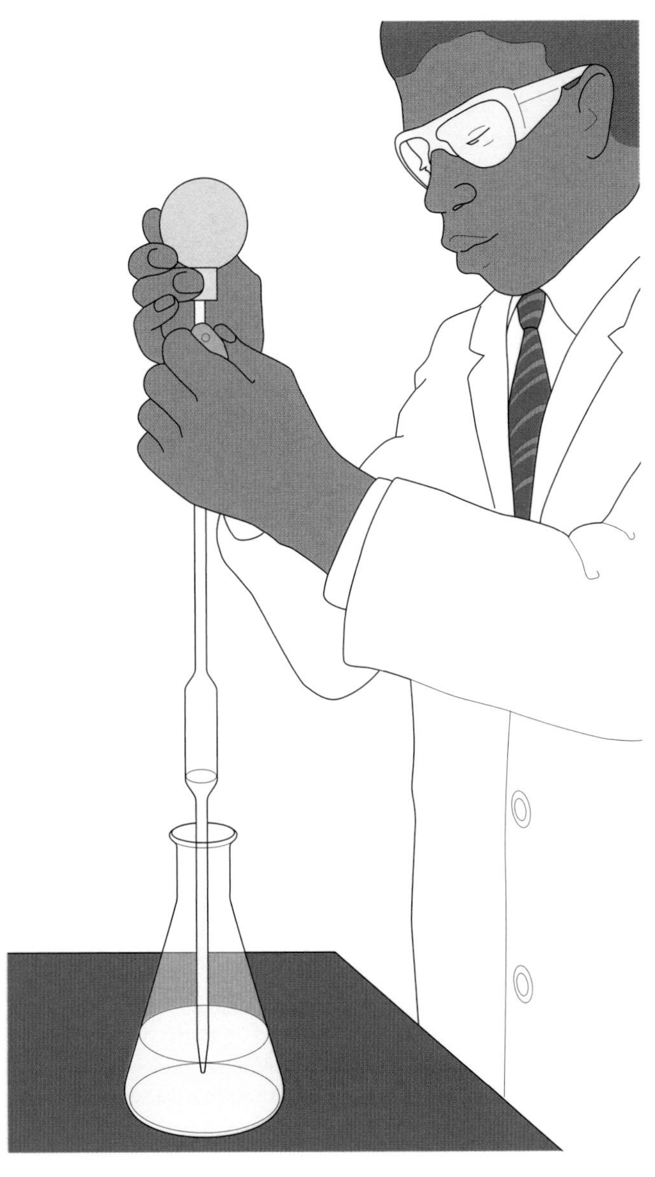

**Figure 7.24** Exactly 25.0 $cm^3$ of sodium hydroxide solution is pipetted into a conical flask.

2 A 0.10 mol $dm^{-3}$ solution of hydrochloric acid is placed in the burette using a filter funnel until it is filled up exactly to the zero mark (Figure 7.25).

3 The filter funnel is now removed.

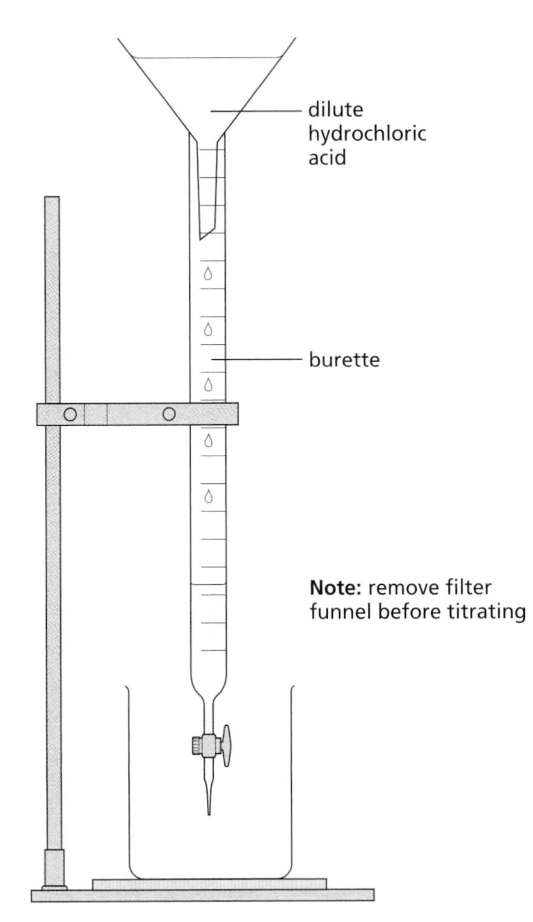


**Figure 7.25** The burette is filled up to the zero mark with a 0.10 mol $dm^{-3}$ solution of hydrochloric acid.

4 The hydrochloric acid is added to the sodium hydroxide solution in small quantities – usually no more than 0.5 $cm^3$ at a time (Figure 7.26). The contents of the flask must be swirled after each addition of acid for thorough mixing.

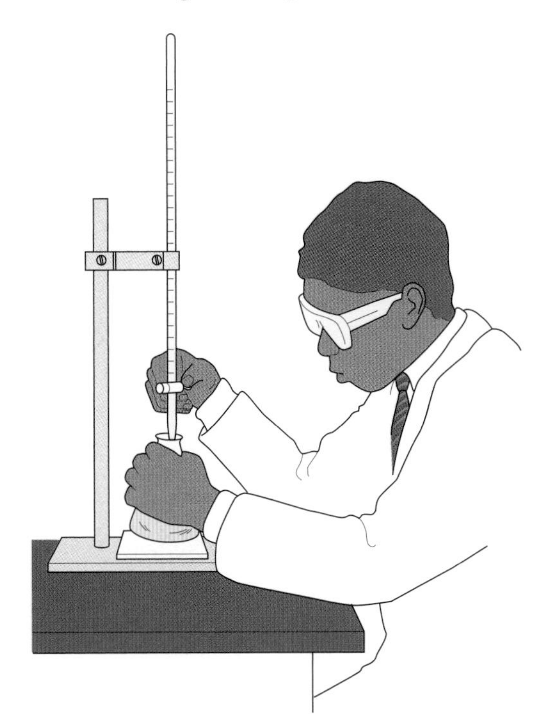

**Figure 7.26** The titration is carried out accurately.

5 The acid is added until the alkali has been neutralised completely. This is shown by the pink colour of the indicator just disappearing.
6 The final reading on the burette at the end-point is recorded and further titrations carried out until consistent results are obtained (within $0.1\,cm^3$ of each other). Some sample data are shown below.

Volume of sodium hydroxide solution
$= 25.0\,cm^3$
Average volume of $0.10\,mol\,dm^{-3}$ solution of hydrochloric acid added
$= 21.0\,cm^3$

The neutralisation reaction which has taken place is:

| hydrochloric acid | + | sodium hydroxide | → | sodium chloride | + | water |
|---|---|---|---|---|---|---|
| $HCl(aq)$ | + | $NaOH(aq)$ | → | $NaCl(aq)$ | + | $H_2O(l)$ |

From this equation it can be seen that 1 mole of hydrochloric acid neutralises 1 mole of sodium hydroxide.

Now you can work out the number of moles of the acid using the formula given in Chapter 5, p. 73.

$$\text{moles} = \frac{\text{volume}}{1000} \times \text{concentration}$$

$$= \frac{21.0 \times 0.10}{1000}$$

$$= 2.1 \times 10^{-3}$$

$$\text{number of moles of hydrochloric acid} = \text{number of moles of sodium hydroxide}$$

Therefore, the number of moles of sodium hydroxide $= 2.1 \times 10^{-3}$

$2.1 \times 10^{-3}$ moles of sodium hydroxide is present in $25.0\,cm^3$ of solution.

Therefore, in $1\,cm^3$ of sodium hydroxide solution we have

$$\frac{2.1 \times 10^{-3}}{25.0}\ \text{moles}$$

Therefore, in 1 litre of sodium hydroxide solution we have

$$\frac{2.1 \times 10^{-3}}{25.0} \times 1000 = 0.084\,\text{mole}$$

The concentration of sodium hydroxide solution is $0.084\,mol\,dm^{-3}$ (0.084 M).

You can simplify the calculation by substituting in the following mathematical equation:

$$\frac{M_1 V_1}{M_{acid}} = \frac{M_2 V_2}{M_{alkali}}$$

where:

$M_1$ = molarity (concentration) of the acid used
$V_1$ = volume of acid used ($cm^3$)
$M_{acid}$ = number of moles of acid shown in the chemical equation
$M_2$ = molarity (concentration) of the alkali used
$V_2$ = volume of the alkali used ($cm^3$)
$M_{alkali}$ = number of moles of alkali shown in the chemical equation

In the example:

$M_1 = 0.10\,mol\,dm^{-3}$
$V_1 = 21.0\,cm^3$
$M_{acid} = 1$ mole
$M_2$ = unknown
$V_2 = 25.0\,cm^3$
$M_{alkali} = 1$ mole

Substituting in the equation:

$$\frac{0.10 \times 21.0}{1} = \frac{M_2 \times 25.0}{1}$$

Rearranging:

$$M_2 = \frac{0.10 \times 21.0 \times 1}{1 \times 25.0}$$

$$M_2 = 0.084$$

The molarity of the sodium hydroxide solution = 0.084 M. It is a solution of concentration $0.084\,mol\,dm^{-3}$.

## Questions

1 $24.2\,cm^3$ of a solution containing $0.20\,mol\,dm^{-3}$ of hydrochloric acid just neutralised $25.0\,cm^3$ of a potassium hydroxide solution. What is the concentration of this potassium hydroxide solution?

2 $22.4\,cm^3$ of a solution containing $0.10\,mol\,dm^{-3}$ of sulphuric acid just neutralised $25.0\,cm^3$ of a sodium hydroxide solution. What is the concentration of this sodium hydroxide solution?

## Checklist

**After studying Chapter 7 you should know and understand the following terms.**

- **Acid** A substance which dissolves in water, producing $H^+(aq)$ ions as the only positive ions.
- **Acid salt** A substance formed when only some of the replaceable hydrogen of an acid is replaced by metal ions or the ammonium ion ($NH_4^+$).
- **Alkali** A soluble base which produces $OH^-(aq)$ ions in water.
- **Base** A substance which neutralises an acid, producing a salt and water as the only products.
- **Deliquescence** The process during which a substance absorbs water vapour from the atmosphere and eventually forms a very concentrated solution.
- **Double decomposition** The process by which an insoluble salt is prepared from solutions of two suitable soluble salts.
- **Efflorescence** The process during which a substance loses water of crystallisation to the atmosphere.
- **Hygroscopic** The ability to absorb water vapour from the atmosphere without forming solutions or changing state, for example concentrated sulphuric acid.
- **Indicator** A substance used to show whether a substance is acidic or alkaline (basic), for example phenolphthalein.
- **Ionic equation** The simplified equation of a reaction which we can write if the chemicals involved are ionic substances.
- **Neutralisation** The process in which the acidity or alkalinity of a substance is destroyed. Destroying acidity means removing $H^+(aq)$ by reaction with a base, carbonate or metal. Destroying alkalinity means removing the $OH^-(aq)$ by reaction with an acid.

$$H^+(aq) + OH^-(aq) \rightarrow H_2O(l)$$

- **Normal salt** A substance formed when all the replaceable hydrogen of an acid is completely replaced by metal ions or the ammonium ion ($NH_4^+$).
- **pH scale** A scale running from 0 to 14, used for expressing the acidity or alkalinity of a solution.
- **Salt hydrates** Salts containing water of crystallisation.
- **Solubility** The solubility of a solute in a solvent at a given temperature is the number of grams of that solute which can dissolve in 100 g of solvent to produce a saturated solution at that temperature.
- **Solubility curve** This is a graph of solubility against temperature.
- **Strong acid** One which produces a high concentration of $H^+(aq)$ ions in water solution, for example hydrochloric acid.
- **Strong alkali** One which produces a high concentration of $OH^-(aq)$ ions in water solution, for example sodium hydroxide.
- **Testing for a carbonate** If effervescence occurs when an acid is added to the suspected carbonate and the gas produced tests positively for carbon dioxide, the substance is a carbonate.
- **Testing for a chloride** If a white precipitate is produced when dilute nitric acid and silver nitrate solution are added to the suspected chloride, the solution contains a chloride.
- **Testing for a sulphate** If a white precipitate is produced when dilute hydrochloric acid and barium chloride solution are added to the suspected sulphate, the solution contains a sulphate.
- **Titration** A method of volumetric analysis in which a volume of one reagent (for example an acid) is added to a known volume of another reagent (for example an alkali) slowly from a burette until an end-point is reached. If an acid and alkali are used, then an indicator is used to show that the end-point has been reached.
- **Water of crystallisation** Water incorporated into the structure of substances as they crystallise, for example copper(II) sulphate pentahydrate ($CuSO_4.5H_2O$).
- **Weak acid** One which produces a low concentration of $H^+(aq)$ in water solution, for example ethanoic acid.
- **Weak alkali** One which produces a low concentration of $OH^-(aq)$ in water solution, for example ammonia solution.

# Acids, bases and salts

# *Additional questions*

1 Descibe briefly the contribution to the theory of acids and bases by:
   a Antoine Lavoisier (1743–1794)
   b Sir Humphry Davy (1778–1829)
   c Justus von Liebig (1803–1873)
   d Svante Arrhenius (1859–1927)
   e Johannes Brønsted (1879–1947) and Thomas Lowry (1874–1936).

2 a Copy out and complete the table, which covers the different methods of preparing salts.

| Method of preparation | Name of salt prepared | Two substances used in the preparation |
|---|---|---|
| Acid + alkali | Potassium sulphate | .............. and .............. |
| Acid + metal | ............ | .............. and dilute hydrochloric acid |
| Acid + insoluble base | Magnesium sulphate | .............. and .............. |
| Acid + carbonate | Copper ............ | .............. and .............. |
| Precipitation | Lead iodide | .............. and .............. |

   b Write word and balanced chemical equations for each reaction shown in your table. Also write ionic equations where appropriate.

3 Study the following scheme.

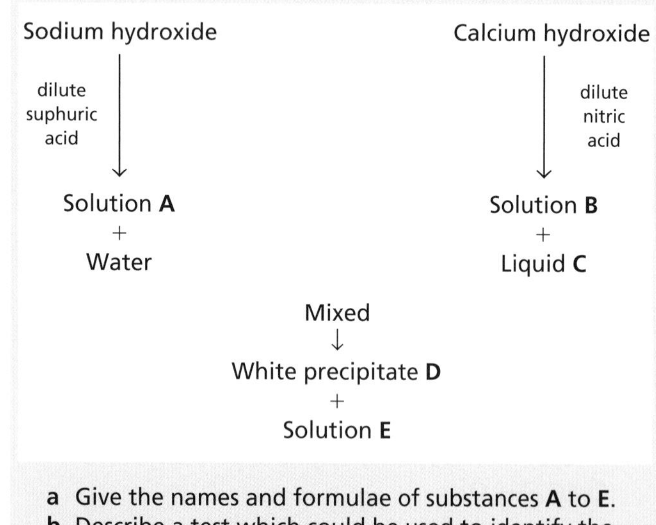


   a Give the names and formulae of substances **A** to **E**.
   b Describe a test which could be used to identify the presence of water.
   c Which indicator is suitable for the initial reaction between the hydroxides and the dilute acids shown?
   d Write balanced chemical equations for the reactions taking place in the scheme.
   e Write an ionic equation for the production of the white precipitate **D**.

4 In a titration involving 24.0 $cm^3$ potassium hydroxide solution against a solution containing 1 mol $dm^{-3}$ of sulphuric acid, 28.0 $cm^3$ of the acid was found to just neutralise the alkali completely.
   a Write a word and balanced chemical equation for the reaction.
   b Name a suitable indicator for the titration and state the colour change you would observe.
   c Calculate the concentration of the alkali in mol $dm^{-3}$.
   d Describe a chemical test which you could use to identify the type of salt produced during the reaction.

5 Explain the following, with the aid of examples:
   a neutralisation
   b titration
   c soluble salt
   d insoluble salt.

6 Ammonium nitrate is used in fertilisers. The table below shows the solubility of this substance in water at various temperatures.

| Temperature/°C | 0 | 10 | 20 | 30 | 40 | 50 | 60 | 70 |
|---|---|---|---|---|---|---|---|---|
| Solubility/g per 100 g $H_2O$ | 28 | 36 | 44 | 52 | 60 | 70 | 82 | 96 |

   a Plot these results as a solubility curve for ammonium nitrate.
   b Use your graph to answer the following questions.
   (i) What is the solubility of ammonium nitrate at 8 °C, 27 °C, 45 °C and 66 °C?
   (ii) Using your answers to (i), calculate the amount of ammonium nitrate that will crystallise out of solution when it is cooled from 45 °C to 8 °C.
   (iii) Calculate the amount of ammonium nitrate which would crystallise out of solution if 20 g of saturated solution was cooled from 63 °C to 28 °C.

7 Read the following passage and then answer the questions which follow.
Sodium carbonate decahydrate *effloresces* quite readily. With some substances, such as solid sodium hydroxide, the reverse of efflorescence occurs – they *deliquesce*. There are some substances, such as concentrated sulphuric acid, which when left open to the atmosphere are diluted – they are *hygroscopic*.
   a What are the meanings of the terms in italics?
   b What precautions should be taken to ensure that these substances are not involved in the processes described above?
   c Which of the other salts shown in Table 7.6 on p. 104 are likely to effloresce? Give a reason for your answer.

**8** Copper(II) sulphate crystals exist as the *pentahydrate*, $CuSO_4.5H_2O$. It is a salt *hydrate*. If it is heated quite strongly, the *water of crystallisation* is driven off and the *anhydrous* salt remains.

**a** Explain the meaning of the terms shown in italics.

**b** Describe the experiment you would carry out to collect a sample of the water given off when the salt hydrate was heated strongly. Your description should include a diagram of the apparatus used and a chemical equation to represent the process taking place.

**c** Describe a chemical test you could carry out to show that the colourless liquid given off was water.

**d** Describe one other test you could carry out to show that the colourless liquid obtained in this experiment was **pure** water.

**e** Sometimes it is necessary to work out the percentage by mass of water of crystallisation as well as the number of moles of water present in a hydrated crystal.

(i) Use the information given below to calculate the percentage, by mass, of water of crystallisation in a sample of hydrated magnesium sulphate.

Mass of crucible = 14.20 g
Mass of crucible + hydrated $MgSO_4$ = 16.66 g
Mass after heating = 15.40 g

(ii) Calculate the number of moles of water of crystallisation driven off during the experiment as well as the number of moles of anhydrous salt remaining. ($A_r$: H = 1; O = 16; Mg = 24; S = 32)

(iii) Using the information you have obtained in (ii), write down, in the form $MgSO_4.xH_2O$, the formula of hydrated magnesium sulphate.

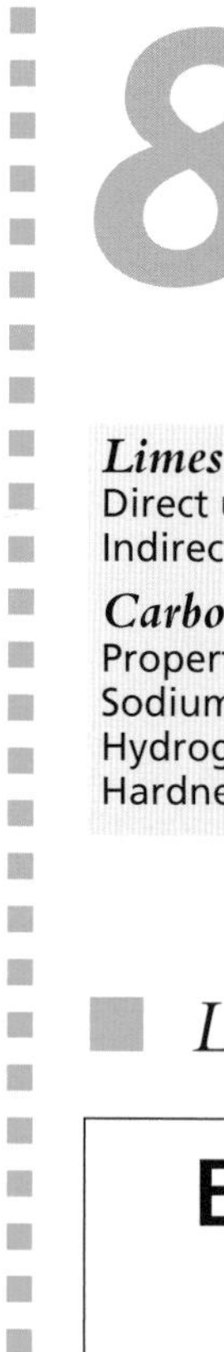

# 8 Inorganic carbon chemistry



## Limestone

**Environmental Problems in the Peak District**

Local residents have been protesting against proposals to site a new limestone quarry in the beautiful Derbyshire Peak District. However, limestone is such a useful and sought-after mineral that demand has encouraged mining in National Park areas.

**Figure 8.1** This gorge in the Peak District is made from limestone.

As the newspaper article on the left says, limestone is found in the Peak District, an example being this gorge on the River Wye (Figure 8.1). In England, it is also found in the Yorkshire Dales, the Sussex Downs and the Mendip Hills. As well as giving rise to beautiful and varied countryside, limestone is a very useful raw material.

Limestone is composed of calcium carbonate ($CaCO_3$) in the form of the mineral calcite (Figure 8.2). Chalk and marble are also made of calcite which is the second most abundant mineral in the Earth's crust after the different types of silicates (which include clay, granite and sandstone).

Chalk is made of the 'shells' of marine algae (that is, plants). It is a form of limestone. Most other limestones are formed from the debris of animal structures, for example brachiopods and crinoids.

Marble is a metamorphic rock made of calcium carbonate. It is formed when limestone is subjected to high pressures or high temperatures, or sometimes both acting together, to create crystals of calcium carbonate in the rock.

In a typical year, about 80 million tonnes of limestone are quarried in Great Britain. Although it is cheap to quarry, as it is found near the surface, there are some environmental costs in its extraction. List the environmental issues which could arise through the quarrying of limestone.

**Figure 8.2** Chalk, calcite and marble are all forms of calcium carbonate.

## Direct uses of limestone

Limestone has a variety of uses in, for example, the making of cement, road building, glass making and the extraction of iron (Figure 8.3).

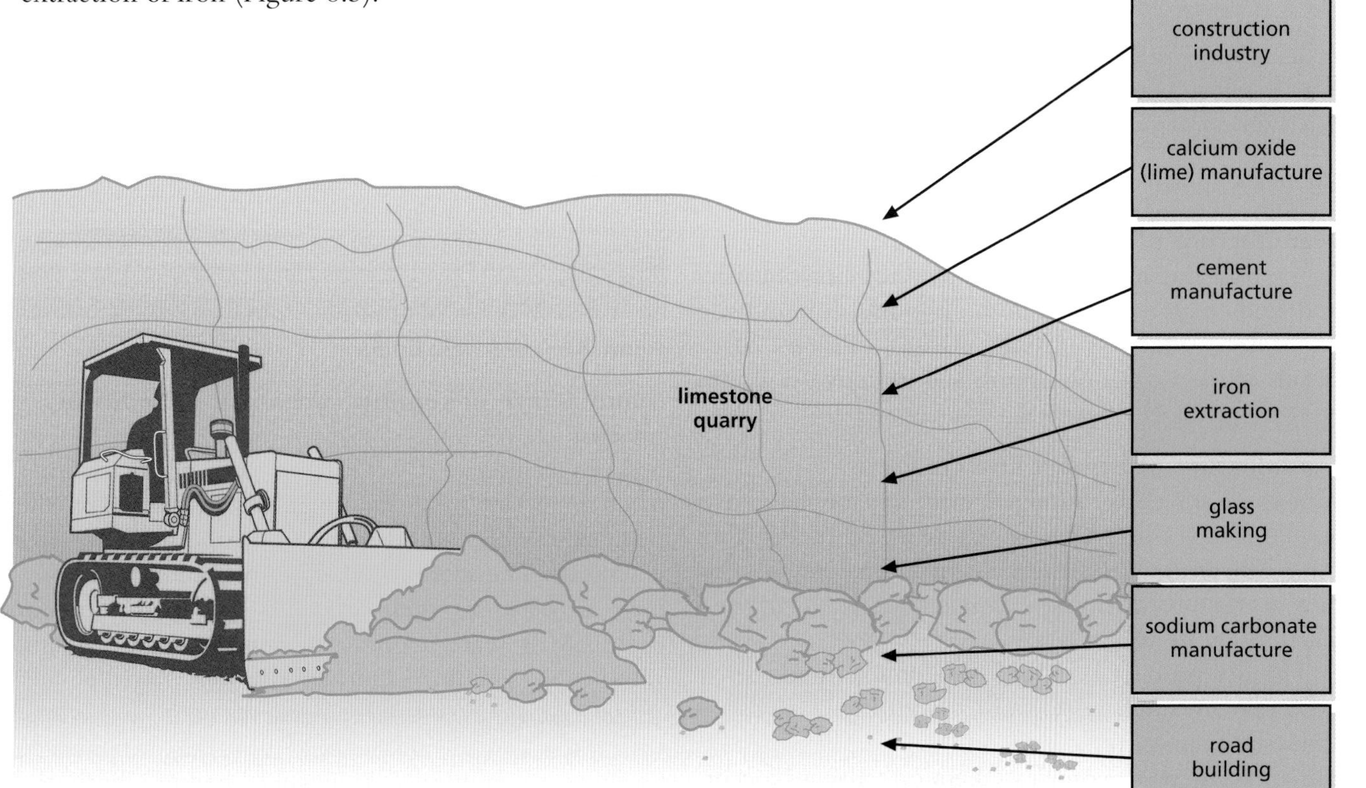


**Figure 8.3** Uses of limestone.

## Neutralisation of acid soil

Powdered limestone is most often used to neutralise acid soil (Figure 8.4) because it is cheaper than any form of lime (calcium oxide), which has to be produced by heating limestone (see p. 115). The reaction of limestone with acidic soil can be shown by the following ionic equation.

carbonate ion (from $CaCO_3$) + hydrogen ion (from acid soils) → carbon dioxide + water

$$CO_3^{2-}(s) + 2H^+(aq) \rightarrow CO_2(g) + H_2O(l)$$

Figure 8.4 Spreading limestone on to soil.

## Manufacture of iron and steel

In the blast furnace, limestone is used to remove earthy and sandy materials found in the iron ore. A liquid slag is formed, which is mainly calcium silicate. More details of the extraction of iron and its conversion into steel are given in Chapter 9.

## Manufacture of cement and concrete

Limestone (or chalk) is mixed with clay (or shale) in a heated rotary kiln, using coal or oil as the fuel (Figure 8.5). The material produced is called cement. It contains a mixture of calcium aluminate ($Ca(AlO_2)_2$) and calcium silicate ($CaSiO_3$).

The dry product is ground to a powder and then a little calcium sulphate ($CaSO_4$) is added to slow down the setting rate of the cement. When water is added to the mixture slow complex chemical changes occur, resulting in the formation of a hard interlocking mass of crystals of hydrated calcium aluminate and silicate.

Figure 8.5 A modern rotary kiln.

Concrete is a mixture of cement with stone chippings and sand, which help to give it body. After the ingredients have been mixed with water they are poured into wooden moulds and allowed to set hard. Reinforced concrete is made by allowing concrete to set around steel rods or mesh to give it greater tensile strength, which is required for the construction of large bridges (Figure 8.6) and tall buildings.

Figure 8.6 The Severn Bridge.

## Manufacture of sodium carbonate – the Solvay process

Sodium carbonate ($Na_2CO_3$) is one of the world's most important industrial chemicals. It is used in the manufacture of soaps, detergents, dyes, drugs and other chemicals. The world production of sodium carbonate now exceeds 30 million tonnes per year. Sodium carbonate is manufactured ingeniously using calcium carbonate, sodium chloride, carbon dioxide and ammonia in a continuous process, with the Solvay tower the centre of reactions. Sodium carbonate is made as economically as possible because both the carbon dioxide and the ammonia components are recycled continuously.

## Indirect uses of limestone

### Lime manufacture

When calcium carbonate is heated strongly it thermally dissociates (breaks up reversibly) to form calcium oxide (lime) and carbon dioxide.

$$\text{calcium carbonate} \rightleftharpoons \text{calcium oxide} + \text{carbon dioxide}$$
$$CaCO_3(s) \rightleftharpoons CaO(s) + CO_2(g)$$

This reaction can go in either direction, depending on the temperature and pressure used. This reaction is an important industrial process and takes place in a lime kiln (Figure 8.7). Two million tonnes of calcium oxide are produced in the UK every year.

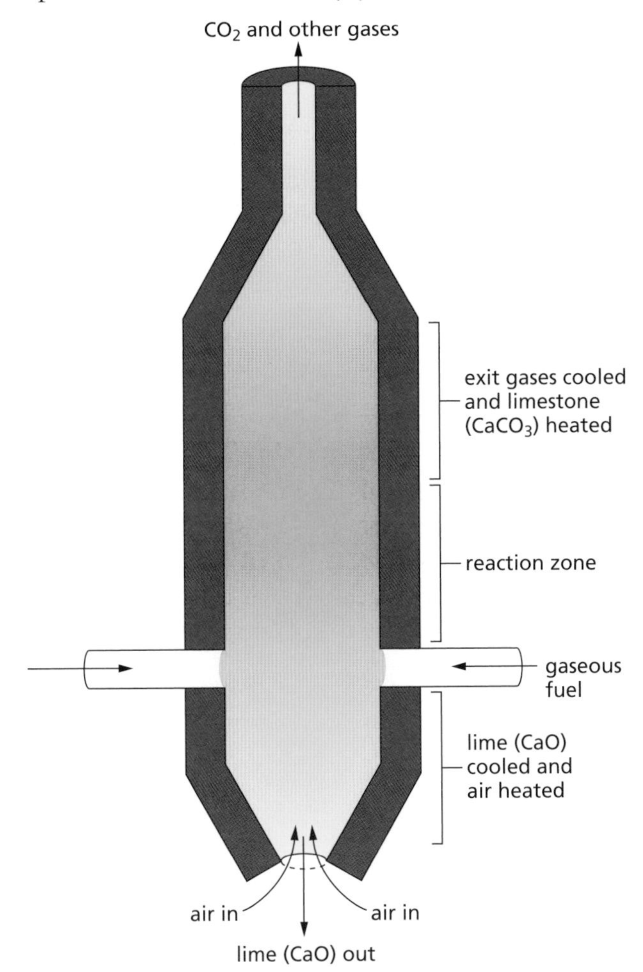


**Figure 8.7** The calcium oxide produced from this process is known as quicklime or lime and is used in large quantities in the manufacture of soda glass.

Calcium oxide (CaO) is a base and is still used by some farmers to spread on fields to neutralise soil acidity and to improve drainage of water through soils that contain large amounts of clay. It also has a use as a drying agent in industry. Soda glass is made by heating sand with soda (sodium carbonate, $Na_2CO_3$) and lime (Figure 8.8). For further discussion of glasses see Chapter 4, p. 66.

Large amounts of calcium oxide are also converted into calcium hydroxide ($Ca(OH)_2$) which is called **slaked lime**.

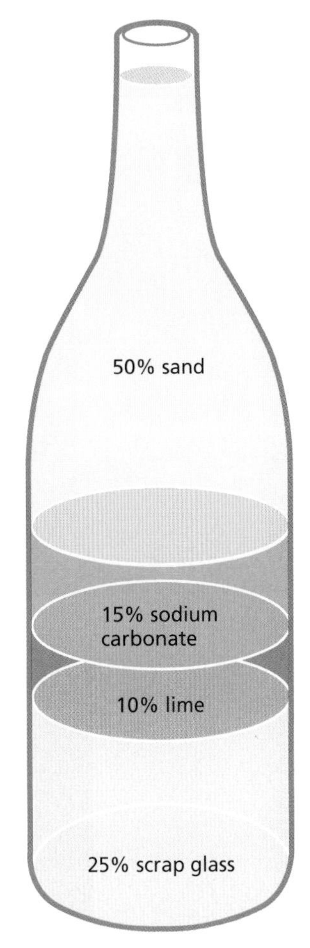


**Figure 8.8** The composition of glass.

### Manufacture of calcium hydroxide – slaked lime

Calcium hydroxide is a cheap industrial alkali (Figure 8.9). It is used in large quantities to make bleaching powder, by some farmers to reduce soil acidity, in the manufacture of whitewash, in glass manufacture and in water purification. Calcium hydroxide, in its white powder form, is produced by adding an equal amount of water to calcium oxide in a carefully controlled reaction. The control is needed because it is a very exothermic reaction.

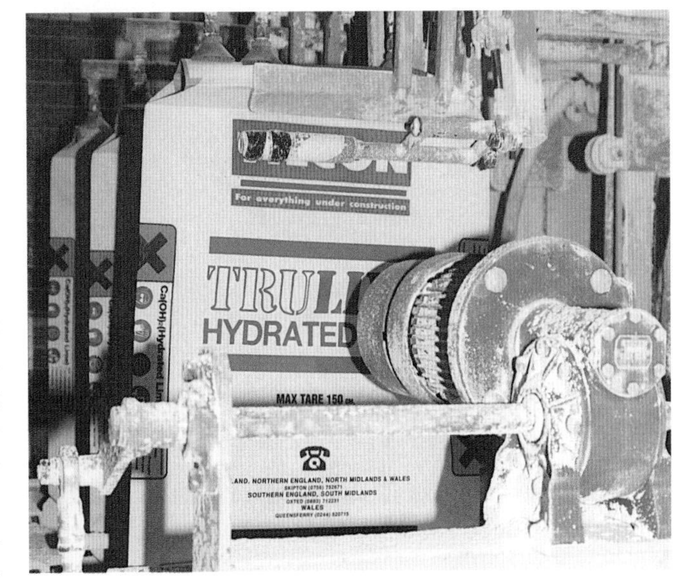


**Figure 8.9** Large-scale production of calcium hydroxide.

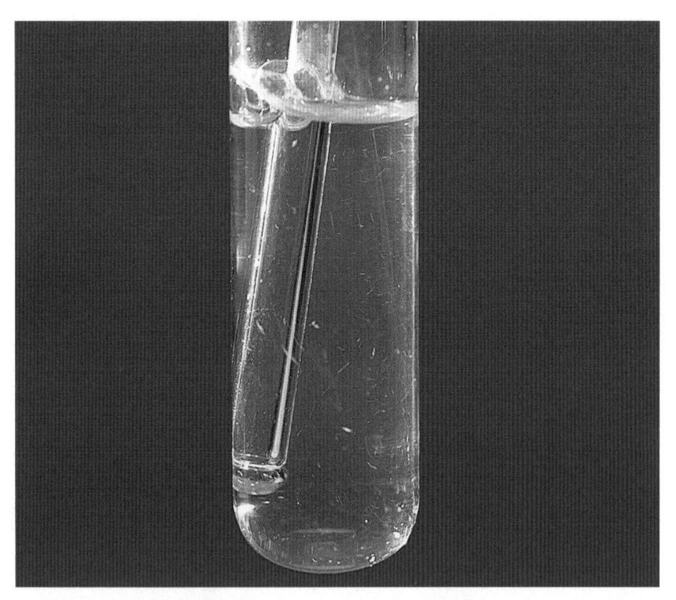

**a** At the start of the test.

**b** The limewater turns cloudy.

**c** If the $CO_2$ is bubbled for a longer length of time the white precipitate dissolves.

**Figure 8.10** Testing for carbon dioxide.

This process can be shown on the small scale in the laboratory by heating a lump of limestone very strongly to convert it to calcium oxide. Water can then be carefully added dropwise to the calcium oxide. An exothermic reaction takes place as the water and calcium oxide react together in this slaking process to form calcium hydroxide.

calcium oxide + water → calcium hydroxide

$$CaO(s) + H_2O(l) \rightarrow Ca(OH)_2(s)$$

A weak solution of calcium hydroxide in water is called **limewater**. It is used to test for carbon dioxide gas, as a white solid of calcium carbonate is formed if carbon dioxide gas is mixed with it (Figure 8.10):

calcium hydroxide + carbon dioxide → calcium carbonate + water

$$Ca(OH)_2(aq) + CO_2(g) \rightarrow CaCO_3(s) + H_2O(l)$$

If carbon dioxide is bubbled for a further length of time then the white precipitate of calcium carbonate dissolves and a solution of calcium hydrogencarbonate is produced:

calcium carbonate + carbon dioxide + water → calcium hydrogen-carbonate

$$CaCO_3(s) + CO_2(g) + H_2O(l) \rightarrow Ca(HCO_3)_2(aq)$$

Calcium hydroxide (slaked lime) is mixed with sand to give mortar. When it is mixed with water and allowed to set a strongly bonded material is formed, which is used to hold bricks together. The hardening of mortar takes place as the following reaction occurs.

calcium hydroxide + carbon dioxide → calcium carbonate + water

$$Ca(OH)_2(aq) + CO_2(g) \rightarrow CaCO_3(s) + H_2O(l)$$

## Questions

**1** Some suggested building alternatives to limestone are sandstone and granite. What would be the benefits of using either of these two materials instead of limestone? For further information, consult Chapter 17 as well as available Internet sites.

**2** Devise an experiment that you could carry out in the laboratory to establish whether or not powdered limestone is better at curing soil acidity than calcium oxide.

**3** Why is calcium oxide described as a 'base' in the reaction which occurs in question **2**?

# *Carbonates and hydrogencarbonates*

Carbonates form an important range of compounds. They are all salts of carbonic acid ($H_2CO_3$) and contain the carbonate ion ($CO_3^{2-}$). Many of them occur naturally in rock formations. For example, in addition to the calcium carbonate in limestone, chalk and marble, malachite is copper(II) carbonate and dolomite is magnesium carbonate (Figure 8.11).

**Figure 8.11** Three naturally occurring carbonates – dolomite (left) with magnesite in front and malachite (right).

## Properties of carbonates

- Most metal carbonates thermally decompose when heated to form the metal oxide and carbon dioxide gas. For example,

copper(II) carbonate → copper(II) oxide + carbon dioxide

$$CuCO_3(s) \rightarrow CuO(s) + CO_2(g)$$

Group 1 metal carbonates, except for lithium carbonate, do not dissociate on heating. It is generally found that the carbonates of the more reactive metals are much more difficult to dissociate than, for example, copper(II) carbonate.

- Carbonates are generally insoluble in water except for those of sodium, potassium and ammonium.
- Carbonates react with acids to form salts, carbon dioxide and water (Chapter 7, p. 101). For example, calcium carbonate reacts with dilute hydrochloric acid to form calcium chloride, carbon dioxide and water.

calcium carbonate + hydrochloric acid → calcium chloride + carbon dioxide + water

$$CaCO_3(s) + 2HCl(aq) \rightarrow CaCl_2(aq) + CO_2(g) + H_2O(l)$$

- This reaction is used in the laboratory preparation of carbon dioxide. It is also used as a test for a carbonate because the reaction produces carbon dioxide which causes effervescence and if bubbled through limewater turns it chalky white.

## Sodium carbonate – an important industrial chemical

Approximately 1 million tonnes of sodium carbonate are manufactured in the UK annually, mainly by the Solvay process (see p. 114). Sodium carbonate is used for glass manufacture, brine treatment and water purification, as well as in the manufacture of detergents and textiles (Figure 8.12).

**Figure 8.12** Sodium carbonate is an important chemical. It has many uses, for example in the manufacture of glass.

## Hydrogencarbonates

Almost all the hydrogencarbonates are known only in solution because they are generally too unstable to exist as solids. They contain the hydrogencarbonate ion ($HCO_3^-$).

Sodium hydrogencarbonate ($NaHCO_3$) is the most common solid hydrogencarbonate. It is found in indigestion remedies because it reacts with the excess stomach acid (HCl) which causes indigestion.

sodium hydrogencarbonate + hydrochloric acid → sodium chloride + water + carbon dioxide

$$NaHCO_3(s) + HCl(aq) \rightarrow NaCl(aq) + H_2O(l) + CO_2(g)$$

Sodium hydrogencarbonate is the substance which is put into 'plain' flour to make it into 'self-raising' flour. When heated, the sodium hydrogencarbonate in the flour decomposes to give carbon dioxide gas, which makes the bread or cakes rise.

sodium hydrogencarbonate → sodium carbonate + water + carbon dioxide

$$2NaHCO_3(s) \rightarrow Na_2CO_3(s) + H_2O(l) + CO_2(g)$$

Calcium hydrogencarbonate and magnesium hydrogencarbonate are two of the hydrogencarbonates found only in solution. They are the substances mainly responsible for the hardness in water.

## Hardness in water

Rainwater dissolves carbon dioxide as it falls through the atmosphere. A small fraction of this dissolved carbon dioxide reacts with the water to produce carbonic acid, which is a weak acid.

water + carbon dioxide ⇌ carbonic acid

$$H_2O(l) + CO_2(g) \rightleftharpoons H_2CO_3(aq)$$

As this solution passes over and through rocks containing limestone and dolomite, the weak acid in the rain attacks these rocks and very slowly dissolves them. The dissolved substances are called calcium and magnesium hydrogencarbonates.

calcium carbonate + carbonic acid → calcium hydrogencarbonate

$$CaCO_3(s) + H_2CO_3(aq) \rightarrow Ca(HCO_3)_2(aq)$$

Some of the rock strata may contain gypsum (calcium sulphate, $CaSO_4.2H_2O$), anhydrite ($CaSO_4$) or kieserite ($MgSO_4.H_2O$), which are very sparingly soluble in water. The presence of these dissolved substances causes the water to become **hard**. Hardness in water can be divided into two types – **temporary** and **permanent**. Temporary hardness is so called because it can be removed easily by boiling. Permanent hardness is so called because it is much more difficult to remove and certainly cannot be removed by boiling.

**Figure 8.13** Evaporating pure water leaves no deposit, while temporary or permanent hard water leaves white deposits behind.

Temporary hardness is caused by the presence of dissolved calcium or magnesium hydrogencarbonates. Permanent hardness is caused by the presence of dissolved calcium or magnesium sulphates. When water containing these substances is evaporated, a white solid deposit is left behind (Figure 8.13). These white deposits are calcium or magnesium sulphates and/or calcium carbonate. Calcium carbonate causes the 'furring' in kettles that occurs in hard water areas (Figure 8.14a). This furring may be removed by the addition of a dilute acid:

hydrogen ions + carbonate ion → carbon dioxide + water

$$2H^+(aq) + CO_3^{2-}(aq) \rightarrow CO_2(g) + H_2O(l)$$

Also, blockages in hot water pipes are caused by a similar process to the furring of kettles. A thick deposit of limescale builds up.

**a** The deposit in kettles ('furring') is caused by calcium carbonate from hard water.

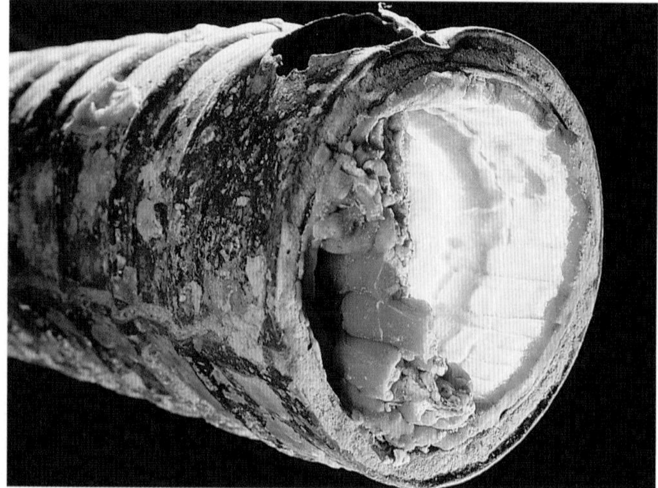

**b** The thick deposits of limescale in the pipe have resulted from heating hot water.

**Figure 8.14**

Stalactites and stalagmites are found in underground caverns in limestone areas. They are formed from the slow decomposition of water containing calcium or magnesium hydrogencarbonates (Figure 8.15).

calcium hydrogencarbonate → calcium carbonate + carbon dioxide + water

$$Ca(HCO_3)_2(aq) \rightarrow CaCO_3(s) + CO_2(g) + H_2O(l)$$

**Figure 8.15** Stalactites and stalagmites have formed over hundreds of thousands of years.

## Effect of hard water on soap

If you live in one of the hard water areas shown in Figure 8.16 then you may have noticed that it is difficult to make the soap lather. Instead, the water becomes cloudy. This cloudiness is caused by the presence of a solid material (a precipitate) formed by the reaction of the dissolved substances in the water with soap (sodium stearate). This white precipitate is known as scum (Figure 8.17).

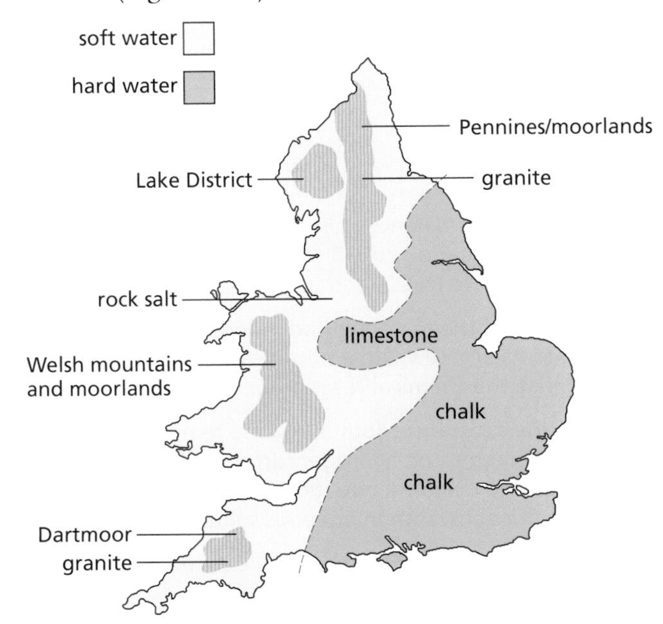


**Figure 8.16** The hard and the soft water areas in England and Wales.

**Figure 8.17** Soap and hard water form scum.

sodium stearate (soap) + calcium hydrogencarbonate → calcium stearate (scum) + sodium hydrogencarbonate

$$2NaSt(aq) + Ca(HCO_3)_2(aq) \rightarrow Ca(St)_2(s) + 2NaHCO_3(aq)$$

St = stearate $\quad NaSt = C_{17}H_{35}COO^-Na^+$

The amount of soap required to just produce a lather can be used to estimate the hardness in water.

To overcome the problem of scum formation, soapless detergents have been produced which do not produce a scum because they do not react with the substances in hard water. For further discussion of soapless detergents see Chapter 14, p. 210.

## Removal of hardness

Temporary hardness is easily removed from water by boiling. When heated the calcium hydrogencarbonate decomposes, producing insoluble calcium carbonate.

calcium hydrogencarbonate $\xrightarrow{heat}$ calcium carbonate + water + carbon dioxide

$$Ca(HCO_3)_2(aq) \rightarrow CaCO_3(s) + H_2O(l) + CO_2(g)$$

The substances in permanently hard water are not decomposed when heated and therefore cannot be removed by boiling. Both types of hardness can be removed by any of the following.

- **Addition of washing soda ($Na_2CO_3.10H_2O$) crystals.** In each case the calcium or magnesium ion, which actually causes the hardness, is removed as a precipitate and can, therefore, no longer cause hardness.

calcium ions (from hard water) + carbonate ions (from washing soda) → calcium carbonate

$$Ca^{2+}(aq) + CO_3^{2-}(aq) \rightarrow CaCO_3(s)$$

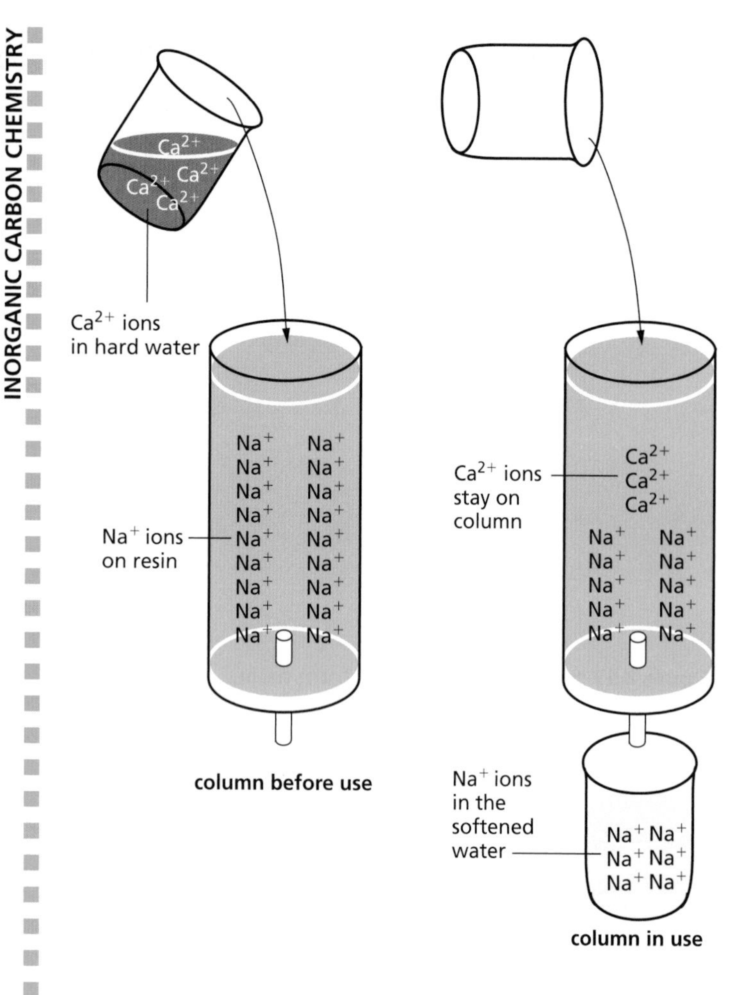


a The ion exchange process.

b Ion exchange resin.

**Figure 8.18**

- **Ion exchange.** The water is passed through a container filled with a suitable resin containing sodium ions. The calcium or magnesium ions causing the hardness are exchanged for the sodium ions in the resin (Figure 8.18).

$$\text{calcium ion} + \text{sodium–resin} \rightarrow \text{calcium–resin} + \text{sodium ion}$$

$$Ca^{2+}(aq) + Na_2{-}R(s) \rightarrow Ca^{2+}{-}R(s) + 2Na^{+}(aq)$$

When all the sodium ions have been removed from the resin, it can be regenerated by pouring a solution of a suitable sodium salt through it.

- **Distillation.** The water is distilled away from the dissolved substances. This method, however, is far too expensive to be used on a large scale.

## Advantages and disadvantages of hard water

As we mentioned earlier, there are some problems associated with hard water; however, there are advantages as well. Both are detailed in Table 8.1.

**Table 8.1** Disadvantages and advantages of hard water.

| Disadvantages | Advantages |
|---|---|
| Wastes soap | Has a nice taste |
| Causes kettles to fur | Calcium ions in hard water are required by the body for bones and teeth |
| Can cause hot water pipes to block | Coats lead pipes with a thin layer of lead(II) sulphate or lead(II) carbonate and cuts down the possibility of lead poisoning |
| Can spoil the finish of some fabrics | Is good for brewing beer |

## Questions

1 Write word and balanced chemical equations to show the effect of heat on cobalt carbonate ($CoCO_3$) and nickel carbonate ($NiCO_3$).

2 Make a list of the main methods of softening hard water. In each case write a suitable equation(s) to summarise the chemical reactions involved.

3 One of the substances found in some temporary hard waters is magnesium hydrogencarbonate. Write word and balanced chemical equations to show the effect of heat on this substance in aqueous solution.

4 Write word and balanced chemical equations for the reaction of dilute hydrochloric acid with:
 a copper(II) carbonate
 b sodium carbonate.

## Carbon dioxide

Carbon forms two oxides – carbon monoxide (CO) and carbon dioxide ($CO_2$). Carbon dioxide is the more important of the two, and in industry large amounts of carbon dioxide are obtained from the liquefaction of air. Air contains approximately 0.03% by volume of carbon dioxide. This value has remained almost constant for a long period of time and is maintained via the **carbon cycle** (Figure 8.19). However, scientists have recently detected a slight increase in the amount of carbon dioxide in the atmosphere.

Carbon dioxide is produced by burning fossil fuels. It is also produced by all living organisms through **aerobic respiration**. Animals take in oxygen and breathe out carbon dioxide.

$$\text{glucose} + \text{oxygen} \rightarrow \text{carbon dioxide} + \text{water} + \text{energy}$$
$$C_6H_{12}O_6(aq) + 6O_2(g) \rightarrow 6CO_2(g) + 6H_2O(l)$$

Carbon dioxide is taken in by plants through their leaves and used together with water, taken in through their roots, to synthesise sugars. This is the process of **photosynthesis**, and it takes place only in sunlight and only in green leaves, as they contain **chlorophyll** (the green pigment) which catalyses the process.

$$\text{carbon dioxide} + \text{water} \xrightarrow[\text{chlorophyll}]{\text{sunlight}} \text{glucose} + \text{oxygen}$$
$$6CO_2(g) + 6H_2O(l) \longrightarrow C_6H_{12}O_6(aq) + 6O_2(g)$$

This cycle has continued in this manner for millions of years. However, scientists have detected an imbalance in the carbon cycle due to the increase in the amount of carbon dioxide produced through burning fossil fuels and the deforestation of large areas of tropical rain forest. The Earth's climate is affected by the levels of carbon dioxide (and water vapour) in the atmosphere. If the amount of carbon dioxide, in particular, builds up in the air, it is thought that the average temperature of the Earth will rise. This effect is known as the **greenhouse effect** (Figure 8.20).

Some energy from the Sun is absorbed by the Earth and its atmosphere. The remainder is reflected back into space. The energy that is absorbed helps to heat up the Earth. The Earth radiates some heat energy back into space but the 'greenhouse gases', including carbon dioxide, prevent it from escaping. This effect is similar to that observed in a greenhouse where sunlight (visible/ultraviolet radiation) enters through the glass panes but heat (infrared radiation) has difficulty escaping through the glass.

The long-term effect of the higher temperatures will be the gradual melting of ice caps and consequent flooding in low-lying areas of the Earth. There will also be changes in the weather patterns which would affect agriculture worldwide.

These problems have been recognised by nations worldwide. Recent agreements under the Kyoto Accord between nations mean that there will be some reduction in the amount of carbon dioxide (and other greenhouse gases) produced over the next few years. However, there is still a long way to go.

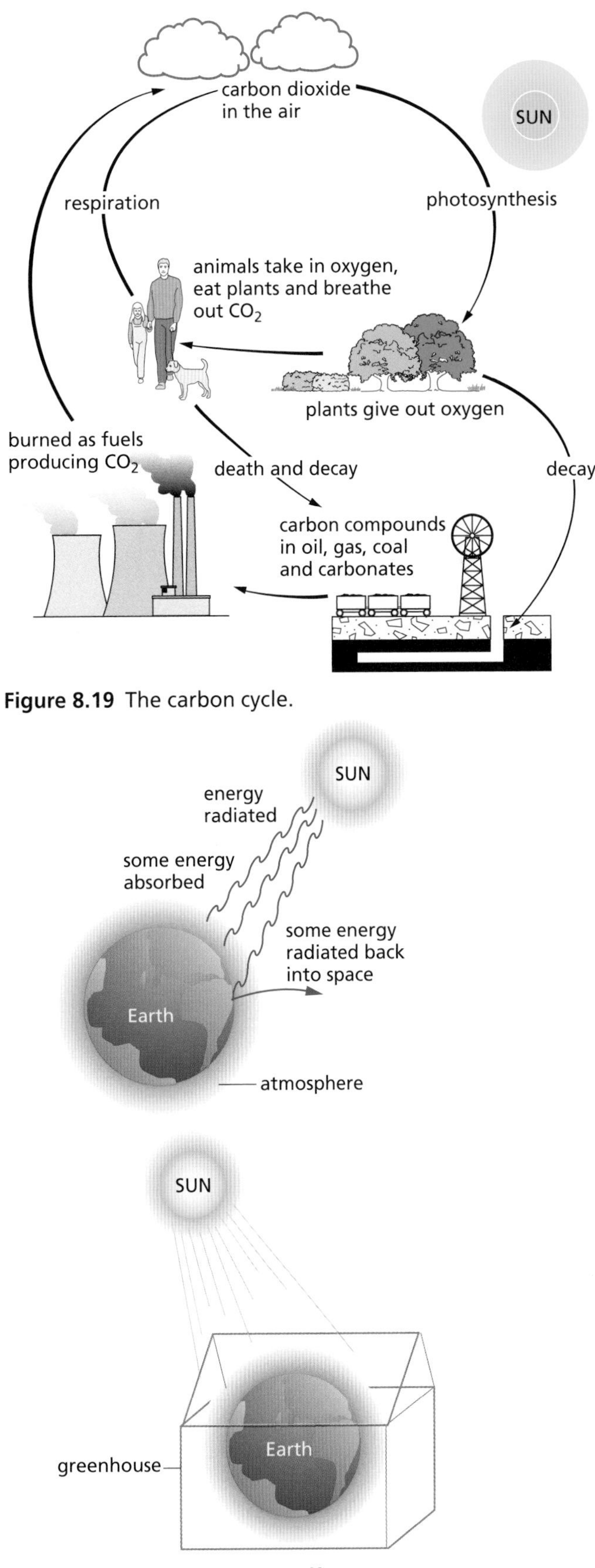


**Figure 8.19** The carbon cycle.



**Figure 8.20** The greenhouse effect.

## Uses of carbon dioxide

Carbon dioxide has some important uses.

- **Carbonated drinks.** Large quantities are used to make soda and mineral waters as well as beer. The carbon dioxide gas is bubbled into the liquid under pressure, which increases its solubility.
- **Fire extinguishers.** It is used in extinguishers for use on electrical fires. Carbon dioxide is denser than air and forms a layer around the burning material. It covers the fire and starves it of oxygen. Carbon dioxide does not burn and so the fire is put out (Figure 8.21).

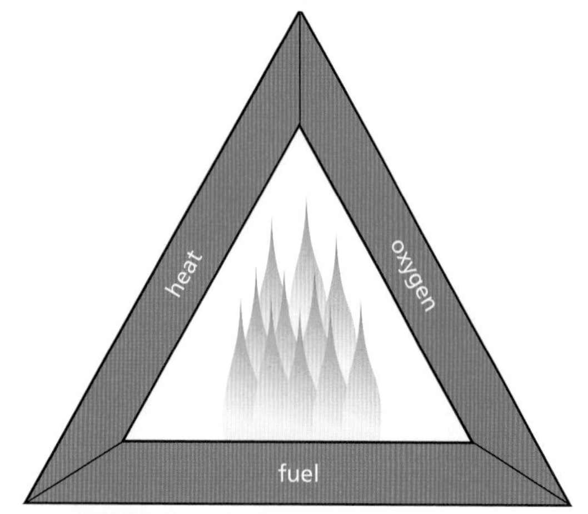


**a** Any fire needs fuel, oxygen and heat. If any of these is removed, the fire triangle is destroyed and the fire will be extinguished.

**b** Carbon dioxide fire extinguishers starve the fire of oxygen. They are used mainly for electrical fires.

**Figure 8.21**

- **Refrigerants.** Solid carbon dioxide (dry ice) is used for refrigerating ice cream, meat and soft fruits. It is used for this purpose because it is colder than ice and it sublimes (p. 5), and so it does not pass through a potentially damaging liquid stage.
- **Special effects.** Carbon dioxide is used to create the 'smoke' effect you may see at pop concerts and on television. Dry ice is placed in boiling water and it forms thick clouds of white 'smoke'. It stays close to the floor due to the fact that carbon dioxide is denser than air.
- **Heat transfer agents.** Carbon dioxide gas is used for transferring heat in some nuclear power stations.

## Laboratory preparation of carbon dioxide gas

In the laboratory the gas is made by pouring dilute hydrochloric acid on to marble chips ($CaCO_3$).

calcium carbonate + hydrochloric acid → calcium chloride + water + carbon dioxide

$$CaCO_3(s) + 2HCl(aq) \rightarrow CaCl_2(aq) + H_2O(l) + CO_2(g)$$

A suitable apparatus for preparing carbon dioxide is shown in Figure 8.22.

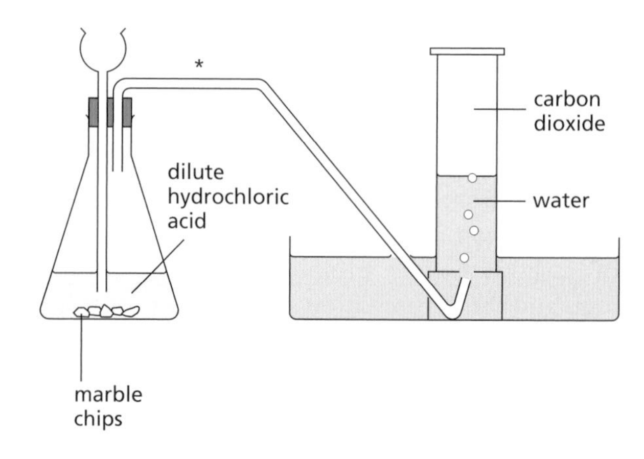


If dry gas is required, then it is passed through concentrated sulphuric acid (to dry it) and then collected as shown below

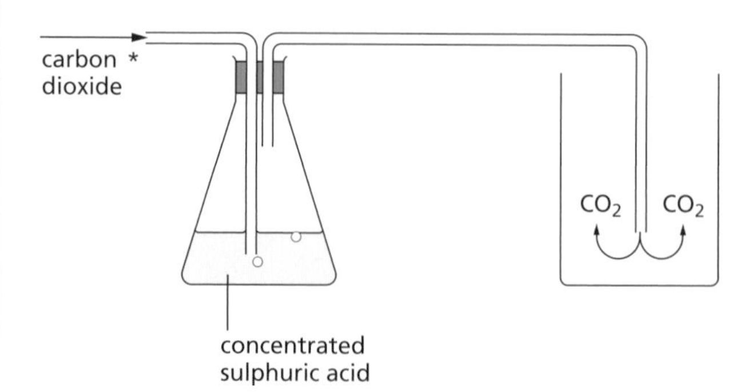


**Figure 8.22** Preparation and collection of carbon dioxide gas.

## Properties of carbon dioxide gas

### Physical properties

Carbon dioxide is:

- a colourless gas
- sparingly soluble in water
- denser than air.

### Chemical properties

- When bubbled into water it dissolves slightly and some of the carbon dioxide reacts, forming a solution of the weak acid carbonic acid which shows a pH of 4 or 5.

$$\text{water} + \text{carbon dioxide} \rightleftharpoons \text{carbonic acid}$$
$$H_2O(l) + CO_2(g) \rightleftharpoons H_2CO_3(aq)$$

- It will support the combustion only of strongly burning substances such as magnesium. This burning reactive metal decomposes the carbon dioxide to provide oxygen for its continued burning in the gas. This reaction is accompanied by much crackling (Figure 8.23).

$$\text{magnesium} + \text{carbon dioxide} \rightarrow \text{magnesium oxide} + \text{carbon}$$
$$2Mg(s) + CO_2(g) \rightarrow 2MgO(s) + C(s)$$

**Figure 8.23** When magnesium burns in carbon dioxide gas, magnesium oxide (white) and carbon (black) are produced.

This reaction is a good example of a **redox** process (Chapter 2, p. 17). Which of these substances has been oxidised and which has been reduced? Which of these substances is the oxidising agent and which is the reducing agent?

- When carbon dioxide is bubbled through limewater (calcium hydroxide solution), a white precipitate is formed. This white solid is calcium carbonate ($CaCO_3$). This reaction is used as a test to show that a gas is carbon dioxide.

$$\text{carbon dioxide} + \text{calcium hydroxide} \rightarrow \text{calcium carbonate} + \text{water}$$
$$CO_2(g) + Ca(OH)_2(aq) \rightarrow CaCO_3(s) + H_2O(l)$$

If carbon dioxide is bubbled through this solution continuously then it will eventually become clear. This is because of the formation of soluble calcium hydrogencarbonate solution.

$$\text{calcium carbonate} + \text{water} + \text{carbon dioxide} \rightarrow \text{calcium hydrogencarbonate}$$
$$CaCO_3(s) + H_2O(l) + CO_2(g) \rightarrow Ca(HCO_3)_2(aq)$$

- Carbon dioxide reacts with strong alkalis, such as sodium hydroxide, to form carbonates. A solution of sodium hydroxide can be used to absorb carbon dioxide from the air. If excess carbon dioxide is bubbled through a solution of an alkali then a white precipitate may be obtained. When using sodium hydroxide this is due to the formation and precipitation of the sodium hydrogencarbonate:

$$\text{sodium carbonate} + \text{water} + \text{carbon dioxide} \rightarrow \text{sodium hydrogencarbonate}$$
$$Na_2CO_3(aq) + H_2O(l) + CO_2(g) \rightarrow 2Na(HCO_3)_2(s)$$

## Questions

1. List the important uses of carbon dioxide and for each use you have given in your answer explain why carbon dioxide is used.
2. Describe an experiment that you could carry out to show that carbonic acid is a weak acid.
3. Why is it not possible to use dilute sulphuric acid to make carbon dioxide from limestone in the laboratory?
4. When carbon dioxide is 'poured' from a gas jar on to a burning candle the candle goes out. What properties of carbon dioxide does this experiment show?

## *Checklist*

**After studying Chapter 8 you should know and understand the following terms.**

- **Carbonate** A salt of carbonic acid containing the carbonate ion, $CO_3^{-2}$, for example $CuCO_3$.
- **Carbon dioxide** A colourless, odourless gas, soluble in water, producing a weak acid called carbonic acid. It makes up 0.03% of air. It is produced by respiration in all living things and by the burning of fossil fuels. It is consumed by plants in photosynthesis.
- **Greenhouse effect** The absorption of reflected infrared radiation from the Earth by gases in the atmosphere such as carbon dioxide (a greenhouse gas) leading to atmospheric warming.
- **Hardness of water** This is caused by the presence of calcium (or magnesium) ions in water, which form a 'scum' with soap and prevent the formation of a lather. There are two types of hardness:
  - temporary hardness – caused by the presence of dissolved calcium (or magnesium) hydrogencarbonate
  - permanent hardness – this results mainly from dissolved calcium (or magnesium) sulphate.
- **Lime** A white solid known chemically as calcium oxide (CaO). It is produced by heating limestone. It is used to counteract soil acidity and to manufacture calcium hydroxide (slaked lime). It is also used as a drying agent in industry.
- **Limestone** A form of calcium carbonate ($CaCO_3$). Other forms include chalk, calcite and marble. It is used directly to neutralise soil acidity and in the manufacture of iron and steel, glass, cement, concrete, sodium carbonate and lime.
- **Photosynthesis** The chemical process by which green plants synthesise their carbon compounds from atmospheric carbon dioxide using light as the energy source and chlorophyll as the catalyst.
- **Removal of hardness** Temporary hardness is removed by boiling. Both temporary and permanent hardness are removed by:
  - addition of washing soda (sodium carbonate)
  - ion exchange
  - distillation.
- **Thermal decomposition** The breakdown of a substance under the influence of heat.

# Inorganic carbon chemistry

# *Additional questions*

**1** This question is about the limestone cycle.

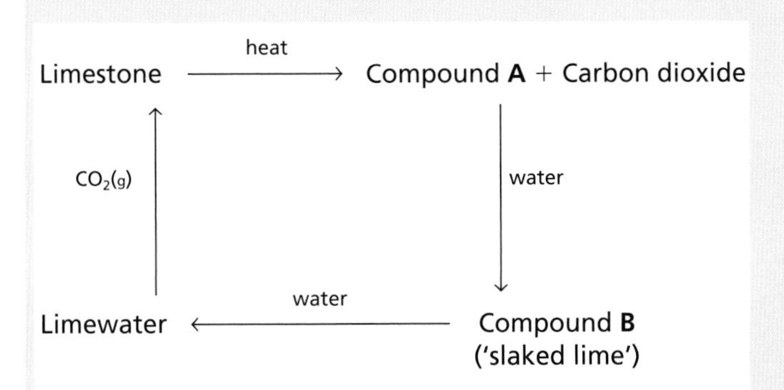


**a** Name and give the formula of:
(i) compound **A**
(ii) compound **B**.

**b** Write balanced chemical equations for the formation of both compounds **A** and **B**.

**c** Name and give the symbol/formula for the ions present in limewater.

**d** Describe with the aid of a balanced chemical equation what happens if carbon dioxide is bubbled through limewater until there is no further change.

**2** The diagram shown below is a simplified version of the carbon cycle.

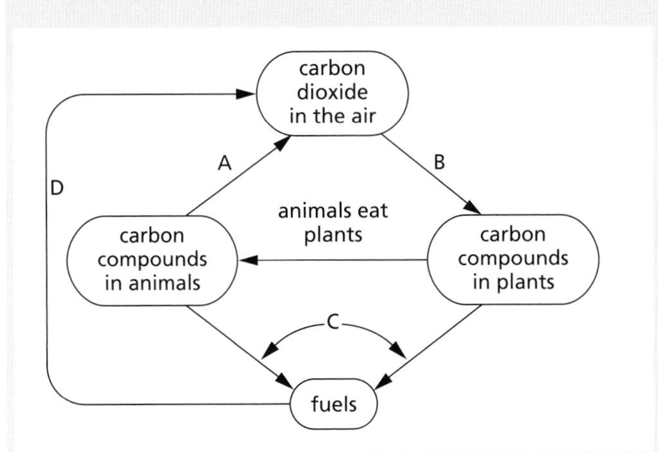


**a** Name the processes **A**, **B**, **C** and **D**.

**b** Write balanced chemical equations to represent the processes taking place in **A** and **B**.

**c** There are a number of fuels which could be placed in the 'fuels' box. Name three such fuels.

**d** 'The air contains approximately 0.03% of carbon dioxide and this amount is almost constant.' Explain why this is a true statement.

**e** (i) What is the 'greenhouse effect'?
(ii) Discuss the possible consequences of the greenhouse effect if nothing is done to counteract it.

**3** Limestone is an important raw material used in many different industries.

**a** One of the properties of limestone is that it reacts with acids.
(i) Why do farmers spread powdered limestone on their fields?
(ii) How can buildings made of limestone be affected by 'acid rain'?
(iii) Write an ionic equation which would represent the reactions taking place in both (i) and (ii).

**b** Limestone is used in the manufacture of iron (Chapter 9, pp. 133–4).
(i) Why is it added to the blast furnace along with coke and haematite?
(ii) Write chemical equations for the reactions it is involved in, in this process.

**c** (i) Name a building material made by heating a mixture of limestone and clay in a rotary kiln.
(ii) What substance is the dry product produced in the rotary kiln added to?
Explain why this substance is added.

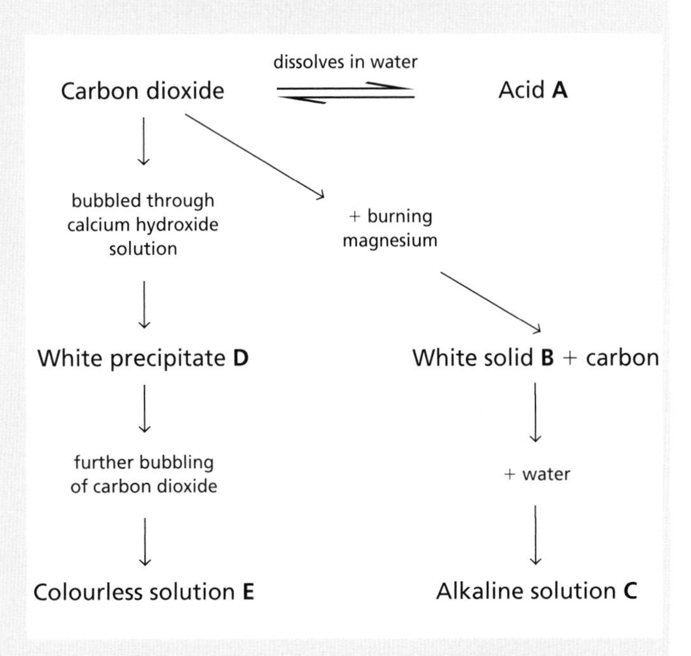


**4** **a** Name and give the formulae of substances **A** to **E**.

**b** Write balanced chemical equations for the reactions in which compounds **B**, **C** and **E** were formed.

**c** Where would you expect to find acid **A**?

**d** Universal indicator solution was added to solution **C**. What colour did it go?

**e** Upon addition of dilute hydrochloric acid to solution **C**, a neutralisation reaction took place.
(i) Write a balanced chemical equation for the reaction taking place.
(ii) Name the salt produced in this reaction.

**5** The following question is about carbon dioxide.

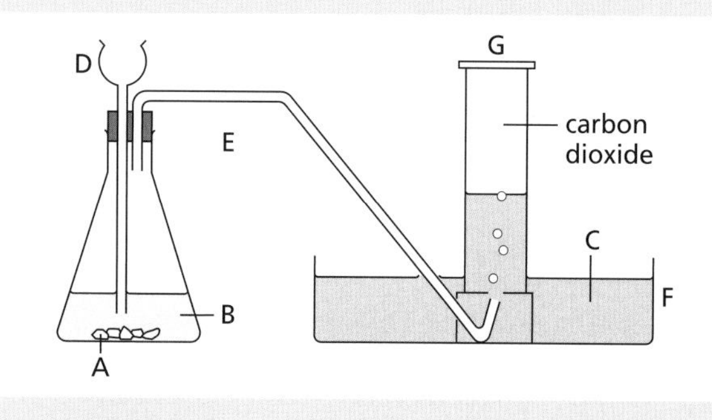


**a** Name and give the formula of each of the substances **A, B** and **C**.

**b** Identify by name the different pieces of apparatus **D, E, F** and **G**.

**c** Draw and label the apparatus that should be used if a dry sample of carbon dioxide gas was required.

**d** When a gas jar containing carbon dioxide is held over a burning wooden splint and the cover removed, the flame goes out. State two properties of carbon dioxide illustrated by this observation.

**e** Carbon dioxide is also produced when zinc carbonate is heated strongly.

(i) Write a balanced chemical equation for the reaction taking place.

(ii) Name the process which is taking place as the zinc carbonate is heated.

(iii) Calculate the volume of carbon dioxide that would be produced (measured at room temperature and pressure (rtp)) if 12.5 g of zinc carbonate were heated strongly.

(One mole of any gas occupies 24 $dm^3$ at rtp.
$A_r$: Zn = 65; C = 12; O = 16)

**6** Explain the following.

**a** Industry normally requires water which has been softened.

**b** Hard water is good for the promotion of healthy bones and teeth.

**c** Hard water causes kettles to fur. This 'fur' can be removed by using a dilute acid.

**d** Hard water wastes soap.

**e** Hard water can coat lead pipes and reduce the possibility of lead poisoning.

**7** The results of testing five samples of water from different parts of the country are shown in the table below. The soap solution was gradually added to 25 $cm^3$ of each sample of water with shaking until a permanent lather (one which lasts for 20 seconds) was obtained.

| Water sample/25 $cm^3$ | Volume of soap solution added/$cm^3$ | |
|---|---|---|
| | **Before boiling** | **After boiling** |
| A | 12 | 1 |
| B | 13 | 6 |
| C | 11 | 11 |
| D | 14 | 3 |
| E | 16 | 16 |

**a** (i) Which samples are permanently hard?

(ii) Which samples are temporarily hard?

(iii) Which sample contains both temporary and permanent hardness?

**b** Name a compound which could be present in sample **D** but not in sample **E**.

**c** Name a compound which could be present in sample **E** but not in sample **D**.

**d** Explain how the compound you have named in **c** gets into the water.

**e** Sample **E** was distilled. The water collected was tested with soap solution. What volume of soap solution might you expect to be added to produce a permanent lather? Comment on your answer.

**8** Lime (calcium oxide) is produced in very large quantities in a lime kiln. The equation for the reaction is:

$$CaCO_3(s) \rightarrow CaO(s) + CO_2(g)$$

**a** How much limestone would be needed to produce 61.60 tonnes of lime?
($A_r$: C = 12; O = 16; Ca = 40)

**b** Why is the carbon dioxide swept out of the lime kiln?

**c** Give three uses of lime.

**d** (i) What problems are associated with the large-scale quarrying of limestone?

(ii) What steps have been taken to overcome or reduce the problems you have outlined in (i)?

# 9 Metal extraction and chemical reactivity



You have already seen in Chapter 2, p. 14, that metals usually have similar physical properties. However, they differ in other ways. Look closely at the three photographs in Figure 9.1.

Sodium is soft and reacts violently with both air and water. Iron also reacts with air and water but much more slowly, forming rust. Gold, however, remains totally unchanged after many hundreds of years. Sodium is said to be more reactive than iron and, in turn, iron is said to be more reactive than gold.

**a** Sodium burning in air/oxygen.

**b** Iron rusts when left unprotected.

**c** Gold is used in leaf form on this giant Buddha as it is unreactive.

**Figure 9.1**

## Metal reactions

By carrying out reactions in the laboratory with other metals and air, water and dilute acid, it is possible to produce an order of reactivity of the metals.

### With acid

Look closely at the photograph in Figure 9.2 showing magnesium metal reacting with dilute hydrochloric acid. You will notice effervescence, which is caused by bubbles of hydrogen gas being formed as the reaction between the two substances proceeds. The other product of this reaction is the salt, magnesium chloride.

magnesium + hydrochloric acid → magnesium chloride + hydrogen

$$Mg(s) + 2HCl(aq) \rightarrow MgCl_2(aq) + H_2(g)$$

Figure 9.2 Effervescence occurs when magnesium is put into acid.

If a metal reacts with dilute hydrochloric acid then hydrogen and the metal chloride are produced.

If similar reactions are carried out using other metals with acid, an order of reactivity can be produced. This is known as a **reactivity series**. An order of reactivity, giving the most reactive metal first, using results from experiments with dilute acid, is shown in Table 9.1. The table also shows how the metals react with air/oxygen and water/steam, and, in addition, the ease of extraction of the metal.

**Table 9.1** Order of reactivity of metals.

| Reactivity series | Reaction with dilute acid | Reaction with air/oxygen | Reaction with water | Ease of extraction |
|---|---|---|---|---|
| Potassium (K)<br>Sodium (Na) | Produce $H_2$ with decreasing vigour ↓ | Burn very brightly and vigorously | Produce $H_2$ with decreasing vigour with cold water | Difficult to extract |
| Calcium (Ca)<br>Magnesium (Mg) | | Burn to form an oxide with decreasing vigour ↓ | React with steam with decreasing vigour ↓ | Easier to extract ↓ |
| Aluminium (Al*) | | | | |
| Zinc (Zn) | | | | |
| Iron (Fe) | | | | |
| Lead (Pb) | | React slowly to form the oxide | Do not react with cold water or steam ↓ | |
| Copper (Cu) | Do not react with dilute acids ↓ | | | |
| Silver (Ag) | | Do not react ↓ | | Found as the element (native) ↓ |
| Gold (Au) | | | | |
| Platinum (Pt) | | | | |

↑ Increasing reactivity of metal

* Because aluminium reacts so readily with the oxygen in the air, a protective oxide layer is formed on its surface. This often prevents any further reaction and disguises aluminium's true reactivity. This gives us the use of a light and strong metal.

## With air/oxygen

Many metals react directly with oxygen to form oxides. For example, magnesium burns brightly in oxygen to form the white powder magnesium oxide.

$$\text{magnesium} + \text{oxygen} \rightarrow \text{magnesium oxide}$$
$$2Mg(s) + O_2(g) \rightarrow 2MgO(s)$$

## With water/steam

Reactive metals such as potassium, sodium and calcium react with cold water to produce the metal hydroxide and hydrogen gas. For example, the reaction of sodium with water produces sodium hydroxide and hydrogen.

$$\text{sodium} + \text{water} \rightarrow \text{sodium hydroxide} + \text{hydrogen}$$
$$2Na(s) + 2H_2O(l) \rightarrow 2NaOH(aq) + H_2(g)$$

The moderately reactive metals, magnesium, zinc and iron, react slowly with water. They will, however, react more rapidly with steam (Figure 9.3). In their reaction with steam, the metal oxide and hydrogen are formed. For example, magnesium produces magnesium oxide and hydrogen gas.

$$\text{magnesium} + \text{steam} \rightarrow \text{magnesium oxide} + \text{hydrogen}$$
$$Mg(s) + H_2O(g) \rightarrow MgO(s) + H_2(g)$$

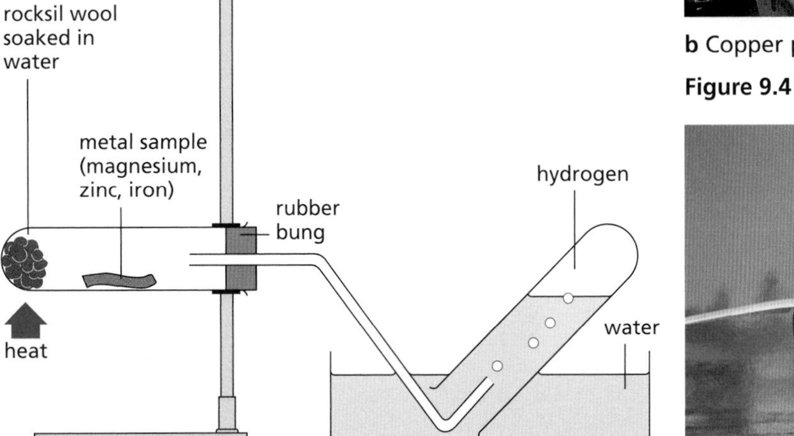


**Figure 9.3** Apparatus used to investigate how metals such as magnesium react with steam.

Generally, it is the unreactive metals that we find the most uses for; for example, the metals iron and copper can be found in many everyday objects (Figure 9.4). However, magnesium is one of the metals used in the construction of Concorde (Figure 9.5).

Both sodium and potassium are so reactive that they have to be stored under oil to prevent them from coming into contact with water or air. However, because they have low melting points and are good conductors of heat, they are used as coolants for nuclear reactors.

**a** This stove is made of iron.

**b** Copper pots.

**Figure 9.4**

**Figure 9.5** Concorde is made of an alloy which contains magnesium and aluminium.

## *Questions*

**1** Write balanced chemical equations for the reactions between:
- **a** iron and dilute hydrochloric acid
- **b** zinc and oxygen
- **c** calcium and water.

**2** Make a list of six things you have in your house made from copper or iron.
Give a use for each of the other unreactive metals shown in the reactivity series.

## Using the reactivity series

What predictions can be made using the reactivity series? It is useful in predicting how metals react.

### Competition reactions in the solid state

If a more reactive metal is heated with the oxide of a less reactive metal, then it will remove the oxygen from it (as the oxide anion). You can see from the reactivity series that iron is less reactive than aluminium (p.128). If iron(III) oxide is mixed with aluminium and the mixture is heated using a magnesium fuse (Figure 9.6), a very violent reaction occurs as the competition between the aluminium and the iron for the oxygen takes place.

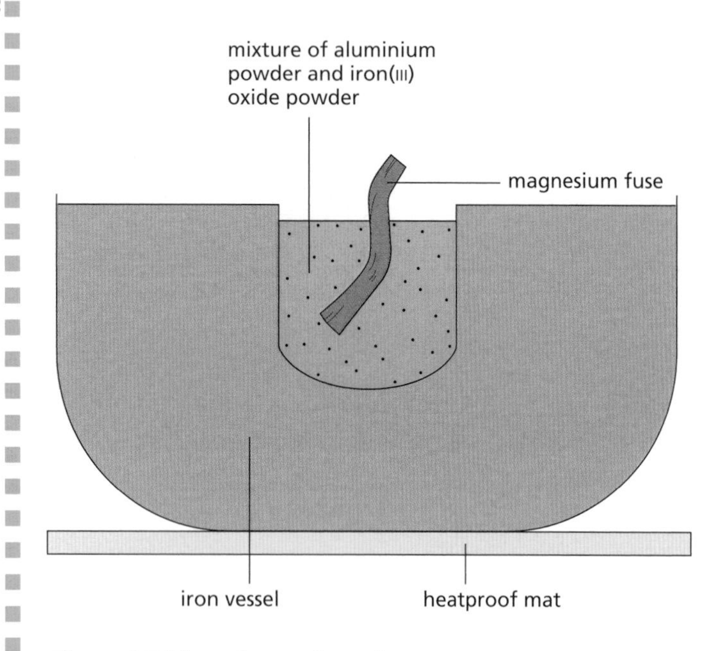


**Figure 9.6** Thermit reaction mixture apparatus.

The aluminium, being the more reactive metal, takes the oxygen from the less reactive iron. It is a very exothermic reaction. When the reaction is over, a solid lump of iron is left along with a lot of white aluminium oxide powder.

$$\text{iron(III) oxide} + \text{aluminium} \xrightarrow{\text{heat}} \text{aluminium oxide} + \text{iron}$$

$$Fe_2O_3(s) + 2Al(s) \rightarrow Al_2O_3(s) + 2Fe(s)$$

This is a redox reaction (see Chapter 2, p. 17 for a discussion of this type of reaction).

This particular reaction is known as the **Thermit reaction** (Figure 9.7). Since large amounts of heat are given out and the iron is formed in a molten state, this reaction is used to weld together damaged railway lines. It is also used in incendiary bombs (Figure 9.8).

Some metals, such as the transition metals chromium and titanium, are prepared from their oxides using this type of competition reaction.

**Figure 9.7** The Thermit reaction is used to weld damaged railway lines.

**Figure 9.8** Incendiary bombs were used during World War II.

### Competition reactions in aqueous solutions

In another reaction, metals compete with each other for other anions. This type of reaction is known as a **displacement reaction**. As in the previous type of competitive reaction, the reactivity series can be used to predict which of the metals will 'win'.

In a displacement reaction, a more reactive metal will displace a less reactive metal from a solution of its salt. Zinc is above copper in the reactivity series. Figure 9.9 shows what happens when a piece of zinc metal is left to stand in a solution of copper(II) nitrate.

The copper(II) nitrate slowly loses its blue colour as the zinc continues to displace the copper from the solution and eventually becomes colourless zinc nitrate.

zinc + copper(II) nitrate → zinc nitrate + copper

$$Zn(s) + Cu(NO_3)_2(aq) \rightarrow Zn(NO_3)_2(aq) + Cu(s)$$

The ionic equation for this reaction is:

zinc + copper ions → zinc ions + copper

$$Zn(s) + Cu^{2+}(aq) \rightarrow Zn^{2+}(aq) + Cu(s)$$

This is also a redox reaction involving the transfer of two electrons from the zinc metal to the copper ions. The zinc is oxidised to zinc ions in aqueous solution, while the copper ions are reduced. (See Chapter 6, p. 82, for a discussion of oxidation and reduction in terms of electron transfer.) It is possible to confirm the reactivity series for metals using competition reactions of the types discussed in this section.

**Figure 9.9** Zinc displaces copper.

## Questions

1 Predict whether or not the following reactions will take place:
   a magnesium + copper(II) oxide
   b iron + aluminium oxide
   c calcium + magnesium oxide.
   Complete the word equations, and write balanced chemical and ionic equations for those reactions which do take place.

2 Predict whether or not the following reactions will take place:
   a magnesium + calcium nitrate solution
   b iron + copper(II) nitrate solution
   c copper + silver nitrate solution.
   Complete the word equations, and write balanced chemical and ionic equations for those reactions which do take place.

## *Identifying metal ions*

When an alkali dissolves in water, it produces hydroxide ions. It is known that most metal hydroxides are insoluble. So if hydroxide ions from a solution of an alkali are added to a solution of a metal salt, an insoluble, often coloured, metal hydroxide is precipitated from solution (Figure 9.10).

Let's take the example of iron(III) chloride with sodium hydroxide solution:

iron(III) chloride + sodium hydroxide → iron(III) hydroxide + sodium chloride

$$FeCl_3(aq) + 3NaOH(aq) \rightarrow Fe(OH)_3(s) + 3NaCl(aq)$$

The ionic equation for this reaction is:

iron(III) ions + hydroxide ions → iron(III) hydroxide

$$Fe^{3+}(aq) + 3OH^-(aq) \rightarrow Fe(OH)_3(s)$$

**Figure 9.10** **a** Iron(III) hydroxide is precipitated. **b** Copper(II) hydroxide is precipitated.

Table 9.2 shows some of the colours of insoluble metal hydroxides. The colours of these insoluble metal hydroxides can be used to identify the metal cations present in solution.

**Table 9.2** Some of the colours of insoluble metal hydroxides.

| Name | Formula | Colour of hydroxide |
|---|---|---|
| Aluminium hydroxide | $Al(OH)_3$ | White |
| Calcium hydroxide | $Ca(OH)_2$ | White |
| Copper(II) hydroxide | $Cu(OH)_2$ | Blue |
| Iron(II) hydroxide | $Fe(OH)_2$ | Green |
| Iron(III) hydroxide | $Fe(OH)_3$ | Brown (rust) |
| Zinc hydroxide | $Zn(OH)_2$ | White |

## Amphoteric hydroxides

The hydroxides of metals are basic and they react with acids to form salts (Chapter 7, p. 99). The hydroxides of some metals, however, will also react with strong bases, such as sodium hydroxide, to form soluble salts. Hydroxides of this type are said to be **amphoteric**. For example,

zinc hydroxide + hydrochloric acid → zinc chloride + water

$$Zn(OH)_2(aq) + 2HCl(aq) \rightarrow ZnCl_2(aq) + 2H_2O(l)$$

and

zinc hydroxide + sodium hydroxide → sodium zincate

$$Zn(OH)_2(aq) + 2NaOH(aq) \rightarrow Na_2Zn(OH)_4(aq)$$

Other amphoteric hydroxides are lead hydroxide ($Pb(OH)_2$) and aluminium hydroxide ($Al(OH)_3$). We can use this sort of behaviour to help identify metal cations, as their hydroxides are soluble in strong bases.

Both aluminium and zinc metals will also react readily with moderately concentrated acids and alkalis. For example:

zinc + hydrochloric acid → zinc chloride + hydrogen

$$Zn(s) + 2HCl(aq) \rightarrow ZnCl_2(aq) + H_2(g)$$

zinc + sodium hydroxide + water → sodium zincate + hydrogen

$$Zn(s) + 2NaOH(aq) + 2H_2O(l) \rightarrow Na_2Zn(OH)_4(aq) + H_2(g)$$

It should be noted that the oxides of the metals used as examples above are also amphoteric. Aluminium oxide and zinc oxide will react with both acids and alkalis.

zinc oxide + hydrochloric acid → zinc chloride + water

$$ZnO(s) + 2HCl(aq) \rightarrow ZnCl_2(aq) + H_2O(l)$$

zinc oxide + sodium hydroxide + water → sodium zincate

$$ZnO(s) + 2NaOH(aq) + H_2O(l) \rightarrow Na_2Zn(OH)_4(aq)$$

### Questions

1 Write equations for the reaction between:
  a aluminium and moderately concentrated hydrochloric acid
  b aluminium and moderately concentrated sodium hydroxide (producing sodium aluminate, $NaAl(OH)_4$).

2 Write ionic equations for the reactions which take place to produce the metal hydroxides shown in Table 9.2 on p.131.

3 Describe what you would see when sodium hydroxide is added slowly to a solution containing iron(II) nitrate.

## *Discovery of metals*

Metals have been used since prehistoric times. Many primitive iron tools have been excavated. These were probably made from small amounts of native iron found in rock from meteorites. It was not until about 2500 BC that iron became more widely used. This date marks the dawn of the iron age, when people learned how to get iron from its **ores** in larger quantities by reduction using charcoal. An ore is a naturally occurring mineral from which a metal can be extracted.

Over the centuries other metals, which like iron are also relatively low in the reactivity series, were isolated in a similar manner. These included copper, lead, tin and zinc. However, due to the relatively low abundance of the ores containing these metals, they were not extracted and used in large amounts.

Metals high in the reactivity series have proved very difficult to isolate. It was not until more recent times, through Sir Humphry Davy's work on electrolysis, that potassium (1807), sodium (1807), calcium (1808) and magnesium (1808) were isolated. Aluminium, the most plentiful reactive metal in the Earth's crust, was not extracted from its ore until 1827, by Friedrich Wöhler (p. 81), and the extremely reactive metal rubidium was not isolated until 1861 by Robert Bunsen and Gustav Kirchhoff.

## *Extraction of metals from their ores*

The majority of metals are too reactive to exist on their own in the Earth's crust, and they occur naturally in rocks as compounds in ores (Figure 9.11). These ores are usually carbonates, oxides or sulphides of the metal, mixed with impurities.

**Figure 9.11** Metal ores – chalcopyrite (left) and galena.

Some metals, such as gold and silver, occur in a native form as the free metal (Figure 9.12). They are very unreactive and have withstood the action of water and the atmosphere for many thousands of years without reacting to become compounds.

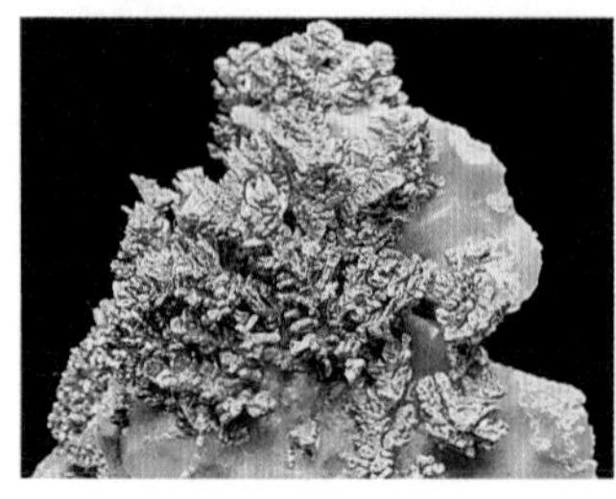

**Figure 9.12** Gold crystals.

Table 9.3 Some common ores.

| Metal | Name of ore | Chemical name of compound in ore | Formula | Usual method of extraction |
|---|---|---|---|---|
| Aluminium | Bauxite | Aluminium oxide | $Al_2O_3.2H_2O$ | Electrolysis of oxide dissolved in molten cryolite |
| Copper | Copper pyrites | Copper iron sulphide | $CuFeS_2$ | The sulphide ore is roasted in air |
| Iron | Haematite | Iron(III) oxide | $Fe_2O_3$ | Heat oxide with carbon |
| Sodium | Rock salt | Sodium chloride | NaCl | Electrolysis of molten sodium chloride |
| Zinc | Zinc blende | Zinc sulphide | ZnS | Sulphide is roasted in air and the oxide produced is heated with carbon |

Some of the common ores are shown in Table 9.3.

Large lumps of the ore are first crushed and ground up by very heavy machinery. Some ores are already fairly concentrated when mined. For example, in some parts of the world, haematite contains over 80% $Fe_2O_3$. However, other ores, such as copper pyrites, are often found to be less concentrated, with only 1% or less of the copper compound, and so they have to be concentrated before the metal can be extracted. The method used to extract the metal from its ore depends on the position of the metal in the reactivity series.

## Extraction of reactive metals

Because reactive metals, such as sodium, hold on to the element(s) they have combined with, they are usually difficult to extract. For example, sodium chloride (as rock salt) is an ionic compound with the $Na^+$ and $Cl^-$ ions strongly bonded to one another. The separation of these ions and the subsequent isolation of the sodium metal is therefore difficult.

Electrolysis of the molten, purified ore is the method used in these cases. During this process, the metal is produced at the cathode while a non-metal is produced at the anode. As you might expect, extraction of metal by electrolysis is expensive. In order to keep costs low, many metal smelters using electrolysis are situated in regions where there is hydroelectric power. Hydroelectric power is discussed further in Chapter 13, p. 187.

For further discussion of the extraction of sodium and aluminium, see Chapter 6, pp. 81–85.

## Extraction of fairly reactive metals

Metals towards the middle of the reactivity series, such as iron and zinc, may be extracted by reducing the metal oxide with the non-metal carbon.

### Iron

Iron is extracted mainly from its oxides, haematite ($Fe_2O_3$) and magnetite ($Fe_3O_4$), in a blast furnace (Figures 9.13 and 9.14). These ores contain at least 60% iron. The iron ores used are a blend of those extracted in Australia, Canada, Sweden, Venezuela and Brazil. The blast furnace is a steel tower approximately 50 m high lined with heat-resistant bricks. It is loaded with the 'charge' of iron ore (usually haematite), coke (made by heating coal) and limestone (calcium carbonate). A blast of hot air is sent in near the bottom of the furnace through holes (tuyères) which makes the 'charge' glow, as the coke burns in the preheated air.

$$\text{carbon} + \text{oxygen} \rightarrow \text{carbon dioxide}$$
$$C(s) + O_2(g) \rightarrow CO_2(g)$$

Figure 9.13 A blast furnace.

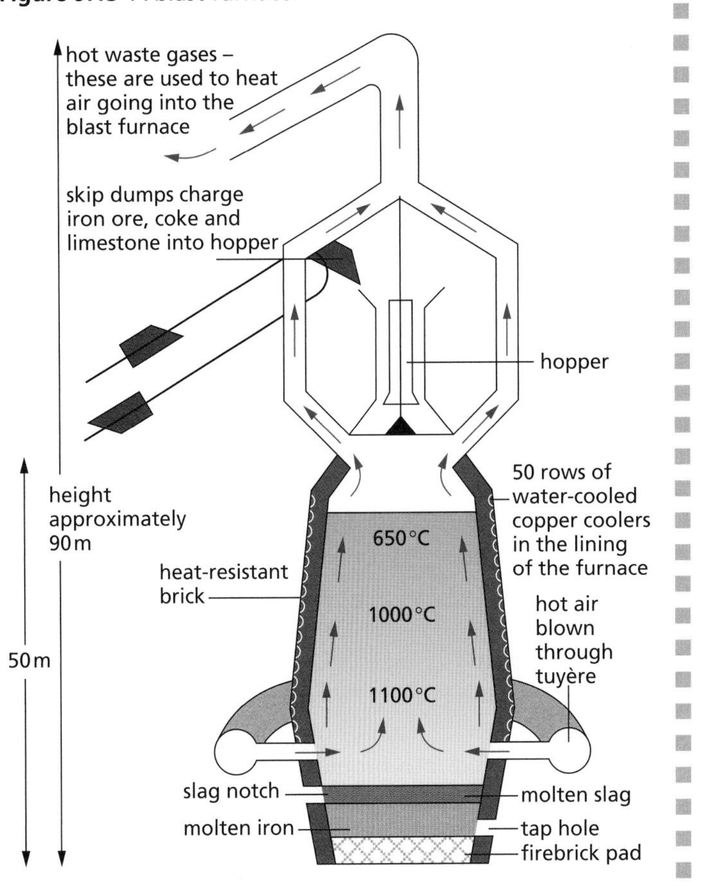


Figure 9.14 Cross-section of a blast furnace.

A number of chemical reactions then follow.

- The limestone begins to decompose:

calcium carbonate → calcium oxide + carbon dioxide

$$CaCO_3(s) \rightarrow CaO(s) + CO_2(g)$$

- The carbon dioxide gas produced reacts with more hot coke higher up in the furnace, producing carbon monoxide in an endothermic reaction.

carbon dioxide + coke → carbon monoxide

$$CO_2(g) + C(s) \rightarrow 2CO(g)$$

- Carbon monoxide is a reducing agent. It rises up the furnace and reduces the iron(III) oxide ore. This takes place at a temperature of around 700 °C:

iron(III) oxide + carbon monoxide → iron + carbon dioxide

$$Fe_2O_3(s) + 3CO(g) \rightarrow 2Fe(l) + 3CO_2(g)$$

The molten iron produced trickles to the bottom of the furnace.

- The calcium oxide formed from the limestone reacts with acidic impurities, for example silicon(IV) oxide (sand) ($SiO_2$), in the iron ore to form a liquid **slag**, which is mainly calcium silicate.

calcium oxide + silicon(IV) oxide → calcium silicate

$$CaO(s) + SiO_2(s) \rightarrow CaSiO_3(l)$$

- This material also trickles to the bottom of the furnace, but because it is less dense than the molten iron, it floats on top of it. The molten iron, as well as the molten slag, may be **tapped off** (run off) at regular intervals.

The waste gases, mainly nitrogen and oxides of carbon, escape from the top of the furnace. They are used in a heat exchange process to heat incoming air and so help to reduce the energy costs of the process. Slag is the other waste material. It is used by builders and road makers (Figure 9.15) for foundations.

The extraction of iron is a continuous process and is much cheaper to run than an electrolytic method.

The iron obtained by this process is known as 'pig' or cast iron and contains about 4% carbon (as well as some other impurities). The name pig iron arises from the fact that if it is not subsequently converted into steel it is poured into moulds called pigs. Because of its brittle and hard nature, the iron produced by this process has limited use. Gas cylinders are sometimes made of cast iron, since they are unlikely to get deformed during their use.

The majority of the iron produced in the blast furnace is converted into different steel alloys (p. 141) such as manganese and tungsten steels as well as the well-known example of stainless steel (p. 140).

The annual production of iron worldwide is 560 million tonnes, 13.2 million tonnes of which are produced in the UK. The largest blast furnace in the UK is at Redcar in the north-east of England (Figure 9.16). This furnace is capable of producing 10 000 tonnes of iron per day.

**Figure 9.15** Slag is used in road foundations.

**Figure 9.16** Redcar steelworks at Teeside, North Yorkshire.

## Titanium

Titanium is the fourth most abundant structural metal after aluminium, iron and magnesium. It occurs naturally in the ores rutile ($TiO_2$) and ilmenite $FeTiO_3$. However, the metal is mainly extracted from rutile. Titanium metal cannot be extracted successfully by electrolysis. The element of choice to reduce the oxide to the metal would be carbon since it is plentiful and cheap. However, rutile cannot be reduced by carbon or carbon monoxide or even hydrogen. Instead, $TiO_2$ is first converted to titanium(IV) chloride, $TiCl_4$, which is covalently bonded and so cannot conduct electricity. It is then extracted by reaction with a more reactive metal, such as sodium. This makes titanium a very expensive metal since the sodium is produced by electrolysis and during the process it is converted to inexpensive sodium chloride.

The annual production of titanium is approximately 66 000 tonnes, of which 6000 tonnes are produced in the UK. Titanium is a hard, silvery metal of low density. It is corrosion resistant (Figure 9.17). Titanium has many uses in the manufacture of strong, light alloys for use in aircraft, missile manufacture and car engines.

The metal is extracted by the following processes.

- The ore is processed to obtain pure titanium(IV) oxide, $TiO_2$.
- The titanium(IV) oxide is converted to titanium(IV) chloride, $TiCl_4$, by reaction with hydrochloric acid.
- The titanium(IV) chloride is purified by fractional distillation.
- Titanium metal is obtained from the reduction of titanium in titanium(IV) chloride by sodium or magnesium metals, at a temperature of approximately 1000 °C in an argon atmosphere:

$$\text{titanium chloride} + \text{sodium} \rightarrow \text{sodium chloride} + \text{titanium}$$

$$TiCl_4(l) + 4Na(l) \rightarrow 4NaCl(s) + Ti(s)$$

- The reaction vessel or reactor is kept at the operating temperature for about four days before going through a cooling process for a further four days. The resulting mixture of titanium and sodium chloride is then crushed and mixed with dilute sulphuric acid. The sodium dissolves in the acid and can therefore be removed, leaving free titanium.

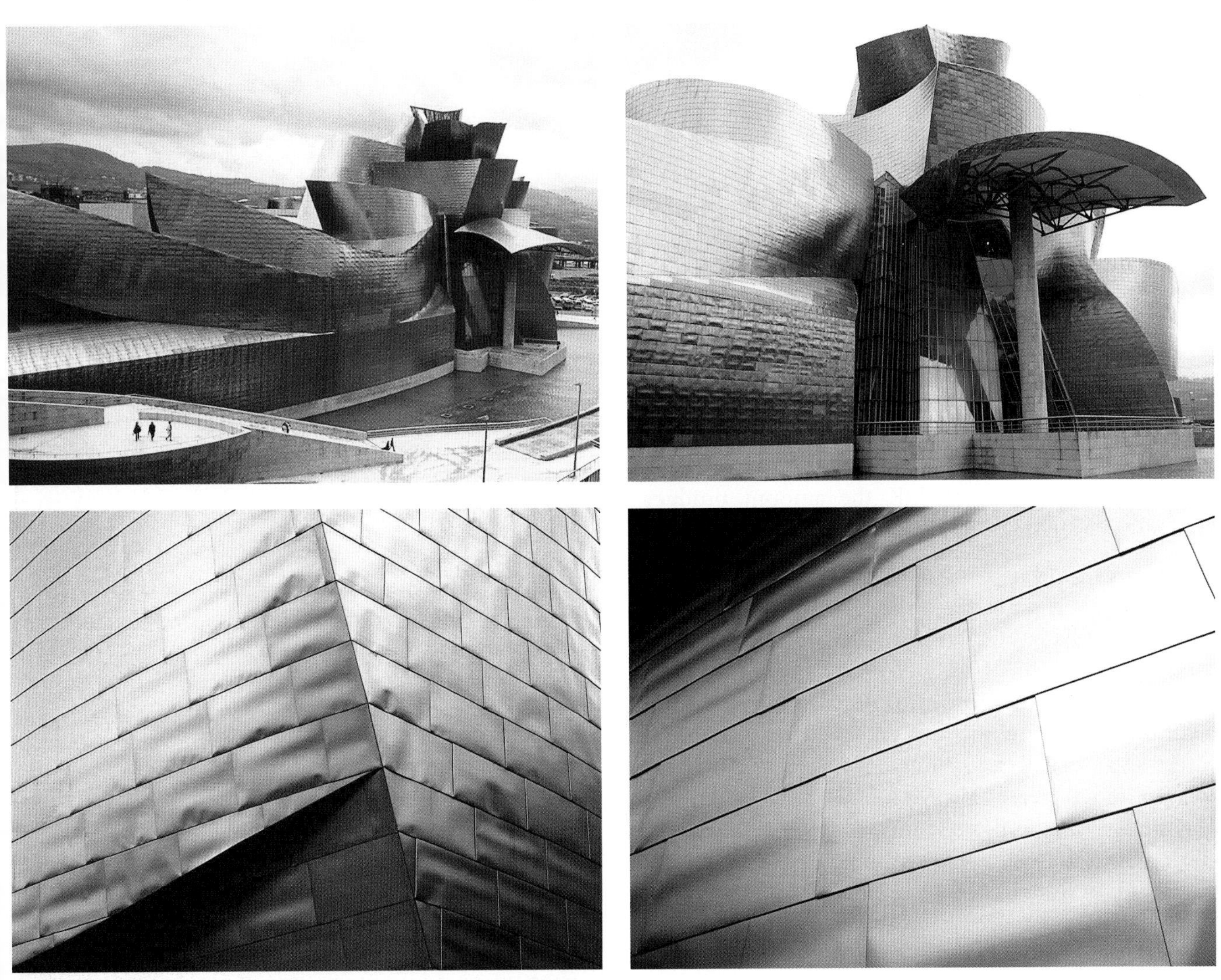

**Figure 9.17** Titanium has a low density, resistance to corrosion and good mechanical strength. Titanium has been used to clad part of the Guggenheim Museum in Bilbao, Spain.

## Extraction of unreactive metals

### Copper

Copper is quite a long way down the reactivity series. It can be found as the free metal element or 'native' in the US. It is principally extracted, however, from copper pyrites, $CuFeS_2$. The crushed ore is concentrated by froth flotation. A chemical known as a collector is added to an ore/water mixture and sticks to the surface of the copper pyrites particles, giving them a water-repellant coating. Detergent is added and air is blown into the mixture to make it froth. The copper pyrites particles are concentrated in the froth and can be removed easily. They are then roasted in a limited supply of air to ensure conversion of copper pyrites to copper(I) sulphide:

copper pyrites + oxygen → copper(I) sulphide + sulphur dioxide + iron(II) oxide

$$2CuFeS_2(s) + 4O_2(g) \rightarrow Cu_2S(s) + 3SO_2(g) + 2FeO(s)$$

Silicon is then added and the mixture is heated in the absence of air. The iron(II) oxide is converted into iron(II) silicate ($FeSiO_3$), which is run off. The remaining copper(I) sulphide is then reduced to copper by heating in a controlled amount of air.

copper(I) sulphide + oxygen → copper + sulphur dioxide

$$Cu_2S(s) + O_2(g) \rightarrow 2Cu(s) + SO_2(g)$$

Copper is then refined by electrolysis (Chapter 6, p. 88) to give a product which is at least 99.92% pure. The purified copper is easily drawn into wires (it is highly ductile), which makes it useful for electrical wiring. It is also used in alloys, such as bronze and brass (Figure 9.18). Copper is also used to make water and central heating pipes, as well as steam boilers (it is a good conductor of heat).

The annual production of copper worldwide is 13.8 million tonnes; 1.8 million tonnes of this total is from recycling copper. In the UK the amount produced annually is 57 000 tonnes, which is all recycled copper.

### Silver and gold

These are very unreactive metals. Silver exists mainly as silver sulphide, $Ag_2S$ (silver glance). The extraction involves treatment of the pulverised ore with sodium cyanide. Zinc is then added to displace the silver from solution. The pure metal is obtained by electrolysis. Silver also exists to a small extent native in the Earth's crust. Gold is nearly always found in its native form (Figure 9.19). It is also obtained in significant amounts during both the electrolytic refining of copper and the extraction of lead.

Silver and gold, because of their resistance to corrosion, are used to make jewellery. Both of these metals are also used in the electronics industry because of their high electrical conductivity.

**Figure 9.18** Copper is alloyed to produce brass.

**Figure 9.19** This man is panning for gold.

## Questions

1. How does the method used for extracting a metal from its ore depend on the metal's position in the reactivity series?
2. 'It is true to say that almost all the reactions by which a metal is extracted from its ore are reduction reactions.' Discuss this statement with respect to the extraction of iron, aluminium and zinc.
3. Zinc is extracted from zinc blende (ZnS). It is extracted in a furnace in a similar way to iron. Use your research skills, including the Internet, to suggest the detail of the extraction process.
4. In the reaction shown for the extraction of titanium from titanium(IV) chloride, write a word and symbol equation for the reduction of $TiCl_4$ with magnesium.

# *Recycling metals*

Recycling 'banks' have become commonplace in recent years (Figure 9.20). Why should we really want to recycle metals? Certainly, if we extract fewer metals from the Earth then the existing reserves will last that much longer. Also, recycling metals prevents the creation of a huge environmental problem (Figure 9.21). However, one of the main considerations is that it saves money.

The main metals which are recycled include aluminium and iron. Aluminium is saved by many households as drinks cans and milk bottle tops, to be melted down and recast. Iron is collected at local authority tips in the form of discarded household goods and it also forms a large part of the materials collected by scrap metal dealers. Iron is recycled to steel. Many steel-making furnaces run mainly on scrap iron.

Aluminium is especially easy to recycle at low cost. Recycling uses only 5% of the energy needed to extract the metal by electrolysis from bauxite. Approximately 60% of the European need for aluminium is obtained by recycling.

**Figure 9.20** Aluminium can recycling.

**Figure 9.21** If we did not recycle metals, then this sight would be commonplace.

# *Rusting of iron*

After a period of time, objects made of iron or steel will become coated with rust. The rusting of iron is a serious problem and wastes enormous amounts of money in the UK each year. It is estimated that upwards of £500 million a year is spent on replacing iron and steel structures.

Rust is an orange–red powder consisting mainly of hydrated iron(III) oxide ($Fe_2O_3.xH_2O$). Both water and oxygen are essential for iron to rust, and if one of these two substances is not present then rusting will not take place. The rusting of iron is encouraged by salt. Figure 9.22 shows an experiment to show that oxygen (from the air) and water are needed for iron to rust.

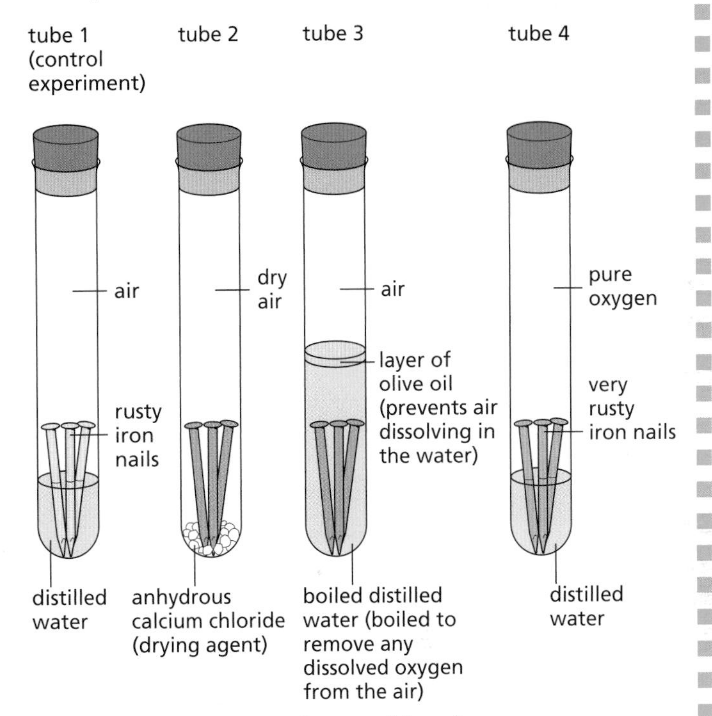


**Figure 9.22** Rusting experiment with nails.

## Rust prevention

To prevent iron rusting, it is necessary to stop oxygen (from the air) and water coming into contact with it. There are several ways of doing this.

### Painting

Ships, lorries, cars, bridges and many other iron and steel structures are painted to prevent rusting (Figure 9.23, overleaf). However, if the paint is scratched, the iron beneath it will start to rust (Figure 9.24, overleaf) and corrosion can then spread under the paintwork which is still sound. This is why it is essential that the paint is kept in good condition and checked regularly.

### Oiling/greasing

The iron and steel in the moving parts of machinery are coated with oil to prevent them from coming into contact with air or moisture. This is the most common way of protecting moving parts of machinery, but the protective film must be renewed.

Figure 9.23 Painting keeps the air and water away from the steel structures of the Forth Rail Bridge.

Figure 9.24 A brand new car is protected against corrosion (top). However, if the paintwork is damaged, then rusting will result.

### Coating with plastic

The exteriors of refrigerators, freezers and many other items are coated with plastic, such as PVC, to prevent the steel structure rusting (Figure 9.25).

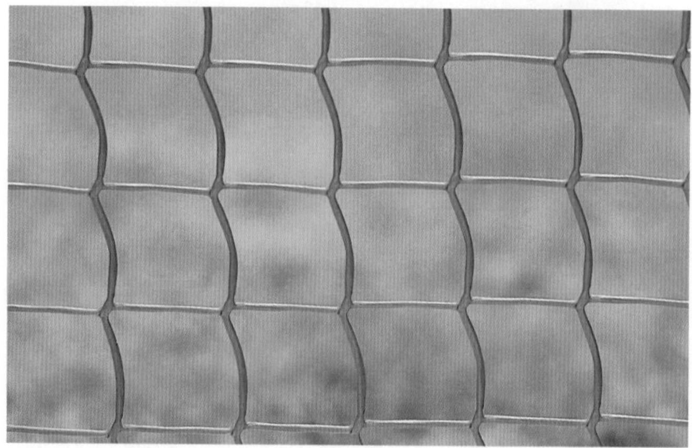

Figure 9.25 The coating of plastic stops the object coming into contact with oxygen or water.

## Plating

Cans for food can be made from steel coated with tin. The tin is deposited on to the steel used to make food cans by dipping the steel into molten tin. Some car bumpers, as well as bicycle handlebars, are electroplated with chromium to prevent rusting. The chromium gives a decorative finish as well as protecting the steel beneath.

## Galvanising

Some steel girders, used in the construction of bridges and buildings, are **galvanised**. Coal bunkers and steel dustbins are also galvanised. This involves dipping the object into molten zinc (Figure 9.26). The thin layer of the more reactive zinc metal coating the steel object slowly corrodes and loses electrons to the iron, thereby protecting it. This process continues even when much of the layer of zinc has been scratched away, so the iron continues to be protected.

**Figure 9.26** A metal object being galvanised.

**a** Bars of zinc on an oil rig.

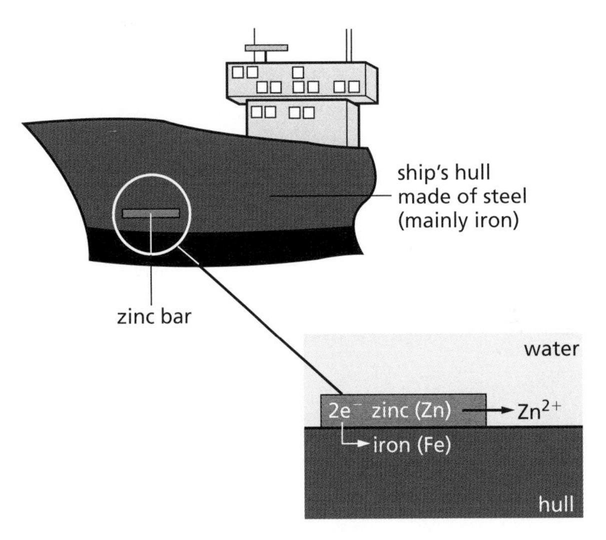


**b** The zinc is sacrificed to protect the steel. Electrons released from the dissolving zinc attract the $Fe^{2+}$ ions back to the steel hull.

**Figure 9.27** Examples of sacrificial protection.

## Sacrificial protection

Bars of zinc are attached to the hulls of ships and to oil rigs (as shown in Figure 9.27a). Zinc is above iron in the reactivity series and will react in preference to it and so is corroded. As long as some of the zinc bars remain in contact with the iron structure, the structure will be protected from rusting. When the zinc runs out, it must be renewed. Gas and water pipes made of iron and steel are connected by a wire to blocks of magnesium to obtain the same result. In both cases, as the more reactive metal corrodes it loses electrons to the iron and so protects it (Figure 9.27b).

# Corrosion

Rusting is the most common form of corrosion. Corrosion is the name given to the process which takes place when metals and alloys are chemically attacked by oxygen, water or any other substances found in their immediate environment. The metals in the reactivity series will corrode to a greater or lesser extent. Generally, the higher the metal is in the reactivity series, the more rapidly it will corrode. If sodium and potassium were not stored under oil they would corrode very rapidly indeed. Magnesium, calcium and aluminium are usually covered by a thin coating of oxide after initial reaction with oxygen in the air. Freshly produced copper is pink in colour (see Figure 6.15 on p. 88). However, it soon turns brown due to the formation of copper(II) oxide on the surface of the metal.

In more exposed environments, copper roofs and pipes quickly become covered in verdigris. Verdigris is green in colour (Figure 9.28) and is composed of copper salts formed on copper. The composition of verdigris varies depending on the atmospheric conditions, but includes mixed copper(II) carbonate and copper(II) hydroxide ($CuCO_3.Cu(OH)_2$).

Gold and platinum are unreactive and do not corrode, even after thousands of years.

**Figure 9.28** Verdigris soon covers copper roofs in exposed environments.

## Questions

1. What is rust? Explain how rust forms on structures made of iron or steel.
2. Rusting is a redox reaction. Explain the process of rusting in terms of oxidation and reduction (Chapter 6, p. 82).
3. Design an experiment to help you decide whether steel rusts faster than iron.
4. Why do car exhausts rust faster than other structures made of steel?

## *Alloys*

The majority of the metallic substances used today are **alloys**. Alloys are mixtures of two or more metals and are formed by mixing molten metals thoroughly. It is generally found that alloying produces a metallic substance that has more useful properties than the original pure metals it was made from. Steel, which is a mixture of the metal iron and the non-metal carbon, is also considered to be an alloy.

Of all the alloys we use, steel is perhaps the most important. Many steels have been produced; they contain not only iron but also carbon and other metals. For example, nickel and chromium are the added metals when stainless steel is produced (Figure 9.29). The chromium prevents the steel from rusting while the nickel makes it harder.

**Figure 9.29** A stainless steel exhaust system. Why do you think more people are buying these exhaust systems?

### Production of steel

The 'pig iron' obtained from the blast furnace contains between 3% and 5% of carbon and other impurities, such as sulphur, silicon and phosphorus. These impurities make the iron hard and brittle. In order to improve the quality of the metal, most of the impurities must be removed and in doing this, steel is produced.

The impurities are removed in the **basic oxygen process**, which is the most important of the steel-making processes. In this process:

- Molten pig iron from the blast furnace is poured into the basic oxygen furnace (Figure 9.30).
- A water-cooled 'lance' is introduced into the furnace and oxygen at 5–15 atm pressure is blown onto the surface of the molten metal.
- Carbon is oxidised to carbon monoxide and carbon dioxide, while sulphur is oxidised to sulphur dioxide. These escape as gases.
- Silicon and phosphorus are oxidised to silicon(IV) oxide and phosphorus pentoxide, which are solid oxides.
- Some calcium oxide (lime) is added to remove these solid oxides as slag. The slag may be skimmed or poured off the surface.
- Samples are continuously taken and checked for carbon content. When the required amount of carbon has been reached, the blast of oxygen is turned off.

The basic oxygen furnace can convert up to 300 tonnes of pig iron to steel per hour.

Worldwide production by this process is 430 million tonnes. The UK produces 14 million tonnes.

There are various types of steel that differ only in their carbon content. The differing amounts of carbon present confer different properties on the steel and they are used for different purposes (Table 9.4). If other types of steel are required then up to 30% scrap steel is added, along with other metals (such as tungsten), and the carbon is burned off.

## Steel recycling

The recycling of scrap steel contributes 310 million tonnes to the world supply of the alloy – 750 million tonnes. It has been calculated that the energy savings are equivalent to 160 million tonnes of coal. Also, it has been calculated that the raw materials conserved are equivalent to 200 million tonnes of iron ore.

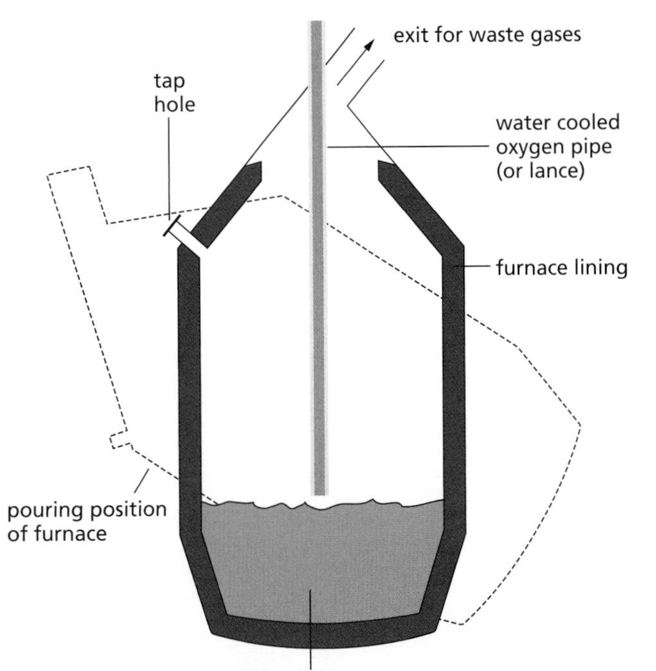


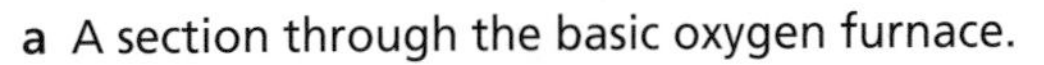
**a** A section through the basic oxygen furnace.

**b** Basic oxygen furnace.

**Figure 9.30**

**Table 9.4** Different types of steel.

| Steel | Typical composition | Properties | Uses |
|---|---|---|---|
| Mild steel | 99.5% iron, 0.5% carbon | Easily worked<br>Lost most of brittleness | Car bodies, large structures |
| Hard steel | 99% iron, 1% carbon | Tough and brittle | Cutting tools, chisels, razor blades |
| Manganese steel | 87% iron, 13% manganese | Tough, springy | Drill bits, springs |
| Stainless steel | 74% iron, 18% chromium, 8% nickel | Tough, does not corrode | Cutlery, kitchen sinks, surgical instruments |
| Tungsten steel | 95% iron, 5% tungsten | Tough, hard, even at high temperatures | Edges of high-speed cutting tools |

## Alloys to order

Just as the properties of iron can be changed by alloying, so the same can be done with other useful metals. Metallurgists have designed alloys to suit a wide variety of different uses. Many thousands of alloys are now made, with the majority being 'tailor-made' to do a particular job (Figure 9.31).

Table 9.5 shows some of the more common alloys, together with some of their uses.

## Questions

1 Calcium oxide is a base. It combines with solid, acidic oxides in the basic oxygen furnace.
Write a chemical equation for one of these oxides reacting with the added lime.

2 'Many metals are more useful to us when mixed with some other elements.' Discuss this statement with respect to stainless steel.

**a** Bronze is often used in sculptures.

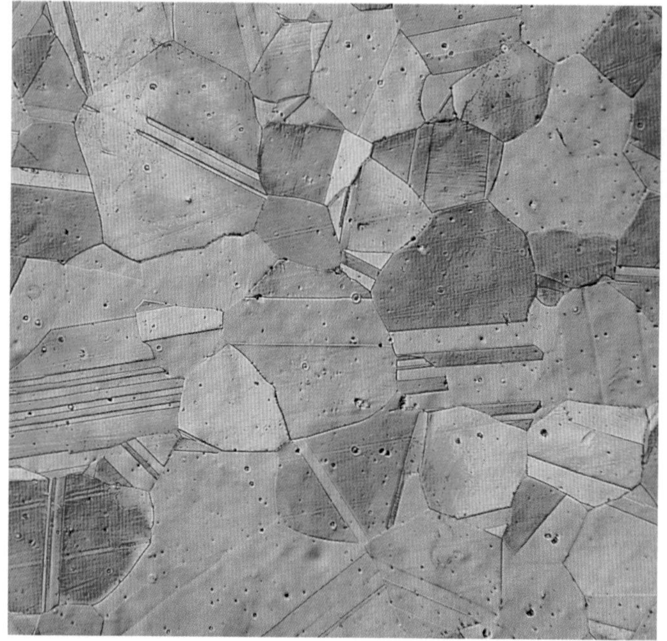

**b** A polarised light micrograph of brass showing the distinct grain structure of this alloy.

**Figure 9.31**

**Table 9.5** Uses of common alloys.

| Alloy | Composition | Use |
|---|---|---|
| Brass | 65% copper, 35% zinc | Jewellery, machine bearings, electrical connections, door furniture |
| Bronze | 90% copper, 10% tin | Castings, machine parts |
| Cupro-nickel | 30% copper, 70% nickel | Turbine blades |
| | 75% copper, 25% nickel | Coinage metal |
| Duralumin | 95% aluminium, 4% copper, 1% magnesium, manganese and iron | Aircraft construction, bicycle parts |
| Magnalium | 70% aluminium, 30% magnesium | Aircraft construction |
| Pewter | 30% lead, 70% tin, a small amount of antimony | Plates, ornaments and drinking mugs |
| Solder | 70% lead, 30% tin | Connecting electrical wiring |

## Checklist

**After studying Chapter 9 you should know and understand the following terms.**

- **Alloy** Generally, a mixture of two or more metals (for example, brass is an alloy of zinc and copper) or a metal and a non-metal (for example, steel is an alloy of iron and carbon, sometimes with other metals included). They are formed by mixing the molten substances thoroughly. Generally, it is found that alloying produces a metallic substance which has more useful properties than the original pure metals it was made from.
- **Amphoteric hydroxide** A hydroxide which can behave as an acid (react with an alkali) or a base (react with an acid), for example zinc hydroxide.
- **Blast furnace** A furnace for smelting iron ores such as haematite ($Fe_2O_3$) and magnetite ($Fe_3O_4$) to produce pig (or cast) iron. In a modified form it can be used to extract metals such as zinc.
- **Competition reactions** Reactions in which metals compete for oxygen or anions. The more reactive metal:
  - takes the oxygen from the oxide of a less reactive metal
  - displaces the less reactive metal from a solution of that metal salt – this type of competition reaction is known as a displacement reaction.
- **Corrosion** The name given to the process that takes place when metals and alloys are chemically attacked by oxygen, water or any other substances found in their immediate environment.
- **Metal extraction** The method used to extract a metal from its ore depends on the position of the metal in the reactivity series.
  - Reactive metals are usually difficult to extract. The preferred method is by electrolysis of the molten ore (electrolytic reduction); for example, sodium from molten sodium chloride.
  - Moderately reactive metals (those near the middle of the reactivity series) are extracted using a chemical reducing agent (for example carbon) in a furnace; for example, iron from haematite in the blast furnace.
  - Unreactive metals, for example gold and silver, occur in an uncombined (native) state as the free metal.
- **Metal ion precipitation** These are reactions in which certain metal cations form insoluble hydroxides. The colours of these insoluble hydroxides can be used to identify the metal cations which are present; for example, copper(II) hydroxide is a blue precipitate.
- **Ore** A naturally occurring mineral from which a metal can be extracted.
- **Reactivity series of metals** An order of reactivity, giving the most reactive metal first, based on results from experiments with oxygen, water and dilute hydrochloric acid.
- **Recycling metals** Metal drink cans such as those made of aluminium are collected in large 'banks' for the sole purpose of recycling them. Reusing the metal in this way saves money.
- **Rust** A loose, orange–brown, flaky layer of hydrated iron(III) oxide found on the surface of iron or steel. The conditions necessary for rusting to take place are the presence of oxygen and water. The rusting process is encouraged by other substances such as salt. It is an oxidation process.
- **Rust prevention** To prevent iron rusting it is necessary to stop oxygen and water coming into contact with it. The methods employed include painting, oiling/greasing, coating with plastic, plating, galvanising and sacrificial protection.

# Metal extraction and chemical reactivity

# *Additional questions*

1 Use the following list of metals to answer the questions **a** to **i**: iron, calcium, potassium, gold, aluminium, magnesium, sodium, zinc, platinum.
  a Which of the metals is found native?
  b Which of the metals is found in nature as the ore:
    (i) rock salt?
    (ii) rutile?
  c Which metal has a carbonate found in nature called chalk?
  d Which of the metals will not react with oxygen to form an oxide?
  e Which of the metals will react violently with cold water?
  f Choose one of the metals in your answer to **e** and write a balanced chemical equation for the reaction which takes place.
  g Which of the metals has a protective oxide coating on its surface?
  h Which of the metals reacts very slowly with cold water but extremely vigorously with steam?
  i Which of the metals is used to galvanise iron?

2

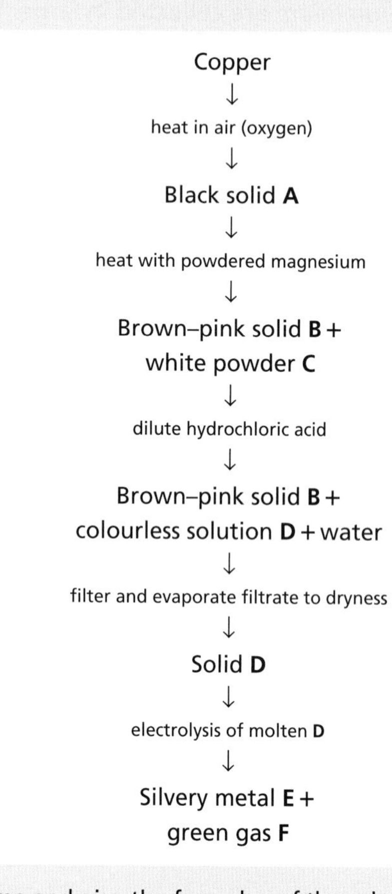


  a Name and give the formulae of the substances **A** to **F**.
  b Write balanced chemical equations for the reactions in which:
    (i) black solid **A** was formed
    (ii) white powder **C** and brown–pink solid **B** were formed
    (iii) colourless solution **D** was formed.
  c The reaction between black solid **A** and magnesium is a redox reaction. With reference to this reaction, explain what you understand by this statement.
  d Write anode and cathode reactions for the processes which take place during the electrolysis of molten **D**.
  e Suggest a use for:
    (i) brown–pink solid **B**
    (ii) silvery metal **E**
    (iii) green gas **F**.

3 Explain the following:
  a metals such as gold and silver occur native in the Earth's crust
  b the parts of shipwrecks made of iron rust more slowly in deep water
  c zinc bars are attached to the structure of oil rigs to prevent them from rusting
  d copper roofs quickly become covered with a green coating when exposed to the atmosphere
  e recycling metals can save money.

4 Iron is extracted from its ores haematite and magnetite. Usually it is extracted from haematite (iron(III) oxide). The ore is mixed with limestone and coke and reduced to the metal in a blast furnace. The following is a brief outline of the reactions involved.

coke + oxygen → gas **X**
gas **X** + coke → gas **Y**
iron(III) oxide + gas **Y** → iron + gas **X**

  a Name the gases **X** and **Y**.
  b Give a chemical test to identify gas **X**.
  c Write balanced chemical equations for the reactions shown above.
  d The added limestone is involved in the following reactions:

limestone → calcium oxide + gas **X**
calcium oxide + silicon(IV) oxide → slag

    (i) Give the chemical names for limestone and slag.
    (ii) Write balanced chemical equations for the reactions shown above.
    (iii) Why is the reaction between calcium oxide and silicon(IV) oxide called an acid–base reaction?
    (iv) Describe what happens to the liquid iron and slag when they reach the bottom of the furnace.
  e Why do you think that the furnace used in the extraction of iron is called a blast furnace?

5 The iron obtained from the blast furnace is known as pig or cast iron. Because of the presence of impurities, such as carbon, it has a hard and brittle nature. Most of this type of iron is therefore converted into steel in the basic oxygen process. During this process either all or some of the carbon is removed. Calculated quantities of other elements are then added to produce the required type of steel.
  a Explain the meaning of the term alloy as applied to steel.
  b Name two impurities, other than carbon, which are present in cast iron and which are removed completely during the steel manufacture.
  c Describe the method of steel manufacture used which removes the impurities referred to in **b**.
  d Name two metallic elements which may be added to the basic oxygen furnace to produce different varieties of steel.
  e Give two uses of stainless steel.
  f Give two advantages of stainless steel compared with cast iron.

6 The metal titanium is extracted from its ore rutile ($TiO_2$).
  a Describe the processes involved in the extraction of titanium from rutile.
  b Name the area of the periodic table in which titanium is found.
  c The final part of the process involves the following reaction:

$$TiCl_4(l) + 4Na(l) \rightarrow 4NaCl(s) + Ti(s)$$

  (i) Calculate the quantity of titanium produced from 19 tonnes of titanium(IV) chloride ($TiCl_4$).
  (ii) How much sodium would have to be used to extract the titanium from the 19 tonnes of $TiCl_4$ used in (i)?
  d Give two uses of titanium metal.

7 Zinc can be reacted with steam using the apparatus shown below. When gas **A** is collected, mixed with air and ignited it gives a small pop. A white solid **B** remains in the test tube when the reaction has stopped and the apparatus cooled down.

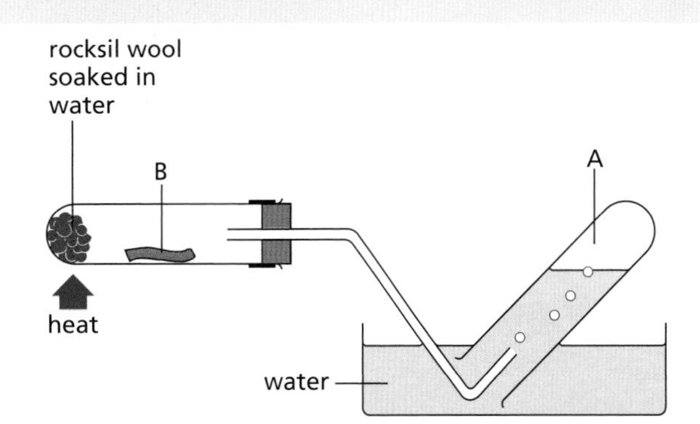


  a Name and give the formula of gas **A**.
  b (i) Name the product formed when gas **A** burns in air.
  (ii) Write a balanced chemical equation for this reaction.
  c (i) Name white solid **B**.
  (ii) Write a balanced chemical equation to represent the reaction between magnesium and steam.
  d Name two other metals which could be safely used to replace zinc and produce another sample of gas **A**.
  e When zinc reacts with dilute hydrochloric acid, gas **A** is produced again. Write a balanced chemical equation to represent this reaction and name the other product of this reaction.

8 Copper is extracted in the final stage by roasting copper(I) sulphide ($Cu_2S$) in a current of air. This converts the sulphide to the metal.

$$Cu_2S(s) + O_2(g) \rightarrow 2Cu(s) + SO_2(g)$$

  a (i) What is the name of the chemical process by which copper(I) sulphide is converted to copper?
  (ii) At what temperature does this process take place?
  b It is a very expensive business to maintain the high temperature needed to efficiently convert copper(I) sulphide to copper. How are the heating costs offset?
  c Calculate the quantity of copper(I) sulphide required to produce 1270 tonnes of copper. ($A_r$: O = 16; S = 32, Cu = 63.5)
  d For use in electrical wiring, the copper must be 99.99% pure. How is the copper purified to obtain this high purity?

# 10 Atmosphere and oceans



## The developing atmosphere

About 4500 million years ago the planets in the solar system were formed. Each planet had a thick layer of gases, mainly hydrogen and helium, surrounding its core. This layer of gas is known as the **primary atmosphere** (Figure 10.1). Over a period of time intense solar activity caused these lighter gases to be removed from this primary atmosphere of planets nearest to the Sun. During this time the Earth was cooling to become a molten mass upon which a thin crust formed.

a Lightning flashes.

b Aurora borealis.

c A tornado or 'twister'.

**Figure 10.1** The Earth's atmosphere formed over many millions of years and involved various atmospheric phenomena.

Volcanic activity through the crust pushed out huge quantities of gases, such as ammonia, nitrogen, methane, carbon monoxide, carbon dioxide and a small amount of sulphur dioxide, which formed a secondary atmosphere around the Earth (Figure 10.2a).

About 3800 million years ago, when the Earth had cooled below 100 °C, the water vapour in this secondary atmosphere condensed and fell as rain. This caused the formation of the first oceans, lakes and seas on the now rapidly cooling Earth. The structure of the surface of the Earth, as we know it today, has evolved as a result of the presence of these large expanses of water (Figure 10.2b). The oceans are even today an important reservoir for carbon dioxide gas, removing it from the atmosphere.

Eventually, early forms of life developed in these oceans, lakes and seas at depths which prevented potentially harmful ultraviolet light from the Sun affecting them. About 3000 million years ago the first forms of bacteria appeared. These were the predecessors of algae-like organisms which used the light from the Sun, via photosynthesis, to produce their own food, and oxygen was released into the atmosphere as a waste product. This process also acted to reduce the amount of $CO_2$ in the atmosphere.

The process of photosynthesis can be described by the following equation:

$$\text{carbon dioxide} + \text{water} \xrightarrow[\text{chlorophyll}]{\text{sunlight}} \text{glucose} + \text{oxygen}$$

$$6CO_2(g) + 6H_2O(l) \longrightarrow C_6H_{12}O_6(aq) + 6O_2(g)$$

The ultraviolet radiation broke down some of the oxygen molecules, which had been released, into oxygen atoms.

$$\text{oxygen molecules} \xrightarrow{\text{UV light}} \text{oxygen atoms}$$

$$O_2(g) \longrightarrow 2O(g)$$

Some of the highly reactive oxygen atoms reacted with molecules of oxygen to form ozone molecules, $O_3(g)$.

$$\text{oxygen atom} + \text{oxygen molecule} \rightarrow \text{ozone}$$

$$O(g) + O_2(g) \rightarrow O_3(g)$$

Ozone is an unstable molecule which readily decomposes, under the action of ultraviolet radiation, into single oxygen atoms and oxygen molecules.

$$\text{ozone} \xrightarrow{\text{UV light}} \text{oxygen atom} + \text{oxygen molecule}$$

$$O_3(g) \longrightarrow O(g) + O_2(g)$$

However, the single oxygen atoms would react quickly, re-forming ozone molecules.

**a** Volcanic activity expelled gases through the crust to form a secondary atmosphere.

**b** The Seven Sisters in Sussex. The Earth's surface has evolved as a result of lakes and other large expanses of water.

**Figure 10.2**

Ozone is an important gas in the atmosphere (stratosphere). It prevents harmful ultraviolet radiation from reaching the Earth. Over many millions of years the amount of ultraviolet radiation was reduced significantly. About 400 million years ago the first land plants appeared on the Earth, and so the amount of oxygen and hence ozone increased.

Oxygen is a reactive gas, and over millions of years organisms have adapted to make use of it. The oxygen from the atmosphere was used, along with the carbon they obtained from their food, to produce energy in a process known as **respiration**. The process of respiration can be represented as:

glucose + oxygen → carbon dioxide + water + energy

$$C_6H_{12}O_6(aq) + 6O_2(g) \rightarrow 6CO_2(g) + 6H_2O(l) + \text{energy}$$

Over recent years, scientists have become aware of a reduction in the amount of ozone in our atmosphere and of the formation of ozone holes (Figure 10.3). The reduction of ozone in our atmosphere will lead to an increased risk of skin cancer as more harmful ultraviolet radiation reaches the surface of the Earth.

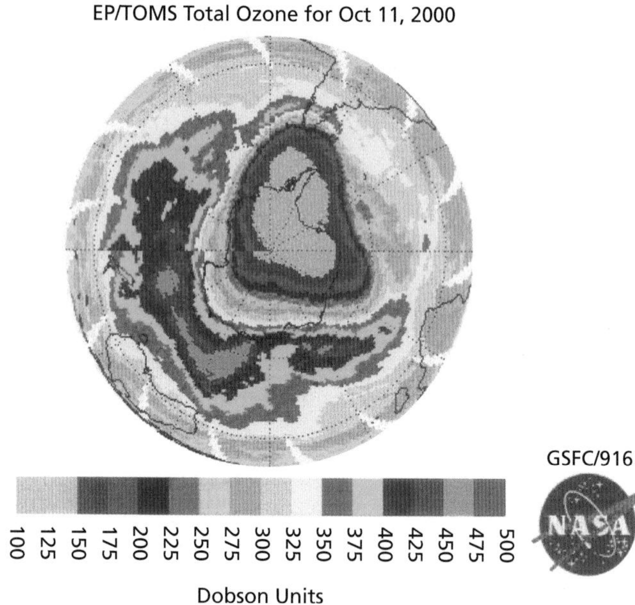


**Figure 10.3** This diagram comes from NASA's ozone monitoring programme TOMS (Total Ozone Mapping Spectrometer). You can view diagrams similar to this at http://toms.gsfc.nasa.gov/. The ozone hole over the Antarctic (shown in purple and pink on the diagram) is largest in the Antarctic spring. Note: Dobson Units are a measure of the total amount of ozone in a vertical column from the ground to the top of the atmosphere.

## Question

**1** What precautions are necessary to prevent an increase in skin cancers?

# *The structure of the atmosphere*

The gases in the atmosphere are held in an envelope around the Earth by its gravity. The atmosphere is 80 km thick (Figure 10.4) and it is divided into four layers:

- troposphere
- stratosphere
- mesosphere
- thermosphere.

About 75% of the mass of the atmosphere is found in the layer nearest the Earth called the troposphere. Beyond this layer the atmosphere reaches into space but becomes extremely thin beyond the mesosphere.

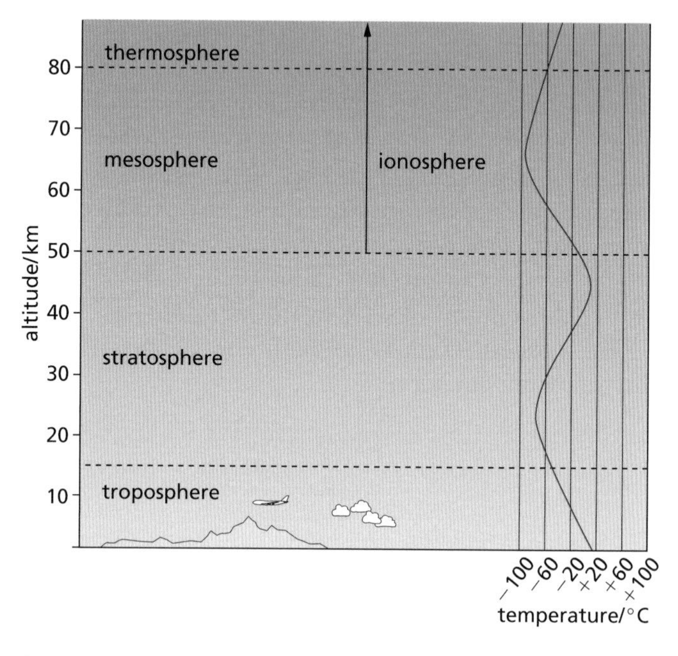


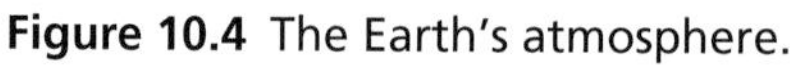

**Figure 10.4** The Earth's atmosphere.

## The composition of the atmosphere

If a sample of dry, unpolluted air was taken from any location in the troposphere and analysed, the composition by volume of the sample would be similar to that shown in Table 10.1.

**Table 10.1** Composition of the atmosphere.

| Component | % |
|---|---|
| Nitrogen | 78.08 |
| Oxygen | 20.95 |
| Argon | 0.93 |
| Carbon dioxide | 0.03 |
| Neon | 0.002 |
| Helium | 0.0005 |
| Krypton | 0.0001 |
| Xenon<br>plus minute amounts of other gases | 0.00001 |

## Measuring the percentage of oxygen in the air

When 100 $cm^3$ of air is passed backwards and forwards over heated copper turnings it is found that the amount of gas decreases (Figure 10.5). This is because the reactive part of the air, the oxygen gas, is reacting with the copper to form black copper(II) oxide (Figure 10.6). In such an experiment the volume of gas in the syringe decreases from 100 $cm^3$ to about 79 $cm^3$, showing that the air contained 21 $cm^3$ of oxygen gas. The percentage of oxygen gas in the air is:

$$\frac{21}{100} \times 100 = 21\%$$

The composition of the atmosphere is affected by the following factors:

- respiration
- photosynthesis
- volcanic activity
- radioactive decay, in which helium is formed
- human activity, involving burning of fossil fuels, in which carbon dioxide and water vapour are produced as well as other gases (p. 185).

Compare the components of our atmosphere with those of the other planets in the solar system (Table 10.2).

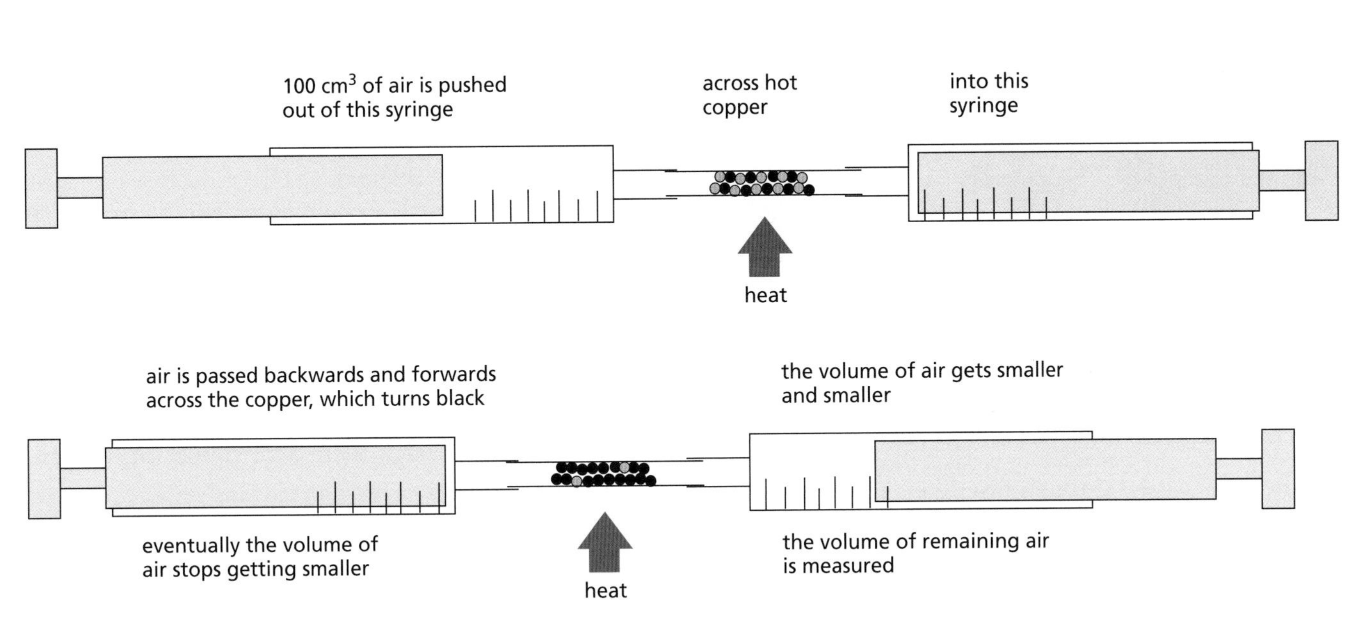


**Figure 10.5** This apparatus can be used to find out the volume of oxygen gas in the air.

**Figure 10.6** Copper turnings before and after reaction.

**Table 10.2** Atmospheres of the other planets in the solar system.

| Planet | Atmosphere |
|---|---|
| Mercury | No atmosphere – the gases were burned off by the heat of the Sun |
| Venus | Carbon dioxide and sulphur dioxide |
| Mars | Mainly carbon dioxide |
| Jupiter | Ammonia, helium, hydrogen, methane |
| Saturn | Ammonia, helium, hydrogen, methane |
| Uranus | Ammonia, helium, hydrogen, methane |
| Neptune | Helium, hydrogen, methane |
| Pluto | No atmosphere; it is frozen |

## *Questions*

1. Draw a pie chart to show the data given in Table 10.1.
2. Is air a compound or a mixture? Explain your answer.
3. Design an experiment to find out how much oxygen there is in exhaled air.

## Fractional distillation of liquid air

Air is the major source of oxygen, nitrogen and the noble gases. The gases are obtained by fractional distillation of liquid air but it is a complex process, involving several different steps (Figure 10.7).

**Figure 10.7** BOC produces large amounts of gases obtained from the fractional distillation of liquid air.

- The air is passed through fine filters to remove dust.
- The air is cooled to about −80 °C to remove water vapour and carbon dioxide as solids. If these are not removed, then serious blockages of pipes can result.
- Next, the cold air is compressed to about 100 atm of pressure. This warms up the air, so it is passed into a heat exchanger to cool it down again.
- The cold, compressed air is allowed to expand rapidly and that cools it still further.
- The process of compression followed by expansion is repeated until the air reaches a temperature below −200 °C. At this temperature the majority of the air liquefies (Table 10.3).
- The liquid air is passed into a fractionating column and it is fractionally distilled.
- The gases are then stored separately in large tanks and cylinders.

**Table 10.3** Boiling points of atmospheric gases.

| Gas | Boiling point/°C |
|---|---|
| Helium | −269 |
| Neon | −246 |
| Nitrogen | −196 |
| Argon | −186 |
| Oxygen | −183 |
| Krypton | −157 |
| Xenon | −108 |

It should be noted that the noble gases neon, argon, krypton and xenon are obtained by this method; however, helium is more profitably obtained from natural gas.

### Question

1 Use information given in the text to construct a flow chart to show the processes involved in the extraction of gases from air.

## Uses of the gases

### Oxygen

- Large quantities are used in industry to convert pig iron into steel (Chapter 9, p. 141) and for producing very hot flames for welding by mixing with gases such as ethyne (acetylene).
- It is used in hospitals to help patients with breathing difficulties (Figure 10.8).
- People such as mountaineers and divers use oxygen.
- It is carried in space rockets so that the hydrogen and kerosene fuels can burn.
- Space shuttles use oxygen gas in fuel cells which convert chemical energy into electrical energy.
- Astronauts must carry their own supply of oxygen, as do fire-fighters.
- It is used to restore life to polluted lakes and rivers and in the treatment of sewage.

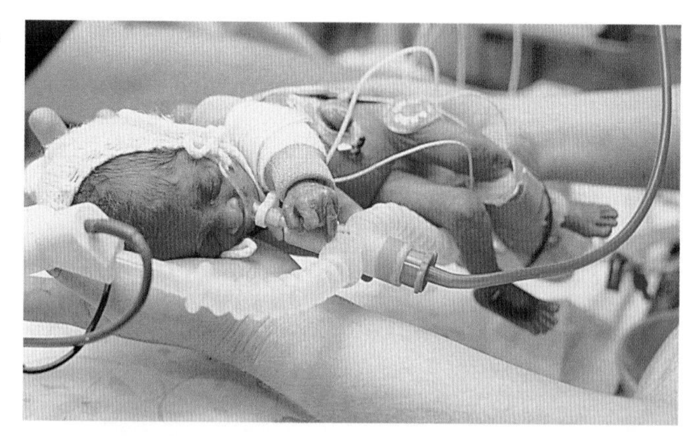

**Figure 10.8** The incubator has its own oxygen supply.

### Nitrogen

- Nitrogen is used in large quantities in the production of ammonia gas (Chapter 15, p. 219), which is used to produce nitric acid. Nitric acid is used in the manufacture of dyes, explosives and fertilisers.
- Liquid nitrogen is used as a refrigerant. Its low temperature makes it useful for freezing food quickly.
- Because of its unreactive nature, nitrogen is used as an inert atmosphere for some processes and chemical reactions. For example, empty oil tankers are filled with nitrogen to prevent fires.

- It is used in food packaging, for example in crisp packets, to keep the food fresh and in this case to prevent the crisps being compressed (Figure 10.9).

**Figure 10.9** Inert nitrogen gas is used in food packaging.

## Noble gases

Argon is used:

- to fill ordinary and long-life light bulbs to prevent the tungsten filament from reacting with oxygen in the air and forming the oxide (Figure 10.10)
- to provide an inert atmosphere in arc welding and in the production of titanium metal.

Neon is used:

- in advertising signs, because it glows red when electricity is passed through it
- in the helium–neon gas laser (Figure 10.11)
- in Geiger–Müller tubes, which are used for the detection of radioactivity.

Helium is used:

- to provide an inert atmosphere for welding
- as a coolant in nuclear reactors
- with 20% oxygen as a breathing gas used by deep-sea divers
- to inflate the tyres of large aircraft
- to fill airships and weather balloons (Figure 10.12)
- in the helium–neon laser
- in low-temperature research, because of its low boiling point.

Krypton and xenon are used:

- in lamps used in photographic flash units, in stroboscopic lamps and in lamps used in lighthouses.

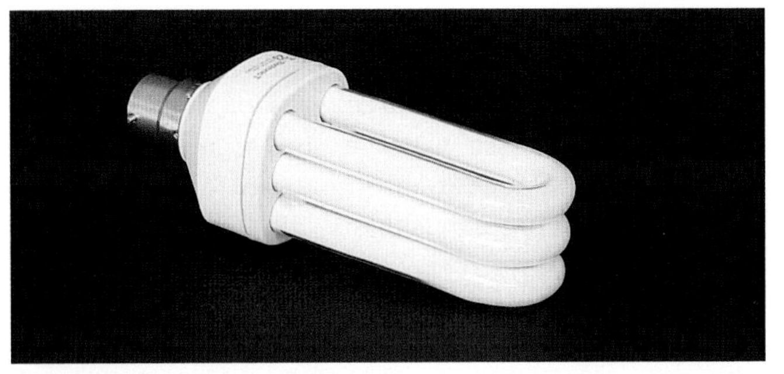

**Figure 10.10** This long-life bulb contains argon.

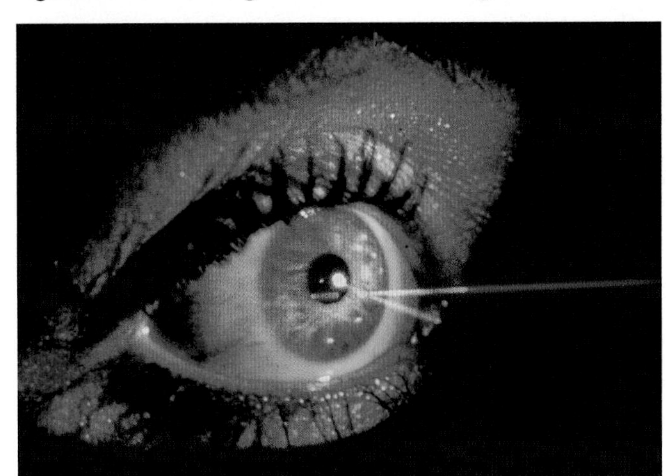

**Figure 10.11** A helium–neon laser used in eye surgery.

**Figure 10.12** Helium is used to fill this airship as it has a low density and is unreactive.

## *Questions*

1. How does oxygen help to restore life to polluted lakes?
2. Why is it important to have nitrogen in fertilisers?
3. Why is helium needed to produce an inert atmosphere for welding?

## Resources in the oceans

In some hot, arid countries the sea is the main source of pure water for drinking. The water is obtained by a process known as desalination (Chapter 2, p. 23).

Sea water contains about 35 g of dissolved substances in each kilogram. A typical analysis of the varied elements present in sea water is shown in Table 10.4. Some elements, such as bromine, are extracted from sea water on a commercial basis.

**Table 10.4** Elements present in sea water.

| Element | Concentration/$g\,dm^{-3}$ of sea water |
|---|---|
| Bromine | 0.07 |
| Calcium | 0.4 |
| Chlorine | 19.2 |
| Magnesium | 1.3 |
| Potassium | 0.4 |
| Sodium | 10.7 |
| Sulphur | 0.9 |
| Other elements | 1.4 |

### Bromine

Eighty per cent of the world's supply of bromine is extracted from sea water. Bromine is extracted in this way in a chemical plant at Amlwch, Anglesey, Wales, but one of the biggest factories for extracting bromine is to be found on the south-west shores of the Dead Sea in Israel. The Dead Sea is a particularly rich source of bromine as it contains 5.2 g of the element per $dm^3$ of sea water.

Worldwide 55 000 tonnes of bromine are harvested from the sea each year.

The extraction of bromine from sea water begins with the evaporation of the sea water using energy from the Sun, forming a more concentrated solution. The bromine is present in solution as bromide ions ($Br^-(aq)$), which can be converted to bromine via a displacement reaction with chlorine gas. The process is as follows.

- Chlorine gas is bubbled through the warm, partially evaporated Dead Sea water and displacement takes place.

$$\underset{\text{gas}}{\text{chlorine}} + \underset{\text{ions}}{\text{bromide}} \rightarrow \underset{\text{gas}}{\text{bromine}} + \underset{\text{ions}}{\text{chloride}}$$

$$Cl_2(g) + 2Br^-(aq) \rightarrow Br_2(aq) + 2Cl^-(aq)$$

- Steam is blown through the resulting solution and the volatile bromine is given off as bromine vapour mixed with steam.
- Bromine is not very soluble in water and therefore two separate layers form when the bromine and steam condense. The less dense water floats on top of the denser bromine layer. The bromine layer is then run off.
- The impure bromine is then purified by distillation and dried.

Half a tonne of chlorine is required for each tonne of bromine manufactured. Bromine is used in the manufacture of the fuel additive tetraethyllead(IV) (TEL), although to a decreasing extent. Bromine compounds are also used in photography (AgBr), medicines (KBr) and herbicides as well as in the manufacture of flame-retardant materials.

**Figure 10.13** The Dead Sea is a particularly rich source of salts, including bromides from which bromine is extracted.

## Magnesium

Magnesium can also be extracted from sea water. The sea water is first treated with calcium hydroxide, and magnesium hydroxide is precipitated and filtered off.

| magnesium ions | + hydroxide ions | → magnesium hydroxide |
|---|---|---|
| $Mg^{2+}(aq)$ | $+ 2OH^-(aq)$ | $\rightarrow Mg(OH)_2(s)$ |

a Magnesium (as the element) is used in alloys to build space shuttles and racing cycles.

b As a compound it is used in the manufacture of indigestion remedies and toothpaste.

**Figure 10.14**

The magnesium hydroxide is then dissolved in hydrochloric acid and the resulting solution is evaporated to produce magnesium chloride. This is then fused, at about 700 °C, with additives to lower its melting point and is electrolysed with a graphite anode and a steel cathode. The magnesium produced is liberated at the cathode and is 99.9% pure. Around 250 000 tonnes are produced worldwide each year.

| magnesium ions | + electrons | → magnesium |
|---|---|---|
| $Mg^{2+}(aq)$ | $+ \quad 2e^-$ | $\rightarrow \quad Mg(s)$ |

Magnesium is used in alloys which are used in the space and aviation industries, and in racing cycles (Figure 10.14a). Compounds of magnesium ($Mg(OH)_2$ and $MgSO_4.7H_2O$) are also used in medicines, for example indigestion remedies (Figure 10.14b), and in toothpastes.

## Sodium chloride

As you might guess, salt is the most abundant resource in sea water. There is about 25 g of sodium chloride per $dm^3$ of sea water. It is extracted in several areas of the world by evaporation, for example in France, Saudi Arabia and Australia.

The sea water is kept in shallow ponds until all the water has been evaporated by the heat of the Sun. The salt is then harvested.

Sodium chloride is used to flavour food, in the manufacture of sodium carbonate and sodium hydrogencarbonate and as the raw material for the chlor-alkali industry (Chapter 6, p. 86).

## *Questions*

1 Use the information in the text and any other information you can obtain from other sources to construct a flow diagram to show the processes involved in the extraction of bromine from sea water.

2 In the extraction of magnesium from sea water, the magnesium hydroxide precipitate is treated with dilute hydrochloric acid. Write both a word and a balanced chemical equation to describe this reaction. In addition, write an electrode equation to show the production of chlorine gas at the anode.

## The water cycle

Figure 10.15 illustrates the water cycle, which shows how water circulates around the Earth. The driving force behind the water cycle is the heat of the Sun.

- Heat from the Sun causes evaporation from oceans, seas and lakes. Water vapour is also formed from the evaporation of water from leaves (transpiration), through respiration and through combustion. The water vapour rises and cools, and condenses to form tiny droplets of water. These droplets form clouds.
- The clouds are moved along by air currents. As they cool, the tiny droplets join to form larger droplets, which fall as rain when they reach a certain size.
- The water that falls as rain runs into streams and rivers and then on into lakes, seas and oceans.

### Question

1 Construct a simplified version of the water cycle using 'key words' in boxes and the 'processes involved' over linking arrows.

## Pollution

The two major resources considered in this chapter, water and air, are essential to our way of life and our very existence. However, we are continually guilty of polluting these resources. We now look at the effects of the various sources of pollution and the methods used to control or eliminate them.

### Water pollution

Water is very good at dissolving substances. It is, therefore, very unusual to find really pure water on this planet. As water falls through the atmosphere and down on to and through the surface of the Earth, it dissolves a tremendous variety of substances. Chemical fertilisers washed off surrounding land will add nitrate ions ($NO_3^-$) and phosphate ions ($PO_4^{3-}$) to the water, owing to the use of artificial fertilisers such as ammonium nitrate and ammonium phosphate. It may also contain human waste as well as insoluble impurities such as grit and bacteria and oil and lead 'dust' (to a decreasing extent) from the exhaust fumes of lorries and cars (Figure 10.16).

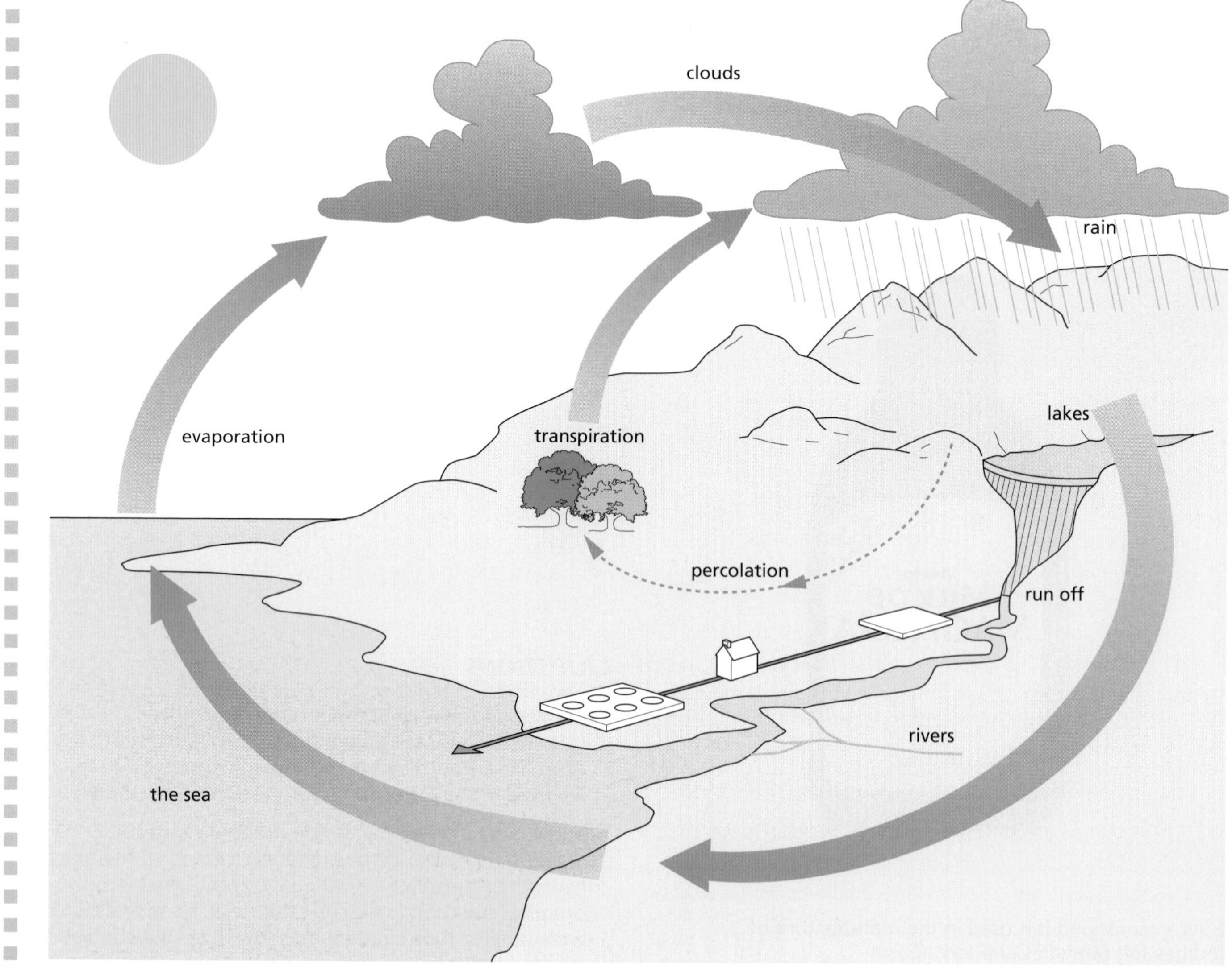


Figure 10.15 The water cycle.

Figure 10.16 A badly polluted river.

All these artificial as well as natural impurities must be removed from the water before it can be used. Recent European Union (EU) regulations have imposed strict guidelines on the amounts of various substances allowed in drinking water (Figure 10.17).

Figure 10.17 A water treatment works.

Most drinking water in the UK is obtained from lakes and rivers where the pollution levels are low. Undesirable materials removed from water include:

- colloidal clay (clay particles in the water)
- bacteria
- chemicals which cause the water to be coloured and foul tasting
- acids, which are neutralised.

The process of water treatment involves both filtration and chlorination and is summarised in Figure 10.18.

1. Impure water is first passed through screens to filter out floating debris.
2. Aluminium sulphate is added to coagulate small particles of clay so that they form larger clumps, which settle more rapidly.
3. Filtration through coarse sand traps larger, insoluble particles. The sand also contains specially grown microbes which remove some of the bacteria.
4. A sedimentation tank has chemicals known as flocculants, for example aluminium sulphate, added to it to make the smaller particles (which remain in the water as colloidal clay) stick together and sink to the bottom of the tank.
5. These particles are removed by further filtration through fine sand. Sometimes a carbon slurry is used to remove unwanted tastes and odours and a lime slurry is used to adjust the acidity.
6. Finally, a little chlorine gas is added, which kills any remaining bacteria. This sterilises the water. Excess chlorine can be removed by the addition of sulphur dioxide gas. The addition of chlorine gas makes the water more acidic and so appropriate amounts of sodium hydroxide solution are added.

Fluoride is added to water in some local supplies if there is insufficient occurring naturally, as it helps to prevent tooth decay.

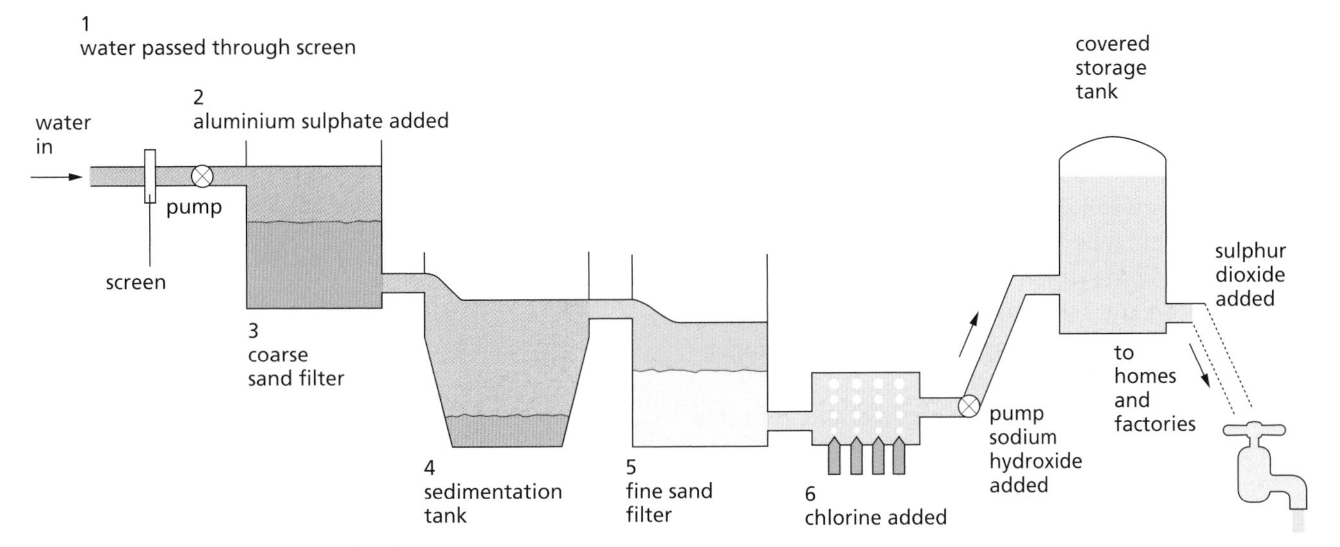


Figure 10.18 The processes involved in water treatment.

### The 'iron problem'

If the acidity level of the treated water is not controlled then problems occur due to the precipitation of iron(III) hydroxide. These include:

- vegetables turning brown
- tea having an inky appearance and a bitter taste
- clothes showing rusty stains after washing.

### Sewage treatment

After we have used water it must be treated again before it can be returned to rivers, lakes and seas. This multi-stage process known as sewage treatment is shown in Figure 10.19.

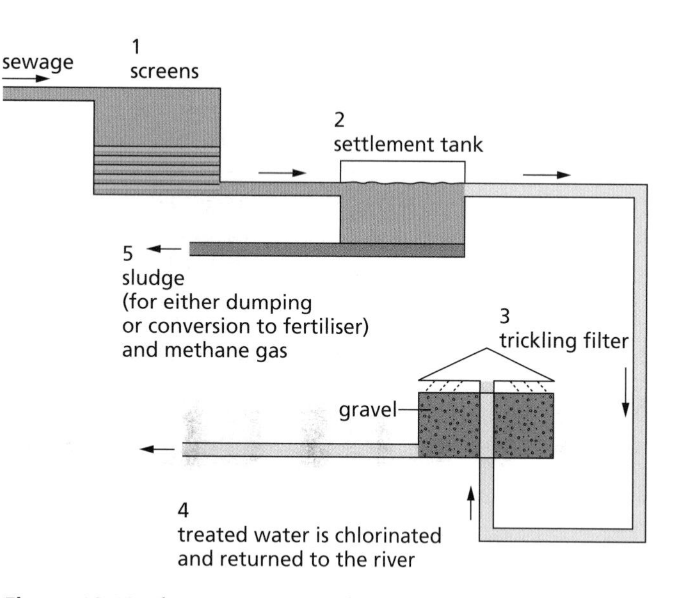


**Figure 10.19** The processes involved in sewage treatment.

Used water, sewage, contains waste products such as human waste and washing-up debris as well as everything else that we put down a drain or sink. The processes that are involved are as follows.

1. Large screens remove large pieces of rubbish.
2. Sand and grit are separated on large sedimentation tanks. The process is speeded up by adding aluminium sulphate, which helps the solids to coagulate into larger particles that separate more rapidly. The sand and grit often contain large amounts of useful chemicals which, by the action of selected microbes, can be used as fertilisers.
3. The impure water is then removed and sent to a trickling filter, where it is allowed to drain through gravel on which microbes have been deposited. These kill off any remaining bacteria in the water by aerobic processes. This stage is known as biological filtration.
4. The treated water is then chlorinated and returned to a river, after checking.
5. Anaerobic bacteria digest what remains from the other stages. Methane gas is produced, which can be used as a fuel.

## Atmospheric pollution

Air pollution is all around us. Concentrations of gases in the atmosphere such as carbon monoxide, sulphur dioxide and nitrogen oxides are increasing with the increasing population. As the population rises there is a consequent increase in the need for energy, industries and motor vehicles. These gases are produced primarily from the combustion of the fossil fuels coal, oil and gas, but they are also produced by the smoking of cigarettes.

Motor vehicles are responsible for much of the air pollution in large towns and cities. They produce four particularly harmful pollutants:

- carbon monoxide
- sulphur dioxide
- hydrocarbons
- oxides of nitrogen.

Concern about pollution due to cars has led to the introduction of strict regulations by the EU and now all new cars must have a device known as a catalytic converter fitted to eliminate the production of some of these gases.

The catalytic converter acts as a device to speed up reactions which involve the pollutant gases, converting them to less harmful products, such as nitrogen and carbon dioxide. It should be noted that catalytic converters can only be used with unleaded petrol as the lead 'poisons' the catalyst, preventing it from catalysing the reactions. For a further discussion of catalytic converters see Chapter 11, p. 166.

Another method that has been introduced to reduce the amount of pollutants is that of the 'lean burn' engine. Although this type of engine reduces the amounts of carbon monoxide and oxides of nitrogen produced, it actually increases the amount of hydrocarbons in the exhaust gases.

A further method of regulating pollutant gases is to convert petrol burning engines to LPG (liquid petroleum gas) engines, whilst retaining the ability to burn petrol. These cars are known as dual-fuel cars. Much research has also been carried out to produce efficient electric motors that can be fitted in place of the petrol engine in a car. This development is moving forward at a pace.

Power stations are a major source of sulphur dioxide, a pollutant formed by the combustion of coal, oil and gas, which contain small amounts of sulphur.

$$\text{sulphur} + \text{oxygen} \rightarrow \text{sulphur dioxide}$$
$$S(s) + O_2(g) \rightarrow SO_2(g)$$

This sulphur dioxide gas dissolves in rainwater to form the weak acid, sulphurous acid ($H_2SO_3$).

$$\text{sulphur dioxide} + \text{water} \rightleftharpoons \text{sulphurous acid}$$
$$SO_2(g) + H_2O(l) \rightleftharpoons H_2SO_3(aq)$$

**Figure 10.20** Sulphur dioxide is a major pollutant produced by industry.

A further reaction occurs in which the sulphurous acid is oxidised to sulphuric acid.

Solutions of these acids are the principal contributors to acid rain. For a further discussion of acid rain, see Chapter 16, p. 230.

Units called flue gas desulphurisation (FGD) units are being fitted to some power stations to prevent the emission of sulphur dioxide gas. Here, the sulphur dioxide gas is removed from the waste gases by passing it through calcium hydroxide slurry. This not only removes the sulphur dioxide but also creates calcium sulphate, which can be sold to produce plasterboard. The FGD units are very expensive and therefore the sale of the calcium sulphate is an important economic part of the process.

## *Questions*

1. Make a list of four major water pollutants and explain where they come from. What damage can these pollutants do?
2. Write a balanced chemical equation to represent the reaction which takes place between sulphur dioxide and calcium hydroxide slurry in the FGD unit of a power station.
3. In the treatment of water for public use, state the purpose of the addition of:
   **a** aluminium sulphate
   **b** chlorine
   **c** sodium hydroxide
   **d** sulphur dioxide.
4. Many industries use water as a coolant. Suggest the sorts of problems that may be created by this 'thermal pollution'.

## Checklist

**After studying Chapter 10 you should know and understand the following terms.**

- **Fractional distillation of air** The process to extract individual gases from the air. Air is a major raw material. The mixture of gases is separated by first liquefying the mixture at low temperature and high pressure. The temperature is then allowed to rise and the gases collected as they boil off. The gases so produced have many and varied uses.
- **Mesosphere** A section of the atmosphere above the stratosphere in which the temperature falls with increasing height.
- **Ozone (trioxygen)** A colourless gas produced in the stratosphere by the action of high-energy ultraviolet radiation on oxygen gas, producing oxygen atoms. These oxygen atoms then react with further oxygen molecules to produce ozone. Its presence in the stratosphere acts as a screen (ozone layer) against dangerous ultraviolet radiation.
- **Pollution** The modification of the environment caused by human influence. It often renders the environment harmful and unpleasant to life. Water pollution is caused by many substances, such as those found in fertilisers and in industrial effluent. Atmospheric pollution is caused by gases such as sulphur dioxide, carbon monoxide and nitrogen oxides being released into the atmosphere by a variety of industries and also by the burning of fossil fuels.
- **Primary atmosphere** The original thick layer of gases, mainly hydrogen and helium, that surrounded the Earth's core soon after the planet was formed 4500 million years ago.
- **Sea water as a resource** Sea water is a major resource. By evaporation (desalination) we can produce water for drinking or irrigation. Also, bromine, magnesium and salt are extracted in large quantities from sea water.
- **Secondary atmosphere** A layer of a mixture of gases created by early volcanic activity. The mixture that formed this atmosphere included ammonia, nitrogen, methane, carbon monoxide, carbon dioxide and sulphur dioxide gases.
- **Stratosphere** A layer of the atmosphere above the troposphere in which the temperature increases with increasing height.
- **Troposphere** A layer of the atmosphere closest to the Earth which contains about 75% of the mass of the atmosphere. The composition of dry air is relatively constant in this layer of the atmosphere.
- **Water cycle** This cycle shows how water circulates around the Earth. The driving force behind the water cycle is the Sun.

# Atmosphere and oceans
# *Additional questions*

1 The apparatus shown on p. 149, Figure 10.5, was used to estimate the proportion of oxygen in the atmosphere. A volume of dry air (200 $cm^3$) was passed backwards and forwards over heated copper until no further change in volume took place. The apparatus was then allowed to cool down to room temperature and the final volume reading was then taken. Some typical results are shown below.

Volume of gas before = 200 $cm^3$
Volume of gas after = 157 $cm^3$

During the experiment the copper slowly turned black.

**a** Why was the apparatus allowed to cool back to room temperature before the final volume reading was taken?

**b** (i) Using the information given above, calculate the volume reduction which has taken place.
(ii) Calculate the percentage reduction in volume.

**c** Explain briefly why there is a change in volume.

**d** What observation given above supports your explanation in **c**? Write a balanced chemical equation for any reaction which has occurred.

**e** Give the name of the main residual gas at the end of the experiment.

**f** Would you expect the copper to have increased or decreased in mass during the experiment? Explain your answer.

2 Explain the following.

**a** Air is a mixture of elements and compounds.

**b** The percentage of carbon dioxide in the atmosphere does not significantly vary from 0.03%.

**c** When liquid air has its temperature slowly raised from −270 °C, helium is the first gas to boil off.

**d** Power stations are thought to be a major cause of acid rain.

3 Air is a raw material from which several useful substances can be separated. They are separated in the following process.
Dry and 'carbon dioxide free' air is cooled under pressure. Most of the gases liquefy as the temperature falls below −200 °C. The liquid mixture is separated by fractional distillation. The boiling points of the gases left in the air after removal of water vapour and carbon dioxide are given in the table below:

| Gas | Boiling point/°C |
|---|---|
| Argon | −186 |
| Helium | −269 |
| Krypton | −157 |
| Neon | −246 |
| Nitrogen | −196 |
| Oxygen | −183 |
| Xenon | −108 |

**a** Why is the air dried and carbon dioxide removed before it is liquefied?

**b** Which of the gases will not become liquid at −200°C?

**c** Which of the substances in the liquid mixture will be the first to change from liquid to gas as the temperature is slowly increased?

**d** Give a use for each of the gases shown in the table.

**e** Use the data given in Table 10.1, p. 148, to calculate the volume of each of the gases found in 1 $dm^3$ of air.

4 Explain what is meant by the term 'pollution' with reference to air and water.

**a** (i) Name an air pollutant produced by the burning of coal.
(ii) Name a different air pollutant produced by the combustion of petrol in a car engine.

**b** Some of our drinking water is obtained by purifying river water.
(i) Would distillation or filtration produce the purest water from river water? Give a reason for your answer.
(ii) Which process, distillation or filtration, is actually used to produce drinking water from river water? Comment on your answer in comparison to your answer in **c** (i).

**c** Power stations produce warm water. This causes thermal pollution as this warm water is pumped into nearby rivers.
(i) Why do power stations produce such large quantities of warm water?
(ii) What effect does this warm water have on aquatic life?

5 In the final stage of the extraction of magnesium from sea water, molten magnesium chloride is electrolysed. Substances are added to bring the working temperature of the electrolysis cell down to 700 °C.

**a** Write equations to represent the reactions taking place at the cathode and anode. State clearly whether oxidation or reduction is taking place.

**b** Why are substances added to the molten magnesium chloride before electrolysis takes place?

**c** The industrial production of magnesium by electrolysis uses a current of 14 000 amps. Calculate the time required to produce 50 kg of magnesium from molten magnesium chloride.
(1 faraday = 96 500 coulombs; $A_r$: Mg = 24)

**d** A large proportion of the magnesium produced (43%) is used to make alloys.
(i) Why has it such a use in the production of magnesium alloys?
(ii) The world production of magnesium is approximately 240 000 tonnes. Calculate the number of tonnes used in the production of magnesium alloys.

6 The developed countries produce about 230 000 tonnes of bromine per year: 80% of this production is from sea water.
   a Calculate the amount of bromine obtained in developed countries from sea water.
   b Sea water contains 0.07 g of bromine per $dm^3$. Using your answer to **a**, calculate the volume of sea water required to obtain that amount of bromine.
   c The main reaction in the extraction process is that involving displacement of bromine using chlorine gas.
      (i) Write an ionic equation for this reaction.
      (ii) Explain why you could not use iodine instead of chlorine in the displacement reaction.
   d Chlorine and bromine are hazardous substances. Describe some of the precautions that have to be taken to ensure the safety of members of the workforce who deal with these two substances.

7 France obtains some of its sodium chloride by evaporation of sea water.
   a If France produces 1.1 million tonnes of salt per year by this method and sea water contains 25 g of sodium chloride per $dm^3$, calculate the volume of sea water required to produce the annual salt production.
   b Give four important uses of sodium chloride.
   c Sodium chloride is an ionic substance.
      (i) Draw a diagram to show the bonding which takes place within sodium chloride.
      (ii) What are the properties of ionic substances such as sodium chloride?

8 In plants, during photosynthesis, carbon dioxide and water are converted into carbohydrates such as glucose ($C_6H_{12}O_6$), and oxygen is released.
   a Write a balanced chemical equation for the reaction taking place during photosynthesis.
   b What conditions are essential for this reaction to take place?
   c Why are animals unable to photosynthesise?
   d Which process occurs if we reverse photosynthesis?
   e Some of the oxygen released during photosynthesis is broken up by ultraviolet radiation. Some of the oxygen atoms produced combine with further oxygen molecules to produce an important allotrope of oxygen.
      (i) Name and give the formula of this allotrope of oxygen.
      (ii) Write a balanced chemical equation for the reaction in which this allotrope is produced.
      (iii) Why is this allotrope of oxygen so important to us?

# 11 Rates of reaction



Figure 11.1 shows some slow and fast reactions. The two photographs on the left show examples of slow reactions. The ripening of apples takes place over a number of weeks, and the making and maturing of cheese may take months. The burning of solid fuels, such as coal, can be said to involve chemical reactions taking place at a medium speed or rate. The other example shows a fast reaction. The chemicals inside explosives, such as TNT, react very rapidly in reactions which are over in seconds or fractions of seconds.

As new techniques have been developed, the processes used within the chemical industry have become more complex. Therefore, chemists and chemical engineers have increasingly looked for ways to control the rates at which chemical reactions take place. In doing so, they have discovered that there are five main ways in which you can alter the rate of a chemical reaction. These ideas are not only incredibly useful to industry but can also be applied to reactions which occur in the school laboratory.

**Figure 11.1** Some slow (ripening fruit and cheese making), medium (coal fire) and fast (explosion) reactions.

## Factors that affect the rate of a reaction

- Surface area of the reactants.
- Concentration of the reactants.
- Temperature at which the reaction is carried out.
- Light.
- Use of a catalyst.

### Surface area

In Chapter 8, we discussed the use of limestone (calcium carbonate) as a substance which can be used to neutralise soil acidity. Powdered limestone is used as it neutralises the acidity faster than if lumps of limestone are used. Why do you think this is the case?

In the laboratory, the reaction between acid and limestone in the form of lumps or powder can be observed in a simple test-tube experiment. Figure 11.2 shows the reaction between dilute hydrochloric acid and limestone in lump and powdered form.

hydrochloric acid + calcium carbonate → calcium chloride + carbon dioxide + water

$$2HCl(aq) + CaCO_3(s) \rightarrow CaCl_2(aq) + CO_2(g) + H_2O(l)$$

The rates at which the two reactions occur can be found by measuring either:

- the volume of the carbon dioxide gas which is produced, or
- the loss in mass of the reaction mixture with time.

These two methods are generally used for measuring the rate of reaction for processes involving the formation of a gas as one of the products.

**Figure 11.2** The powdered limestone (left) reacts faster with the acid than the limestone in the form of lumps.

The apparatus shown in Figure 11.3 is used to measure the loss in mass of the reaction mixture. The mass of the conical flask plus the reaction mixture is measured at regular intervals. The total loss in mass is calculated for each reading of the balance, and this is plotted against time. Some sample results from experiments of this kind have been plotted in Figure 11.4.

**Figure 11.3** After 60 seconds the mass has fallen by 1.24 g.

The reaction is generally at its fastest in the first minute. This is indicated by the slopes of the curves during this time. The steeper the slope, the faster the rate of reaction. You can see from the two traces in Figure 11.4 that the rate of reaction is greater with the powdered limestone than the lump form.

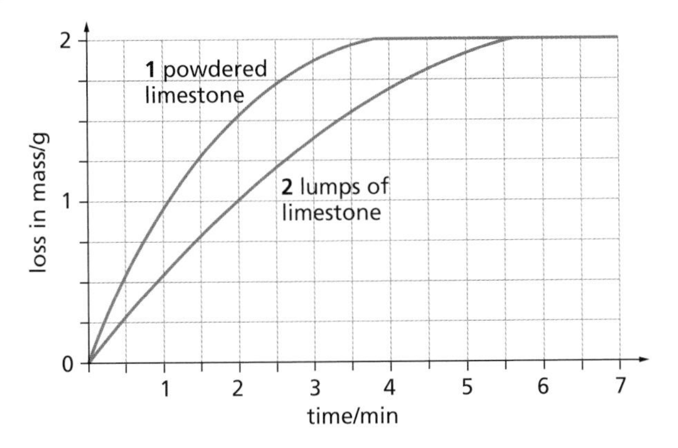


**Figure 11.4** Sample results for the limestone/acid experiment.

The surface area has been increased by powdering the limestone (Figure 11.5). The acid particles now have an increased amount of surface of limestone with which to collide. The products of a reaction are formed when collisions occur between reactant particles. Therefore, the increase in surface area of the limestone increases the rate of reaction.

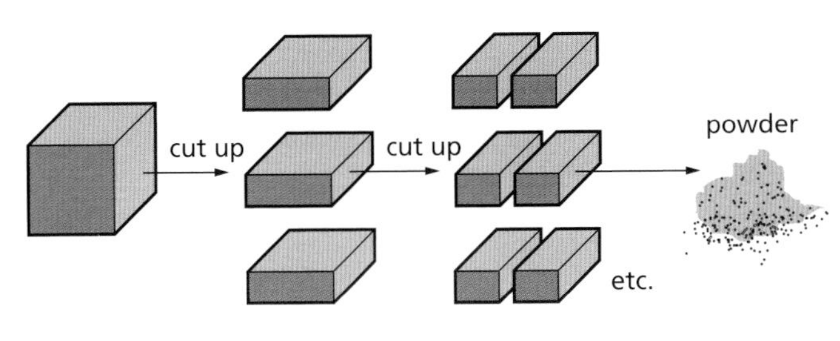


**Figure 11.5** A powder has a larger surface area.

## *Questions*

**1** What apparatus would you use to measure the rate of reaction of the limestone with dilute hydrochloric acid by measuring the volume of carbon dioxide produced?

**2** The following results were obtained from an experiment of the type you were asked to design in question **1**.

| Time/min | 0 | 0.5 | 1.0 | 1.5 | 2.0 | 2.5 | 3.0 | 3.5 | 4.0 | 4.5 | 5.0 |
|---|---|---|---|---|---|---|---|---|---|---|---|
| Total volume of $CO_2$ gas/$cm^3$ | 0 | 15 | 24 | 28 | 31 | 33 | 35 | 35 | 35 | 35 | 35 |

**a** Plot a graph of the total volume of $CO_2$ against time.
**b** At which point is the rate of reaction fastest?
**c** What volume of $CO_2$ was produced after 1 minute 15 seconds?
**d** How long did it take to produce 30 $cm^3$ of $CO_2$?

## Concentration

A yellow precipitate is produced in the reaction between sodium thiosulphate and hydrochloric acid.

sodium thiosulphate + hydrochloric acid → sodium chloride + sulphur + sulphur dioxide + water

$$Na_2S_2O_3(aq) + 2HCl(aq) \rightarrow 2NaCl(aq) + S(s) + SO_2(g) + H_2O(l)$$

The rate of this reaction can be followed by recording the time taken for a given amount of sulphur to be precipitated. This can be done by placing a conical flask containing the reaction mixture on to a cross on a piece of paper (Figure 11.6). As the precipitate of sulphur forms, the cross is obscured and finally disappears from view. The time taken for this to occur is a measure of the rate of this reaction. To obtain sufficient information about the effect of changing the concentration of the reactants, several experiments of this type must be carried out, using different concentrations of sodium thiosulphate or hydrochloric acid.

**Figure 11.6** The precipitate of sulphur obscures the cross.

Some sample results of experiments of this kind have been plotted in Figure 11.7. You will note from the graph that when the most concentrated sodium thiosulphate solution was used, the reaction was at its fastest. This is shown by the shortest time taken for the cross to be obscured.

As discussed earlier, the products of the reaction are formed as a result of the collisions between reactant particles. There are more particles in a more concentrated solution and collisions occur more often. The more often they collide, the greater the chance they have of reacting. This means that the rate of a chemical reaction will increase if the concentration of reactants is increased.

In reactions involving only gases, for example the Haber process (Chapter 15, p. 219), an increase in the overall pressure at which the reaction is carried out increases the rate of the reaction. The increase in pressure results in the gas particles being pushed closer together. This means that they collide more often and so react faster.

time for cross to disappear/s
concentration of sodium thiosulphate/cm$^3$ (0.5 mol dm$^{-3}$ sodium thiosulphate)

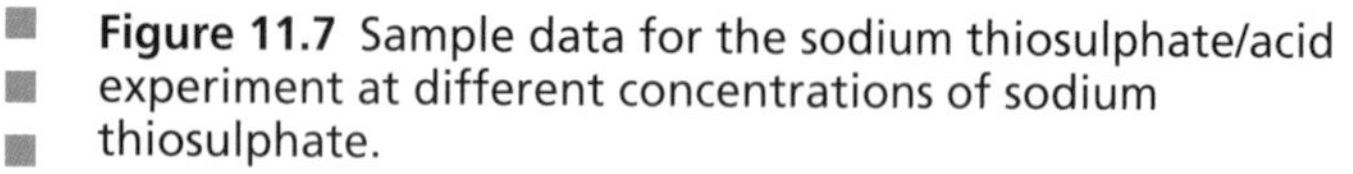

**Figure 11.7** Sample data for the sodium thiosulphate/acid experiment at different concentrations of sodium thiosulphate.

## Question

1 Devise an experiment to show the effect of changing the concentration of dilute acid on the rate of reaction between magnesium and hydrochloric acid.

## Temperature

Why do you think food is stored in a refrigerator? The reason is that the rate of decay is slower at lower temperatures. This is a general feature of the majority of chemical processes.

The reaction between sodium thiosulphate and hydrochloric acid can also be used to study the effect of temperature on the rate of a reaction. Figure 11.8 shows some sample results of experiments with sodium thiosulphate and hydrochloric acid carried out at different temperatures.

You can see from the graph that the rate of the reaction is fastest at high temperatures. When the temperature at which the reaction is carried out is increased, the energy that the particles have also increases – the particles move faster. This increases the number of collisions of sodium thiosulphate and hydrochloric acid particles, and the collisions which occur are more energetic and so more likely to form products. Therefore, if the temperature at which a reaction takes place is increased then the rate of reaction will increase.

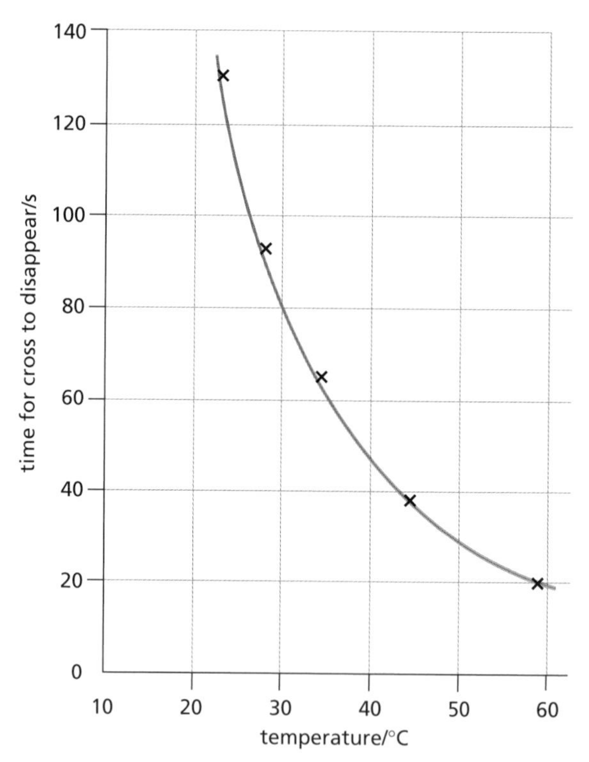


**Figure 11.8** Sample data for the sodium thiosulphate/acid experiment at different temperatures.

## Questions

1 Explain why potatoes cooked in oil cook faster than those cooked in water.

2 Devise an experiment to study the effect of temperature on the reaction between magnesium and hydrochloric acid.

3 Explain why food cooks faster in a pressure cooker.

## Light

Some chemical reactions are affected by light. Photosynthesis is a very important reaction (Chapter 8, p. 121) which occurs only when sunlight falls on leaves containing the green pigment chlorophyll. Another chemical reaction which takes place only when exposed to light is that which occurs when you take a photograph. Photographic film is a transparent plastic strip coated with emulsion: a layer of gelatin throughout which are spread many millions of tiny crystals of silver halides, in particular silver bromide (AgBr). The emulsion used is similar for both black-and-white and colour film. In the case of colour film there are three layers of emulsion with each layer of emulsion containing a different dye.

When light hits a silver bromide crystal, silver cations ($Ag^+$) accept an electron from the bromide ions ($Br^-$) and silver atoms are produced.

silver ion + electron → silver atom

$$Ag^+ + e^- \rightarrow Ag$$

The bromine atom produced in the process is trapped in the gelatin. The more light that falls on the photographic film the greater the amount of silver deposited.

### Question

1 Devise an experiment to show how sunlight affects the rate of formation of silver from the silver salts, silver chloride and silver bromide.

## Catalysts

Over 90% of industrial processes use **catalysts**. A catalyst is a substance which can alter the rate of a reaction without being chemically changed itself. In the laboratory, the effect of a catalyst can be observed using the decomposition of hydrogen peroxide as an example.

hydrogen peroxide → water + oxygen

$$2H_2O_2(aq) \rightarrow 2H_2O(l) + O_2(g)$$

The rate of decomposition at room temperature is very slow. There are substances, however, which will speed up this reaction, one being manganese(IV) oxide. When black manganese(IV) oxide powder is added to hydrogen peroxide solution, oxygen is produced rapidly. The rate at which this occurs can be followed by measuring the volume of oxygen gas produced with time.

Some sample results from experiments of this type have been plotted in Figure 11.9. At the end of the reaction, the manganese(IV) oxide can be filtered off and used again. The reaction can proceed even faster by increasing the amount and surface area of the catalyst. This is because the activity of a catalyst involves its surface. Note that, in gaseous reactions, if dirt or impurities are present on the surface of the catalyst, it will not act as efficiently; it is said to have been 'poisoned'. Therefore, the gaseous reactants must be pure.

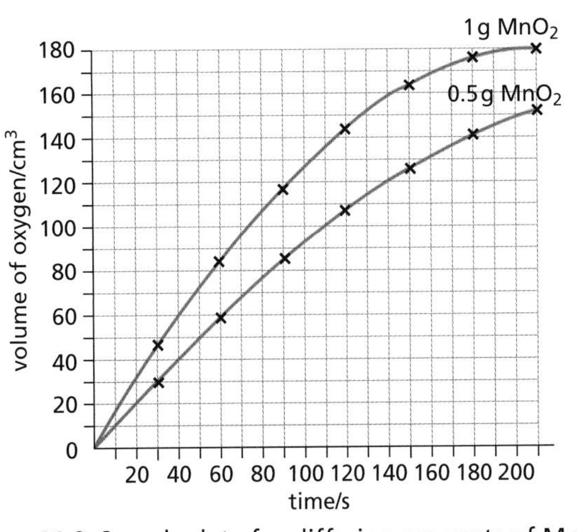


**Figure 11.9** Sample data for differing amounts of $MnO_2$ catalyst.

Chemists have found that:

- a small amount of catalyst will produce a large amount of chemical change
- catalysts remain unchanged chemically after a reaction has taken place, but they can change physically. For example, a finer manganese(IV) oxide powder is left behind after the decomposition of hydrogen peroxide
- catalysts are very specific to a particular chemical reaction.

Some examples of chemical processes and the catalysts used are shown in Table 11.1.

**Table 11.1** Examples of catalysts.

| Process | Catalyst |
|---|---|
| Haber process – for the manufacture of ammonia | Iron |
| Contact process – for the manufacture of sulphuric acid | Vanadium(V) oxide |
| Oxidation of ammonia to give nitric acid | Platinum |
| Fermentation of sugars to produce alcohol | Enzymes (in yeast) |
| Hydrogenation of unsaturated oils to form fats in the manufacture of margarines | Nickel |

A catalyst increases the rate by providing an alternative reaction path with a lower **activation energy**. The activation energy is the energy barrier which reactants must overcome, when their particles collide, to react successfully and form products (Figure 11.10).

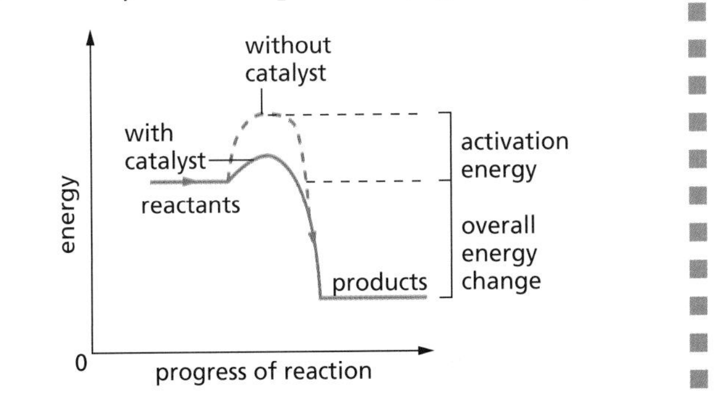


**Figure 11.10** Energy level diagram showing activation energy.

## Catalytic converters

In the previous chapter you saw that European regulations state that all new cars have to be fitted with catalytic converters as part of their exhaust system (Figure 11.11). Car exhaust fumes contain pollutant gases such as carbon monoxide (CO) and nitrogen(II) oxide (NO). The following reactions proceed of their own accord but very slowly under the conditions inside an exhaust.

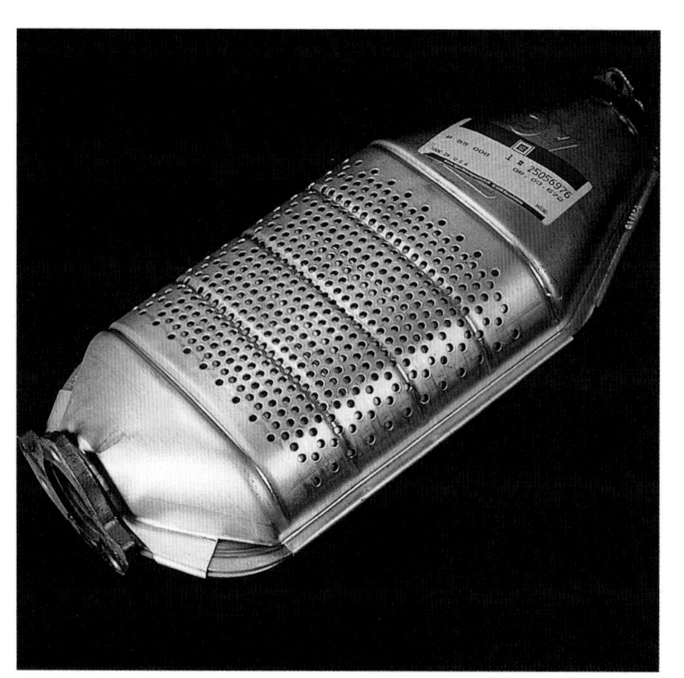

a Catalytic converter.

carbon monoxide + oxygen → carbon dioxide

$$2CO(g) + O_2(g) \rightarrow 2CO_2(g)$$

nitrogen(II) oxide + carbon monoxide → nitrogen + carbon dioxide

$$2NO(g) + 2CO(g) \rightarrow N_2(g) + 2CO_2(g)$$

The catalyst in the converter speeds up these reactions considerably. In these reactions, the pollutants are converted to carbon dioxide and nitrogen, which are naturally present in the air. It should be noted, however, that the catalytic converter can only be used with unleaded petrol and that, due to impurities being deposited on the surface of the catalyst, it becomes poisoned and has to be replaced every five or six years.

## Questions

1 Using a catalysed reaction of your choice, devise an experiment to follow the progress of the reaction and determine how effective the catalyst is.

2 Why do some people consider catalytic converters not to be as environmentally friendly as suggested in their advertising material?

3 Unreacted hydrocarbons such as octane, $C_8H_{18}$ (from petrol), also form part of the exhaust gases. These gases are oxidised in the converter to carbon dioxide and water vapour. Write an equation for the oxidation of octane.

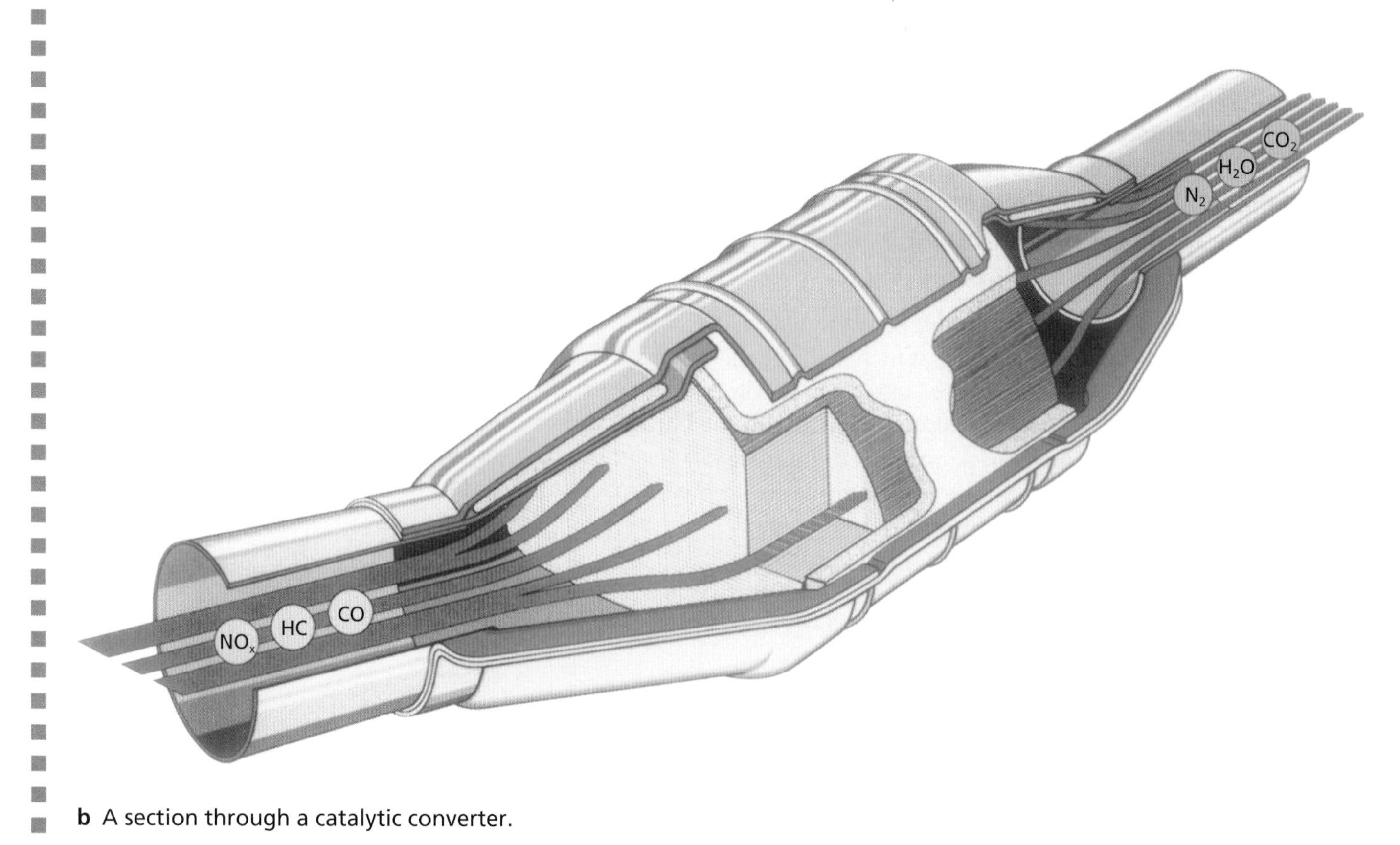


b A section through a catalytic converter.

**Figure 11.11**

## Enzymes

**Enzymes** are protein molecules produced in living cells. These substances are used by living organisms as catalysts to speed up hundreds of different chemical reactions going on inside them. These biological catalysts are very specific in that each chemical reaction taking place has a different enzyme catalyst. You can imagine, therefore, that there are literally hundreds of different kinds of enzyme. For example, hydrogen peroxide is a substance naturally produced within our bodies (a natural metabolic product). However, it is extremely damaging and must be decomposed very rapidly. Catalase is the enzyme which converts hydrogen peroxide into harmless water and oxygen within our livers:

$$\text{hydrogen peroxide} \xrightarrow{\text{catalase}} \text{water} + \text{oxygen}$$

$$2H_2O_2(aq) \xrightarrow{\text{catalase}} 2H_2O(l) + O_2(g)$$

Although many chemical catalysts can work under various conditions of temperature and pressure as well as alkalinity or acidity, biological catalysts operate only under very particular conditions. For example, they operate over a very narrow temperature range and if the temperature becomes too high, they become inoperative. At temperatures above about 45 °C, they denature. This means that the specific shape of the active site of the enzyme molecule changes due to the breaking of bonds. This means that the reactant molecules are no longer able to fit into the active site (Figure 11.12).

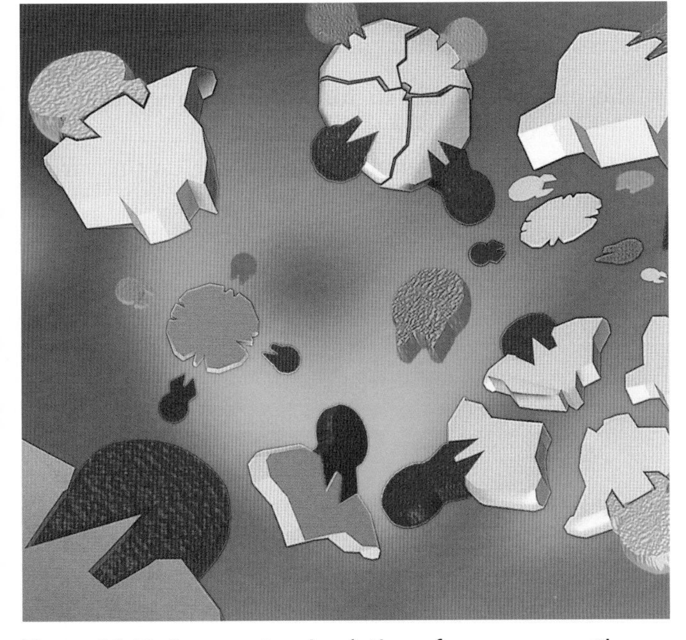

**Figure 11.12** A computer simulation of an enzyme active site. The enzyme molecules (red, pink, green and blue) have an active site that locks on exactly to a particular reactant (yellow). Once locked on, they can work to break up the pieces of the molecules.

A huge multimillion-pound industry has grown up around the use of enzymes to produce new materials. Biological washing powders (Figure 11.13) contain enzymes to break down stains such as sweat, blood and egg, and they do this at the relatively low temperature of 40 °C. This reduces energy costs, because the washing water does not need to be heated as much.

**Figure 11.13** These biological washing powders contain enzymes.

There were problems associated with the early biological washing powders. Some customers suffered from skin rashes, because they were allergic to the enzymes (Figure 11.14). This problem has been overcome to a certain extent by advising that extra rinsing is required. Also, many manufacturers have placed warnings on their packets, indicating that the powder contains enzymes which may cause skin rashes.

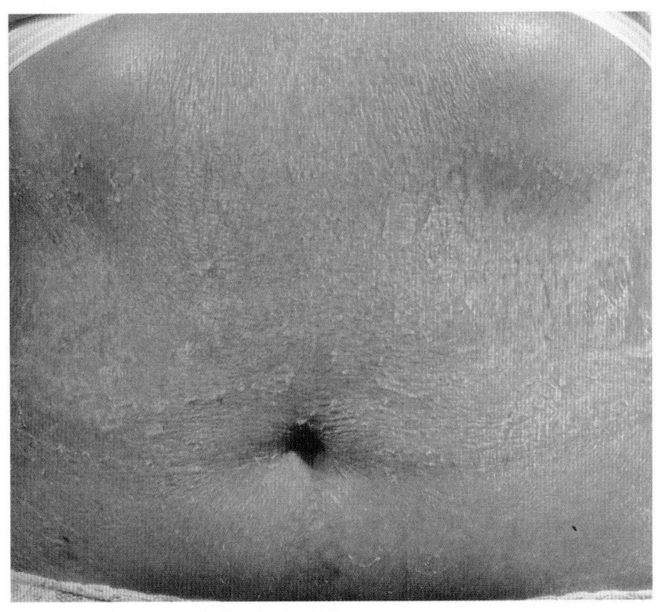

**Figure 11.14** An allergic reaction to a biological detergent.

Other industrial processes make use of enzymes.

- In the manufacture of baby foods, enzymes called proteases are used to 'pre-digest' the protein part of the baby food.
- The enzyme isomerase is used to convert glucose syrup to fructose syrup. Fructose syrup is much sweeter than glucose syrup and can be used as a sweetener in slimming foods as less is needed.
- In the production of yoghurt, milk is initially heated to 90 °C for 15–30 minutes to kill any bacteria in the milk. After cooling to 40 °C, a starter culture of *Lactobacillus* bacteria is added and the mixture incubated at 40 °C for eight hours (Figure 11.15). The bacteria ferment the lactose in the milk to lactic acid, which causes the milk protein to become solid.

**Figure 11.15** Yoghurt is incubated in these tanks, and allowed to mature.

- In cheese making, milk is initially heated to kill bacteria and then cooled. A starter culture of *Streptococcus* bacteria is then added, which coagulates the milk into curds and whey (Figure 11.16). The curds are put into steel or wooden drums and pressed and allowed to dry.

**Figure 11.16** Cheese is separated into curds and whey by the addition of bacteria. The liquid, whey, is separated from the curds which are then pressed.

In industry, enzymes are used to bring about reactions at normal temperatures and pressures that would otherwise require expensive conditions and equipment. Successful processes using enzymes need to ensure that:

- the enzyme is able to function for long periods of time by optimising the environment
- the enzyme is not lost by trapping it on the surface of an inert solid
- continuous processes occur rather than batch processes.

## Questions

1 When using biological washing powders what factors have to be taken into consideration?

2 Enzymes in yeast are used in the fermentation of glucose. Why, when the temperature is raised to 45 °C, is very little ethanol actually produced compared with the amount formed at room temperature?

## Checklist

**After studying Chapter 11 you should know and understand the following terms.**

- **Activation energy** The excess energy that a reaction must acquire to permit the reaction to occur.

- **Catalyst** A substance which alters the rate of a chemical reaction without itself being chemically changed.

- **Catalytic converter** A device for converting dangerous exhaust gases from cars into less harmful emissions. For example, carbon monoxide gas is converted to carbon dioxide gas.

- **Enzymes** Protein molecules produced in living cells. They act as biological catalysts and are specific to certain reactions. They operate only within narrow temperature and pH ranges.

- **Reaction rate** A measure of the change which happens during a reaction in a single unit of time. It may be affected by the following factors:
  - surface area of the reactants
  - concentration of the reactants
  - the temperature at which the reaction is carried out
  - light
  - use of a catalyst.

# Rates of reaction

# *Additional questions*

1 Explain the following statements.

a A car exhaust pipe will rust much faster if the car is in constant use.

b Carrots cook faster when they are chopped up.

c Industrial processes become more economically viable if a catalyst can be found for the reactions involved.

d In fireworks it is usual for the ingredients to be powdered.

e Tomatoes ripen faster in a greenhouse.

f The reaction between zinc and dilute hydrochloric acid is slower than the reaction between zinc and concentrated hydrochloric acid.

2 A student performed two experiments to establish how effective manganese(IV) oxide was as a catalyst for the decomposition of hydrogen peroxide. The results below were obtained by carrying out these experiments with two different quantities of manganese(IV) oxide. The volume of the gas produced was recorded against time.

| Time/s | 0 | 30 | 60 | 90 | 120 | 150 | 180 | 210 |
|---|---|---|---|---|---|---|---|---|
| Volume for 0.3 g/cm$^3$ | 0 | 29 | 55 | 79 | 98 | 118 | 133 | 146 |
| Volume for 0.5 g/cm$^3$ | 0 | 45 | 84 | 118 | 145 | 162 | 174 | 182 |

a Draw a diagram of the apparatus you could use to carry out these experiments.

b Plot a graph of the results.

c Is the manganese(IV) oxide acting as a catalyst in this reaction? Explain your answer.

d (i) At which stage does the reaction proceed most quickly?

(ii) How can you tell this from your graph?

(iii) In terms of particles, explain why the reaction is quickest at the point you have chosen in (i).

e Why does the slope of the graph become less steep as the reaction proceeds?

f What volume of gas has been produced when using 0.3 g of manganese(IV) oxide after 50 s?

g How long did it take for 60 cm$^3$ of gas to be produced when the experiment was carried out using 0.5 g of the manganese(IV) oxide?

h Write a balanced chemical equation for the decomposition of hydrogen peroxide.

3 a Which of the following reaction mixtures will produce hydrogen more quickly at room temperature:

(i) zinc granules + dilute nitric acid?

(ii) zinc powder + dilute nitric acid?

b Give an explanation of your answer to **a**.

c Suggest two other methods by which the speed of this reaction can be altered.

4 A flask containing dilute hydrochloric acid was placed on a digital balance. An excess of limestone chippings was added to this acid, a plug of cotton wool was placed in the neck of the flask and the initial mass was recorded. The mass of the apparatus was recorded every two minutes. At the end of the experiment the loss in mass of the apparatus was calculated and the following results were obtained.

| Time/min | 0 | 2 | 4 | 6 | 8 | 10 | 12 | 14 | 16 |
|---|---|---|---|---|---|---|---|---|---|
| Loss in mass/g | 0 | 2.1 | 3.0 | 3.1 | 3.6 | 3.8 | 4.0 | 4.0 | 4.0 |

a Plot the results of the experiment.

b Which of the results would appear to be incorrect? Explain your answer.

c Write a balanced chemical equation to represent the reaction taking place.

d Why did the mass of the flask and its contents decrease?

e Why was the plug of cotton wool used?

f How does the rate of reaction change during this reaction? Explain this using particle theory.

g How long did the reaction last?

h How long did it take for half of the reaction to occur?

5 a What is a catalyst?

b List the properties of catalysts.

c Name the catalyst used in the following processes:

(i) the Contact process

(ii) the Haber process

(iii) the hydrogenation of unsaturated fats.

d Which series of metallic elements, in the periodic table, do the catalysts you have named in **c** belong to?

e What are the conditions used in the industrial processes named in **c**? The following references will help you: Chapters 12, 14 and 15.

6 This question concerns the reaction of copper(II) carbonate with dilute hydrochloric acid. The equation for the reaction is:

$$CuCO_3(s) + 2HCl(aq) \rightarrow CuCl_2(aq) + CO_2(aq) + H_2O(l)$$

a Sketch a graph to show the rate of production of carbon dioxide when an excess of dilute hydrochloric acid is added. The reaction lasts 40 s and produces 60 cm$^3$ of gas.

b Find on your graph the part which shows:

(i) where the reaction is at its fastest

(ii) when the reaction has stopped.

c Calculate the mass of copper(II) carbonate used to produce 60 cm$^3$ of carbon dioxide. ($A_r$: C = 12; O = 16; Cu = 63.5. One mole of a gas occupies 24 dm$^3$ at room temperature and pressure (rtp).)

**d** Sketch a further graph using the same axes to show what happens to the rate at which the gas is produced if:
(i) the concentration of the acid is decreased
(ii) the temperature is increased.

**7** European regulations state that all new cars have to be fitted with catalytic converters as part of their exhaust system.
**a** Why are these regulations necessary?
**b** Which gases are removed by catalytic converters?
**c** Which metals are often used as catalysts in catalytic converters?
**d** What does the term 'poisoned' mean with respect to catalysts?
**e** The latest converters will also remove unburnt petrol. An equation for this type of reaction is:

$$2C_7H_{14}(g) + 21O_2(g) \rightarrow 14CO_2(g) + 14H_2O(g)$$

(i) Calculate the mass of carbon dioxide produced by 1.96 g of unburnt fuel.
(ii) Convert this mass of carbon dioxide into a volume measured at rtp.
(iii) If the average car produces 7.84 g of unburnt fuel a day, calculate the volume of carbon dioxide produced by the catalytic converter measured at rtp.
($A_r$: H = 1; C = 12; O = 16. One mole of any gas occupies 24 $dm^3$ at rtp.)

**8** **a** Give examples of chemical reactions which happen:
(i) very slowly
(ii) at a moderate rate
(iii) very quickly.
**b** How could you speed up the reaction named in **a** (i)?
**c** How could you slow down the reaction named in **a** (ii)?

# 12 The petroleum industry



## Substances from oil

What do the modes of transport shown in Figure 12.1 have in common? They all use liquids obtained from **crude oil** as fuels.

**Figure 12.1** Modes of transport.

## Oil refining

Crude oil is a complex mixture of compounds known as **hydrocarbons** (Figure 12.2a). Hydrocarbons are molecules which contain only the elements carbon and hydrogen bonded together covalently (Chapter 4, p. 59). These carbon compounds form the basis of a group called **organic compounds**. All living things are made from organic compounds based on chains of carbon atoms similar to those found in crude oil. Crude oil is not only a major source of fuel but is also a raw material of enormous importance. It supplies a large and diverse chemical industry to make dozens of products (Figure 12.2b).

**a** Crude oil is a mixture of hydrocarbons.

Crude oil is not very useful to us until it has been processed. The process, known as **refining**, is carried out at an oil refinery (Figure 12.3).

**Figure 12.3** An oil refinery.

**b** The objects above are made from substances obtained from oil.

**Figure 12.2**

refinery gas used as a fuel
fractionating tower
30 °C
gasoline
used as fuel in cars (petrol)
110 °C
naphtha
used to make chemicals
180 °C
kerosene
used as a fuel in jet engines
260 °C
diesel oil or gas oil
used as a fuel in diesel engines
crude oil
fuel oil
used as a fuel for ships and for home heating systems
lubricating oil
heater
340 °C
residue
used to make bitumen for surfacing roads

**Figure 12.4** Fractional distillation of crude oil in a refinery.

Refining involves separating crude oil into various batches or **fractions**. Chemists use a technique called **fractional distillation** to separate the different fractions. This process works in a similar way as that discussed in Chapter 2, p. 24, for separating ethanol (alcohol) and water. The different components (fractions) separate because they have different boiling points. The crude oil is heated to about 400 °C to vaporise all the different parts of the mixture. The mixture of vapours passes into the fractionating column near the bottom (Figure 12.4). Each fraction is obtained by collecting hydrocarbon molecules which have a boiling point in a given range of temperatures (Figure 12.4). For example, the fraction we know as petrol contains molecules which have boiling points between 30 °C and 110 °C. The molecules in this fraction contain between five and ten carbon atoms. These smaller molecules with lower boiling points condense higher up the tower. The bigger hydrocarbon molecules which have the higher boiling points condense in the lower half of the tower.

The liquids condensing at different levels are collected on **trays**. In this way the crude oil is separated into different fractions. These fractions usually contain a number of different hydrocarbons. The individual single hydrocarbons can then be obtained, again by refining the fraction by further distillation.

It is important to realise that the uses of the fractions depend on their properties. For example, one of the lower fractions, which boils in the range 250–350 °C, is quite thick and sticky and makes a good lubricant. However, the petrol fraction burns very easily and this therefore makes it a good fuel for use in engines.

## *Questions*

1 What do you understand by the term hydrocarbon?

2 All organisms are composed of compounds which contain carbon. Why do you think carbon chemistry is often called 'organic chemistry'?

3 List the main fractions obtained by separating the crude oil mixture and explain how they are obtained in a refinery.

## Alkanes

Most of the hydrocarbons in crude oil belong to the family of compounds called **alkanes**. The molecules within the alkane family contain carbon atoms covalently bonded to four other atoms by single bonds. Because these molecules possess only single bonds they are said to be **saturated**, as no further atoms can be added (Figure 12.5). The physical properties of the first six members of the alkane family are shown in Table 12.1.

**Table 12.1** Some alkanes and their physical properties.

| Alkane | Formula | Melting point/ °C | Boiling point/ °C | Physical state at room temperature |
|---|---|---|---|---|
| Methane | $CH_4$ | −182 | −162 | Gas |
| Ethane | $C_2H_6$ | −183 | −89 | Gas |
| Propane | $C_3H_8$ | −188 | −42 | Gas |
| Butane | $C_4H_{10}$ | −138 | 0 | Gas |
| Pentane | $C_5H_{12}$ | −130 | 36 | Liquid |
| Hexane | $C_6H_{14}$ | −95 | 69 | Liquid |

You will notice from Figure 12.5 and Table 12.1 that the compounds have a similar structure and similar name endings. They also behave chemically in a similar way and the family of compounds can be represented by a general formula. In the case of the alkanes the general formula is:

$$C_nH_{(2n+2)}$$

where $n$ is the number of carbon atoms present.

A family with the above factors in common is called a **homologous series**. As you go up a homologous series, in order of increasing number of carbon atoms, the physical properties of the compounds gradually change. For example, the melting and boiling points of the alkanes shown in Table 12.1 gradually increase. This is due to an increase in the intermolecular forces (van der Waals' forces) as the size and mass of the molecule increases (Chapter 4, p. 61).

Under normal conditions molecules with up to four carbon atoms are gases, those with between five and 16 carbon atoms are liquids, while those with greater than 16 carbon atoms are solids.

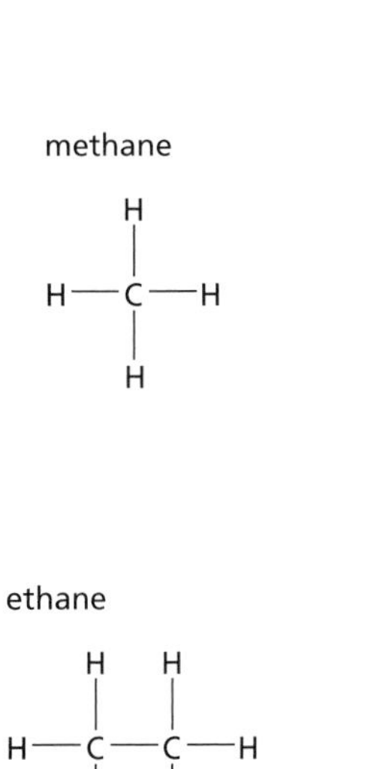

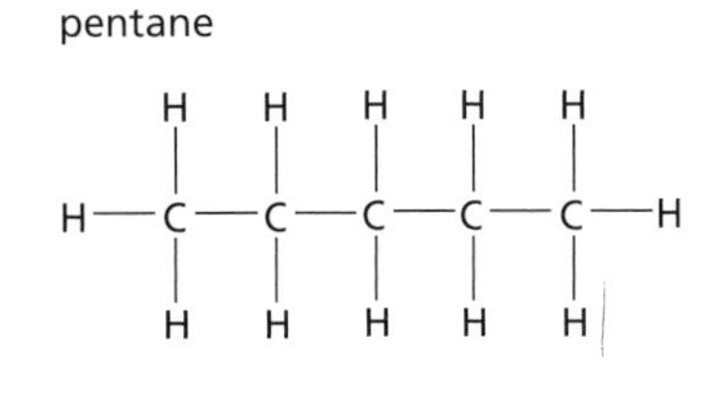


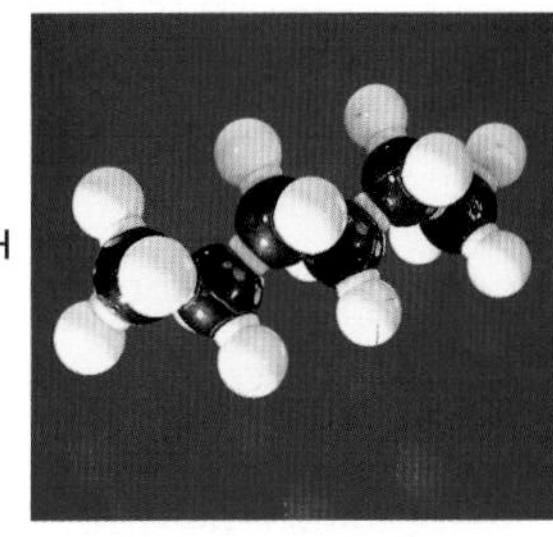

Figure 12.5 The alkane molecules look like the models in the photographs.

### Questions

1. Estimate the boiling points for the alkanes with formulae:
   a $C_7H_{16}$
   b $C_8H_{18}$.
2. Name the alkanes which have the following formulae:
   a $C_7H_{16}$
   b $C_{10}H_{22}$.

## Naming the alkanes

All the alkanes have names ending in *-ane*. The rest of the name tells you the number of carbon atoms present in the molecule. For example, the compound whose name begins with:

- *meth-* has one carbon atom
- *eth-* has two carbon atoms
- *prop-* has three carbon atoms
- *but-* has four carbon atoms
- *pent-* has five carbon atoms

and so on.

## Structural isomerism

Sometimes it is possible to write more than one structural formula to represent a molecular formula. The structural formula of a compound shows how the atoms are joined together by the covalent bonds. For example, there are two different compounds with the molecular formula $C_4H_{10}$. The structural formulae of these two substances along with their names and physical properties are shown in Figure 12.6.

Compounds such as those in Figure 12.6 are known as **isomers**. Isomers are substances which have the same molecular formula but different structural formulae. The different structures of the compounds shown in Figure 12.6 have different melting and boiling points. Molecule (b) contains a branched chain and has a lower melting point than molecule (a), which has no branched chain. All the alkane molecules with four or more carbon atoms possess isomers. Perhaps now you can see why there are so many different organic compounds!

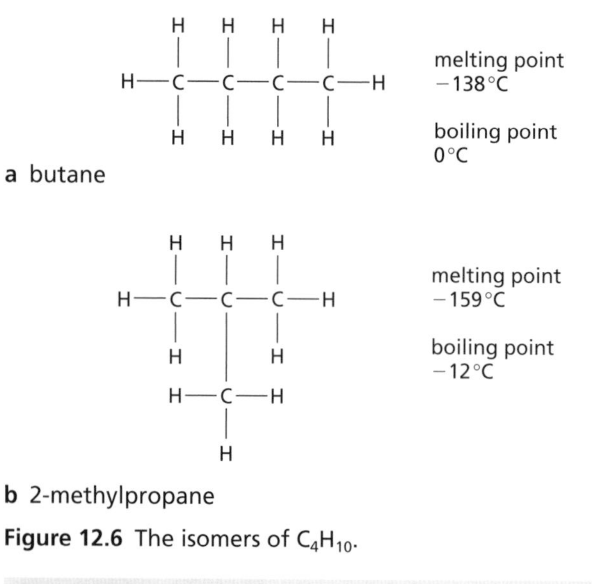


**Figure 12.6** The isomers of $C_4H_{10}$.

### Question

**1** Draw the structural formulae for the isomers of:
**a** $C_5H_{12}$
**b** $C_6H_{14}$.

## The chemical behaviour of alkanes

Alkanes are rather unreactive compounds. For example, they are generally not affected by alkalis, acids or many other substances. Their most important property is that they burn easily.

Gaseous alkanes, such as methane, will burn in a good supply of air, forming carbon dioxide and water as well as plenty of heat energy.

methane + oxygen → carbon dioxide + water + energy

$$CH_4(g) + 2O_2(g) \rightarrow CO_2(g) + 2H_2O(g)$$

The gaseous alkanes are some of the most useful fuels. Methane, better known as natural gas, is used for cooking as well as for heating our offices, schools and homes (Figure 12.7a). Propane and butane burn with very hot flames and they are sold as liquefied petroleum gas (LPG). In rural areas where there is no supply of natural gas, central heating systems can be run on propane gas (Figure 12.7b). Butane, sometimes mixed with propane, is used in portable blowlamps and in gas lighters.

**a** This is burning methane.

**b** Central heating systems can be run on propane.

**Figure 12.7**

Another useful reaction worth noting is that between the alkanes and the halogens. For example, methane and chlorine react in the presence of sunlight (or ultraviolet light). The ultraviolet light splits the chlorine molecules into atoms. When this type of reaction takes place, these atoms are called **free radicals** and they are very reactive.

$$\text{chlorine gas} \xrightarrow{\text{sunlight}} \text{chlorine atoms (free radicals)}$$

$$Cl_2(g) \longrightarrow 2Cl(g)$$

The chlorine atoms then react further with methane molecules, and a hydrogen chloride molecule is produced along with a methyl free radical.

$$\text{chlorine atom} + \text{methane} \rightarrow \text{methyl radical} + \text{hydrogen chloride}$$

$$Cl(g) + CH_4(g) \rightarrow CH_3(g) + HCl(g)$$

The methyl free radical reacts further.

$$\text{methyl radical} + \text{chlorine gas} \rightarrow \text{chloromethane} + \text{chlorine atom}$$

$$CH_3(g) + Cl_2(g) \rightarrow CH_3Cl(g) + Cl(g)$$

This chlorine free radical, in turn, reacts further and the process continues until all the chlorine and the methane have been used up. This type of process is known as a **chain reaction** and it is very fast. The overall chemical equation for this process is:

$$\text{methane} + \text{chlorine} \rightarrow \text{chloromethane} + \text{hydrogen chloride}$$

$$CH_4(g) + Cl_2(g) \rightarrow CH_3Cl(g) + HCl(g)$$

We can see from this final equation that one hydrogen atom of the methane molecule is **substituted** by a chlorine atom. This type of reaction is known as a **substitution reaction**.

Because we cannot control the chlorine free radicals produced in this reaction, we also obtain small amounts of other 'substituted' products – $CH_2Cl_2$ (dichloromethane), $CHCl_3$ (trichloromethane or chloroform) and $CCl_4$ (tetrachloromethane). Many of these so-called **halogenoalkanes** are used as solvents. For example, dichloromethane is used as a solvent in paint stripper (Figure 12.8).

**Figure 12.8** Dichloromethane is used as a solvent in paint stripper.

Early anaesthetics relied upon trichloromethane, $CHCl_3$, or chloroform. Unfortunately, this anaesthetic had a severe problem since the lethal dose was only slightly higher than that required to anaesthetise the patient. In 1956, halothane was discovered by chemists working at ICI. This is a compound containing chlorine, bromine and fluorine. Its formula is $CF_3CHBrCl$. However, even this is not the perfect anaesthetic since evidence suggests that prolonged exposure to this substance may cause liver damage. The search continues for even better anaesthetics.

A group of compounds were discovered in the 1930s and were called the chlorofluorocarbons or CFCs for short. Because of their inertness they found many uses, especially as a propellant in aerosol cans. CFC-12 or dichlorodifluoromethane, $CF_2Cl_2$, was one of the most popular CFCs in use in aerosols. Scientists now believe that the CFCs released from aerosols are destroying the ozone layer.

## The ozone hole problem

Our atmosphere protects us from harmful ultraviolet radiation from the Sun. This damaging radiation is absorbed by the relatively thin ozone layer found in the stratosphere (Figure 12.9).

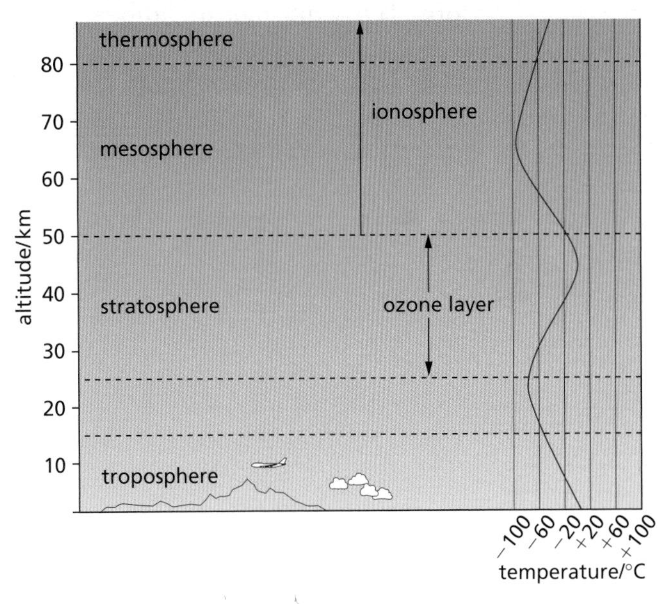


**Figure 12.9** The ozone layer is between 25 km and 50 km above sea level.

Large holes have recently been discovered in the ozone layer over Antarctica, Australasia and Europe. Scientists think that these holes have been produced by CFCs such as CFC-12. CFCs escape into the atmosphere and, because of their inertness, remain without further reaction until they reach the stratosphere and the ozone layer. In the stratosphere the high-energy ultraviolet radiation causes a chlorine atom to split off from the CFC molecule. This chlorine atom, or free radical, then reacts with the ozone.

$$Cl(g) + O_3(g) \rightarrow OCl(g) + O_2(g)$$

This is not the only problem with CFCs. They are also significant 'greenhouse gases' (Chapter 8, p. 121). The ozone depletion and greenhouse effects have become such serious problems that an international agreement known as the *Montreal Protocol on Substances that Deplete the Ozone Layer* was agreed in 1987. The proposed controls were tightened in 1990 by the second meeting of the parties to the *Montreal Protocol*. Modifications have since been made at meetings held in 1992, 1995 and 1996.

Research is now going ahead, with some success, to produce safer alternatives to CFCs. At present, better alternatives called hydrochlorofluorocarbons (HCFCs) have been developed. These substances have lower ozone-depletion effects and are not very effective greenhouse gases.

## Other uses of alkanes

Besides their major use as fuels (Chapter 13, p. 184), some of the heavier alkanes are used as waxes (Figure 12.10), as lubricating oils and in the manufacture of another family of hydrocarbons – the alkenes.

**Figure 12.10** Candles contain a mixture of heavier alkanes.

## Questions

1 Write a balanced chemical equation to represent the combustion of propane.

2 In what mole proportions should chlorine and methane be mixed to produce:
**a** mainly chloromethane?
**b** mainly tetrachloromethane?

3 Describe a method you would use to separate chloromethane from the other possible reaction products when methane reacts with chlorine.

4 Why do you think that CFCs release chlorine atoms into the stratosphere and not fluorine atoms?

# *Alkenes*

Alkenes form another homologous series of hydrocarbons of the general formula:

$$C_nH_{2n}$$

where *n* is the number of carbon atoms. The alkenes are more reactive than the alkanes because they each contain a double covalent bond between the carbon atoms (Figure 12.11). Molecules that possess a double covalent bond of this kind are said to be **unsaturated**, because it is possible to break one of the two bonds to add extra atoms to the molecule.

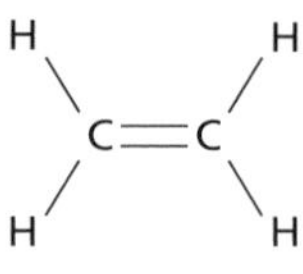

**Figure 12.11** The bonding in ethene, the simplest alkene.

All alkenes have names ending in *-ene*. Alkenes, especially ethene, are very important industrial chemicals. They are used extensively in the plastics industry and in the production of alcohols such as ethanol and propanol. Table 12.2 gives the names, formulae and some physical properties of the first three members of the alkene family. Figure 12.12 shows the structure of these members as well as models of their shape.

**Table 12.2** The first three alkenes and their physical properties.

| Alkene | Formula | Melting point/°C | Boiling point/°C | Physical state at room temperature |
|---|---|---|---|---|
| Ethene | $C_2H_4$ | −169 | −104 | Gas |
| Propene | $C_3H_6$ | −185 | −47 | Gas |
| Butene | $C_4H_8$ | −184 | −6 | Gas |

ethene

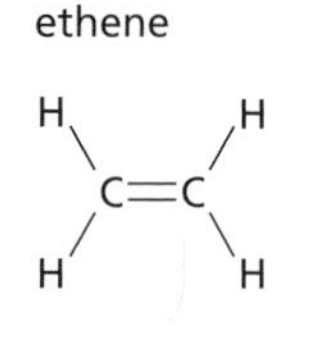

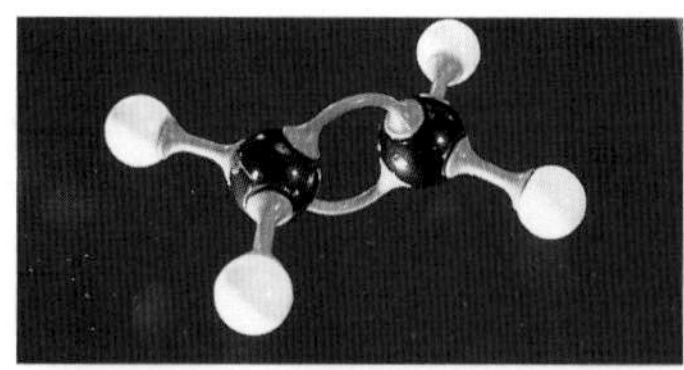

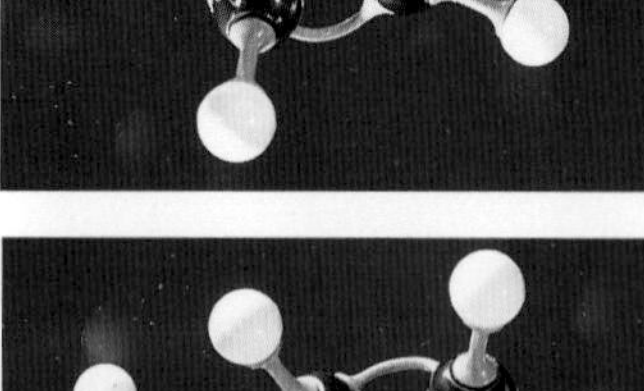

propene

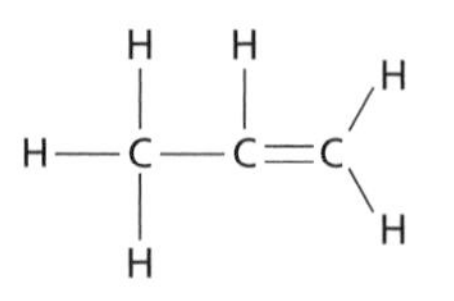

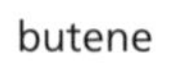

butene

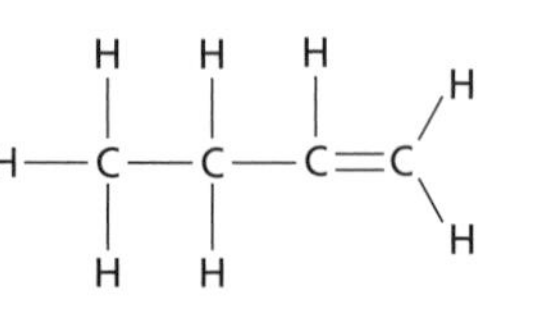

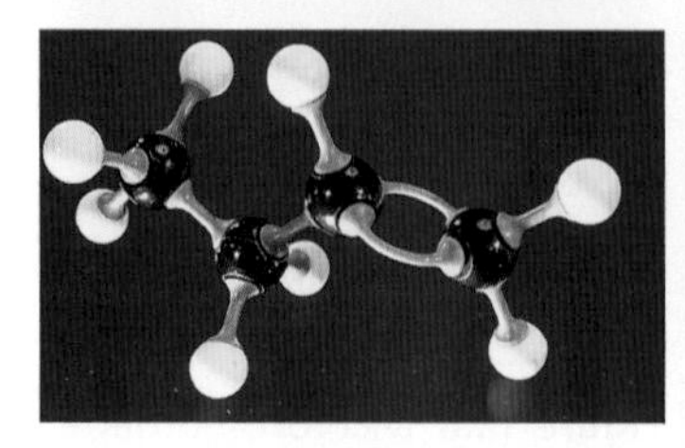

**Figure 12.12** Structure and shape of the first three alkenes.

## Where do we get alkenes from?

Very few alkenes are found in nature. Most of the alkenes used by the petrochemical industry are obtained by breaking up larger, less useful alkane molecules obtained from the fractional distillation of crude oil. This is usually done by a process called **catalytic cracking**. In this process the alkane molecules to be 'cracked' (split up) are passed over a mixture of aluminium and chromium oxides heated to about 500 °C.

| dodecane | → | decane | + ethene |
|---|---|---|---|
| $C_{12}H_{26}(g)$ | → | $C_{10}H_{22}(g)$ | + $C_2H_4(g)$ |
| (found in kerosene) | | shorter alkane | alkene |

Figure 12.13 shows the simple apparatus that can be used to carry out cracking reactions in the laboratory. You will notice that in the laboratory we may use a catalyst of broken, unglazed pottery.

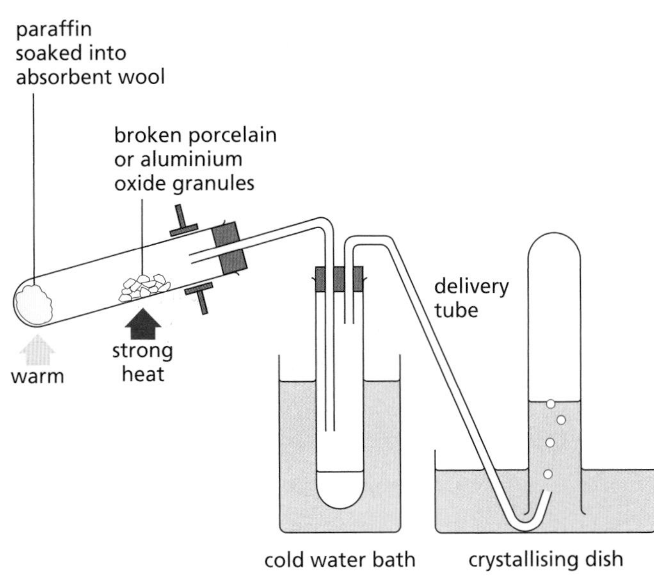


**Figure 12.13** The cracking of an alkane in the laboratory.

## The chemical behaviour of alkenes

The double bond makes alkenes more reactive than alkanes during chemical reactions. For example, hydrogen will add across the double bond of ethene, under suitable conditions, forming ethane (Figure 12.14).

$$\text{ethene} + \text{hydrogen} \xrightarrow[\text{Ni/Pt catalyst}]{150\text{–}300\,°\text{C}} \text{ethane}$$

$$C_2H_4(g) + H_2(g) \longrightarrow C_2H_6(g)$$

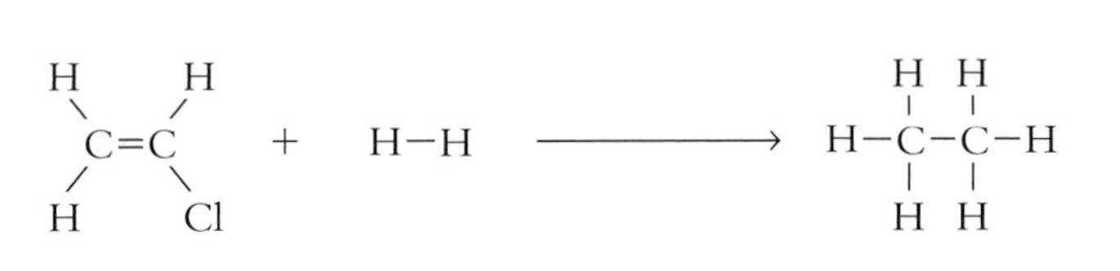


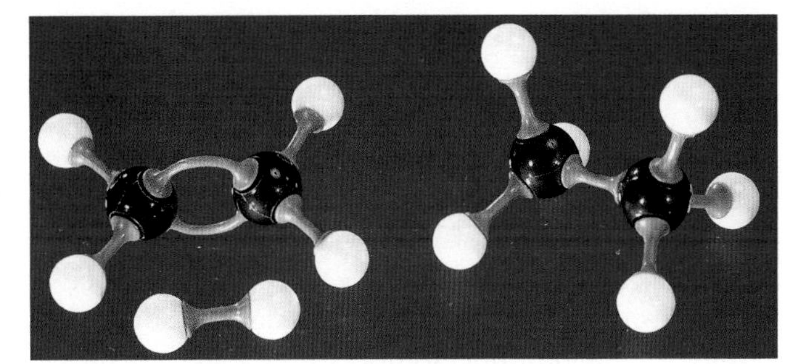

**Figure 12.14** The addition of hydrogen to ethene using molecular models.

This reaction is called **hydrogenation**. Hydrogenation reactions like the one shown with ethene are used in the manufacture of margarines from vegetable oils. Vegetable oils contain fatty acids, such as linoleic acid ($C_{18}H_{32}O_2$). These are unsaturated molecules, containing several double bonds. These double bonds make the molecule less flexible. Hydrogenation can convert these molecules into more saturated ones. Now the molecules are less rigid and can flex and twist more easily, and hence pack more closely together. This in turn causes an increase in the intermolecular forces and so raises the melting point. The now solid margarines can be spread on bread more easily than liquid oils.

There is another side to this process. Many doctors now believe that unsaturated fats are more healthy than saturated ones. Because of this, many margarines are left partially unsaturated. They do not have all the C=C taken out of the fat molecules. However, the matter is far from settled and the debate continues.

Another important **addition reaction** is the one used in the manufacture of ethanol. Ethanol has important uses as a solvent and a fuel (p. 189). It is formed when water (as steam) is added across the double bond in ethene. For this reaction to take place, the reactants have to be passed over a catalyst of phosphoric(V) acid (absorbed on silica pellets) at a temperature of 300 °C and pressure of 60 atmospheres (1 atmosphere = $1 \times 10^5$ pascals).

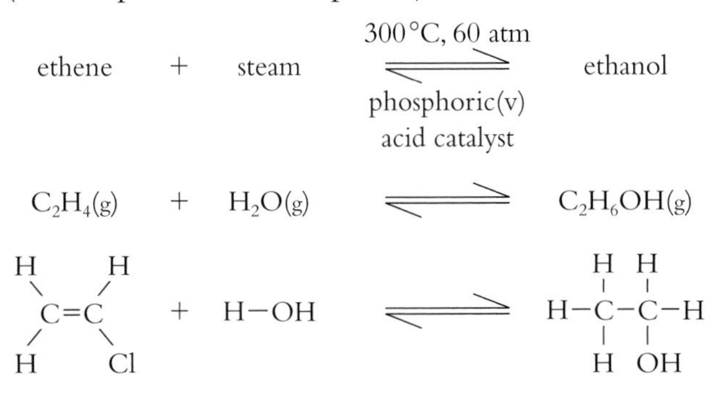

$$\text{ethene} + \text{steam} \underset{\text{phosphoric(V) acid catalyst}}{\overset{300\,°\text{C, 60 atm}}{\rightleftharpoons}} \text{ethanol}$$

$$C_2H_4(g) + H_2O(g) \rightleftharpoons C_2H_5OH(g)$$



This reaction is reversible as is shown by the equilibrium (⇌) sign. The conditions have been chosen to ensure the highest possible yield of ethanol. In other words, the conditions have been chosen so that they favour the forward reaction.

## A test for unsaturated compounds

The addition reaction between bromine dissolved in an organic solvent, or water, and alkenes is used as a chemical test for the presence of a double bond between two carbon atoms. When a few drops of this bromine solution are shaken with the hydrocarbon, if it is an alkene, such as ethene, a reaction takes place in which bromine joins to the alkene double bond. This results in the bromine solution losing its red/brown colour. If an alkane, such as hexane, is shaken with a bromine solution of this type, no colour change takes place (Figure 12.15). This is because there are no double bonds between the carbon atoms of alkanes.

ethane + bromine ⟶ dibromoethane

$$C_2H_4(g) + Br_2(\text{in solution}) \longrightarrow C_2H_4Br_2(\text{in solution})$$

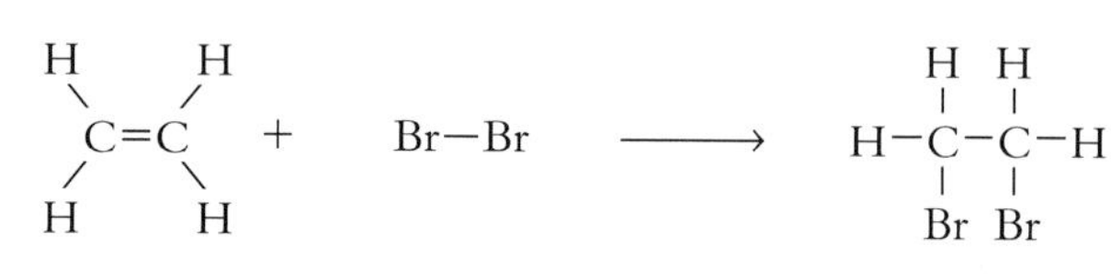

**Figure 12.15** The alkene decolorises bromine in 1,1,1-trichloroethane.

## *Questions*

1. Using the information in Table 12.2 (p. 178), make an estimate of the boiling point of pentene.
2. Write a balanced chemical equation to represent the process that takes place when decane is cracked.
3. What is meant by the term 'addition reaction'?
4. Write a word and balanced chemical equation for the reaction between ethene and hydrogen chloride.
5. Write the structural formula for pentene.

## Checklist

**After studying Chapter 12 you should know and understand the following terms.**

- **Alkanes** A family of saturated hydrocarbons with the general formula $C_nH_{2n+2}$. The term 'saturated', in this context, is used to describe molecules that have only single bonds. The alkanes can only undergo substitution reactions in which there is replacement of one atom in the molecule by another atom.
- **Alkenes** A family of unsaturated hydrocarbons with the general formula $C_nH_{2n}$. The term 'unsaturated', in this context, is used to describe molecules which contain one or more double carbon–carbon bonds. Unsaturated compounds undergo addition reactions across the carbon–carbon double bonds and so produce saturated compounds. The addition of hydrogen across the carbon–carbon double bonds is used to reduce the amount of unsaturation during the production of margarines.
- **Catalytic cracking** The decomposition of higher alkanes into alkenes and alkanes of lower relative molecular mass. The process involves passing the larger alkane molecules over a catalyst of aluminium and chromium oxides, heated to 500 °C.
- **CFC** Abbreviation for chlorofluorocarbon, a type of organic compound in which some or all of the hydrogen atoms of an alkane have been replaced by fluorine and chlorine atoms. These substances are generally unreactive but they can diffuse into the stratosphere where they break down under the influence of ultraviolet light. The products of this photochemical process then react with ozone (in the ozone layer). Because of this, their use has been discouraged. They are now being replaced by hydrochlorofluorocarbons (HCFCs).
- **Chain reaction** A reaction which is self-sustaining owing to the products of one step of the reaction assisting in promoting further reaction.
- **Free radicals** Atoms or groups of atoms with unpaired electrons and are therefore highly reactive. They can be produced by high-energy radiation such as ultraviolet light in photochemical reactions.
- **Halogenoalkanes** Organic compounds in which one or more hydrogen atoms of an alkane have been substituted by halogen atoms such as chlorine.
- **Hydrocarbon** A substance which contains atoms of carbon and hydrogen only.
- **Isomers** Compounds which have the same molecular formula but different structural arrangements of the atoms.
- **Oil refining** The general process of converting the mixture that is collected as crude oil into separate fractions. These fractions, known as petroleum products, are used as fuels, lubricants, bitumens and waxes. The fractions are separated from the crude oil mixture by fractional distillation.
- **Organic chemistry** The branch of chemistry concerned with compounds of carbon found in living organisms.
- **Test for unsaturation** A few drops of bromine dissolved in an organic solvent are shaken with the hydrocarbon. If it is decolorised, the hydrocarbon is unsaturated.

# The petroleum industry
# *Additional questions*

1 Explain the following.
  a Ethene is called an unsaturated hydrocarbon.
  b The cracking of larger alkanes into simple alkanes and alkenes is important to the petrochemical industry.
  c The conversion of ethene to ethanol is an example of an addition reaction.

2 The following question is about some of the reactions of ethene.

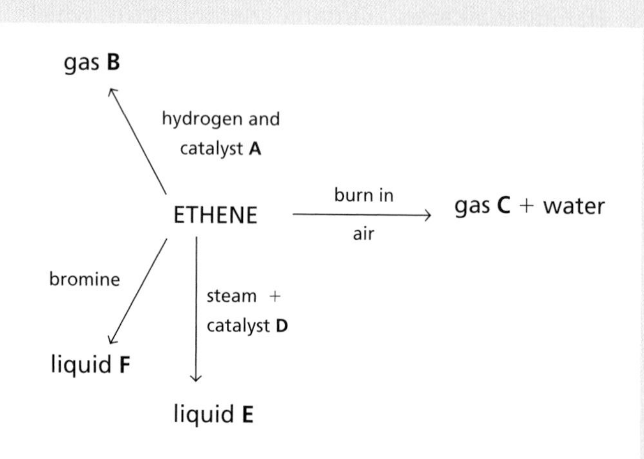


  a Give the names and formulae for substances **A** to **F**.
  b (i) Write a word and balanced chemical equation to represent the reaction in which liquid **E** is formed.
  (ii) What reaction conditions are required for the process to take place?
  (iii) Hydrogen is used in the production of margarine to remove unsaturation. Explain what you understand by this statement.
  c Name the homologous series that gas **B** belongs to.
  d Describe a chemical test which would allow you to identify gas **C**.

3 a Crude oil is a mixture of *hydrocarbons* which belong to the *homologous series* called the *alkanes*. This mixture can be separated into fractions by the process of *fractional distillation*. Some of the fractions obtained are used as *fuels*. Some of the other fractions are subjected to *catalytic cracking* in order to make *alkenes*.
  Explain the meaning of the terms in italics.
  b Alkanes can be converted into substances which are used as solvents. To do this the alkane is reacted with a halogen, such as chlorine, in the presence of ultraviolet light.
  (i) Write a word and balanced chemical equation for the reaction between methane and chlorine.
  (ii) Name the type of reaction taking place.
  (iii) Highly reactive chlorine atoms are produced in the presence of ultraviolet light. When atoms are produced in this way, what are they called?
  (iv) Write a balanced chemical equation for the reaction which takes place between $CHF_3$ and $Cl_2$ to produce a chlorofluorocarbon (CFC).
  (v) Why are CFCs such a problem?

4 Crude oil is a mixture of hydrocarbons. The refining of crude oil produces fractions which are more useful to us than crude oil itself. Each fraction is composed of hydrocarbons which have boiling points within a specific range of temperature. The separation is carried out in a fractionating column, as shown below.

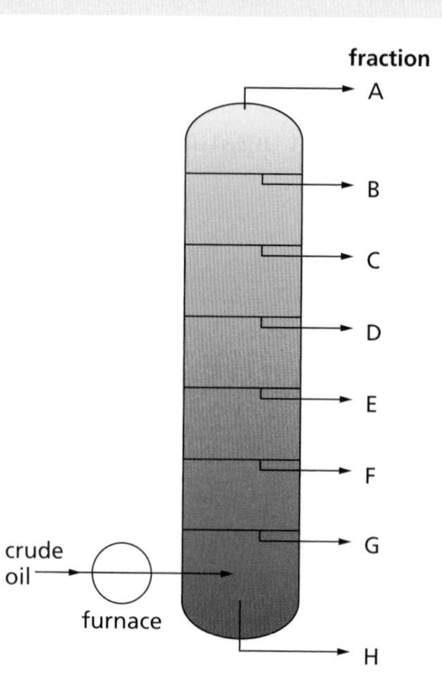


  a Which separation technique is used to separate the fractions?
  b Name each of the fractions **A** to **H** and give a use for each.
  c Why do the fractions come from the fractionating column in this order?
  d What is the connection between your answer to **c** and the size of the molecules in each fraction?
  e Which of the fractions will be the most flammable?

5 Alkanes and alkenes are hydrocarbons. They are composed of molecules which contain covalent bonds. For each of the molecules below, use a dot and cross diagram to show the bonding it contains.
  a Methane, $CH_4$.
  b Propene, $C_3H_6$.
  c Propane, $C_3H_8$.
  d Ethene, $C_2H_4$.

6 Crude oil is an important source of organic chemical fuels. It is refined by fractional distillation. Use the information in the table below to answer the questions which follow.

| Fraction | Boiling point/°C |
|---|---|
| A | 40 |
| B | 80 |
| C | 200 |
| D | 350 |
| E | above 350 |

a For each of the questions that follow, give the letter of the fraction which is most appropriate as an answer. You should also give a reason for your answer in each case.
(i) Which fraction would contain the most volatile substances?
(ii) Which of the fractions would collect at the bottom of the fractionating column?
(iii) Which fraction could be used as a fuel for cars?
(iv) Which fraction would contain the largest molecules?

b Some of the fractions undergo a further process called cracking to produce further substances.
(i) Explain what you understand by the term 'cracking'. What conditions are employed when cracking occurs?
(ii) Write a word and balanced chemical equation to show how octane can be produced by the cracking of $C_{15}H_{32}$.

7 a A hydrocarbon contains 92.3% by mass of carbon. Work out the empirical formula of this hydrocarbon.
b The relative molecular mass of this hydrocarbon was found by mass spectrometry to be 78. Work out its molecular formula. ($A_r$: H = 1, C = 12)

8 a Which of the following formulae represent alkanes, which represent alkenes and which represent neither?

$CH_3$, $C_6H_{12}$, $C_5H_{12}$, $C_6H_6$, $C_9H_{20}$, $C_{12}H_{24}$, $C_{20}H_{42}$, $C_2H_4$, $C_8H_{18}$, $C_3H_7$

b Draw all the possible isomers which have the molecular formula $C_6H_{14}$.

# Energy sources



## Fossil fuels

Coal, oil and natural gas are all examples of **fossil fuels**. The term, fossil fuels, is derived from the fact that they are formed from dead plants and animals which were fossilised over 200 million years ago during the carboniferous era.

Coal was produced by the action of pressure and heat on dead wood from ancient forests which once grew in the swampland in many parts of the world under the prevailing weather conditions of that time. When dead trees fell into the swamps they were buried by mud. This prevented aerobic decay (which takes place in the presence of oxygen). Over millions of years, due to movement of the Earth's crust (Chapter 17, p. 245) as well as to changes in climate, the land sank and the decaying wood became covered by even more layers of mud and sand. Anaerobic decay (which takes place in the absence of oxygen) occurred, and as time passed the gradually forming coal became more and more compressed as other material was laid down above it (Figure 13.1). Over millions of years, as the layers of forming coal were pushed deeper and the pressure and temperature increased, the final conversion to coal took place (Figure 13.2).

Different types of coal were formed as a result of different pressures being applied during its formation. For example, anthracite is a hard coal with a high carbon content, typical of coal produced at greater depths. Table 13.1 shows some of the different types of coal along with their carbon contents.

Oil and gas were formed during the same period as coal. It is believed that oil and gas were formed from the remains of plants, animals and bacteria that once lived in seas and lakes. This material sank to the bottom of these seas and lakes and became covered in mud, sand and silt which thickened with time.

**Figure 13.1** Piece of coal showing a fossilised leaf.

**Figure 13.2** Cutting of coal is extremely mechanised.

**Table 13.1** The different coal types.

| Type of coal | Carbon content/% |
|---|---|
| Anthracite | 90 |
| Bituminous coal | 60 |
| Lignite | 40 |
| Peat | 20 |

Anaerobic decay took place, and, as the mud layers built up, high temperatures and pressures were created which converted the material slowly into oil and gas. As rock formed, earth movements caused it to buckle and split, and the oil and gas were trapped in folds beneath layers of non-porous rock or cap-rock (Figures 13.3 and 13.4).

**Figure 13.3** Oil production in the North Sea.

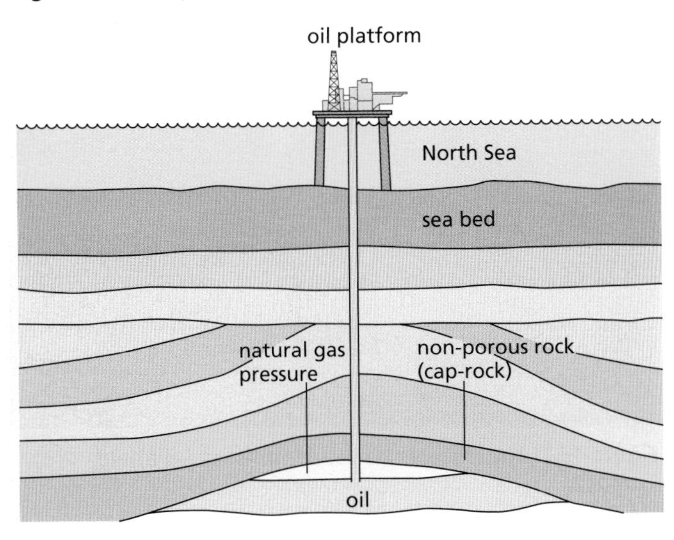


**Figure 13.4** Natural gas and oil are trapped under non-porous rock.

## Questions

1 Coal, oil and natural gas are all termed 'fossil fuels'. Why is the word 'fossil' used in this context?

2 a Name the process by which plants convert carbon dioxide and water into glucose.
  b What conditions are necessary for this process to occur?

3 Draw a flow diagram to represent the formation of coal, oil or gas.

# *What is a fuel?*

A fuel is a substance which can be conveniently used as a source of energy. Fossil fuels produce energy when they undergo **combustion**.

fossil fuel + oxygen → carbon dioxide + water + energy

For example, natural gas burns readily in air (Chapter 12, p. 176).

| methane | + | oxygen | → | carbon dioxide | + | water | + energy |
|---|---|---|---|---|---|---|---|
| $CH_4(g)$ | + | $2O_2(g)$ | → | $CO_2(g)$ | + | $2H_2O(l)$ | |

It should be noted that natural gas, like crude oil, is a mixture of hydrocarbons such as methane, ethane and propane, and may also contain some sulphur. The sulphur content varies from source to source (Chapter 16, p. 228). Natural gas obtained from the North Sea is quite low in sulphur.

The perfect fuel would be:

- cheap
- available in large quantities
- safe to store and transport
- easy to ignite and burn, causing no pollution
- capable of releasing large amounts of energy.

Solid fuels are safer than volatile liquid fuels like petrol and gaseous fuels like natural gas.

## How are fossil fuels used?

A major use of fossil fuels is in the production of electricity. Coal, oil and natural gas are burned in power stations (Figure 13.5) to heat water to produce steam, which is then used to drive large turbines (Figure 13.6). At least 80% of the electricity generated in the UK is generated using fossil fuels. However, it should be noted that the relative importance of the three major fossil fuels is changing. Coal and oil are becoming less important while natural gas is increasingly important.

**Figure 13.5** A power station.

In a power station, the turbine drives a generator to produce electricity which is then fed into the National Grid (Figure 13.6). The National Grid is a system for distributing electricity throughout the country.

Other major uses of the fossil fuels are:

- as a major feedstock (raw material) for the chemicals and pharmaceuticals industries
- for domestic and industrial heating and cooking
- as fuels for various forms of vehicle transport.

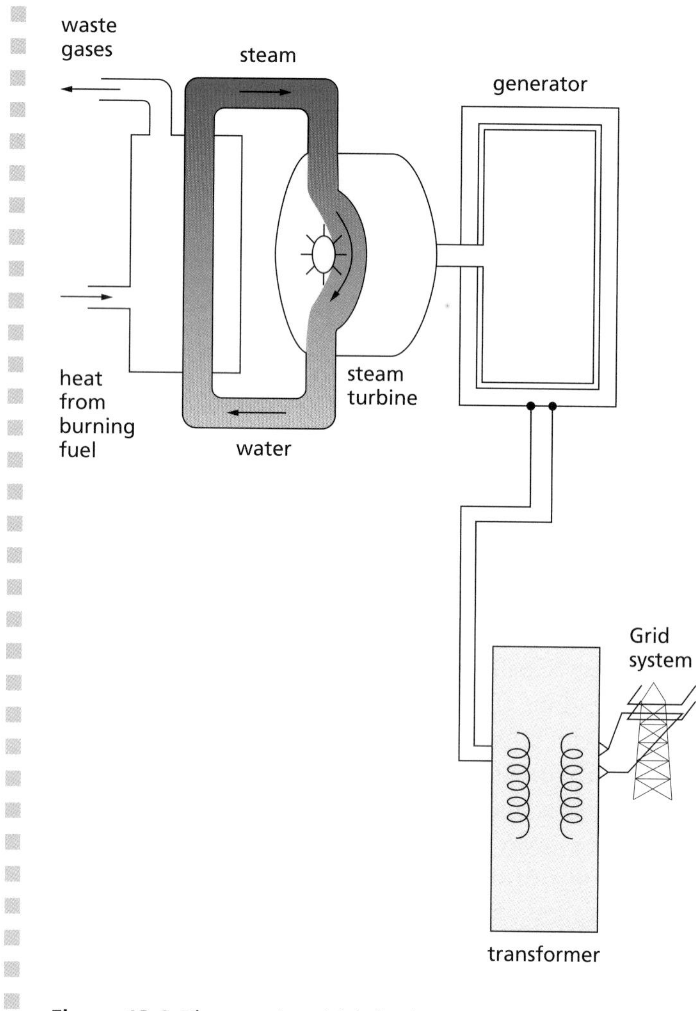


**Figure 13.6** The way in which fuels are used to produce electricity.

## Questions

1 'We have not yet found the perfect fuel.' Discuss this statement.

2 'Fossil fuels are a major feedstock for the chemical and pharmaceutical industries.' With reference to Chapter 12, give examples which support this statement.



## *Alternative sources of energy*

Fossil fuels are an example of **non-renewable** resources, so called because they are not being replaced at the same rate as they are being used up. For example, we have approximately 55 years' supply of crude oil remaining from known reserves if we continue to use it at the current rate as a source of energy and chemicals (Table 13.2). It is important to use non-renewable fuels carefully and to consider alternative **renewable** sources of energy for use in the future.

**Table 13.2** Estimates of how long our fossil fuels will last.

| Fossil fuel | Estimated date it is expected to run out |
|---|---|
| Gas | 2045 |
| Oil | 2055 |
| Coal | 2500 |

### Nuclear power

Calder Hall power station in Cumbria, on the site of the present-day nuclear power complex at Sellafield (Figure 13.7), opened in 1956 and was the first nuclear reactor in the world to produce electricity on an industrial scale.

**Figure 13.7** The nuclear power complex at Sellafield, Cumbria.

Nuclear reactors harness the energy from the fission of uranium-235. **Nuclear fission** occurs when the unstable nucleus of a radioactive isotope splits up, forming smaller atoms and producing a large amount of energy as a result. Scientists believe that the energy comes from the conversion of some of the mass of the isotope.

This fission process begins when a neutron hits an atom of uranium-235, causing it to split and produce three further neutrons. These three neutrons split three more atoms of uranium-235, which produces nine neutrons and so on. This initiates a **chain reaction** (Figure 13.8).

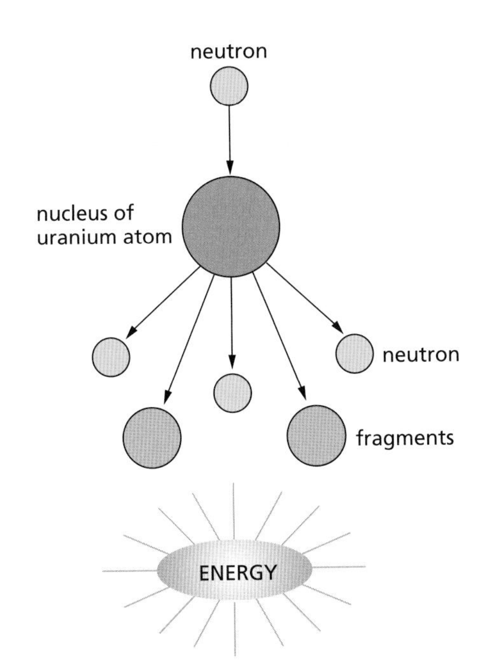


**Figure 13.8** Chain reaction in uranium-235 fission.

In a reactor the fission process cannot be allowed to get out of control as it does in an atomic bomb. To prevent this, boron control rods can be pushed into different positions in the reactor to absorb some of the neutrons which are produced and so slow down the chain reaction. If this is done, the energy released from the reaction is obtained in a more controlled way. The energy is used to produce steam, which in turn is used to generate electricity (Figure 13.6).

However, there are problems. The main problem associated with a nuclear power station is that the reactor produces highly radioactive waste materials. These waste materials are difficult to store and cannot be disposed of very easily. Also, leaks of radioactive material have occurred at various sites throughout the world. Accidents at a small number of nuclear power stations, such as Chernobyl in the Ukraine (Figure 13.9) and Three Mile Island in the US, have led to a great deal of concern about their safety. In the UK the safety record of the nuclear power industry is relatively good because it is subject to strict controls.

**Figure 13.9** A nuclear accident happened at the Chernobyl power station in 1986.

## Hydroelectric power

Hydroelectric power (HEP) is electricity generated from the energy of falling water (Figure 13.10). It is an excellent energy source and electricity has been generated in this way in the mountainous areas of Scotland and Wales for some time. It is a very cheap source of electricity. Once you have built the power station, the energy is absolutely free. In some mountainous areas of the world, such as the Alps, HEP is the main source of electricity. One of the main advantages of this system is that it can be quickly used to supplement the National Grid at times of high demand. A disadvantage of HEP schemes is that they often require valleys to be flooded and communities to be moved.

**Figure 13.10** A hydroelectric power station.

## Geothermal energy

Water is pumped into hot rocks in the Earth's crust far below ground level (Figure 13.11). The internal heat of the rocks converts the water to steam, which is used to drive turbines and hence generate electricity. This is a major source of electrical energy in Iceland.

Geothermal energy is a natural, non-polluting source of energy. Experiments on the viability of geothermal energy have been conducted in Cornwall.

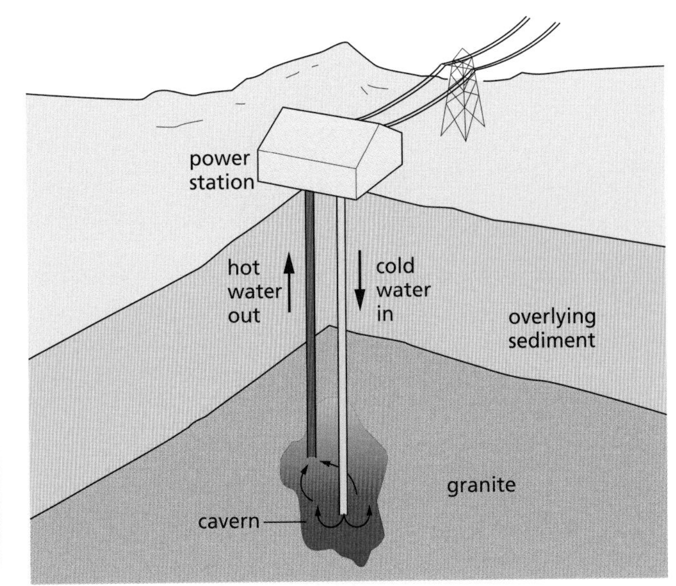


**Figure 13.11** Geothermal plant.

## Wave power

In this method the energy of moving waves is used to generate electricity. Figure 13.12 shows the Salters' 'Duck' wave machine in operation. The vertical motion of the waves is converted to rotary motion, which is used to drive a generator producing electricity. A disadvantage of this method for generating large amounts of electricity is that vast numbers of strings of these ducks would be required.

Wave power is a non-polluting source of energy.

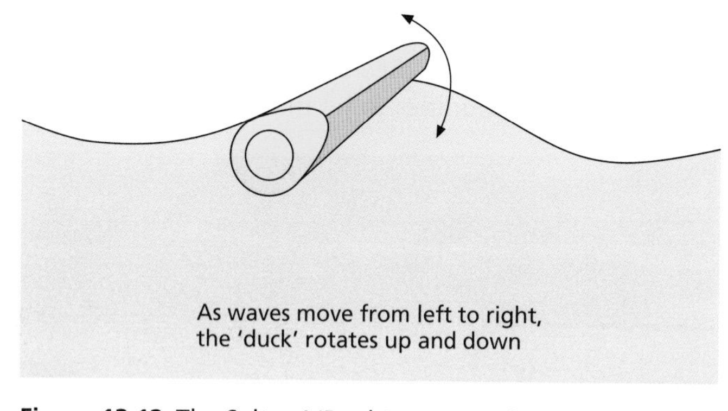

**Figure 13.12** The Salters' 'Duck' wave machine.

## Tidal power

The ebb and flow of the tides drives turbines built into a dam or barrage across an estuary where the height difference between high and low tides is large (Figure 13.13). There is a successful tidal power station across the River Rance near St Malo in northern France. The most likely place in Britain to put a tidal barrage is across the Severn Estuary near Bristol, but there would be environmental disadvantages with such a scheme. For example, there would be a threat to the wildlife around the estuary, since the mud banks are one of the few remaining sites in Britain for wildfowl and wading birds to nest.

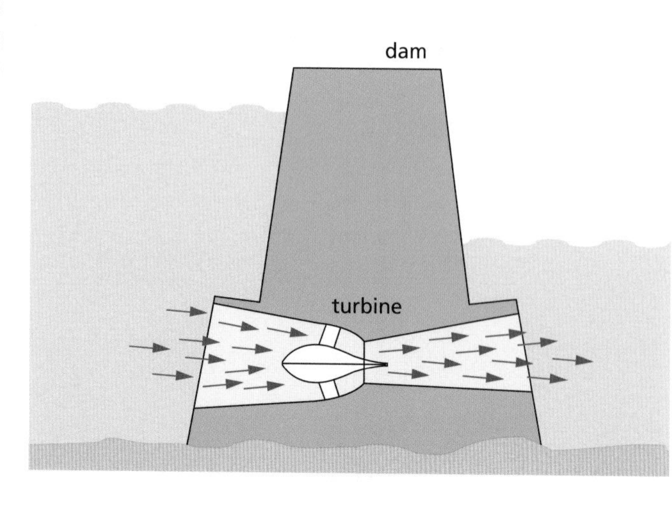


**Figure 13.13** Tidal power – the ebb and flow of the tide drives the turbine set into the barrage or dam.

## Wind power

This method uses the force of the wind to turn generators to produce electricity. Wind machines 24 m high are capable of generating 200 kW of electricity per machine. Large 'wind farms' have been developed in many parts of the world, for example in the US and on the Pennines in England (Figure 13.14).

The disadvantages of wind power are that the wind farms are somewhat unsightly, they require vast amounts of land, as the machines have to be carefully arranged so that their operation is not impaired, and they produce a lot of noise.

**Figure 13.14** Wind farm at Ovenden Moor, Yorkshire.

## Solar energy

Two possible methods exist to use the energy of the Sun. In the first, the Sun's energy can be absorbed on to black-painted collector plates and used to heat water and homes (Figure 13.15). This method is relatively cheap. The second method involves the use of photovoltaic cells or photocells to generate electricity (Figure 13.16). Disadvantages include the initial cost of cells as well as the fact that the Sun does not shine all the time. To solve this problem, solar cells are commonly linked to storage batteries.

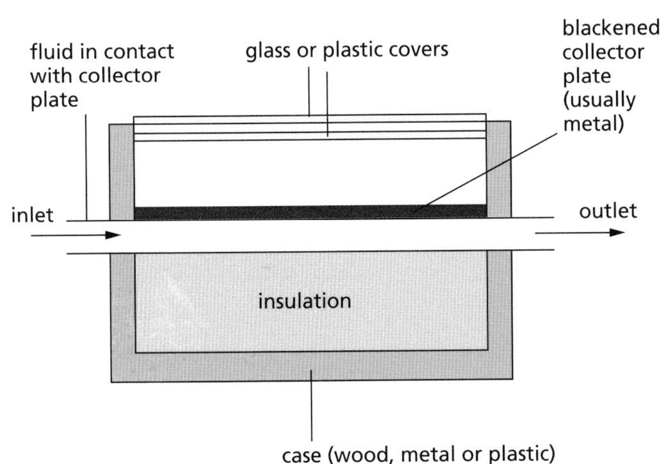

**Figure 13.15** A single solar energy panel.

**Figure 13.16** A bank of photocells.

## Biomass and biogas

When any biological material, whether plant or animal, is converted into energy, this energy is called **biomass** energy. It can be taken from animal or plant materials in different ways:

- by burning it, for example wood (Figure 13.17)
- by pressing out oils that can be burned
- by fermenting it to produce fuels such as ethanol or methane.

**Figure 13.17** Biomass energy is produced by burning wood.

At least 50% of the world's population rely on wood as their main energy source.

In India there are millions of methane generators. Methane generated by the digestion of animal waste is called **biogas**. The biogas produced is used for cooking, heating and lighting. The by-product of this process is an excellent fertiliser.

Some countries have already experimented with ethanol as a fuel for cars. Up to 20% of ethanol can be added to petrol without the need to adjust the carburettor. Brazil, which has few oil reserves, produces ethanol by fermentation (breakdown by enzymes) of sugar cane and grain, and uses it as a petrol additive (Figure 13.18). The Brazilian government has cut down its petrol imports by up to 60% through using this alcohol/petrol mixture.

**Figure 13.18** In Brazil cars use an ethanol/petrol mixture.

## *Questions*

**1** Draw up a table showing the alternative sources of energy along with their advantages and disadvantages.

**2** What is meant by the terms:
- **a** non-renewable energy sources?
- **b** renewable energy sources?

# Chemical energy

We obtain our energy needs from the combustion of fuels, such as hydrocarbons, from the combustion of foods and from many other chemical reactions.

## Combustion

When natural gas burns in a plentiful supply of air it produces a large amount of energy.

methane + oxygen → carbon dioxide + water + heat energy

$$CH_4(g) + 2O_2(g) \rightarrow CO_2(g) + 2H_2O(l) + \text{heat energy}$$

During this process, the **complete combustion** of methane, heat is given out. It is an **exothermic** reaction. If only a limited supply of air is available then the reaction is not as exothermic and the poisonous gas carbon monoxide is produced.

methane + oxygen → carbon monoxide + water + heat energy

$$2CH_4(g) + 3O_2(g) \rightarrow 2CO(g) + 4H_2O(l) + \text{heat energy}$$

This process is known as the **incomplete combustion** of methane.

The energy changes that take place during a chemical reaction can be shown by an **energy level diagram**. Figure 13.19 shows the energy level diagram for the complete combustion of methane.

When any reaction occurs, the chemical bonds in the reactants have to be broken – this requires energy. When the new bonds in the products are formed, energy is given out (Figure 13.20). The **bond energy** is defined as the amount of energy in kilojoules (kJ) associated with the breaking or making of one mole of chemical bonds in a molecular element or compound.

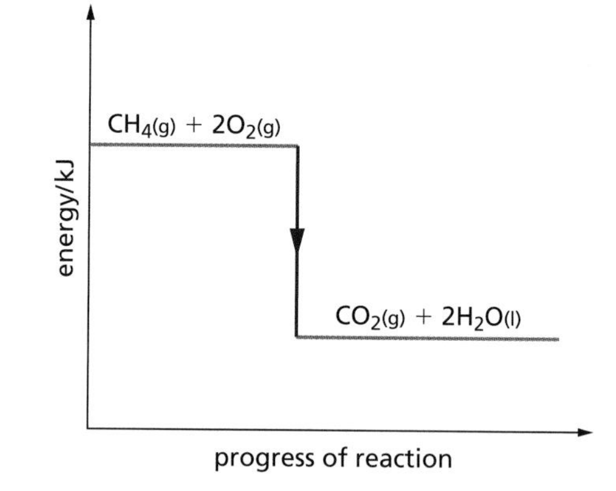


**Figure 13.19** Energy level diagram for the complete combustion of methane.

Using the bond energy data from Table 13.3, which tells us how much energy is needed to break a chemical bond and how much is given out when it forms, we can calculate how much energy is involved in each stage.

### Bond breaking

Breaking 4 C—H bonds in methane requires

$4 \times 435 = 1740\,kJ$

Breaking 2 O=O bonds in oxygen requires

$2 \times 497 = 994\,kJ$

Total $= 2734\,kJ$ of energy

**Table 13.3** Bond energy data.

| Bond | Bond energy ($kJ\,mol^{-1}$) |
|---|---|
| C—H | 435 |
| O=O | 497 |
| C=O | 803 |
| H—O | 464 |
| C—C | 347 |
| C—O | 358 |

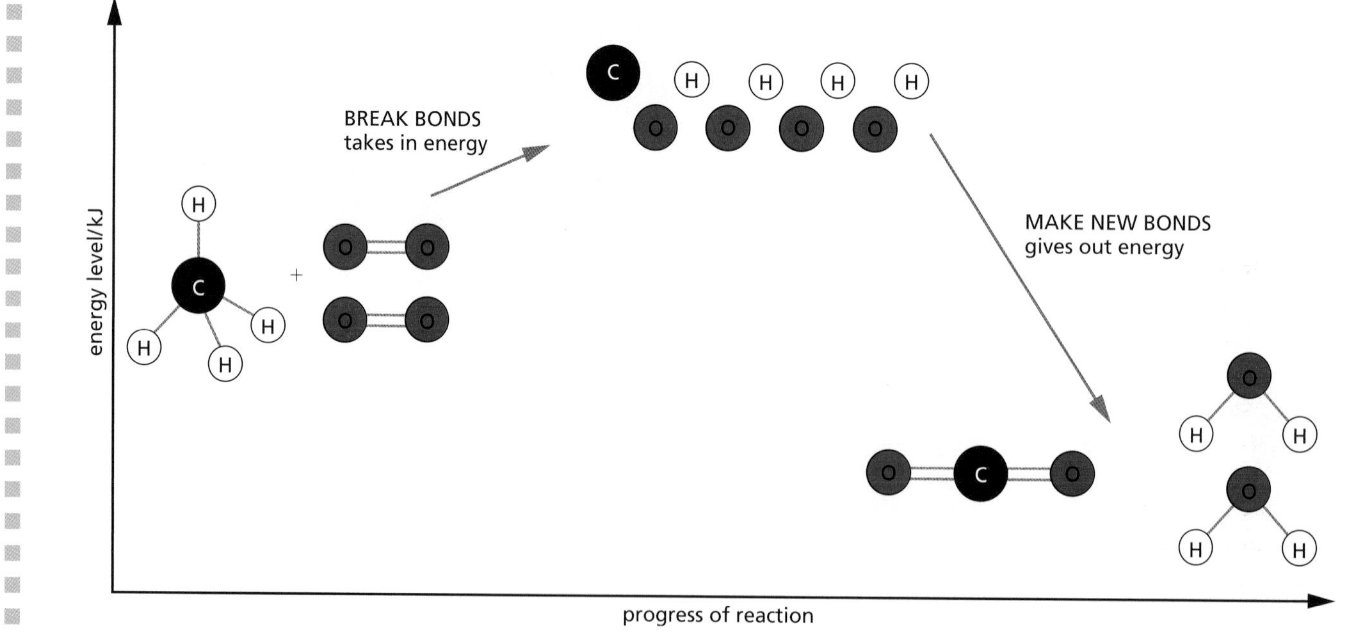


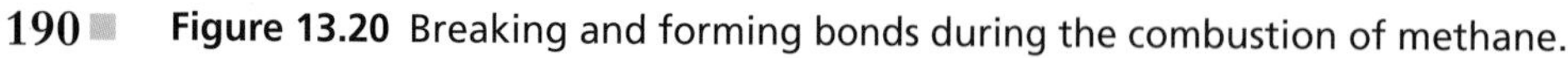

**Figure 13.20** Breaking and forming bonds during the combustion of methane.

**Making bonds**

Making 2 C=O bonds in carbon dioxide gives out

$2 \times 803 = 1606\,\text{kJ}$

Making 4 O—H bonds in water gives out

$4 \times 464 = 1856\,\text{kJ}$

Total $= 3462\,\text{kJ}$ of energy

Energy difference

= energy required to break bonds − energy given out when bonds are made

= 2734 − 3462

= −728 kJ

The negative sign shows that the chemicals are losing energy to the surroundings, that is, it is an exothermic reaction. A positive sign would indicate that the chemicals are gaining energy from the surroundings. This type of reaction is called an **endothermic** reaction.

The energy stored in the bonds is called the **enthalpy** and is given the symbol *H*. The change in energy going from reactants to products is called the **change in enthalpy** and is shown as $\Delta H$ (pronounced 'delta H'). $\Delta H$ is called the **heat of reaction**.

For an exothermic reaction $\Delta H$ is negative and for an endothermic reaction $\Delta H$ is positive.

When fuels, such as methane, are burned they require energy to start the chemical reaction. This is known as the **activation energy, $E_A$** (Figure 13.21).

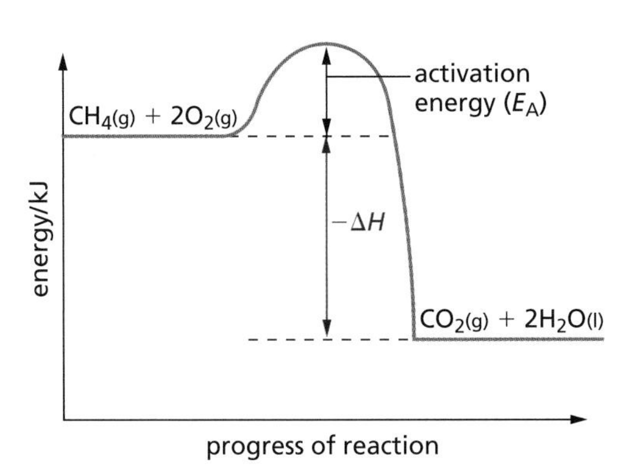


**Figure 13.21** Energy level diagram for methane/oxygen.

In the case of methane reacting with oxygen, it is the energy involved in the initial bond breaking (Figure 13.20). The value of the activation energy will vary from fuel to fuel.

Endothermic reactions are much less common than exothermic ones. In this type of reaction energy is absorbed from the surroundings so that the energy of the products is greater than that of the reactants. The reaction between nitrogen and oxygen gases is endothermic (Figure 13.22).

nitrogen + oxygen → nitrogen(II) oxide

$N_2(g) + O_2(g) \rightarrow 2NO(g)$

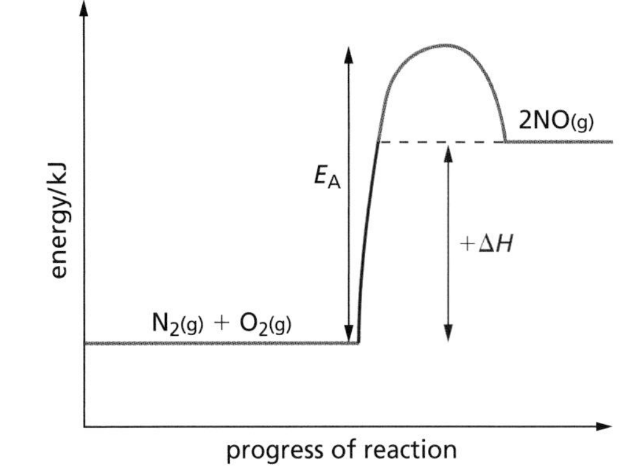

**Figure 13.22** Energy level diagram for nitrogen/oxygen.

Dissolving is often an endothermic process. For example, when ammonium nitrate dissolves in water the temperature of the water falls, indicating that energy is being taken from the surroundings. Photosynthesis and thermal decomposition are other examples of endothermic processes.

In equations it is usual to express the $\Delta H$ value in units of $\text{kJ mol}^{-1}$. For example:

$$CH_4(g) + 2O_2(g) \rightarrow CO_2(g) + 2H_2O(l) \quad \Delta H = -728\,\text{kJ mol}^{-1}$$

This $\Delta H$ value tells us that when 1 mole of methane is burned in oxygen, 728 kJ of energy are released. This value is called the **enthalpy of combustion** of methane (or **molar heat of combustion** of methane).

## Enthalpy of neutralisation (molar heat of neutralisation)

This is the enthalpy change that takes place when 1 mol of hydrogen ions ($H^+(aq)$) is neutralised.

$$H^+(aq) + OH^-(aq) \rightarrow H_2O(l) \quad \Delta H = -57\,\text{kJ mol}^{-1}$$

This process occurs in the titration of an alkali by an acid to produce a neutral solution (Chapter 7, p. 107).

## *Questions*

1 Using the bond energy data given in Table 13.3:
   - **a** Calculate the enthalpy of combustion of ethanol, a fuel added to petrol in some countries.
   - **b** Draw an energy level diagram to represent this combustion process.
   - **c** How does this compare with the enthalpy of combustion of heptane ($C_7H_{14}$), a major component of petrol, of $-4853\,\text{kJ mol}^{-1}$?
   - **d** How much energy is released per gram of ethanol and heptane burned.

2 How much energy is released if:
   - **a** 0.5 mole of methane is burned?
   - **b** 5 moles of methane are burned?
   - **c** 4 g of methane are burned? ($A_r$: C = 12, H = 1)

3 How much energy is released if:
   - **a** 2 moles of hydrogen ions are neutralised?
   - **b** 0.25 mole of hydrogen ions is neutralised?
   - **c** 1 mole of sulphuric acid is completely neutralised?

## Change of state

In Chapter 1, p. 4, we discussed the melting and boiling of a substance. The heating curve for water is shown in Figure 1.11 on p. 5. For ice to melt to produce liquid water, it must absorb energy from its surroundings. This energy is used to break down the weak forces between the water molecules (intermolecular forces) in the ice. This energy is called the **enthalpy of fusion** and is given the symbol $\Delta H_{fusion}$. Similarly, when liquid water changes into steam, the energy required for this process to occur is called the **enthalpy of vaporisation** and is given the symbol $\Delta H_{vap}$. Figure 13.23 shows the energy level diagrams representing both the fusion and the vaporisation processes.

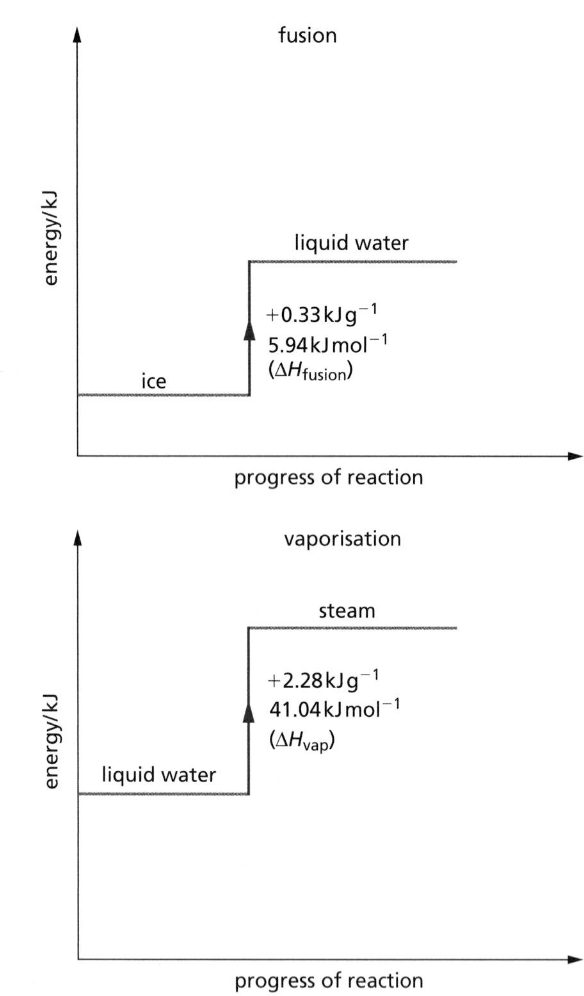


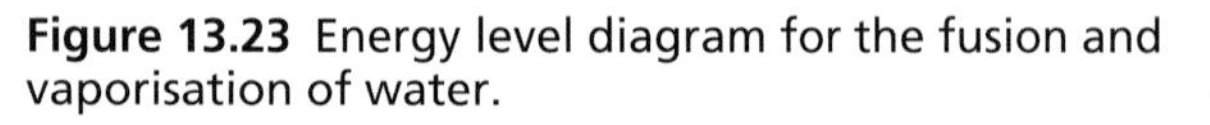

**Figure 13.23** Energy level diagram for the fusion and vaporisation of water.

## Questions

1 Describe the energy changes which take place when the processes described in this section, with water, are reversed.

2 Using the knowledge you have obtained from Chapter 1, p. 5, give a full definition of the enthalpy of fusion and enthalpy of vaporisation for water.

## Cells and batteries

A simple type of chemical cell is that shown in Figure 13.24a. In this cell the more reactive metal zinc dissolves in the dilute sulphuric acid, producing zinc ions ($Zn^{2+}(aq)$) and releasing two electrons.

$$Zn(s) \rightarrow Zn^{2+}(aq) + 2e^-$$

The electrons produced at the zinc electrode flow through the external circuit via the bulb and the bulb glows. Bubbles of hydrogen are seen when the electrons arrive at the copper electrode. The hydrogen gas is produced from the hydrogen ions in the acid, which collect the electrons appearing at the copper electrode.

$$2H^+(aq) + 2e^- \rightarrow H_2(g)$$

Slowly, the zinc electrode dissolves in the acid and the bulb will then go out. If the zinc is replaced by a more reactive metal, such as magnesium, then the bulb glows more brightly. Magnesium loses electrons more easily as it reacts faster with the dilute acid.

The difference in the reactivity between the two metals used in the cell creates a particular voltage reading on the voltmeter shown in Figure 13.24b. The more the two metals differ in reactivity, the larger is the voltage shown and delivered by the cell. This method can be used to confirm the order of reactivity of the metals (Chapter 9, p. 128). Other types of chemical cell in common use are dry cells used in radios, torches, and so on, and lead–acid accumulators used in motor vehicles.

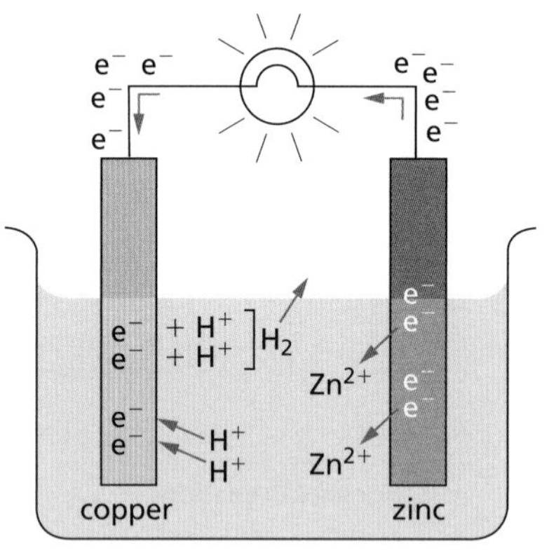


**a** A simple chemical cell.

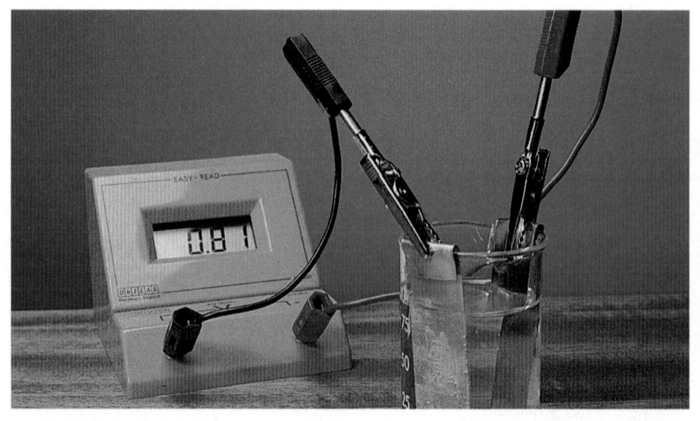


**b** The voltage reflects the difference in reactivity of the metals.

**Figure 13.24**

# Fuel cells

Scientists have found a much more efficient way of changing chemical energy into electrical energy, using a fuel cell (Figure 13.25). Fuel cells are like the chemical cells in the previous section, except that the reagents are supplied continuously to the electrodes. The reagents are usually hydrogen and oxygen. The fuel cell principle was first discovered by Sir William Grove in 1839.

When he was electrolysing water and he switched off the power supply, he noticed that a current still flowed but in the reverse direction. Subsequently, the process was explained in terms of the reactions at the electrodes' surfaces of the oxygen and hydrogen gases which had been produced during the electrolysis.

The hydrogen fuel cells used by NASA in the US space programme are about 70% efficient and, since the only product is water, they are pollution free. The aqueous NaOH electrolyte is kept within the cell by electrodes which are porous, allowing the transfer of $O_2$, $H_2$ and water through them (Figure 13.26). As $O_2$ gas is passed into the cathode region of the cell it is reduced:

$$O_2 + 2H_2O + 4e^- \rightarrow 4OH^-$$

The $OH^-$ ions formed are removed from the fuel cell by reaction with $H_2$:

$$H_2 + 2OH^- \rightarrow 2H_2O + 2e^-$$

The electrons produced by this process pass around an external circuit to the cathode.

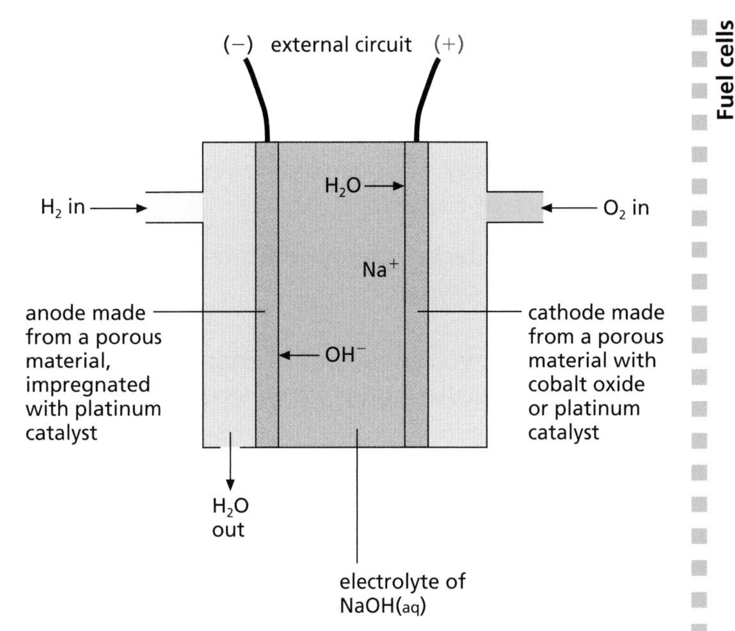


**Figure 13.26** A diagrammatic view of a fuel cell.

## Questions

1 Describe how simple chemical cells can be used to confirm the order of reactivity of the metals in the reactivity series.

2 The fuel cell was discovered during electrolysis experiments with water. It is the reverse process which produces the electricity. Write a balanced chemical equation to represent the overall reaction taking place in a fuel cell.

**Figure 13.25** The space shuttle's computers use electricity produced by fuel cells.

## Checklist

**After studying Chapter 13 you should know and understand the following terms.**

- **Aerobic decay** Decay which takes place in the presence of oxygen.
- **Anaerobic decay** Decay which takes place in the absence of oxygen.
- **Bond energy** An amount of energy associated with a particular bond in a molecular element or compound.
- **Chain reaction** A nuclear reaction which is self-sustaining as a result of one of the products causing further reactions.
- **Chemical cell** A system for converting chemical energy to electrical energy.
- **Combustion** A chemical reaction in which a substance reacts rapidly with oxygen with the production of heat and light.
- **Endothermic reaction** A chemical reaction which absorbs heat energy from its surroundings.
- **Enthalpy** Energy stored in chemical bonds, given the symbol $H$.
- **Enthalpy change** Given the symbol $\Delta H$, it represents the difference between energies of reactants and products.
- **Enthalpy of combustion** The enthalpy change which takes place when one mole of a substance is completely burned in oxygen.
- **Enthalpy of fusion** The enthalpy change that takes place when one mole of a solid is changed to one mole of liquid at the same temperature.
- **Enthalpy of neutralisation** The enthalpy change which takes place when one mole of hydrogen ions is completely neutralised.
- **Enthalpy of vaporisation** The enthalpy change that takes place when one mole of liquid is changed to one mole of vapour at the same temperature.
- **Exothermic reaction** A chemical reaction that releases heat energy into its surroundings.
- **Fossil fuels** Fuels, such as coal, oil and natural gas, formed from the remains of plants and animals.
- **Non-renewable energy sources** Sources of energy, such as fossil fuels, which take millions of years to form and which we are using up at a rapid rate.
- **Nuclear fission** The disintegration of a radioactive nucleus into two or more lighter fragments. The energy released in the process is called nuclear energy.
- **Renewable energy** Sources of energy which cannot be used up or which can be made at a rate faster than the rate of use.

# Energy sources

# *Additional questions*

**1 a** State which of the following processes is endothermic and which is exothermic.
(i) The breaking of a chemical bond.
(ii) The forming of a chemical bond.

**b** The table below shows the bond energy data for a series of covalent bonds.
(i) Use the information given in the table to calculate the overall enthalpy change for the combustion of ethanol producing carbon dioxide and water.
(ii) Is the process in (i) endothermic or exothermic?

| Bond | Bond energy/kJ mol$^{-1}$ |
|---|---|
| C—H | 435 |
| O=O | 497 |
| C=O | 803 |
| H—O | 464 |
| C—C | 347 |
| C—O | 358 |

**2** Explain the following.

**a** Hydroelectric power is a relatively cheap source of electricity.

**b** Geothermal energy is a non-polluting form of energy.

**c** A disadvantage of wind power is that it causes noise pollution.

**d** The by-product from the process by which methane is generated by the digestion of animal waste is an excellent fertiliser.

**e** The fission of uranium-235 in a nuclear reactor is an example of a chain reaction.

**f** Tidal- and wave-generated electricity has a major environmental disadvantage.

**3** One of the first practical chemical cells was the Daniell cell invented by John Daniell in 1836. A diagram of this type of cell is shown below.

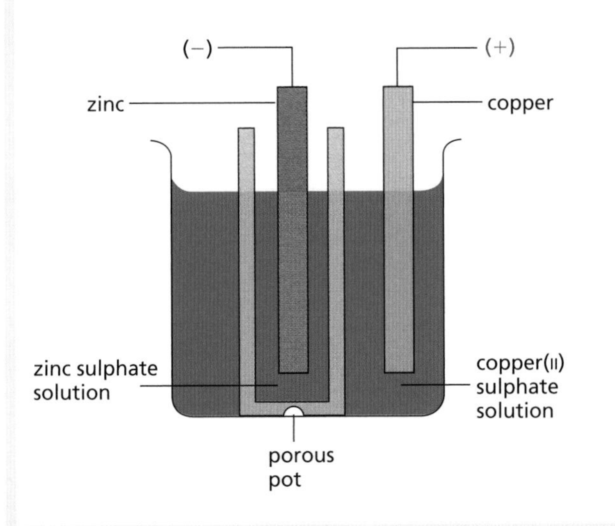


It is capable of generating about 1.1 volts and was used to operate small electrical items such as doorbells.

**a** The electrode reaction taking place at a copper anode is:

$$Cu^{2+}(aq) + 2e^- \rightarrow Cu(s)$$

Write an electrode equation for the process taking place at the cathode.

**b** Which way would the electrons flow in the wire connected to the voltmeter – from 'copper to zinc' or 'zinc to copper'?

**c** Why should copper(II) sulphate crystalise at the bottom of the outer container?

**d** What is the function of the porous pot?

**e** There are problems associated with the Daniell cell which has led to it being replaced by other types of cell. Give two reasons why Daniell cells are no longer in use today.

**4** This question is about endothermic and exothermic reactions.

**a** Explain the meaning of the terms endothermic and exothermic.

**b** (i) Draw an energy level diagram for the reaction:

$$NaOH(aq) + HCl(aq) \rightarrow NaCl(aq) + H_2O(l) \quad \Delta H = -57\,kJ\,mol^{-1}$$

(ii) Is this reaction endothermic or exothermic?
(iii) Calculate the energy change associated with this reaction if 2 moles of sodium hydroxide were neutralised by excess hydrochloric acid.

**c** (i) Draw an energy level diagram for the reaction:

$$2H_2O(l) \rightarrow 2H_2(g) + O_2(g) \quad \Delta H = +575\,kJ\,mol^{-1}$$

(ii) Is this reaction endothermic or exothermic?
(iii) Calculate the energy change for this reaction if only 9 g of water were converted into hydrogen and oxygen.

**5**

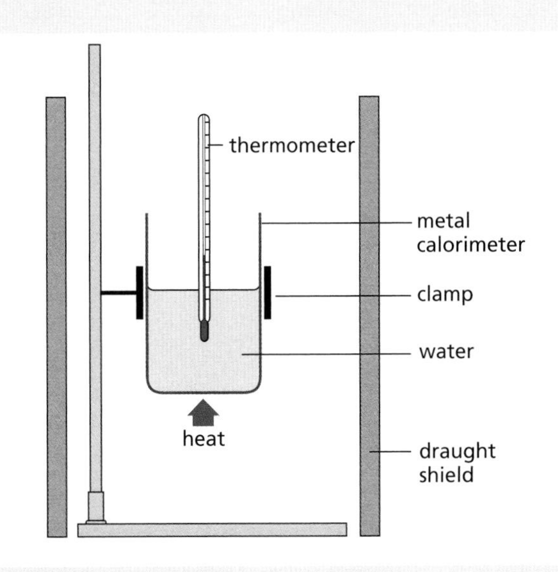

The following results were obtained from an experiment carried out to measure the enthalpy of combustion (heat of combustion) of ethanol. The experiment involved heating a known volume of water with the flame from an ethanol burner. The burner was weighed initially and after the desired temperature rise had been obtained.

Volume of water in glass beaker = $200\,cm^3$
Mass of ethanol burner at start = 85.3 g
Mass of ethanol burner at end = 84.8 g
Temperature rise of water = 12 °C
(Density of water = $1\,g\,cm^{-3}$)

Heat energy given to water = mass of water/g × 4.2 $J\,g^{-1}\,°C^{-1}$ × temperature rise/°C

**a** Calculate the mass of ethanol burned.
**b** Calculate the amount of heat produced, in joules, in this experiment by the ethanol burning.
**c** Convert your answer to **b** into kilojoules.
**d** Calculate the amount of heat produced by 1 g of ethanol burning.
**e** What is the mass of 1 mole of ethanol ($C_2H_5OH$)? ($A_r$: H = 1; C = 12; O = 16)
**f** How much heat would be produced if 1 mole of ethanol had been burned? (This is the heat of combustion of ethanol.)
**g** Compare your value with the actual value of $1371\,kJ\,mol^{-1}$ and suggest two reasons for the difference in values.
**h** Write a balanced chemical equation to represent the combustion of ethanol.

**6** The following results were obtained from a neutralisation reaction between $1\,mol\,dm^{-3}$ hydrochloric acid and $1\,mol\,dm^{-3}$ sodium hydroxide. This experiment was carried out to measure the heat of neutralisation of hydrochloric acid. The temperature rise which occurred during the reaction was recorded.

Volume of sodium hydroxide used = $50\,cm^3$
Volume of acid used = $50\,cm^3$
Temperature rise = 5 °C
(Density of water = $1\,g\,cm^{-3}$)

Heat energy given out during reaction = mass of water/g × 4.2 $J\,g^{-1}\,°C^{-1}$ × temperature rise/°C

**a** Write a balanced chemical equation for the reaction.
**b** What mass of solution was warmed during the reaction?
**c** How much heat energy was produced during the reaction?
**d** How many moles of hydrochloric acid were involved in the reaction?
**e** How much heat would be produced if 1 mole of hydrochloric acid had reacted? (This is the heat of neutralisation of hydrochloric acid.)
**f** The heat of neutralisation of hydrochloric acid is $-57\,kJ\,mol^{-1}$. Suggest two reasons why there is a difference between this and your calculated value.

**7** Write down which factors are most important when deciding on a particular fuel for the purpose given:
**a** fuel for a cigarette lighter
**b** fuel for a camping stove
**c** fuel for an aeroplane
**d** fuel for an underground transport system
**e** fuel for the space shuttle
**f** fuel for domestic heating.

**8** 'Propagas' is used in some central heating systems where natural gas is not available. It burns according to the following equation:

$$C_3H_8(g) + 5O_2(g) \rightarrow 3CO_2(g) + 4H_2O(l)\ \Delta H = -2220\,kJ\,mol^{-1}$$

**a** What are the chemical names for 'propagas' and natural gas?
**b** Would you expect the heat generated per mole of 'propagas' burned to be greater than that for natural gas? Explain your answer.
**c** What is 'propagas' obtained from?
**d** Calculate:
(i) the mass of 'propagas' required to produce 5550 kJ of energy
(ii) the heat energy produced by burning 0.5 mole of 'propagas'
(iii) the heat energy produced by burning 11 g of 'propagas'
(iv) the heat energy produced by burning $2000\,dm^3$ of 'propagas'.
($A_r$: H = 1; C = 12; O = 16. One mole of any gas occupies $24\,dm^3$ at room temperature and pressure.)

# 14 The (wider) organic manufacturing industry



## Plastics and polymers

In 1933, Reginald Gibson and Eric Fawcett carried out a reaction involving ethene and another organic chemical called benzaldehyde ($C_6H_5CHO$) using a pressure of about 2000 atmospheres. They were hoping to make these two chemicals react to produce another organic substance from a homologous series called ketones. The reaction vessel leaked and some air (oxygen) got in and much more ethene had to be added. Upon opening the reaction vessel they found a white waxy solid instead of what they expected. They subsequently discovered that this solid was in fact a new type of substance. It was found to consist of very large molecules, each one made up of many thousands of ethene molecules (Figure 14.1). 'Many' in Greek is *poly*, and so the new substance was called '**poly(ethene)**' (or polythene). The modern plastics industry was born with this accidental discovery.

**Figure 14.1** Molecules of ethene in this poly(ethene) chain are represented by individual poppet beads.

Poly(ethene) has many useful properties.

- It is easily moulded.
- It is an excellent electrical insulator.
- It does not corrode.
- It is tough.
- It is not affected by the weather.
- It is durable.

It was first used to insulate telephone cables and its unique electrical properties were essential during the development of radar. Its properties continued to be exploited, and today it can be found as a substitute for natural materials in plastic bags, sandwich boxes, washing-up bowls, wrapping film, milk-bottle crates and squeezy bottles.

We now manufacture poly(ethene) by heating ethene to a relatively high temperature under a high pressure in the presence of a catalyst.

where $n$ is a very large number. In poly(ethene) the ethene molecules have joined together to form a very long hydrocarbon chain (Figure 14.2). The ethene molecules are able to form chains like this because they possess carbon–carbon double bonds.

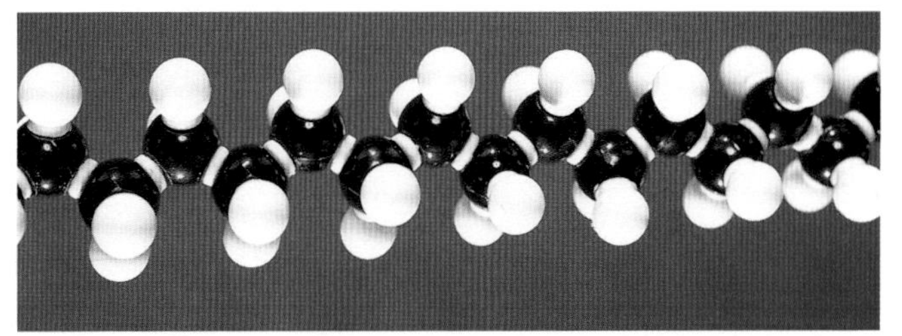

**Figure 14.2** This model shows part of the poly(ethene) polymer chain.

Poly(ethene) is produced in three main forms:

- low density poly(ethene) (LDPE)
- linear low density poly(ethene) (LLDPE)
- high density poly(ethene) (HDPE).

The world production of all types of poly(ethene) is about 52 million tonnes per year, with the UK producing about 0.64 million tonnes.

Other alkene molecules can also produce substances like poly(ethene); for example, propene produces poly(propene), which is used to make ropes and packaging. When small molecules like ethene join together to form long chains of atoms called **polymers**, the process is called **polymerisation**. The small molecules, like ethene, which join together in this way are called **monomers**. A polymer chain often consists of many thousands of monomer units and in any piece of plastic there will be many millions of polymer chains. Since in this polymerisation process the monomer units add together to form only one product, the polymer, the process is called **addition polymerisation**.

## Other addition polymers

Many other addition polymers have been produced. Often the plastics are produced with particular properties in mind, for example PTFE (poly(tetrafluoroethene)) and PVC (polyvinyl chloride or poly(chloroethene)). Both of these plastics have monomer units similar to ethene.

H H
C=C
H Cl

PVC monomer
(vinyl chloride or
chloroethene)

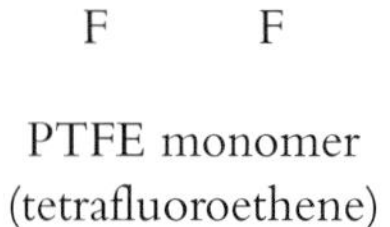


PTFE monomer
(tetrafluoroethene)

If we use chloroethene (Figure 14.3a), the polymer we make is slightly stronger and harder than poly(ethene) and is particularly good for making pipes for plumbing.

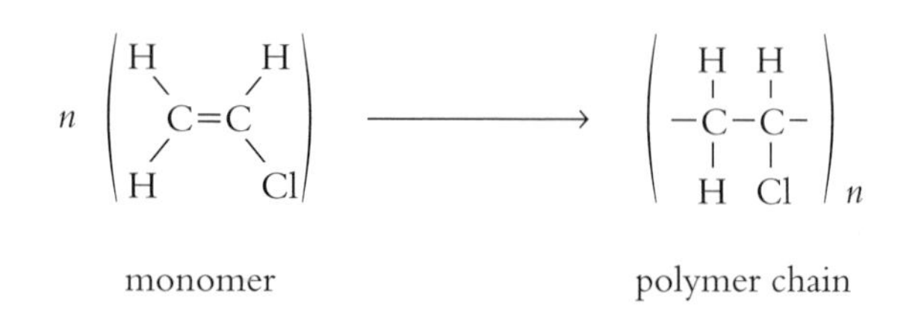


PVC is the most versatile plastic and is the second most widely used, after poly(ethene). Worldwide 27 million tonnes are produced annually, of which 0.47 million tonnes are produced in the UK.

a Model of chloroethene, the PVC monomer.

b Model of part of a PVC polymer chain.

**Figure 14.3**

If we start from tetrafluoroethene (Figure 14.4a) the polymer we make, PTFE, has some slightly unusual properties:

- it will withstand very high temperatures, of up to 260 °C
- it forms a very slippery surface
- it is hydrophobic
- it is highly resistant to chemical attack.

These properties make PTFE an ideal 'non-stick' coating for frying pans and saucepans.

Every year 50 000 tonnes of PTFE are made, of which 3000 tonnes are made in the UK.

monomer

polymer chain

The properties of some addition polymers along with their uses are given in Table 14.1.

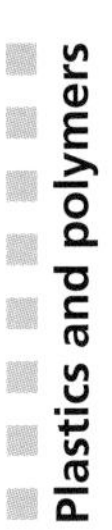

**a** Model of tetrafluoroethene, the PTFE monomer.

**b** Model of part of the PTFE polymer chain.

**Figure 14.4**

**Table 14.1** Some addition polymers.

| Plastic | Monomer | Properties | Uses |
|---|---|---|---|
| Poly(ethene) | $CH_2{=}CH_2$ | Tough, durable | Carrier bags, bowls buckets, packaging |
| Poly(propene) | $CH_3CH{=}CH_2$ | Tough, durable | Ropes, packaging |
| PVC | $CH_2{=}CHCl$ | Strong, hard (less flexible than poly(ethene)) | Pipes, electrical insulation, guttering |
| PTFE | $CF_2{=}CF_2$ | Non-stick surface, withstands high temperatures | Non-stick frying pans, soles of irons |
| Polystyrene | $CH_2{=}CHC_6H_5$ | Light, poor conductor of heat | Insulation, packaging (especially as foam) |
| Perspex | $CH_2{=}C(CO_2CH_3)CH_3$ | Transparent | Used as a glass substitute |

## Condensation polymers

Wallace Carothers discovered a different sort of plastic when he developed nylon in 1935. Nylon is made by reacting two different chemicals together, unlike poly(ethene) which is made only from monomer units of ethene. Poly(ethene), formed by addition polymerisation, can be represented by:

–A–A–A–A–A–A–A–A–A–A–

where A = monomer.

The starting molecules for nylon are more complicated than those for poly(ethene) and are called 1,6-diaminohexane and hexanedioic acid.

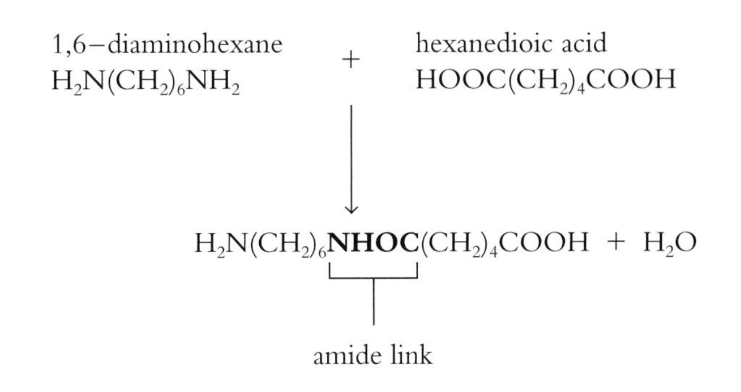


The polymer chain is made up from the two starting molecules arranged alternately (Figure 14.5) as these molecules react and therefore link up. Each time a reaction takes place a molecule of water is lost, and because of this it is called **condensation polymerisation**. Because an amide link is formed during the polymerisation, nylon is known as a **polyamide**. This type of polymerisation, in which two kinds of monomer unit react, results in a chain of the type:

–A–B–A–B–A–B–A–B–A–B–

**Figure 14.5** A nylon polymer chain is made up from the two molecules arranged alternately just like the two different coloured poppet beads in the photo.

When nylon is made in industry, it forms as a solid which is melted and forced through small holes (Figure 4.21, p. 62). The long filaments cool and solid nylon fibres are produced which are stretched to align the polymer molecules and then dried. The resulting yarn can be woven into fabric to make shirts, ties, sheets and parachutes or turned into ropes or racket strings for tennis and badminton rackets. The worldwide production of nylon is 5.4 million tonnes per year.

We can obtain different polymers with different properties if we carry out condensation polymerisation reactions between other monomer molecules. For example, if we react ethane-1,2-diol with benzene-1,4-dicarboxylic acid, then we produce a polymer called terylene.

| ethane-1,2-diol | | benzene-1,4-dicarboxylic acid |
|---|---|---|
| $HO(CH_2)_2OH$ | + | $HOOC(C_6H_4)COOH$ |

$$\downarrow$$

$$HO(CH_2)_2\mathbf{OCO}(C_6H_4)COOH + H_2O$$

ester link

Like nylon, terylene can be turned into yarn, which can then be woven. Terylene clothing is generally softer than that made from nylon but both are hard wearing. Because an ester link is formed during the polymerisation, terylene is known as a **polyester**.

## Thermosoftening and thermosetting plastics

Plastics can be put into one of two categories. If they melt or soften when heated (like poly(ethene), PVC and polystyrene) then they are called **thermoplastics** or **thermosoftening plastics**. If they do not soften on heating but only char and decompose on further heating, they are known as **thermosetting plastics**.

Thermoplastics are easily moulded or **formed** into useful articles. Once they are molten they can be injected or blown into moulds, and a variety of different-shaped items can be produced (Figure 14.6). Thermosetting plastics can be heated and moulded only once, usually by compression moulding (Figure 14.7).

Figure 14.8 shows the different molecular structures for thermosetting and thermosoftening plastics. Thermosetting plastics have polymer chains which are linked or bonded to each other to give a **cross-linked** structure, and so the chains are held firmly in place and no softening takes place on heating. Thermosoftening plastics do not have polymer chains joined in this way, so when they are subjected to heat their polymer chains flow over one another and the plastic softens.

**Figure 14.6** Blow moulding in progress in the recycling of poly(ethene).

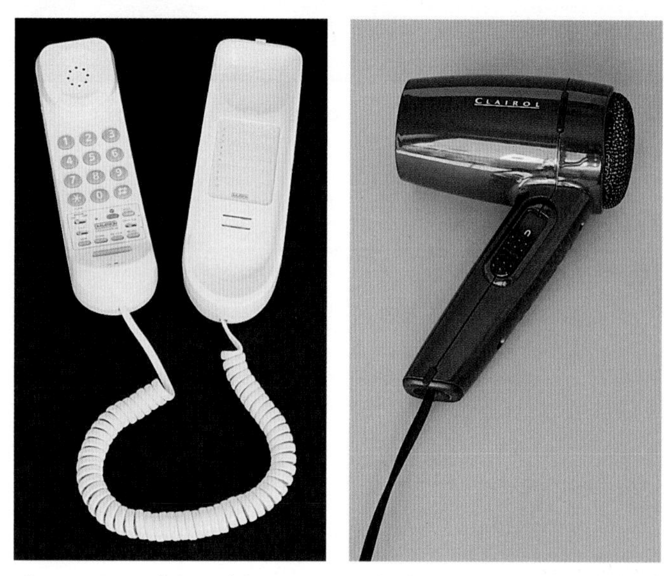


**Figure 14.7** These objects are made from compression-moulded thermosetting plastic.

**a** In thermosetting plastic the chains are cross-linked.

**b** In thermosoftening plastic there is no cross-linking.

**Figure 14.8**

## Disposal of plastics

In the last 30 years plastics have taken over as replacement materials for metals, glass, paper and wood as well as for natural fibres such as cotton and wool. This is not surprising since plastics are light, cheap, relatively unreactive, can be easily moulded and can be dyed bright colours. However, this situation has contributed significantly to the household waste problem. Figure 14.9 shows some EU average figures for solid household waste.

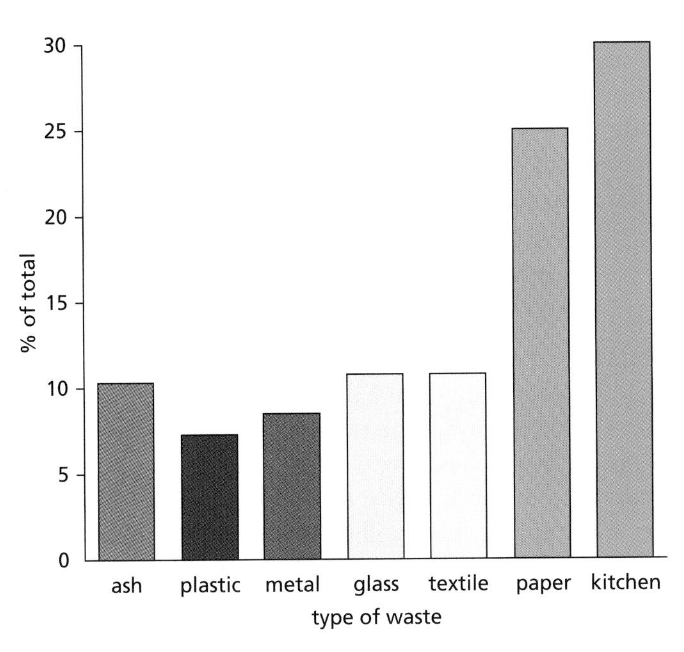


**Figure 14.9** Amount of plastic waste compared with other household waste in EU countries.

In the recent past much of our plastic waste has been used to landfill disused quarries. However, these sites are getting harder to find and it is becoming more and more expensive. The alternatives to dumping plastic waste are certainly more economical and more satisfactory.

- Incineration – schemes have been developed to use the heat generated for heating purposes.
- Recycling – large quantities of black plastic bags and sheeting are produced for resale.
- Biodegradable plastics, as well as those polymers that degrade in sunlight (**photodegradable**, Figure 14.10a), have been developed. Other common categories of degradable plastics include synthetic biodegradable plastics which are broken down by bacteria, as well as plastics which dissolve in water (Figure 14.10b). The property that allows plastic to dissolve in water has been used in relatively new products, such as Ariel Liquitabs and Persil Capsules.

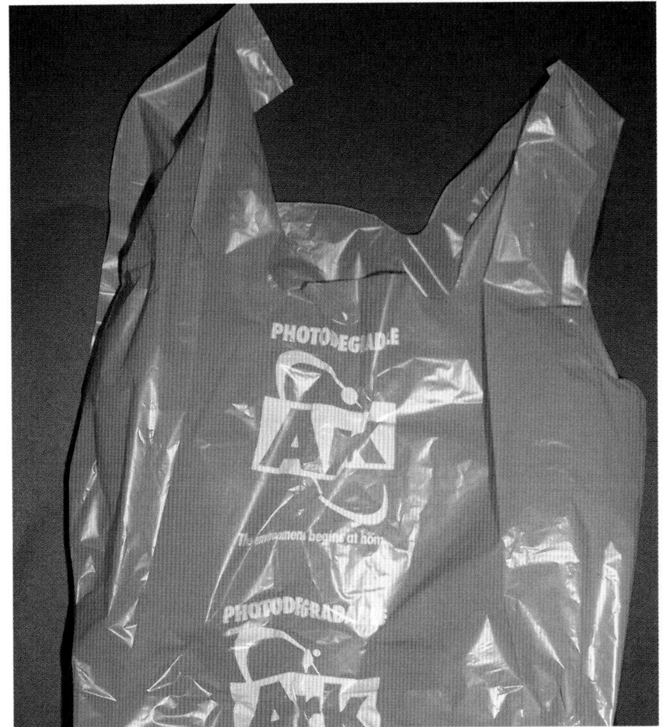

**a** This plastic bag is photodegradable.

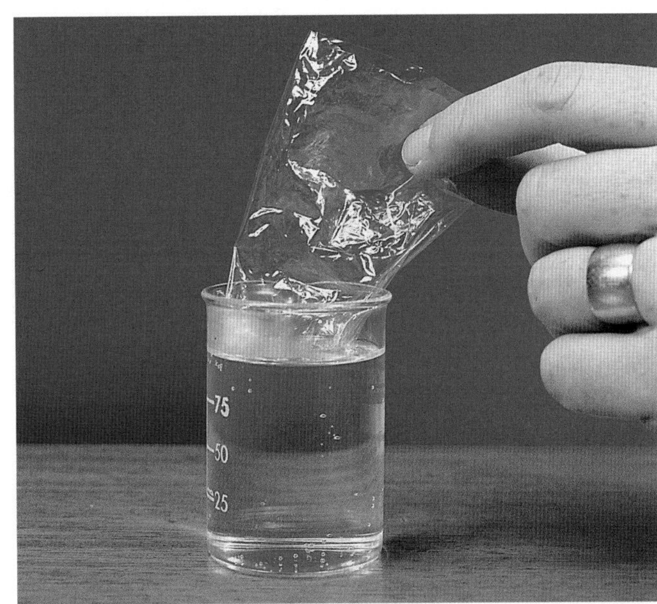


**b** This plastic dissolves in water.

**Figure 14.10**

## Questions

1. Write the general equation to represent the formation of polystyrene from its monomer.
2. Suggest other uses for the polymers shown in Table 14.1.
3. Draw the structure of the repeating units found in:
   a nylon
   b terylene.
4. Explain the differences between an addition polymer and a condensation polymer.
5. Give two advantages and two disadvantages of plastic waste (rubbish).

## *Some biopolymers*

Starch is a **biopolymer** or **natural polymer**. It is a condensation polymer of glucose, a type of sugar. It is often produced as a way of storing energy and is formed as a result of photosynthesis in green plants.

$$\text{carbon dioxide} + \text{water} \xrightarrow[\text{sunlight/ chlorophyll (in green plants)}]{\text{photosynthesis}} \text{glucose} + \text{oxygen}$$

$$6CO_2(g) + 6H_2O(l) \longrightarrow C_6H_{12}O_6(aq) + 6O_2(g)$$

$$\text{glucose} \rightarrow \text{starch} + \text{water}$$

$$nC_6H_{12}O_6(aq) \rightarrow (C_6H_{10}O_5)_n(s) + nH_2O(l)$$

Both starch and glucose are **carbohydrates**, a class of naturally occurring organic compounds which can be represented by the general formula $(CH_2O)_x$. Starch occurs in potatoes, rice and wheat. Glucose, from which starch is polymerised, belongs to a group of simple carbohydrates known as **monosaccharides**. They are sweet to taste and soluble in water. Starch belongs to the more complicated group of carbohydrates known as **polysaccharides**. Starch does not form a true solution and it does not have a sweet taste. With iodine it gives an intense blue colour (nearly black), which is used as a test for starch or iodine itself.

**Figure 14.11** A dark blue-black colour is produced when dilute iodine solution is applied to starch, for example in a potato.

## Hydrolysis of starch

Starch can be broken down in two ways, both of which take place in the presence of water. Hence the reactions are known as **hydrolysis** reactions. Hydrolysis of starch is the key reaction that enables us to use this energy source. If starch is boiled for about one hour with dilute hydrochloric acid, it is broken down into its monomers, glucose molecules.

$$\text{starch} + \text{water} \xrightarrow[\text{heat}]{\text{dilute acid}} \text{glucose}$$

$$(C_6H_{10}O_5)_n(s) + nH_2O(l) \longrightarrow nC_6H_{12}O_6(aq)$$

If starch is mixed with saliva and left to stand for a few minutes, it will break down to maltose, a **disaccharide** (that is two joined monosaccharides). The enzyme present in the saliva, called amylase, catalyses this hydrolysis reaction.

$$\text{starch} + \text{water in saliva} \xrightarrow{\text{amylase}} \text{maltose}$$

$$2(C_6H_{10}O_5)_n(s) + nH_2O(l) \longrightarrow nC_{12}H_{22}O_{11}(aq)$$

Enzymes are very efficient natural catalysts present in plants and animals. They do not require high temperatures to break down the starch to maltose (Chapter 11, p. 167). In humans, a salivary amylase breaks down the starch in our food. If you chew on a piece of bread for several minutes, you will notice a sweet taste in your mouth. The above hydrolysis reactions are summarised in Figure 14.12.

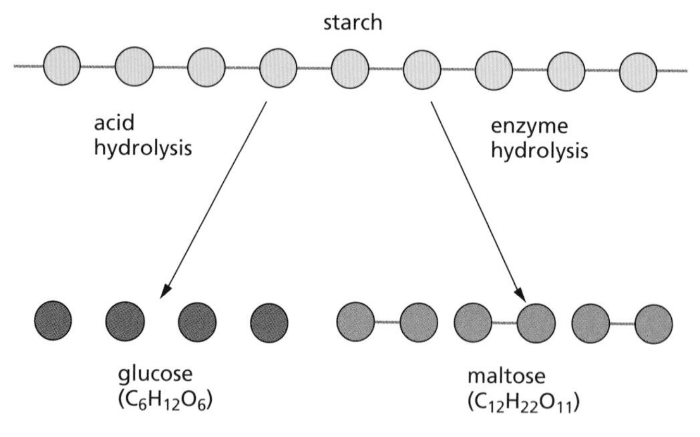


**Figure 14.12** Starch produces glucose or maltose depending on the type of hydrolysis used.

### *Questions*

1 In the hydrolysis of starch, how, using a chemical test, could you tell whether all the starch had been hydrolysed?

2 Describe a method you could possibly use to identify the products of the different types of hydrolysis.

## *Ethanol ($C_2H_5OH$) – an alcohol*

The alcohols form another homologous series with the general formula $C_nH_{2n+1}OH$. All the alcohols possess an –OH as the **functional group**. The functional group is the group of atoms responsible for the characteristic reactions of a compound. Table 14.2 shows the first three members along with their melting and boiling points. Figure 14.13 shows the actual arrangement of the atoms in these first three members of this family.

**Table 14.2** The first three members of the alcohol family.

| Alcohol | Formula | Melting point/ °C | Boiling point/ °C |
|---|---|---|---|
| Methanol | $CH_3OH$ | −94 | 64 |
| Ethanol | $CH_3CH_2OH$ | −117 | 78 |
| Propanol | $CH_3CH_2CH_2OH$ | −126 | 97 |

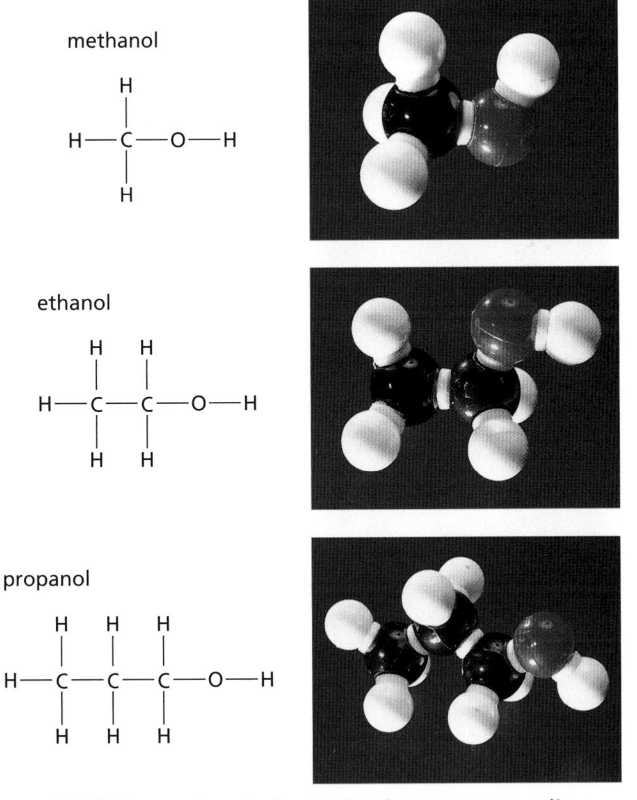


**Figure 14.13** The molecules look like their corresponding models in the photographs.

Ethanol is by far the most important of the alcohols and is often just called 'alcohol'. Ethanol can be produced by fermentation (p. 204) as well as by the hydration of ethene (Chapter 12, p. 179). It is a neutral, colourless, volatile liquid which does not conduct electricity. Ethanol burns quite readily with a clean, hot flame.

ethanol + oxygen → carbon dioxide + water + energy

$$CH_3CH_2OH(l) + 3O_2(g) \rightarrow 2CO_2(g) + 3H_2O(g) + \text{energy}$$

As methylated spirit, or meths, it is used in spirit (camping) stoves. Methylated spirit is ethanol with small amounts of poisonous substances added to stop you drinking it. Also, some countries, like Brazil, are already using ethanol mixed with petrol as a fuel for cars (Chapter 13, p. 189).

Many other materials, such as food flavourings, are made from ethanol. Ethanol is a very good solvent and evaporates easily. It is therefore used extensively as a solvent for paints, glues, aftershave and many other everyday products.

Ethanol can be oxidised to ethanoic acid (an organic acid also called acetic acid) by powerful oxidising agents, such as warm acidified potassium dichromate(VI). During the reaction the orange colour of potassium dichromate(VI) changes to a dark green (Figure 14.14) as the ethanol is oxidised to ethanoic acid.

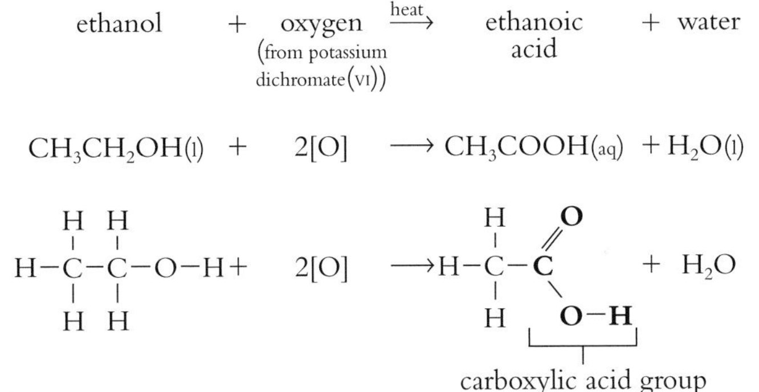

ethanol + oxygen (from potassium dichromate(VI)) $\xrightarrow{\text{heat}}$ ethanoic acid + water

$$CH_3CH_2OH(l) + 2[O] \longrightarrow CH_3COOH(aq) + H_2O(l)$$



A similar oxidation process takes place, but more slowly, if wine or beer is left open. It will eventually turn to 'vinegar' (ethanoic acid) due to bacterial oxidation of the ethanol in the alcoholic drink.

**Figure 14.14** Orange potassium dichromate(VI) slowly turns green as it oxidises ethanol to ethanoic acid.

### *Questions*

1 Write the structural formula for butanol.

2 Write a word and balanced chemical equation for:
  a the combustion of butanol
  b the oxidation of butanol.

## Cholesterol – a steroid

Cholesterol is a naturally occurring and essential chemical. It belongs to a family of chemicals called steroids and also contains an alcohol group (Figure 14.15). Cholesterol is found in almost all of the tissues in the body, including nerve cells. Levels of cholesterol above normal (above 6.5 mmol/l) are associated with an increased risk of heart disease. Cholesterol hardens and blocks off arteries by building up layers of solid material (atheroma) inside the arteries (Figure 14.16). This is particularly serious if the arteries that supply the heart or brain are blocked. Simple tests are now available to monitor cholesterol levels and people with high levels can be treated and can follow special low-fat and low-cholesterol diets.

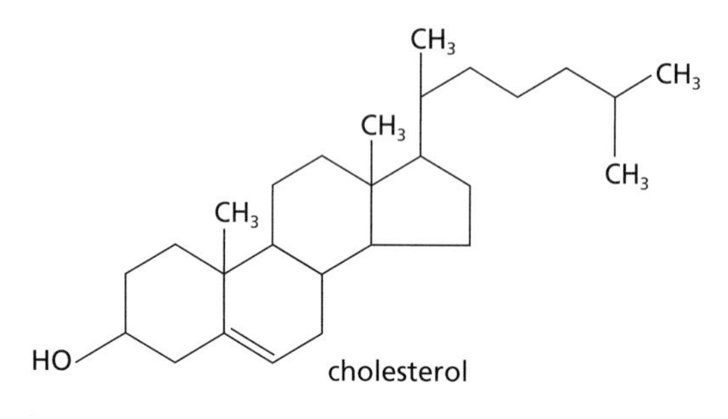


**Figure 14.15** The structure of cholesterol.

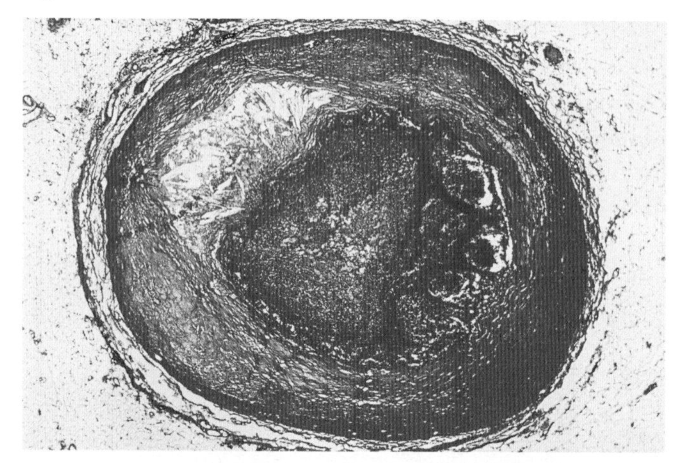

**a** This artery is being blocked by atheroma, which may be related to high levels of cholesterol in the blood.

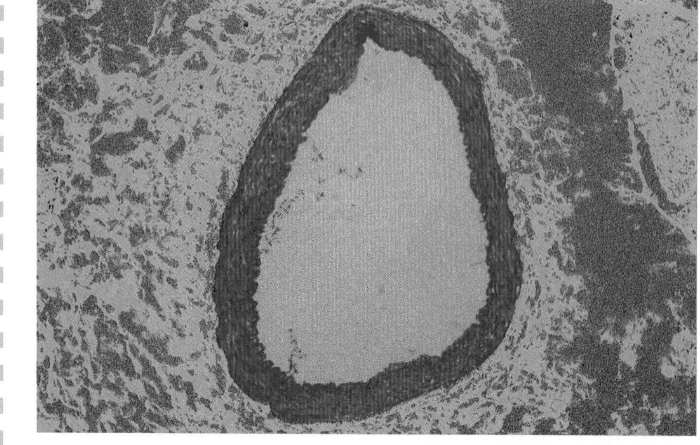

**b** This is a healthy artery.

**Figure 14.16**

## Biotechnology

Biotechnology involves making use of micro-organisms or their components, such as enzymes, for the benefit of humans to produce, for example, foods such as yoghurt and bread. One of the oldest biotechnologies is that of **fermentation**. It involves a series of biochemical reactions brought about by micro-organisms or enzymes. Fermentation is the basic process behind the baking, wine- and beer-making industries.

Fermentation in the laboratory can be carried out using sugar solution. A micro-organism called yeast is added to the solution. The yeast uses the sugar for energy during **anaerobic respiration** (respiration without oxygen), and so the sugar is broken down to give carbon dioxide and ethanol. The best temperature for this process to be carried out is at 37 °C.

$$\text{glucose} \xrightarrow{\text{yeast}} \text{ethanol} + \text{carbon dioxide}$$
$$C_6H_{12}O_6(aq) \longrightarrow 2C_2H_5OH(l) + 2CO_2(g)$$

Figure 14.17 shows a simple apparatus for obtaining ethanol from glucose in the laboratory.

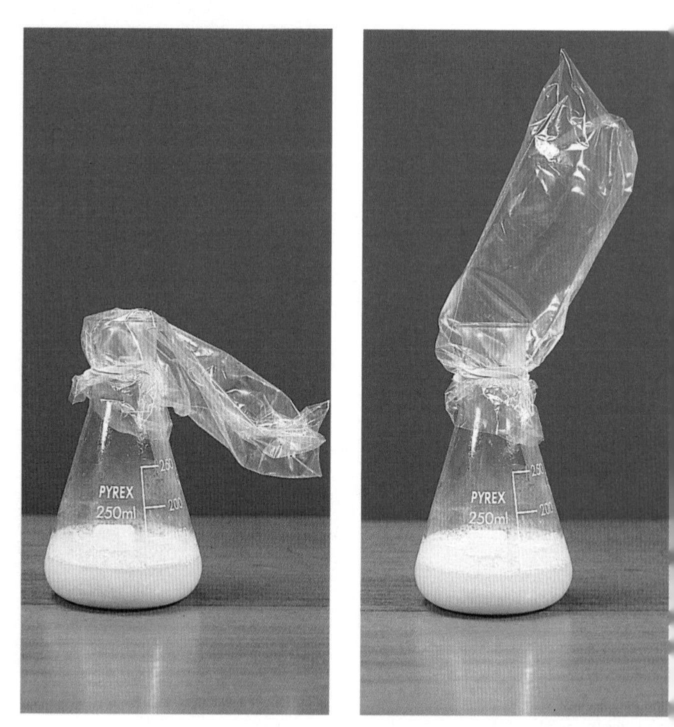


**Figure 14.17** Fermenting glucose and yeast to produce ethanol. The bag is inflated during the experiment by $CO_2$.

Alcoholic drinks such as beer and wine are made on a large scale in vast quantities. Beer is made from barley to which hops are added to produce the distinctive flavour (Figure 14.18a), whereas wine is made by fermenting grape juice, which contains glucose (Figure 14.18b). The micro-organisms in yeast will carry out the fermentation quite successfully in both cases. Beer normally contains only about 4% by volume of ethanol, whereas wine contains about 11%.

a Beer is produced on a large scale.

b Grape picking for the wine-making industry.

**Figure 14.18**

The yeast is killed off if there is much more ethanol than this, so it is not possible to make stronger drinks by fermentation. Some of the stronger drinks in Table 14.3, like sherry, are made up to 'strength' by adding pure ethanol.

**Table 14.3** Ethanol content of different drinks.

| Drink | Percentage ethanol by volume |
|---|---|
| Whisky, brandy | 40 |
| Sherry | 20 |
| Martini | 14 |
| Wine | 11 |
| Beer | 4 |

Spirits such as whisky and brandy are more concentrated forms of alcoholic drink. These higher concentrations of ethanol are produced by distillation after the fermentation process is complete (Figure 14.19).

**Figure 14.19** A whisky distillery.

## Treat alcohol carefully

As you will know, it is the alcohol (ethanol) in these drinks that can make people drunk. It is also the alcohol which causes headaches (the 'hangover'), dizziness and vomiting associated with being drunk. Alcohol can damage vital organs in your body. Excessive intake over an extended period of time can cause irreparable damage to the liver (cirrhosis) and can result in death.

**Figure 14.20** Because alcohol is readily available, it is easy for people to misuse it. Abusing alcohol is a serious problem in our society.

Alcohol is a drug and is addictive (Figure 14.20). There are about half a million alcoholics in the UK. Because of their addiction, alcoholics often have health, family and work problems.

Even small amounts of alcohol in the bloodstream can reduce your judgement and skill. Up to a third of all road accidents are associated with alcohol. The introduction of the breathalyser had a marked effect on road accident figures. In its first year of use there were 40 000 fewer accidents on the roads of the UK. The 'Alcotest 80' breathalyser tubes contain potassium dichromate(VI) crystals (Figure 14.21). When you breathe into a breathalyser, if your exhaled air contains alcohol vapour some of the potassium dichromate(VI) crystals turn green. How far the green colour extends along the tube is a measure of the amount of alcohol in your breath. The extent of the green colour is taken as a measure of the blood alcohol concentration (BAC).

The legal limit on BAC is 80 mg of alcohol per 100 $cm^3$ of blood. If you 'fail' on the 'Alcotest 80' breathalyser, a blood sample is taken to confirm the actual level of alcohol in your blood.

Table 14.4 shows the effects of different blood alcohol levels on an average person. It is important to appreciate that young people may experience these effects after less alcohol than the table shows.

**Table 14.4** Effects of different blood alcohol levels.

| Number of drinks | Approximate BAC level | Effects |
|---|---|---|
| 1 pint of beer | 30 mg | Increased chance of an accident |
| 3 single whiskies | 50 mg | Cheerful, judgement affected, less inhibited |
| 5 single whiskies | 80 mg | This is the legal limit for driving. You are four times more likely to have an accident |

## Baking

A **baker** is involved with biotechnology. To make bread, fresh yeast is mixed with warm sugar solution and the mixture added to the flour. This dough mixture is then put into a warm place to rise. The dough rises due to the production of carbon dioxide from **aerobic respiration** (respiration with oxygen) by the yeast. The products of this style of respiration are different to those of anaerobic respiration (see p. 204).

$$\text{sugar} + \text{oxygen} \xrightarrow{\text{yeast}} \text{carbon dioxide} + \text{water} + \text{energy}$$

$$C_6H_{12}O_6(aq) + 6O_2(g) \longrightarrow 6CO_2(g) + 6H_2O(l) + \text{energy}$$

After the dough has 'risen', it is baked and the heat kills the yeast and the bread stops rising.

**a** A driver's alcohol level being assessed using a breathalyser.

**b** 'Alcotest 80' breathalyser tubes (before, orange, and after, green).

**Figure 14.21**

## New applications

A large number of firms throughout the world are investing large sums of money in the newer biotechnology applications now in use.

- Enzymes can be isolated from micro-organisms and used to catalyse reactions in other processes. For example, proteases are used in biological detergents to digest protein stains such as blood and food. Also, catalase is used in the rubber industry to help convert latex into foam rubber.
- Our ability to manipulate an organism's genes to make it useful to us is called **genetic engineering**. This is being used, for example, to develop novel plants for agriculture as well as making important human proteins such as the hormones insulin and growth hormone.

However, a word of caution is necessary. The new biotechnologies may not be without dangers. For example, new pathogens (organisms that cause disease) might be created accidentally. Also, new pathogens may be created deliberately for use in warfare. As you can imagine, there are very strict guidelines covering these new biotechnologies, especially in the area of research into genetic engineering.

## Questions

1 What do you understand by the term 'biotechnology'? In your answer make reference to the making of beer and bread.

2 With reference to the data shown in Table 14.3 (p. 205), answer the following.
   a Calculate the amount of ethanol (alcohol) in 1 litre of whisky or brandy.
   b An advert for Campari suggested it contained 15% more alcohol by volume than sherry. Calculate the percentage by volume of alcohol in Campari and also calculate the amount of alcohol in a 1 litre bottle of Campari.
   c Low-alcohol beer contains about 0.5% of alcohol by volume. How many litres of this beer contain the same volume of alcohol as 2 litres of normal-strength beer?

## *Ethanoic acid ($CH_3COOH$) – a carboxylic acid*

The carboxylic acids form another homologous series, this time with the general formula $C_nH_{2n-1}COOH$. All the carboxylic acids possess –COOH as their functional group. Table 14.5 shows the first three members of this homologous series along with their melting and boiling points. Figure 14.22 shows the actual arrangement of the atoms in these three members of this family.

**Table 14.5** The first three members of the carboxylic acid series.

| Carboxylic acid | Formula | Melting point/ °C | Boiling point/ °C |
|---|---|---|---|
| Methanoic acid | HCOOH | 9 | 101 |
| Ethanoic acid | $CH_3COOH$ | 17 | 118 |
| Propanoic acid | $CH_3CH_2COOH$ | −21 | 141 |

methanoic acid

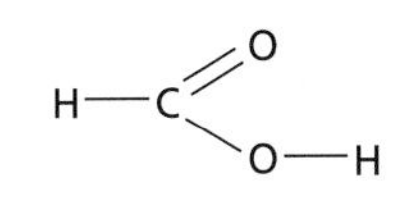

ethanoic acid

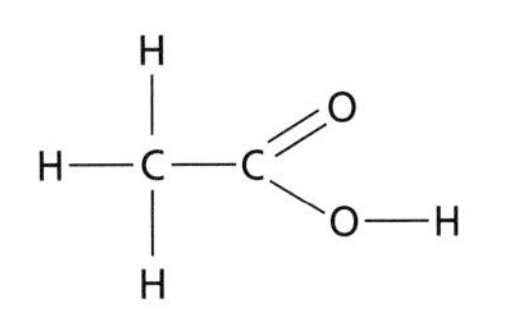

propanoic acid

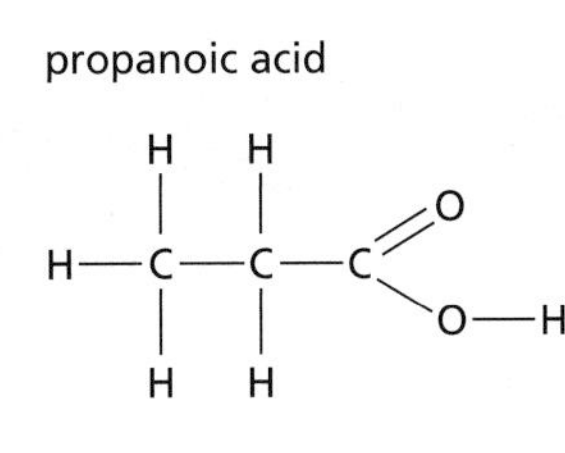

**Figure 14.22** The molecules look like the 3D models in the photographs.

Methanoic acid is present in stinging nettles and ant stings. Ethanoic acid, however, is the most well known as it is the main constituent of vinegar. Like other acids, ethanoic acid affects indicators and will react with metals such as magnesium. However, whereas the mineral acids such as hydrochloric acid are called strong acids, ethanoic acid is a weak acid (Chapter 7, p. 98). Even though it is a weak acid, it will still react with bases to form salts. For example, the salt sodium ethanoate is formed when ethanoic acid reacts with dilute sodium hydroxide.

$$\text{ethanoic acid} + \text{sodium hydroxide} \rightarrow \text{sodium ethanoate} + \text{water}$$

$$CH_3COOH(aq) + NaOH(aq) \rightarrow CH_3COONa(aq) + H_2O(l)$$

Ethanoic acid also undergoes other typical reactions of acids, in that it reacts with indicators, metals and carbonates in the usual way.

Ethanoic acid will react with ethanol, in the presence of a few drops of concentrated sulphuric acid, to produce ethyl ethanoate – an **ester**.

$$\text{ethanoic acid} + \text{ethanol} \underset{H_2SO_4}{\overset{\text{conc}}{\rightleftharpoons}} \text{ethyl ethanoate} + \text{water}$$

$$CH_3COOH(l) + CH_3CH_2OH(l) \rightleftharpoons CH_3COOCH_2CH_3(aq) + H_2O(l)$$

This reaction is called **esterification**.

Members of the 'ester' family have strong and pleasant smells. Many of them occur naturally and are responsible for the flavours in fruits and the smells of flowers. They are used, therefore, in some food flavourings and in perfumes (Figure 14.23).

**Figure 14.23** Perfumes contain esters.

Fats and oils are naturally occurring esters which are used as energy storage compounds by plants and animals (Figure 14.24).

**Figure 14.24** Olive oil is obtained from a plant whereas lard is made from animal fat.

## Questions

1 Write the structural formula for propanoic acid.

2 Write word and balanced chemical equations for the esterification of propanoic acid with ethanol.

# *Other carboxylic acids*

## Aspirin

Aspirin (Figure 14.25) is one of the most frequently used painkillers in the world. It is also able to reduce inflammation and fever and a low dose taken on a daily basis over the age of 50 may prevent heart attacks. It is derived from another acid, salicylic acid, which can be obtained from willow bark. Salicylic acid has the same medicinal properties as aspirin and has been known since 1829. Salicylic acid, however, caused stomach bleeding. The conversion of salicylic acid to aspirin reduced these problems, but aspirin still has some adverse effects on the stomach if taken in excess.

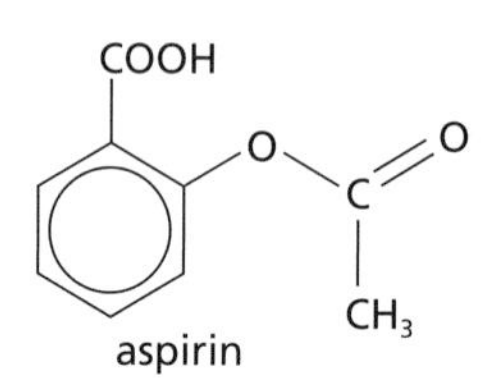


**Figure 14.25** The structure of aspirin.

## Ascorbic acid (vitamin C)

Vitamin C, also known as ascorbic acid, is an essential vitamin (Figure 14.26). Vitamin C is required by the body in very small amounts and it is obtained from foods.

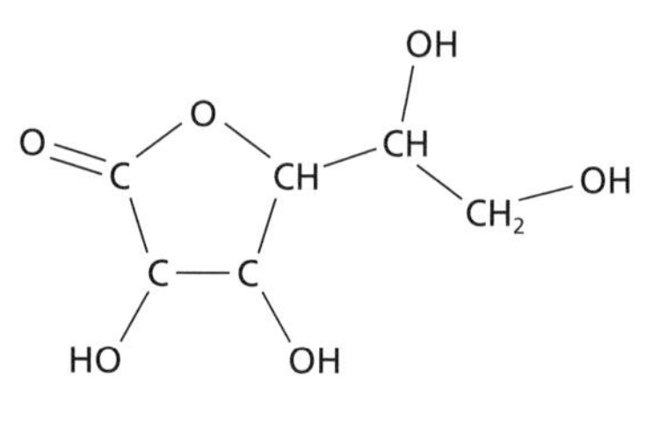


**Figure 14.26** The structure of ascorbic acid (vitamin C).

Vitamin C is important to all animals, including humans, and is vital in the production of collagen. Collagen is important in the formation of connective tissues that give our body shape and help to support vital organs. Vitamin C prevents the disease scurvy. It is found in citrus fruits and brightly coloured vegetables, such as peppers and broccoli. Many people take vitamin C supplements, which are readily available from supermarkets and pharmacies (Figure 14.27). Although vitamin C is destroyed by exposure to air and heat, the average person usually reaches the recommended daily allowance of 60 mg through food.

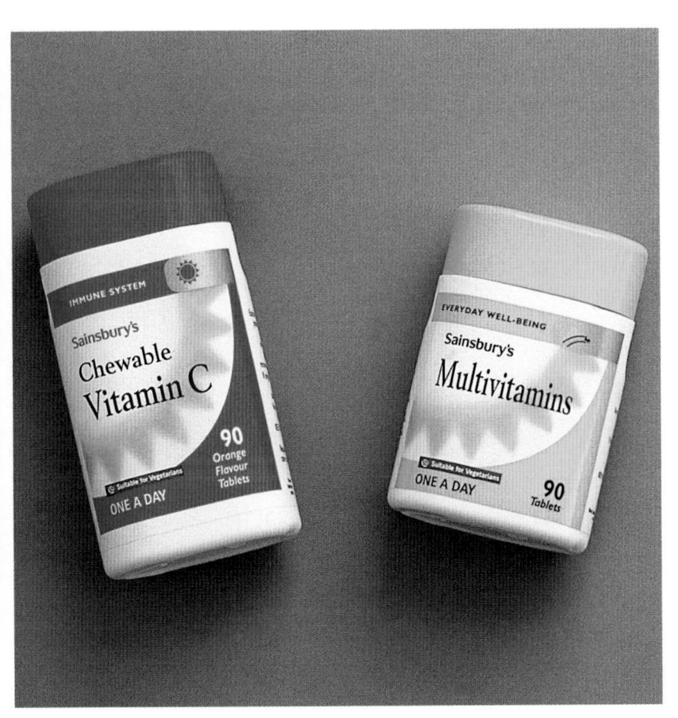


**Figure 14.27** These fruits and supplements all contain vitamin C.

## Citric acid

Citric acid is an example of a tricarboxylic acid, one which contains three —COOH groups (Figure 14.28). It is an important acid and is found in all citrus fruits, for example lemons and oranges.

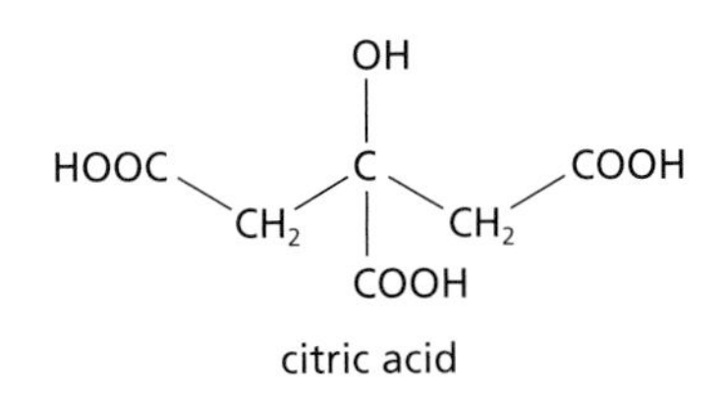


**Figure 14.28** The structure of citric acid.

## *Soaps and detergents*

Millions of tonnes of soaps and soapless detergents are manufactured worldwide every year. Soap is manufactured by heating natural fats and oils of either plants or animals with a strong alkali. These fats and oils, called triglycerides, are complicated ester molecules.

Fat is boiled with aqueous sodium hydroxide to form soap. The esters are broken down in the presence of water – hydrolysed. This type of reaction is called **saponification**. The equation given below is that for the saponification of glyceryl stearate (a fat).

glyceryl stearate + sodium hydroxide → sodium stearate (soap) + glycerol

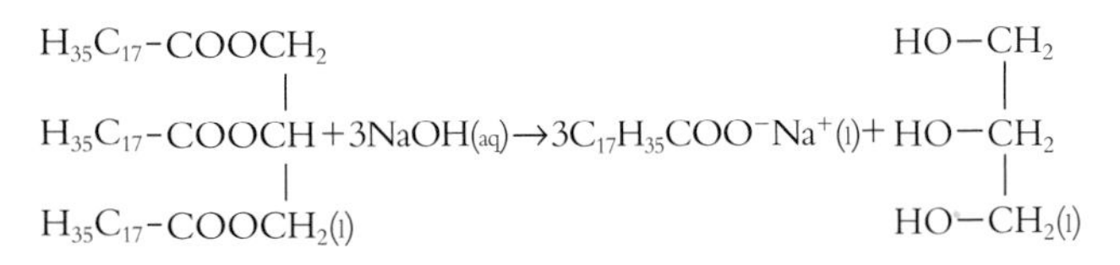


The cleaning properties of the soap depend on its structure and bonding. Sodium stearate consists of a long hydrocarbon chain which is hydrophobic (water hating) attached to an ionic 'head' which is hydrophilic (water loving) (Figure 14.29).

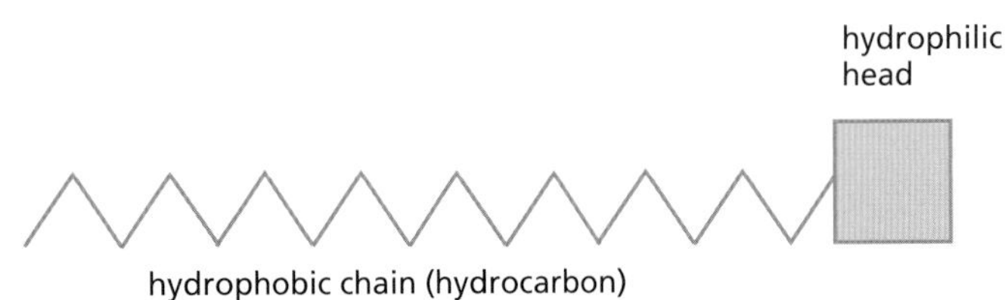


**Figure 14.29** Simplified diagram of a soap molecule.

Covalent compounds are generally insoluble in water but they are more soluble in organic solvents. Ionic compounds are generally water soluble but tend to be insoluble in organic solvents. When a soap is put into water which has a greasy dish (or a greasy cloth) in it, the hydrophobic hydrocarbon chain on each soap molecule becomes attracted to the grease and becomes embedded in it (Figure 14.30).

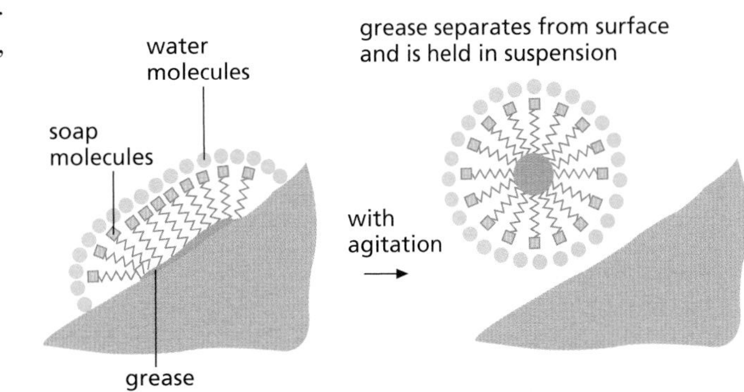


**Figure 14.30** Soap dissolves grease like this.

On the other hand, the hydrophilic ionic head group is not attracted to the grease but is strongly attracted to the water molecules. When the water is stirred, the grease is slowly released and is completely surrounded by the soap molecules. The grease is, therefore, 'solubilised' and removed from the dish. The soap is able to remove the grease because of the combination of the covalent and ionic bonds present.

## Soapless detergents

In Chapter 8, p. 119, we discussed the way in which, in hard water areas, an insoluble scum forms when soap is used. This problem has been overcome by the development of synthetic **soapless detergents**. These new substances do not form scum with hard water since they do not react with $Ca^{2+}$ and $Mg^{2+}$ present in such water. Furthermore, these new soapless detergent molecules have been designed so that they are biodegradable. Bacteria readily break down these new molecules so that they do not persist in the environment.

Sodium alkyl benzene sulphonates were developed in the early 1970s. The structure of sodium 3-dodecylbenzene sulphate, $C_{18}H_{29}SO_3Na$, is given below.

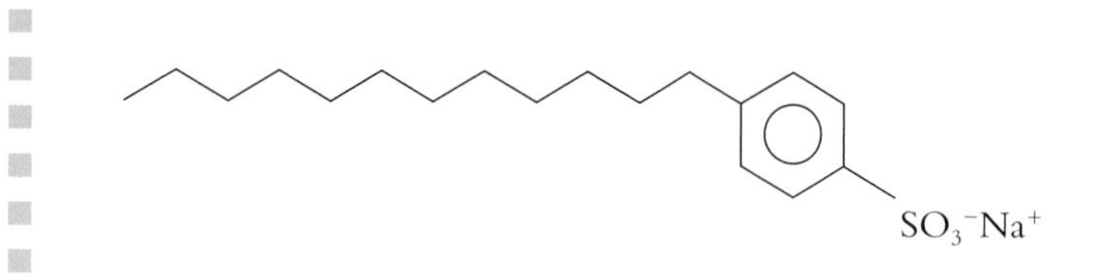


The calcium and magnesium salts of this detergent molecule are water soluble, so the problem of scum is solved. Very many of our washing powders (and liquids) contain this type of substance. The manufacture of soapless detergents is explained in Chapter 16, p. 234.

## Questions

1 What class of organic compound do substances like glyceryl stearate belong to?

2 What do you understand by the terms:
  **a** hydrophobic?
  **b** hydrophilic?
  **c** saponification?

3 What is the main advantage of detergents over soaps?

# *Amino acids*

There are 20 different amino acids and they each possess two functional groups. One is the carboxylic acid group, –COOH. The other is the amine group, $-NH_2$. The two amino acids shown below are glycine and alanine.

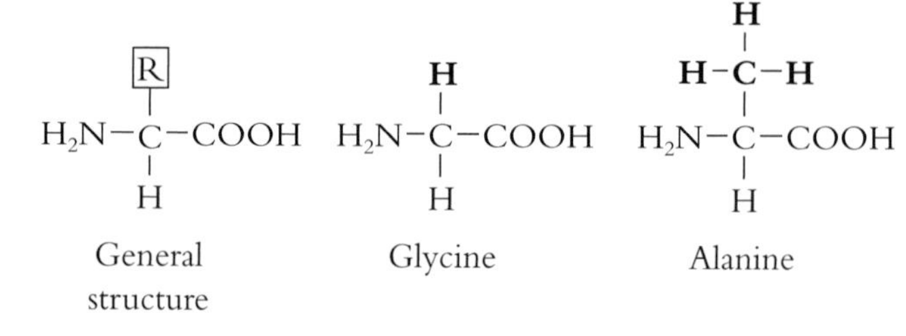


Amino acids are the building blocks of proteins. Similar to nylon (see p. 199) proteins are polyamides, as they contain the –CONH– group, which is called the amide or, in the case of proteins, the peptide link. Proteins are formed by condensation polymerisation.

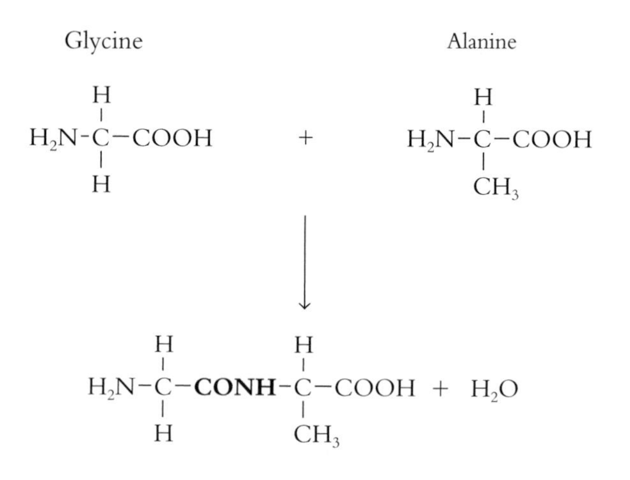


A **dipeptide**
(composed of two amino acids joined together)

Further reaction with many more amino acids takes place at opposite ends of each molecule to produce the final protein (Figure 14.31). For a molecule to be a protein, there must be at least 100 amino acids involved. Below this number, they are called polypeptides. Proteins make up some 15% of our body weight.

Proteins fall broadly into two groups: they can be fibrous or globular.

- Fibrous proteins – these have linear molecules, are insoluble in water and resistant to alkalis and acids. Collagen (in tendons and muscles), keratin (in nails, hair, horn and feathers) and elastin (in arteries) are all fibrous proteins.
- Globular proteins – these have complicated three-dimensional structures and are soluble in water. They are easily affected by acids, alkalis and temperature increase, when they are said to be denatured. Casein (in milk), albumen (in egg white) and enzymes are examples of globular proteins.

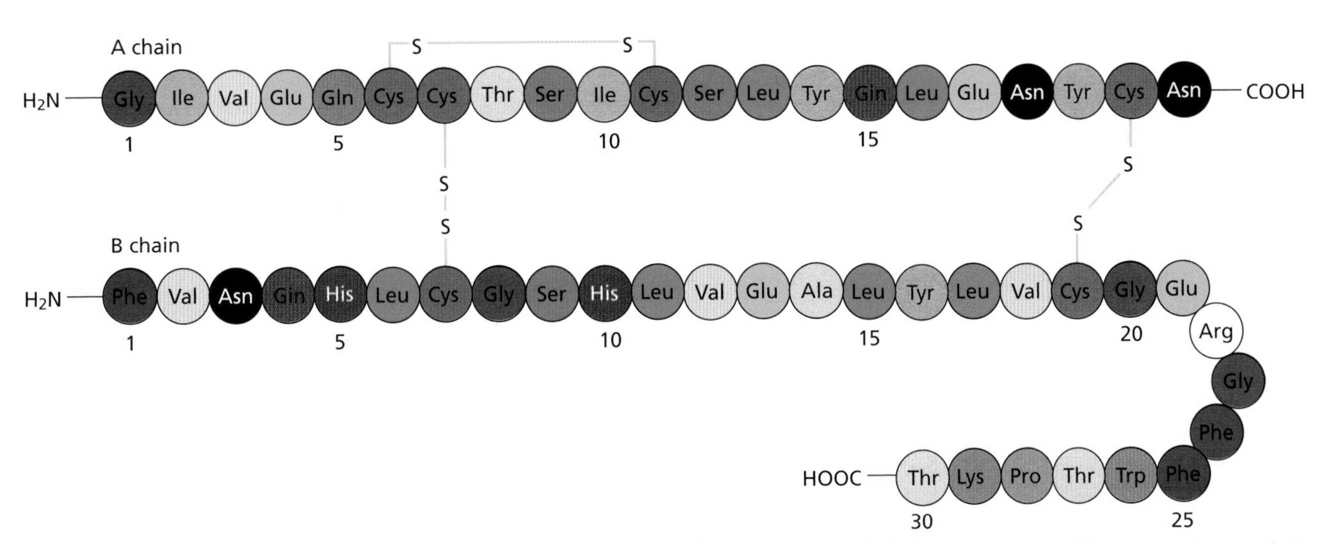


**Figure 14.31** The structure of a protein – human insulin (the different coloured circles represent different amino acids in this protein).

## Protein analysis

How can you determine which amino acids are present in a particular protein? This involves hydrolysis of the peptide (amide) bonds in the protein so that the individual amino acids are released. This can only be done by heating the protein with dilute hydrochloric acid. The mixture of amino acids is then separated by thin layer chromatography (TLC) (Figure 14.32) or electrophoresis. In both cases, a locating agent (Chapter 2, p. 26), such as ninhydrin, is used. This ensures that the spots of acid are visible.

**Figure 14.32** Amino acids can be separated and identified by TLC.

If you are trying to show only the presence of a protein, a quick test to carry out is known as the **Biuret test**. A mixture of dilute sodium hydroxide and 1% copper(II) sulphate solution is shaken with a sample of the material under test. If a protein is present, a purple colour appears after about three minutes (Figure 14.33).

**a** Testing for a protein.

**b** Adding dilute sodium hydroxide.

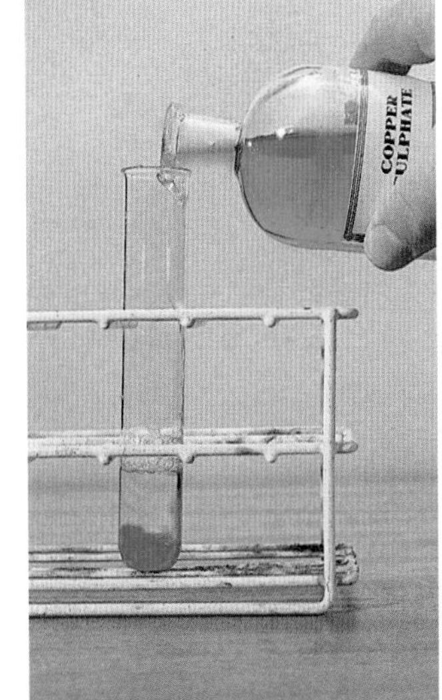


**c** Adding 1% copper(II) sulphate.

**d** The purple colour shows the presence of a protein.

**Figure 14.33**

## *DNA*

**Deoxyribonucleic acid (DNA)** belongs to a group of chemicals called the nucleic acids (Figure 14.34). They are also biopolymers. DNA controls the protein synthesis within your cells. When you eat a food containing proteins, such as meat or cheese, your digestive enzymes break down the proteins present into individual amino acids. The DNA in your cells controls the order in which the amino acids are repolymerised to make the proteins you need!

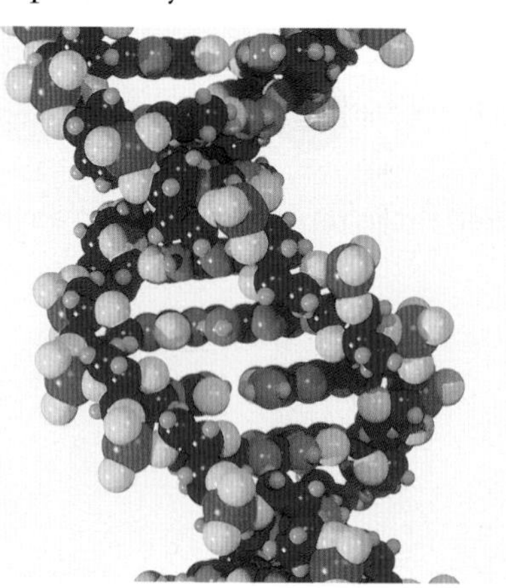

**Figure 14.34** Crick and Watson based this model for DNA on X-ray studies and chemical analysis.

Genes are the units of heredity that control the characteristics of organisms. A gene is made of DNA. No two individuals have the same DNA sequence. DNA 'fingerprinting' has become a very powerful forensic science tool in the investigation of crime (Figure 14.35).

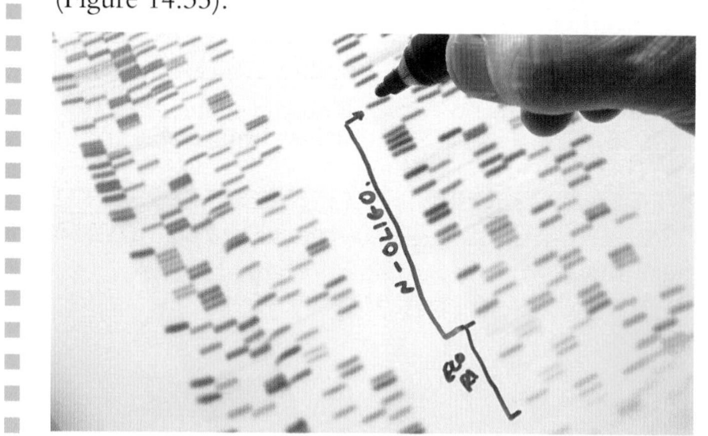

**Figure 14.35** Identical patterns shown by DNA fingerprinting can identify the criminal.

### Questions

1. Which two functional groups do amino acids possess?
2. How many amino acids have to be involved before the biopolymer is called a protein?
3. Name the process by which the individual amino acids in a protein are released by reaction with a dilute acid.
4. Explain how DNA fingerprinting may be used in paternity suits.

## *Pharmaceuticals*

Pharmaceuticals are chemicals called **drugs** that are prepared and sold with the intention of treating illness. A drug is any substance, natural or synthetic, which alters the way in which the body works. There are many categories of drugs. The following are some examples.

- Anaesthetics – these induce loss of feeling and/or consciousness, for example flurothane.
- Analgesics – these relieve pain, for example aspirin.
- Antibiotics – these are substances, for example penicillin, originally produced by micro-organisms, which are used to kill bacteria. However, most antibiotics are now made in chemical laboratories, for example carbenicillin.
- Sedatives – these induce sleep, for example barbiturates.
- Tranquillisers – these will give relief from anxiety, for example Valium.

There are, of course, many other types of drug available which have very specific uses. For example, methyldopa was developed to relieve hypertension (high blood pressure), and antihistamines were developed to help control travel sickness, hayfever and allergic reactions.

The pharmaceutical industry is one of the most important parts of the chemical industry and is a major consumer of the products of the petrochemical industry. It is a high-profit industry but with very high research and development costs. For example, it costs in excess of £100 million to discover, test and get a single drug on to the market.

Today, the pharmaceutical industry could be called the 'medicines by design' industry. Companies such as GlaxoSmithKline have teams of chemists and biochemists working almost around the clock to discover, test, check for safety and produce drugs that can deal with almost every known illness.

Table 14.6 (opposite) shows the structures of a selection of some of the more common drugs available at the present time, along with their uses. The common names for these drugs are used, since their systematic, theoretical names are extremely complex.

**Table 14.6** Commonly available drugs.

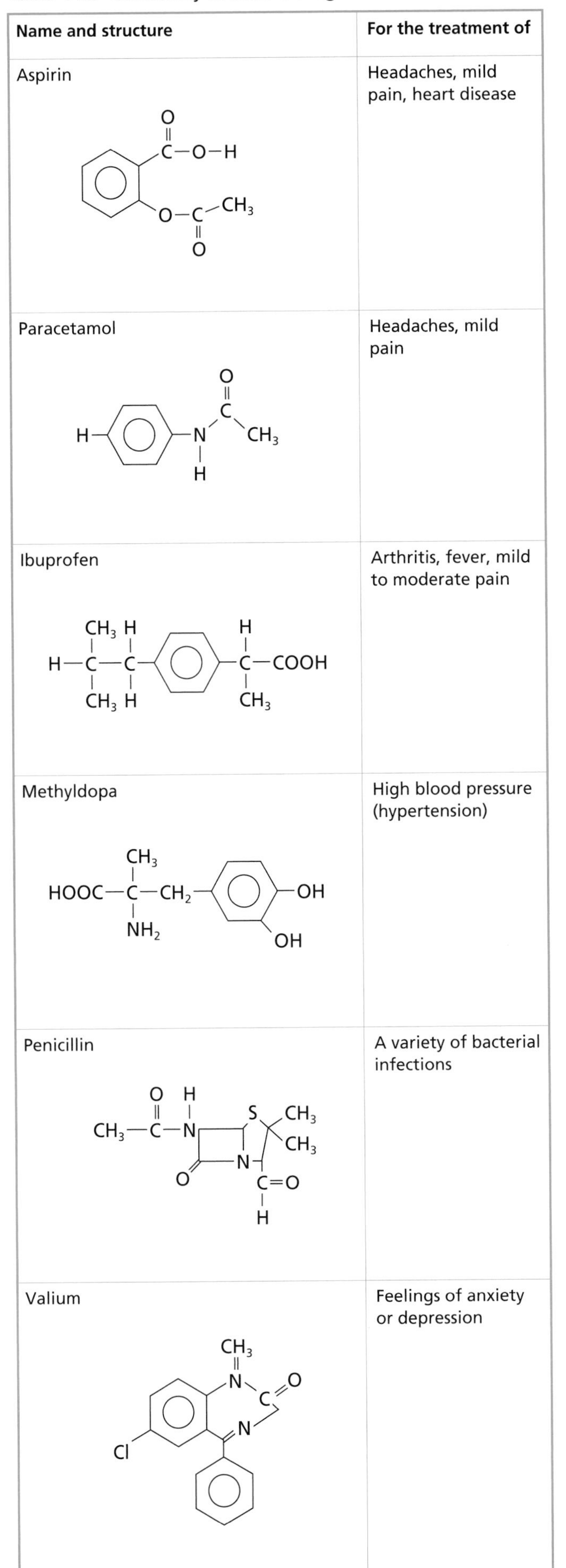

| Name and structure | For the treatment of |
|---|---|
| Aspirin | Headaches, mild pain, heart disease |
| Paracetamol | Headaches, mild pain |
| Ibuprofen | Arthritis, fever, mild to moderate pain |
| Methyldopa | High blood pressure (hypertension) |
| Penicillin | A variety of bacterial infections |
| Valium | Feelings of anxiety or depression |

## Drug abuse

Some of the very useful drugs developed by chemists can be habit forming. For example, barbiturates (in sleeping tablets) and amphetamines (stimulants) fall into this category. Another drug that has created problems in the past is Valium, which is not itself addictive but when used in the long term makes people dependent on it. Severe psychological and physiological problems can arise.

It should be noted, however, that it is the opiates which cause **addiction** (Figure 14.36). Cocaine and heroin are just two examples of such substances. Consequences of the addiction include personal neglect, both of nutritional needs and of hygiene. For a short-term feeling of well-being ('fix'), the addict is prepared to do almost anything. Addicts often turn to a life of crime to fulfil their cravings for the opiates. Addicts, especially those injecting drugs, are at a high risk of HIV (human immunodeficiency virus) infection as they often share needles with other drug addicts, who may be HIV positive. Public awareness campaigns aim to educate everyone on the dangers in society, including drug abuse and its related risks.

**Figure 14.36** Increasing public awareness of the risks associated with drug abuse is very important.

## *Questions*

**1** Using the data given in Table 14.6, suggest which of the pharmaceuticals contains:
- **a** sulphur
- **b** an –OH group
- **c** an $-NH_2$ group
- **d** an ester group.

**2** Drug abuse is a rapidly growing problem. Using the information given in this section as well as your research skills, make a list of the addictive drugs. Also explain the problems that drug abuse can cause.

## *Checklist*

**After studying Chapter 14 you should know and understand the following terms.**

- **Addition polymer** A polymer formed by an addition reaction. For example, poly(ethene) is formed from ethene.
- **Alcohols** Organic compounds containing the –OH group. They have the general formula $C_nH_{2n+1}OH$. Ethanol is by far the most important of the alcohols and is often just called 'alcohol'.
- **Amino acids** These naturally occurring organic compounds possess both an $-NH_2$ group and a –COOH group on adjacent carbon atoms. There are 20 naturally occurring amino acids, of which glycine is the simplest.
- **Biodegradable plastics** Plastics designed to degrade (decompose) under the influence of bacteria.
- **Biopolymers** Natural polymers such as starch and proteins.
- **Biotechnology** Making use of micro-organisms in industrial and commercial processes. For example, the process of fermentation is brought about by the enzymes in yeast.
- **Biuret test** The test for proteins. A mixture of dilute sodium hydroxide and 1% copper(II) sulphate solution is shaken with the material under test. A purple colour appears after about three minutes if a protein is present.
- **Carbohydrates** A group of naturally occurring organic compounds which can be represented by the general formula $(CH_2O)_x$.
- **Carboxylic acids** A family of organic compounds containing the functional group –COOH. They have the general formula $C_nH_{2n-1}COOH$. The most important and well known of these acids is ethanoic acid, which is the main constituent in vinegar. Ethanoic acid is produced by the oxidation of ethanol.
- **Condensation polymer** A polymer formed by a condensation reaction (one in which water is given out). For example, nylon is produced by the condensation reaction between 1,6-diaminohexane and hexanedioic acid.
- **Cross-linking** The formation of side covalent bonds linking different polymer chains and therefore increasing the rigidity of, say, a plastic. Thermosetting plastics are usually heavily cross-linked.
- **DNA** Abbreviation for deoxyribonucleic acid. It belongs to a group of biopolymers called the nucleic acids. It is involved in the polymerisation of amino acids in a specific order to form the particular protein required by a cell.
- **Drug** Any substance, natural or synthetic, that alters the way in which the body works.
- **Drug abuse** This term usually applies to the misuse of addictive drugs, which include barbiturates and amphetamines, as well as the opiates, cocaine and heroin. These drugs create severe psychological and physiological problems through self-indulgence. This leads to a variety of personal problems for the user.
- **Esters** A family of organic compounds formed by the reaction of an alcohol with a carboxylic acid in the presence of concentrated $H_2SO_4$. This type of reaction is known as esterification. Esters are characterised by a strong and pleasant smell (many occur in nature and account for the smell of flowers).
- **Functional group** The atom or group of atoms responsible for the characteristic reactions of a compound.
- **HIV** Short for human immunodeficiency virus, from which AIDS (acquired immunodeficiency syndrome) can develop.
- **Hydrolysis** A chemical reaction involving the reaction of a compound with water. Acid hydrolysis usually involves dilute hydrochloric acid, and enzyme hydrolysis involves enzymes such as amylase.
- **Monomer** A simple molecule, such as ethene, which can be polymerised.
- **Monosaccharides** A group of simple carbohydrates. They are sweet to taste and are water soluble (for example glucose).

*continued*

- **Pharmaceuticals** These are chemicals called drugs that are prepared and sold with the intention of treating disease (for example methyldopa).
- **Photodegradable plastics** Plastics designed to degrade under the influence of sunlight.
- **Polymer** A substance possessing very large molecules consisting of repeated units or monomers. Polymers therefore have a very large relative molecular mass.
- **Polymerisation** The chemical reaction in which molecules (monomers) join together to form a polymer.
- **Polysaccharides** A group of more complicated carbohydrates. They generally do not form true solutions and do not have a sweet taste (for example starch).
- **Proteins** Polymers of amino acids formed by condensation reactions. They fall broadly into the two categories: fibrous proteins (for example keratin and collagen) and globular proteins (for example casein and albumen).
- **Soapless detergents** These are soap-like molecules which do not form a scum with hard water. These substances have been developed from petrochemicals. Their calcium and magnesium salts are water soluble and they are biodegradable.
- **Soaps** Substances formed by saponification. In this reaction, the oil or fat (glyceryl ester) is hydrolysed by aqueous sodium hydroxide to produce the sodium salt of the fatty acid, particularly sodium stearate (from stearic acid). Soap will dissolve grease because of the dual nature of the soap molecule. It has a hydrophobic part (the hydrocarbon chain) and a hydrophilic part (the ionic head) and so will involve itself with both grease and water molecules. However, it forms a scum with hard water by reacting with the $Ca^{2+}$ (or $Mg^{2+}$) present.
- **Thermoplastics** Plastics which soften when heated (for example polythene, PVC).
- **Thermosetting plastics** Plastics which do not soften on heating but only char and decompose (for example melamine).

# The (wider) organic manufacturing industry

## *Additional questions*

**1** Explain the following.

- **a** The problem of plastic waste has been overcome.
- **b** The majority of detergents produced today are biodegradable.
- **c** In bread making, yeast is added to the mix and the dough left to stand for a period of time.
- **d** When wine is left open it eventually turns to vinegar.
- **e** Poly(ethene) is a thermoplastic.

**2** **a** A detergent molecule may be represented by the following simplified diagram.

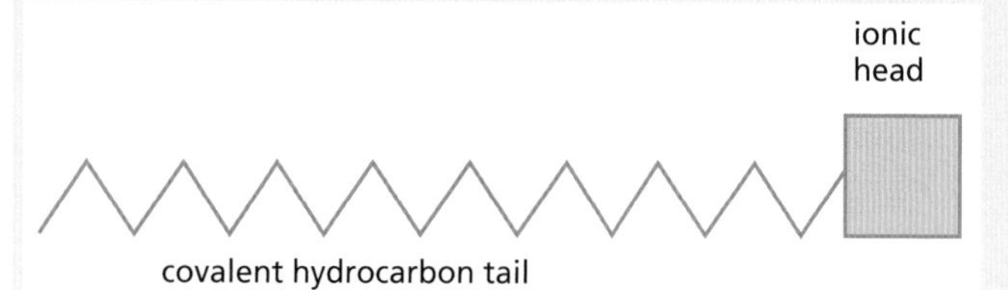


Use this representation of a detergent molecule in a series of labelled diagrams to show how detergents can remove grease from a piece of greasy cloth.

- **b** Explain why detergents do not form a scum with hard water, whereas soaps do.
- **c** The modern detergents are biodegradable.
  - (i) Explain what this statement means.
  - (ii) Why is it necessary for detergents to be biodegradable?

**3** A piece of cheese contains protein. Proteins are natural polymers made up of amino acids. There are 20 naturally occurring amino acids. The structures of two amino acids are shown below.

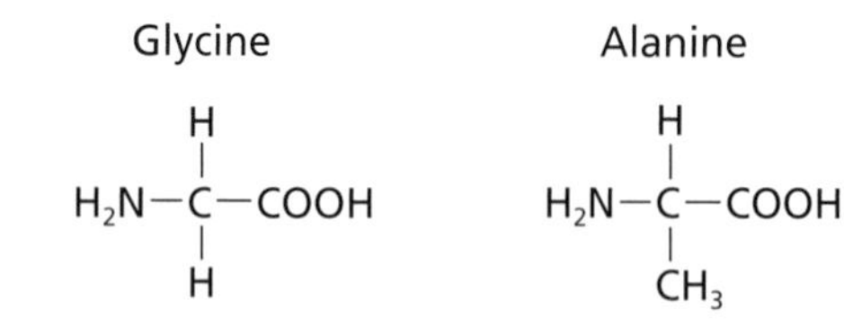


- **a** Name the type of polymerisation involved in protein formation.
- **b** Draw a structural formula to represent the part of the protein chain formed by the reaction between the amino acids shown above.
- **c** What is the name given to the common linkage present in protein molecules?
- **d** Why is there such a huge variety of proteins?
- **e** Name and describe the features of the two broad groups of proteins.

**4**

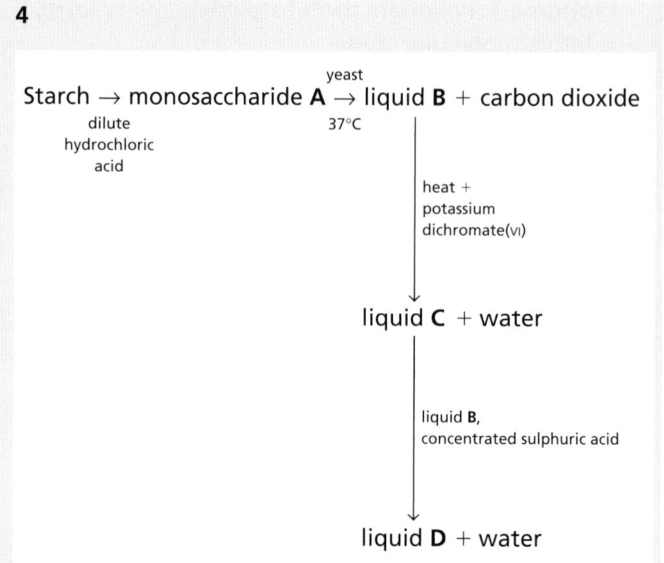


- **a** Name, give the formula and one use of each of the substances **A** to **D**.
- **b** Write word and balanced chemical equations for the reactions involved in the formation of liquids **B**, **C** and **D**.
- **c** Starch is classified as a natural polymer or 'biopolymer'. Explain the meaning of this statement.
- **d** Name the processes by which:
  - (i) starch is broken down
  - (ii) liquid **B** is formed
  - (iii) liquid **C** is formed
  - (iv) liquid **D** is formed.

**5** **a** Copy the following table and complete it by writing the structural formulae for methanol and methanoic acid.

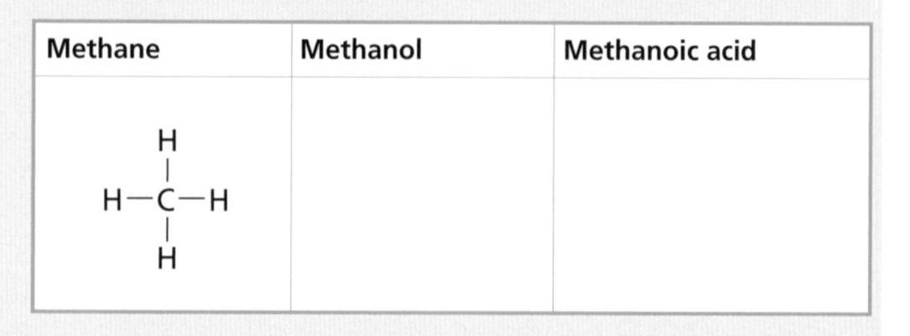

| Methane | Methanol | Methanoic acid |
|---|---|---|
| H<br>\|<br>H—C—H<br>\|<br>H | | |

- **b** Describe a simple chemical test that could be used to distinguish methanol from methanoic acid.
- **c** (i) Name the class of compound produced when methanol reacts with methanoic acid.
  - (ii) Name the type of reaction taking place.
  - (iii) Write a word and balanced chemical equation for this reaction.
  - (iv) Give two uses related to the class of compound formed in this reaction.

**d** The following reaction takes place when methanol is burned:

$$2CH_3OH(l) + 3O_2(g) \rightarrow 2CO_2(g) + 4H_2O(g) \; \Delta H = -1452\,kJ\,mol^{-1}$$

(i) How much heat energy would be liberated by burning:
0.5 mol of methanol?
4.0 mol of methanol?
4 g of methanol?
(ii) Calculate the volume of carbon dioxide produced at room temperature and pressure (rtp) when 16 g of methanol are burned.
($A_r$: H = 1; C = 12; O = 16. One mole of any gas at rtp occupies 24 $dm^3$.)

**6** This question requires you to use the information in Table 14.6 (p. 213).

**a** Which grouping of atoms is common to all three painkillers?

**b** Which of the pharmaceuticals contains:
(i) sulphur?
(ii) the greatest number of nitrogen atoms?
(iii) the carboxylic acid functional group?
(iv) chlorine?
(v) the greatest number of $-CH_3$ groups?

**c** Which of the pharmaceuticals is used to combat:
(i) bacterial infections?
(ii) depression?
(iii) heart disease?

**d** The pharmaceuticals industry has often been termed the 'medicines by design' industry. Explain the meaning of this statement.

**7** When a bottle of wine is left open to the air it turns sour. This is due to the oxidation, by the oxygen in the air, of the ethanol in the wine to ethanoic acid.

**a** Write an equation to represent this oxidation process.

**b** What would you expect to happen to the pH of the wine over a number of days?

**c** Devise a method which would allow you to calculate the amount of ethanoic acid in the wine.

**8** Why is it safe for us to use vinegar, which contains ethanoic acid, on food while it would be extremely dangerous for us to use dilute nitric acid for the same purpose?

# 15 Nitrogen



## Nitrogen – an essential element

Nitrogen gas is all around us. It makes up 78% of the air. Each day each individual breathes thousands of litres of nitrogen into his/her lungs and then exhales it without any chemical change occurring.

Nitrogen gas has no smell, is colourless and is a very unreactive gas.

The nitrogen gas molecule comprises two nitrogen atoms covalently bonded together by a triple bond (Figure 15.1). This triple bond is very strong, requiring a large amount of energy to break it so that nitrogen can take part in chemical reactions. Only at very high temperatures or if an electrical spark is passed through nitrogen gas will it react with oxygen. This occurs in an internal combustion engine, where the spark from the spark plug is sufficient to convert some of the nitrogen to toxic nitrogen oxides ($NO_x$) which are then emitted as car exhaust fumes (Chapter 11, p. 166). Catalytic converters prevent the emission of nitrogen oxides by converting them back to nitrogen gas before they leave the exhaust pipe.

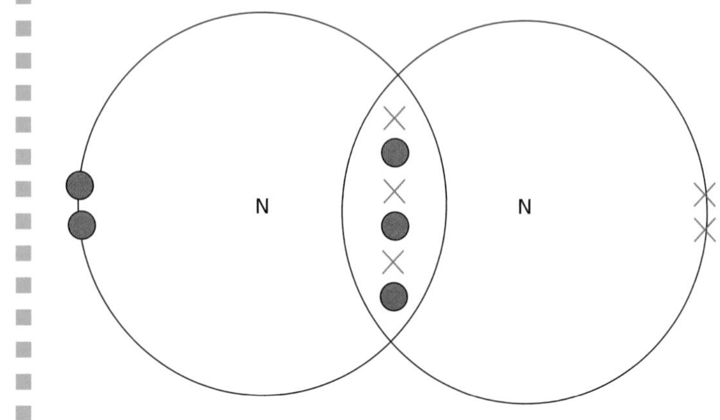


**Figure 15.1** The nitrogen molecule.

Nitrogen gas can be obtained industrially by the fractional distillation of liquid air. It is produced on a very large scale, since the gas has many very important industrial uses. The UK alone produces about 4 million tonnes of nitrogen each year.

Nitrogen is an essential element necessary for the well-being of animals and plants. It is present in proteins, which are found in all living things. Proteins are essential for healthy growth. Animals obtain the nitrogen they need for protein production by feeding on plants and other animals. Most plants obtain the nitrogen they require from the soil.

Nitrogen is found in nitrogen-containing compounds called **nitrates**, which are produced as a result of the effect of lightning and from dead plants and animals as they decay. Nitrates are soluble compounds and can be absorbed by the plants through their roots. Nitrogen in the air is also converted into nitrates by some forms of bacteria. Some of these bacteria live in the soil. A group of plants called **leguminous** plants have these bacteria in nodules on their roots. These bacteria are able to take nitrogen from the atmosphere and convert it into a form in which it can be used to make proteins. This process is called **fixing nitrogen** and is carried out by plants such as beans and clover.

The vital importance of nitrogen to both plants and animals can be summarised by the **nitrogen cycle** (Figure 15.2). If farm crops are harvested from the land rather than left to decay the soil becomes deficient in this important element. The nitrogen is removed in the harvested crops rather than remaining as the plants decay. In addition, nitrates can be washed from the soil by the action of rain (leaching). For the soil to remain fertile for the next crop, the nitrates need to be replaced. The natural process is by decay or by the action of lightning on atmospheric nitrogen. Without

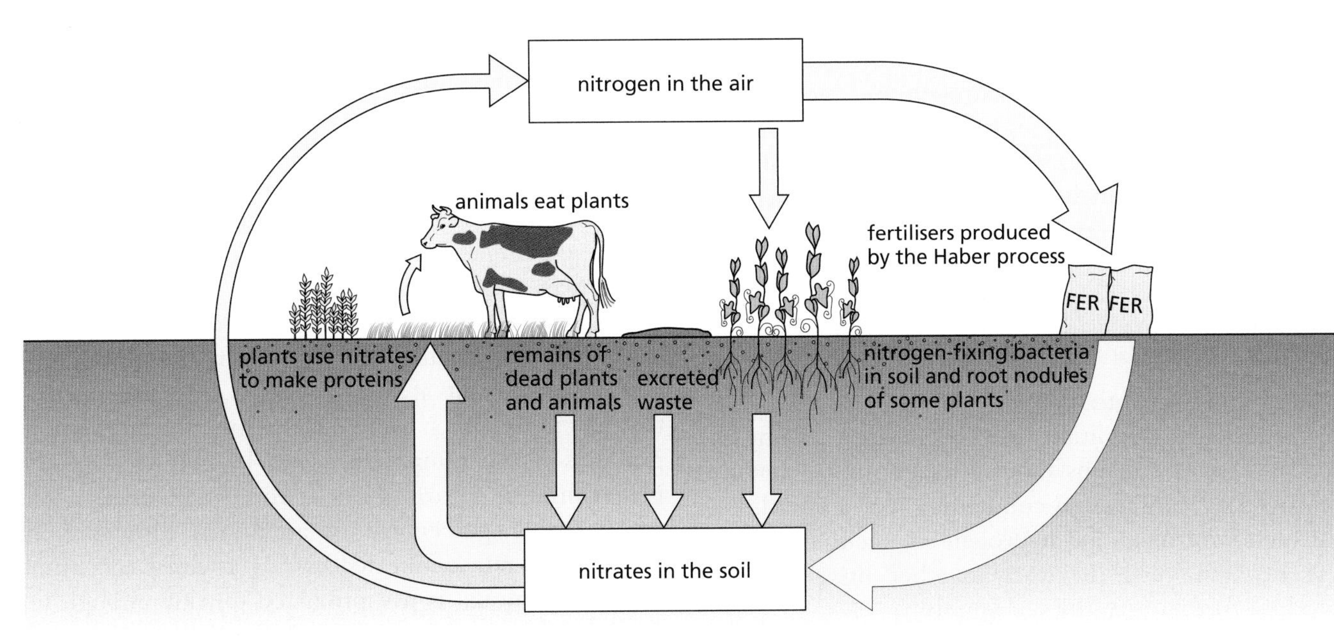


**Figure 15.2** The nitrogen cycle.

the decay, however, the latter process is not efficient enough to produce nitrates on the scale required. Farmers often need to add substances containing these nitrates. Such substances include:

- farmyard manure – this is very rich in nitrogen-containing compounds which can be converted by bacteria in the soil to nitrates
- artificial fertilisers – these are manufactured compounds of nitrogen which are used on an extremely large scale to enable farmers to produce ever-bigger harvests for the increasing population of the world. One of the most commonly used artificial fertilisers is **ammonium nitrate**, which is made from ammonia gas and nitric acid, both nitrogen-containing compounds.

## *Questions*

1 Why can most plants not use nitrogen gas directly from the atmosphere?

2 Nitrogen is essential in both plants and animals to form which type of molecule? Explain why these molecules are essential to life.

3 Explain why natural sources of the element nitrogen are not sufficient to produce the world's annual crop requirement.

# *The Haber process*

The Haber process forms the basis of the artificial fertiliser industry, as it is used to produce ammonia gas. The process was developed by the German scientist Fritz Haber in 1913. He was awarded a Nobel Prize in 1918 for his work. The process involves reacting nitrogen and hydrogen. It was first developed to satisfy the need for explosives during World War I, as explosives can be made from ammonia.

## Obtaining nitrogen

The nitrogen needed in the Haber process is obtained from the atmosphere.

## Obtaining hydrogen

The hydrogen needed in the Haber process is obtained from the reaction between methane and steam.

methane + steam ⇌ hydrogen + carbon monoxide

$$CH_4(g) + H_2O(g) \rightleftharpoons 3H_2(g) + CO(g)$$

This process is known as **steam re-forming**. This reaction is a **reversible** reaction and special conditions are employed to ensure that the reaction proceeds to the right (the forward reaction), producing hydrogen and carbon monoxide. The process is carried out at a temperature of 750 °C, at a pressure of 30 atmospheres with a catalyst of nickel. These conditions enable the maximum amount of hydrogen to be produced at an economic cost.

The carbon monoxide produced is then allowed to reduce some of the unreacted steam to produce more hydrogen gas.

carbon monoxide + steam ⇌ hydrogen + carbon dioxide

$$CO(g) + H_2O(g) \rightleftharpoons H_2(g) + CO_2(g)$$

## Making ammonia

In the Haber process itself, nitrogen and hydrogen in the correct proportions (1:3) are pressurised to approximately 200 atmospheres and passed over a catalyst of freshly produced, finely divided iron at a temperature of between 350 °C and 500 °C. The reaction in the Haber process is:

nitrogen + hydrogen ⇌ ammonia

$$N_2(g) + 3H_2(g) \rightleftharpoons 2NH_3(g) \quad \Delta H = -92\,kJ\,mol^{-1}$$

The reaction is exothermic.

Under these conditions the gas mixture leaving the reaction vessel contains about 15% ammonia, which is removed by cooling and condensing it as a liquid. The unreacted nitrogen and hydrogen are recirculated into the reaction vessel to react together once more to produce further quantities of ammonia.

The 15% of ammonia produced does not seem a great deal. The reason for this is the reversible nature of the reaction. Once the ammonia is made from nitrogen and hydrogen, it decomposes to produce nitrogen and hydrogen. There comes a point when the rate at which the nitrogen and hydrogen react to produce ammonia is equal to the rate at which the ammonia decomposes. This situation is called a **chemical equilibrium**. Because the processes continue to happen, the equilibrium is said to be **dynamic**. The conditions used ensure that the ammonia is made economically. The diagram below shows how the percentage of ammonia produced varies with the use of different temperatures and pressures (Figure 15.3).

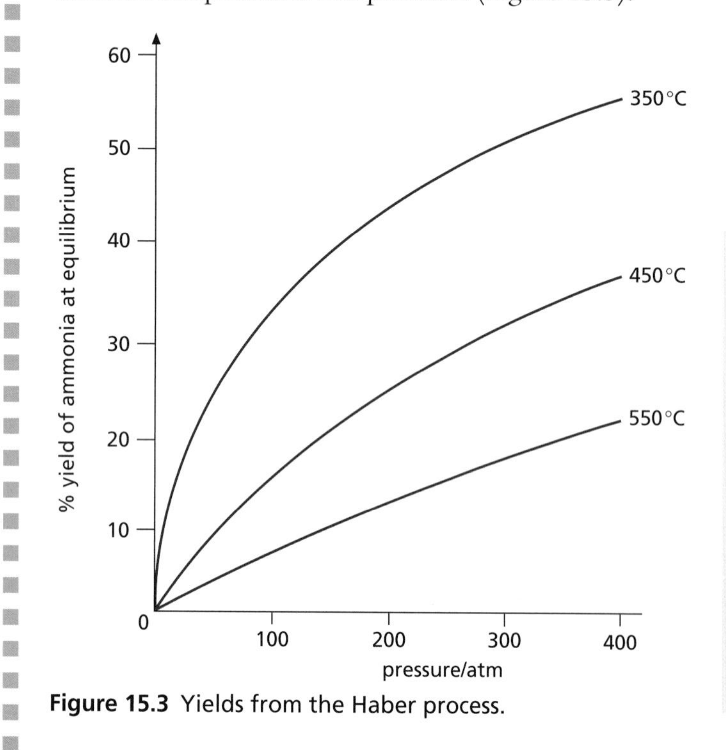


**Figure 15.3** Yields from the Haber process.

You will notice that the higher the pressure and the lower the temperature used, the more ammonia is produced. Relationships such as this were initially observed by Henri Le Chatelier, a French scientist, in 1888. He noticed that if the pressure was increased in reactions involving gases, the reaction which produced the fewest molecules of gas was favoured. If you look at the reaction for the Haber process you will see that, going from left to right, the number of molecules of gas goes from four to two. This is why the Haber process is carried out at high pressures. He also noticed that reactions which were exothermic produced more products if the temperature was low. Indeed, if the Haber process is carried out at room temperature you get a higher percentage of ammonia. However, in practice the rate of the reaction is lowered too much and the ammonia is not produced quickly enough for the process to be economical. An **optimum temperature** is used to produce enough ammonia at an acceptable rate. It should be noted, however, that the increased pressure used is very expensive in capital terms and so alternative, less expensive routes involving biotechnology (Chapter 14, p. 204) are being sought at the present time.

Worldwide, 140 million tonnes of ammonia are produced by the Haber process each year. 1 300 000 tonnes are produced in the UK each year.

### Questions

1 Use the information given above and any other sources you may have to produce a flow diagram of the Haber process. Indicate the flow of the gases, and write equation(s) to show what happens at each stage. You may have to look up the boiling points of the gases involved for the stage in which the ammonia is separated from the reaction vessel mixture.

2 What problems do the builders of a chemical plant to produce ammonia have to consider when they start to build such a plant?

3 What problems are associated with building a plant which uses such high pressures as those required in the Haber process?

# *Ammonia gas*

## Making ammonia in a laboratory

Small quantities of ammonia gas can be produced by heating any ammonium salt, such as ammonium chloride, with an alkali, such as calcium hydroxide.

calcium hydroxide + ammonium chloride → calcium chloride + water + ammonia

$$Ca(OH)_2(s) + 2NH_4Cl(s) \rightarrow CaCl_2(s) + 2H_2O(g) + 2NH_3(g)$$

Water vapour is removed from the ammonia gas by passing the gas formed through a drying tower containing calcium oxide (Figure 15.4).

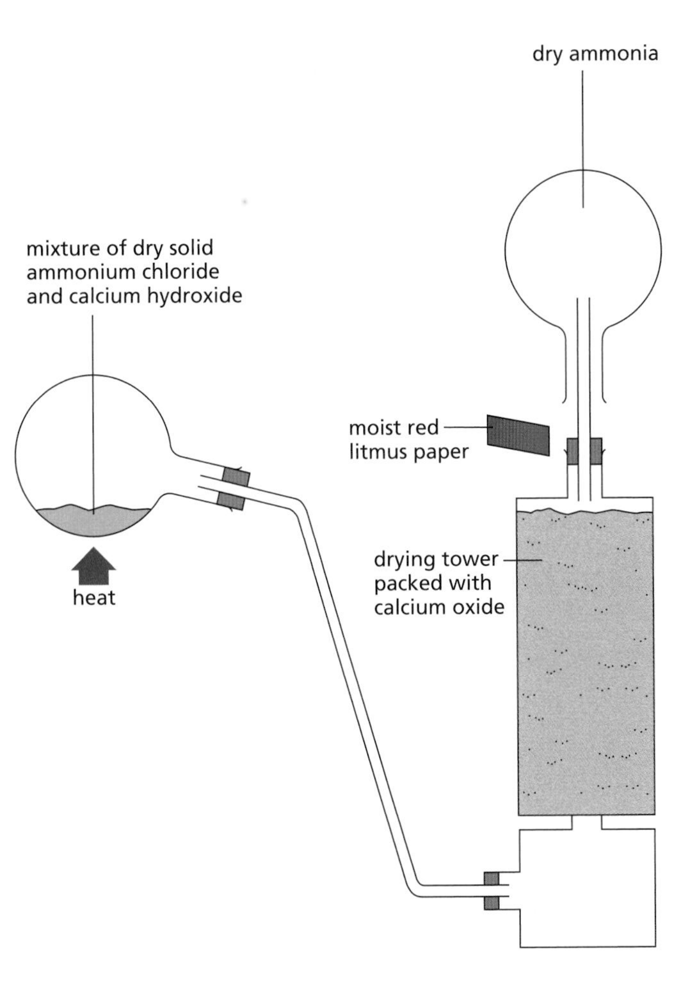


**Figure 15.4** Laboratory production of ammonia gas.

This reaction forms the basis of a chemical test to show that a compound contains the ammonium ion ($NH_4^+$). If any compound containing the ammonium ion is heated with sodium hydroxide, ammonia gas is given off which turns damp red litmus paper blue.

## Physical properties of ammonia

Ammonia (Figure 15.5):

- is a colourless gas
- is less dense than air
- has a sharp or pungent smell
- is very soluble in water with about 680 $cm^3$ of ammonia in each 1 $cm^3$ of water (at 20 °C).

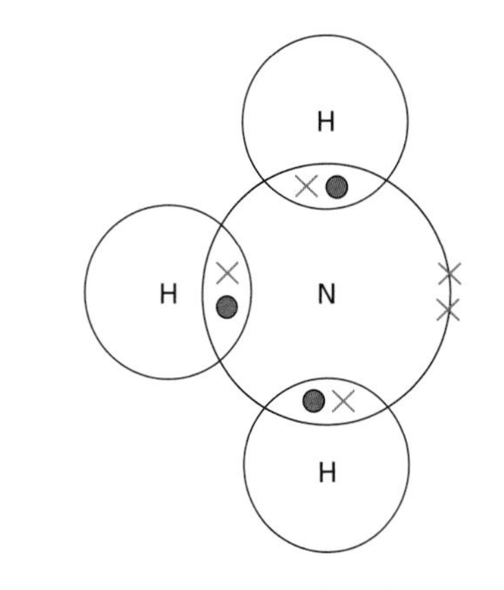


**Figure 15.5** The ammonia molecule.

## Chemical properties of ammonia

The reason ammonia is so soluble in water is that some of it reacts with the water. The high solubility can be shown by the 'fountain flask experiment' (Figure 15.6, overleaf). As the first drop of water reaches the top of the tube all the ammonia gas in the flask dissolves, creating a much reduced pressure. Water then rushes up the tube to fill the space once occupied by the dissolved gas. This creates the fountain.

If the water initially contained some universal indicator, then you would also see a change from green to blue when it comes into contact with the dissolved ammonia. This shows that ammonia solution is a weak alkali, although dry ammonia gas is not. This is because a little of the ammonia gas has reacted with the water, producing ammonium ions and hydroxide ions. The hydroxide ions produced make the solution of ammonia alkaline.

ammonia + water ⇌ ammonium ions + hydroxide ions

$$NH_3(g) + H_2O(l) \rightleftharpoons NH_4^+(aq) + OH^-(aq)$$

The solution is only weakly alkaline because of the reversible nature of this reaction, which results in a relatively low concentration of hydroxide ions. Ammonia gas dissolved in water is usually known as aqueous ammonia.

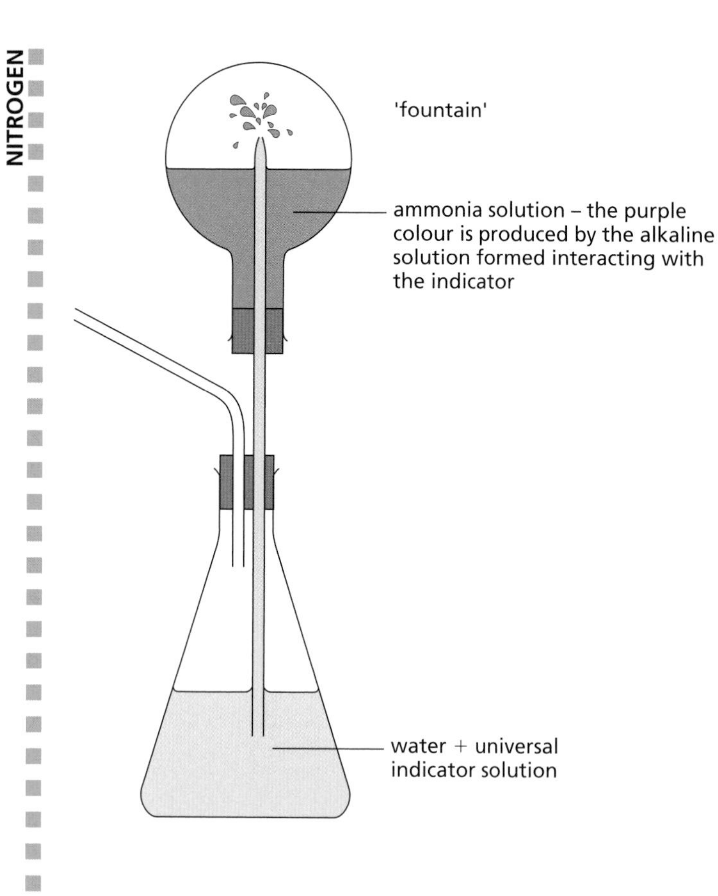

**Figure 15.6** The fountain flask experiment.

Aqueous ammonia can be used to identify salts of $Cu^{2+}$, $Fe^{2+}$, $Fe^{3+}$, $Al^{3+}$, $Zn^{2+}$ and $Mg^{2+}$ ions. The colour of the precipitate or solution formed identifies the metal present (Table 15.1).

**Table 15.1** Identifying metals ions using aqueous ammonia.

| Metal ion | With a few drops of ammonia solution | With excess ammonia solution |
|---|---|---|
| $Cu^{2+}$(aq) | Gelatinous blue precipitate | Precipitate dissolves to give a deep blue solution |
| $Fe^{2+}$(aq) | Dirty green precipitate | Dirty green precipitate remains |
| $Fe^{3+}$(aq) | Rust brown precipitate | Rust brown precipitate remains |
| $Al^{3+}$(aq) | White precipitate | White precipitate remains |
| $Zn^{2+}$(aq) | White precipitate | White precipitate dissolves to give a colourless solution |
| $Mg^{2+}$(aq) | White precipitate | White precipitate remains |

## Questions

1 Calcium oxide is used to dry ammonia gas in its laboratory preparation. Write a word and balanced chemical equation to show how calcium oxide can react with the water vapour to remove it from damp ammonia gas.

2 Explain why ammonia gas only acts as a weak alkali in the presence of water.

# Nitric acid

## Manufacture of nitric acid

One of the largest uses of ammonia is in the production of nitric acid. The process, which was invented by Wilhelm Ostwald, has three stages.

1 A mixture of air and ammonia is heated to about 230 °C and is passed through a metal gauze made of platinum (90%) and rhodium (10%) (Figure 15.7). The reaction produces a lot of heat energy. This energy is used to keep the reaction vessel temperature at around 800 °C. The reaction produces nitrogen monoxide (NO) and water.

ammonia + oxygen → nitrogen monoxide + water

$$4NH_3(g) + 5O_2(g) \rightarrow 4NO(g) + 6H_2O(g)$$
$$\Delta H = -904\,kJ\,mol^{-1}$$

2 The colourless nitrogen monoxide gas produced from the first stage is then reacted with oxygen from the air to form brown nitrogen dioxide gas ($NO_2$).

nitrogen monoxide + oxygen → nitrogen dioxide

$$2NO(g) + O_2(g) \rightarrow 2NO_2(g)$$
$$\Delta H = -114\,kJ\,mol^{-1}$$

3 The nitrogen dioxide is then dissolved in water to produce nitric acid.

nitrogen dioxide + water → nitric acid + nitrogen monoxide

$$3NO_2(g) + H_2O(l) \rightarrow 2HNO_3(aq) + NO(g)$$

**Figure 15.7** Platinum–rhodium gauze is used as a catalyst in the production of nitric acid.

A small amount of nitrogen dioxide cannot be turned into nitric acid and so it is expelled into the atmosphere through a very tall chimney. This can cause pollution because it is an acidic gas.

The nitric acid produced is used in the manufacture of the following:

- artificial fertilisers, such as ammonium nitrate
- explosives, such as 2,4,6-trinitrotoluene (TNT)
- dyes
- artificial fibres, such as nylon.

It is also used in the treatment of metals.

Sixty million tonnes of nitric acid are produced each year. Of this figure 20 million tonnes are produced annually in Europe.

## Properties of nitric acid

### Dilute nitric acid

Nitric acid is a strong acid. As a dilute acid it shows most of the properties of an acid.

- It forms an acidic solution in water and will turn pH paper red.
- It will react with bases such as sodium hydroxide and zinc oxide to give salts, called **nitrates**, and water.

sodium hydroxide + nitric acid → sodium nitrate + water

$$NaOH(aq) + HNO_3(aq) \rightarrow NaNO_3(aq) + H_2O(l)$$

zinc oxide + nitric acid → zinc nitrate + water

$$ZnO(s) + 2HNO_3(aq) \rightarrow Zn(NO_3)_2(aq) + H_2O(l)$$

- It reacts with carbonates to give a salt, water and carbon dioxide gas.

sodium carbonate + nitric acid → sodium nitrate + water + carbon dioxide

$$Na_2CO_3(s) + 2HNO_3(aq) \rightarrow 2NaNO_3(aq) + H_2O(l) + CO_2(g)$$

- Most acids will react with metals to give a salt and hydrogen gas. However, since nitric acid is an oxidising agent it behaves differently and hydrogen is rarely formed. The exceptions to this are the reactions of cold dilute nitric acid with magnesium and calcium.

magnesium + nitric acid → magnesium nitrate + hydrogen

$$Mg(s) + 2HNO_3(aq) \rightarrow Mg(NO_3)_2(aq) + H_2(g)$$

Dilute nitric acid reacts with copper, for example, to produce nitrogen monoxide instead of hydrogen.

copper + nitric acid → copper(II) nitrate + water + nitrogen monoxide

$$3Cu(s) + 8HNO_3(aq) \rightarrow 3Cu(NO_3)_2(aq) + 4H_2O(l) + 2NO(g)$$

The nitrogen monoxide produced quickly reacts with oxygen from the air to give brown fumes of nitrogen dioxide gas.

### Concentrated nitric acid

Concentrated nitric acid is a powerful **oxidising agent**. It will oxidise both metals and non-metals.

- It will oxidise carbon to carbon dioxide:

carbon + conc nitric acid → carbon dioxide + nitrogen(IV) oxide + water

$$C(s) + 4HNO_3(l) \rightarrow CO_2(g) + 4NO_2(g) + 2H_2O(l)$$

- It will oxidise copper to copper(II) nitrate (Figure 15.8).

copper + conc nitric acid → copper(II) nitrate + nitrogen dioxide + water

$$Cu(s) + 4HNO_3(l) \rightarrow Cu(NO_3)_2(aq) + 2NO_2(g) + 2H_2O(l)$$

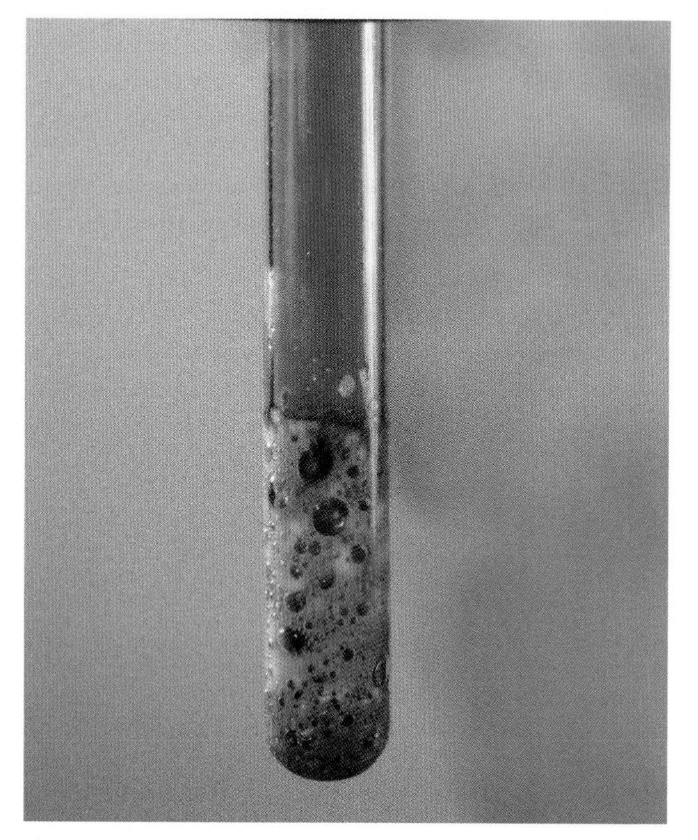

**Figure 15.8** Copper reacting with concentrated nitric acid.

## Questions

1 What do you think happens to the nitrogen monoxide produced in the third step of the industrial preparation of nitric acid?

2 What is the maximum amount of nitric acid that could be produced from 500 tonnes of ammonia gas?

3 In reality the amount of nitric acid produced would be less than the amount calculated above. Explain why this would be so.

4 Write a word and balanced chemical equation to show the neutralisation reaction between nitric acid and ammonium hydroxide.

## *Artificial fertilisers*

The two processes so far described, the production of ammonia and nitric acid, are extremely important in the production of many artificial fertilisers. As was explained earlier, the use of artificial fertilisers is essential if farmers are to produce sufficient crops to feed the ever-increasing population. Crops remove nutrients from the soil as they grow; these include nitrogen, phosphorus and potassium. Artificial fertilisers are added to the soil to replace these nutrients and others, such as calcium, magnesium, sodium, sulphur, copper and iron. Examples of nitrogenous fertilisers (those which contain nitrogen) are shown in Table 15.2.

**Table 15.2** Some nitrogenous fertilisers.

| Fertiliser | Formula |
|---|---|
| Ammonium nitrate | $NH_4NO_3$ |
| Ammonium phosphate | $(NH_4)_3PO_4$ |
| Ammonium sulphate | $(NH_4)_2SO_4$ |
| Urea | $CO(NH_2)_2$ |

Artificial fertilisers can also make fertile land which was once unable to support crop growth. The fertilisers which add the three main nutrients (N, P and K) are called NPK fertilisers. They contain ammonium nitrate ($NH_4NO_3$), ammonium phosphate ($(NH_4)_3PO_4$) and potassium chloride (KCl) in varying proportions.

### Manufacture of ammonium nitrate

Ammonium nitrate (Nitram) is probably the most widely used nitrogenous fertiliser. It is manufactured by reacting ammonia gas and nitric acid.

ammonia + nitric acid → ammonium nitrate

$$NH_3(g) + HNO_3(aq) \rightarrow NH_4NO_3(aq)$$

### Problems with fertilisers

If artificial fertilisers are not used correctly, problems can arise. If too much fertiliser is applied to the land, rain washes the fertiliser off the land and into rivers and streams. This leaching encourages the growth of algae and marine plants. As the algae die and decay oxygen is removed from the water, leaving insufficient amounts for fish and other organisms to survive.

## Questions

1 Calculate the percentage of nitrogen in each of the four fertilisers in Table 15.2.
($A_r$: N = 14; H = 1; P = 31; O = 16; S = 32)

2 Write down a method that you could carry out in a school laboratory to prepare a sample of ammonium sulphate fertiliser.

## Checklist

**After studying Chapter 15 you should know and understand the following terms.**

- **Artificial fertiliser** A substance added to soil to increase the amount of elements such as nitrogen, potassium and phosphorus. This enables crops grown in the soil to grow more healthily and to produce higher yields.

- **Chemical equilibrium** A dynamic state. The concentrations of the reactants and products remain constant because the rate at which the forward reaction occurs is the same as that of the back reaction.

- **Nitrogen cycle** The system by which nitrogen and its compounds, both in the air and in the soil, are interchanged.

- **Nitrogen fixation** The direct use of atmospheric nitrogen in the formation of important compounds of nitrogen. Bacteria present in root nodules of certain plants are able to take nitrogen directly from the atmosphere to form essential protein molecules.

- **Optimum temperature** A compromise temperature used in industry to ensure that the yield of product and the rate at which it is produced make the process as economical as possible.

- **Reversible reaction** A chemical reaction which can go both ways. This means that once some of the products have been formed they will undergo a chemical change once more to re-form the reactants. The reaction from left to right, as the equation for the reaction is written, is known as the forward reaction and the reaction from right to left is known as the back reaction.

# Nitrogen
# *Additional questions*

**1** Study the following reaction scheme.

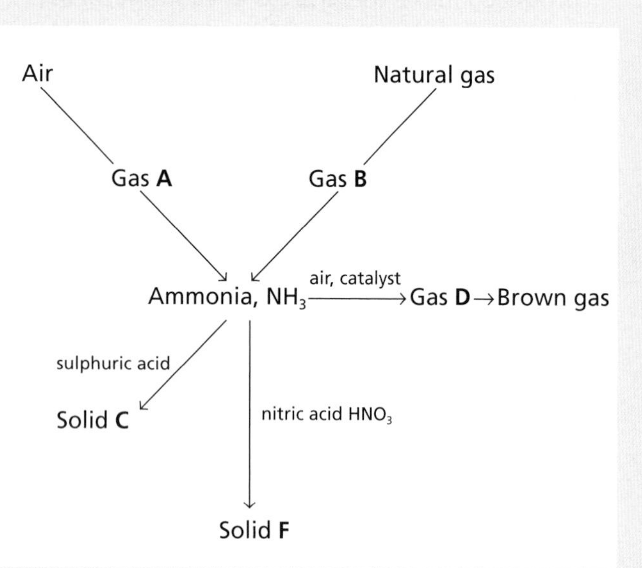


**a** Identify the substances **A** to **F** by giving their names and chemical formulae.
**b** How is gas **A** obtained from the air?
**c** Write a word and balanced chemical equation for the formation of ammonia gas from gases **A** and **B**.
**d** Write an equation for the formation of solid **C** from ammonia.
**e** Give a use for:
(i) solid **C**
(ii) nitric acid
(iii) solid **F**.

**2 a** Write word and balanced chemical equations for the reaction of dilute nitric acid with:
(i) sodium hydroxide
(ii) copper(II) oxide
(iii) magnesium carbonate.
**b** Describe how you would obtain dry crystals of potassium nitrate from dilute nitric acid and potassium hydroxide solution.

**3** An international chemical firm has decided that it needs to build a new plant to produce the fertiliser ammonium sulphate, starting from the raw materials.
**a** What raw materials would be needed for this process?
**b** What factors would a chemical engineer look for before she decided where to build the new plant?
**c** At this type of plant there is always the danger of the leakage of some ammonia gas. What effect would a leak of ammonia gas have on:
(i) people living in the surrounding area?
(ii) the pH of the soil in the surrounding area?
(iii) the growth of crops in the surrounding area?

**4**

| Fertiliser | Formula | % nitrogen |
|---|---|---|
| Ammonia solution | $NH_3$ | 82.4 |
| Calcium nitrate | $Ca(NO_3)_2$ | 17.1 |
| Nitram | $NH_4NO_3$ | 35.0 |
| Sodium nitrate | $NaNO_3$ | |
| Potassium nitrate | $KNO_3$ | |

**a** Copy and complete the above table by calculating the percentage of nitrogen in the fertilisers sodium nitrate and potassium nitrate.
($A_r$: N = 14; H = 1; O = 16; K = 39; Ca = 40; Na = 23)
**b** Including the data you have just calculated, which of the fertilisers contains:
(i) the largest percentage of nitrogen?
(ii) the smallest percentage of nitrogen?
**c** Give the chemical name for the fertiliser that goes by the name Nitram.
**d** Ammonia can be used directly as a fertiliser but not very commonly. Think of two reasons why ammonia is not often used directly as a fertiliser.
**e** Nitram fertiliser is manufactured by the reaction of nitric acid with ammonia solution according to the equation:

$$NH_3(aq) + HNO_3(aq) \rightarrow NH_4NO_3(aq)$$

A bag of Nitram may contain 50 kg of ammonium nitrate. What mass of nitric acid would be required to make it?

**5** Ammonia gas is made industrially by the Haber process, which involves the reaction between the gases nitrogen and hydrogen. The amount of ammonia gas produced from this reaction is affected by both the temperature and the pressure at which the process is run. The graph below shows how the amount of ammonia produced from the reaction changes with both temperature and pressure. The percentage yield of ammonia indicates the percentage of the nitrogen and hydrogen gases that are actually changed into ammonia gas.

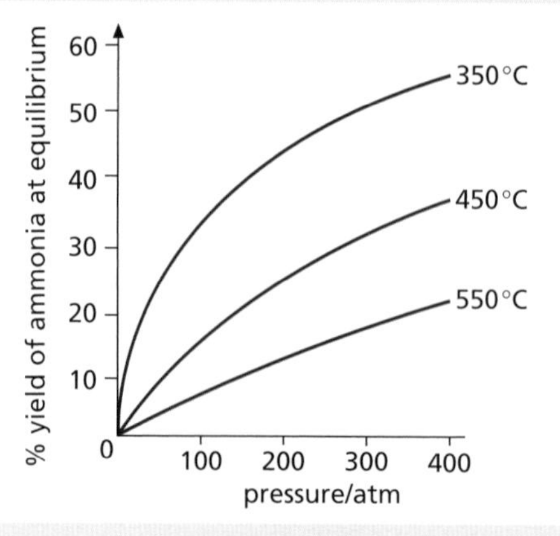

**a** Write a word and balanced chemical equation for the reversible reaction between nitrogen and hydrogen to produce ammonia using the Haber process.
**b** What is meant by the term a 'reversible reaction'?
**c** Use the graphs to say whether more ammonia is produced at:
(i) higher or lower temperatures
(ii) higher or lower pressures.
**d** What is the percentage yield of ammonia if the conditions used to run the process are:
(i) a temperature of 350 °C and a pressure of 100 atmospheres?
(ii) a temperature of 550 °C and a pressure of 350 atmospheres?
**e** The conditions in industry for the production of ammonia are commonly of the order of 200 atmospheres and 450 °C. What is the percentage yield of ammonia using these conditions?
**f** Why does industry use the conditions stated in part **e** if it is possible to obtain a higher yield of ammonia using different conditions?

**6** The following results were obtained from a neutralisation reaction between potassium hydroxide and nitric acid. The dilute nitric acid had been found in a cupboard in the school laboratory. The student carrying out the experiment was interested in finding out the concentration of the nitric acid.
The student used 25 $cm^3$ of 0.15 $mol\,dm^{-3}$ potassium hydroxide solution in a conical flask to which was added the indicator phenolphthalein. The dilute nitric acid was added from a burette until the indicator just changed colour. The student repeated the experiment four times. Her results are shown below.

| | **Rough** | **1** | **2** | **3** |
|---|---|---|---|---|
| **Final burette reading/ $cm^3$** | 10.25 | 13.20 | 13.90 | 12.65 |
| **Initial burette reading/ $cm^3$** | 0.00 | 3.10 | 3.80 | 2.50 |
| **Volume used/$cm^3$** | | | | |

**a** Copy and complete the table above by calculating the volume of dilute nitric acid used in each titration.
**b** From the three most accurate results, calculate the average volume of dilute nitric acid required to neutralise the 25 $cm^3$ of potassium hydroxide solution.
**c** (i) Write a word and balanced chemical equation for the reaction which has taken place.
(ii) Write down the number of moles of nitric acid and potassium hydroxide shown reacting in the equation.
**d** (i) Calculate the number of moles of potassium hydroxide present in 25 $cm^3$ of solution.
(ii) Calculate the number of moles of nitric acid neutralised.
**e** Calculate the molarity of the dilute nitric acid.

**7** Explain the following.
**a** Dry litmus paper does not change colour when added to dry ammonia gas.
**b** Ammonia gas cannot be collected over water but can be collected by downward displacement of air.
**c** Soil often needs artificial fertilisers added to it if it is to continuously support the growth of healthy crops.
**d** When colourless nitrogen(II) oxide gas is formed, it almost instantaneously turns brown.

**8** Nitrogen oxides are emitted from car exhaust pipes. To animals and plants these gases are very harmful.
**a** Normally, nitrogen gas and oxygen gas do not react. Why, in a car engine, do they react to form these dangerous gases?
**b** Where does the nitrogen which forms these nitrogen oxides come from?
**c** How is the production of these nitrogen oxides being minimised in newer cars?

# 16 Sulphur



## *Sulphur – the element*

Sulphur is a non-metallic element which has a very important role in the chemical industry. It is a yellow solid which is found in large quantities but in various forms throughout the world (Figure 16.1).

**Figure 16.1** Sulphur – rhombic (top) and monoclinic.

It is found in metal ores such as copper pyrites ($CuFeS_2$) and zinc blende (ZnS) and in volcanic regions of the world. Natural gas and oil contain sulphur and its compounds, but the majority of this sulphur is removed as it would cause environmental problems. Sulphur obtained from these sources is known as 'recovered sulphur' and it is an important source of the element. It is also found as elemental sulphur in sulphur beds in Poland, Russia and the US (Louisiana). These sulphur beds are typically 200 m below the ground. Sulphur from these beds is extracted using the **Frasch process**, named after its inventor Hermann Frasch.

### The Frasch process

Superheated water at 170 °C and hot compressed air are forced underground through pipes, forcing water and molten sulphur to the surface. Sulphur is insoluble in water and so the two substances emerging from the pipes are easily separated. The sulphur is kept molten and sold in this form. The sulphur obtained from this process is about 99.5% pure and can be used directly (Figure 16.2).

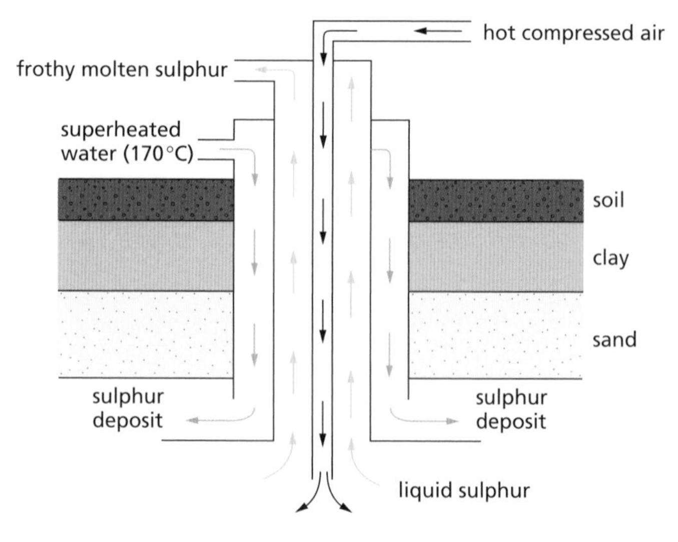


**Figure 16.2** The Frasch process.

## Uses of sulphur

The vast majority of sulphur is used to produce perhaps the most important industrial chemical, sulphuric acid. Sulphur is also used to **vulcanise** rubber, a process which makes the rubber harder and increases its elasticity. Relatively small amounts are used in the manufacture of matches, fireworks and fungicides, as a sterilising agent and in medicines.

## Allotropes of sulphur

Sulphur is one of the few non-metal elements which exist as allotropes (Chapter 4, p. 62). The main allotropes are called rhombic sulphur and monoclinic sulphur. Both of these solid forms of sulphur are made up of $S_8$ molecules (Figure 16.3).

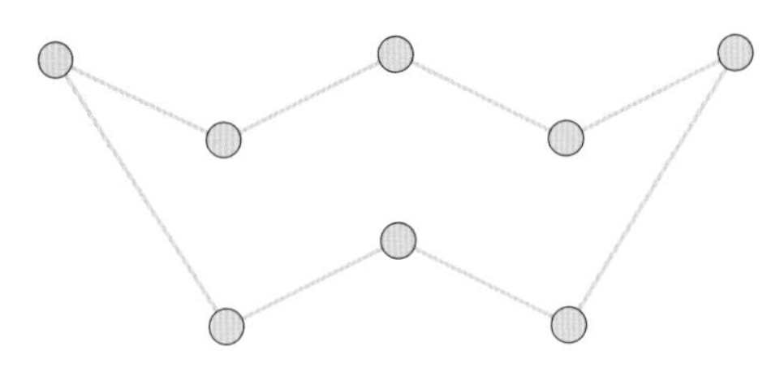

**Figure 16.3** $S_8$ molecule.

The fact that there are two different allotropes of sulphur is due to the way in which these $S_8$ molecules pack together. In rhombic sulphur the molecules are packed more closely than in the monoclinic form (Figure 16.4).

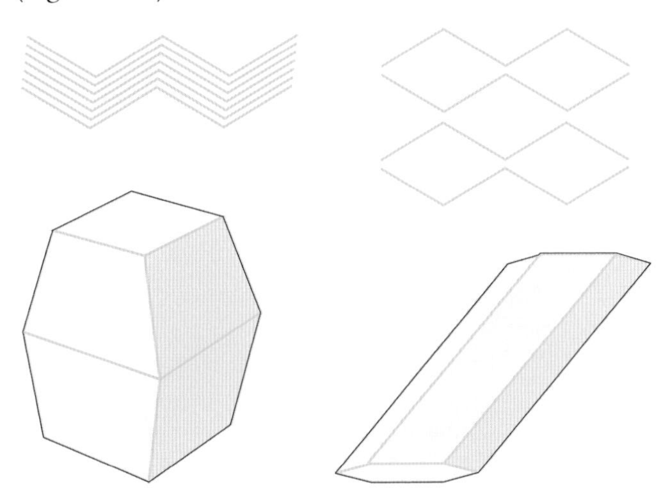

**a** A rhombic crystal. **b** A monoclinic crystal of sulphur.

**Figure 16.4** The packing of $S_8$ molecules

Although sulphur is insoluble in water, it will dissolve in an organic solvent such as methylbenzene. If a solution of sulphur in methylbenzene is heated and allowed to cool then crystals of monoclinic sulphur are produced. When the temperature of the solution falls below 96 °C, rhombic sulphur crystals are produced. Rhombic sulphur is stable below 96 °C and monoclinic sulphur is stable above 96 °C. This temperature is called the **transition temperature**.

When solid sulphur is heated, it melts at 112 °C and forms a runny (mobile) liquid. At this point the $S_8$ molecules are moving freely around each other, as the weak attractive forces between them have been overcome (Figure 16.5).

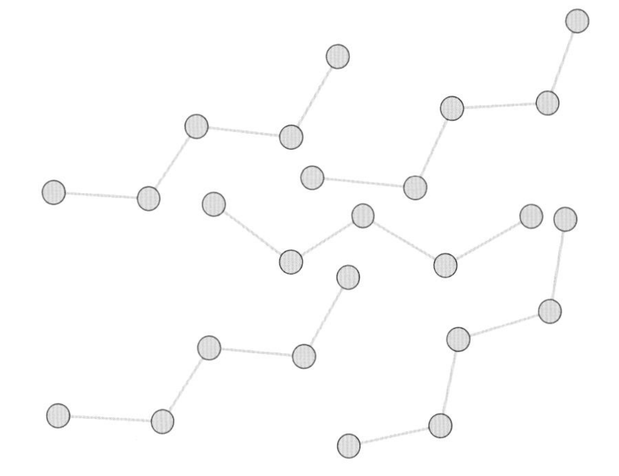

**Figure 16.5** At the melting point $S_8$ rings move freely around one another.

However, if the sulphur is heated further the liquid becomes thicker (viscous). This is because the $S_8$ rings have been broken by the energy given to the sulphur and they bond together, forming long chains of sulphur atoms which become tangled, making the liquid viscous (Figure 16.6). Continued heating, to 444 °C, makes the liquid more mobile once again as the long chains are broken down into smaller ones which move around one another freely.

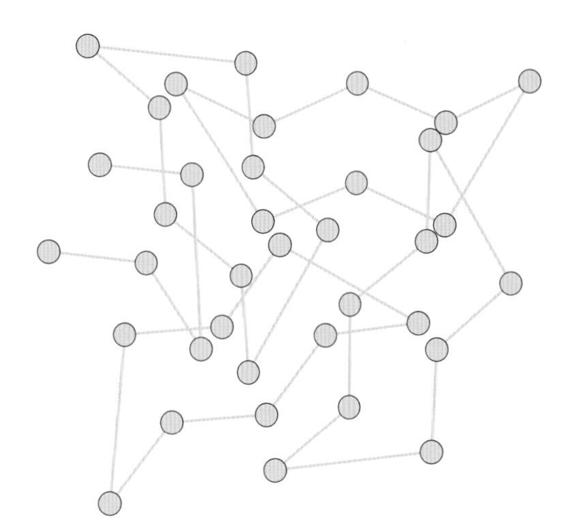

**Figure 16.6** A viscous liquid is produced as long chains of sulphur atoms are formed and get tangled together.

If this liquid is poured into a beaker of cold water, a substance called **plastic sulphur** is formed. This is an elastic, rubber-like substance. In plastic sulphur, the sulphur atoms remain bonded together in the form of chains, very similar to chains of carbon atoms in plastics such as polythene. After a few hours, however, the plastic sulphur loses its elasticity and once again becomes solid as the $S_8$ molecular rings re-form.

## Properties of sulphur

Sulphur:

- is a yellow, brittle solid at room temperature
- does not conduct electricity
- is insoluble in water.

Sulphur will react with both metals and non-metals.

- It reacts with magnesium metal to form magnesium sulphide.

$$\text{magnesium} + \text{sulphur} \rightarrow \text{magnesium sulphide}$$
$$Mg(s) + S(s) \rightarrow MgS(s)$$

- It reacts with oxygen to produce sulphur dioxide gas.

$$\text{sulphur} + \text{oxygen} \rightarrow \text{sulphur dioxide}$$
$$S(s) + O_2(g) \rightarrow SO_2(g)$$

## Sulphur dioxide

Sulphur dioxide is a colourless gas produced when sulphur or substances containing sulphur, for example crude oil or natural gas, are burned in oxygen gas. It has a choking smell and is extremely posionous. The gas dissolves in water to produce an acidic solution of sulphurous acid.

$$\text{sulphur dioxide} + \text{water} \rightleftharpoons \text{sulphurous acid}$$
$$SO_2(g) + H_2O(l) \rightleftharpoons H_2SO_3(aq)$$

It is one of the major pollutant gases and is the gas principally responsible for **acid rain**. However, it does have some uses: as a bleaching agent, in fumigants and in the preservation of food by killing bacteria.

### Questions

1 What is meant by the term 'allotrope'?

2 'Sulphur is a non-metallic element.' Discuss this statement, giving physical and chemical reasons to support your answer.

# *Acid rain*

Rainwater is naturally acidic since it dissolves carbon dioxide gas from the atmosphere as it falls. Natural rainwater has a pH of about 5.7. In recent years, especially in central Europe, the pH of rainwater has fallen to between pH 3 and pH 4.8. This increase in acidity has led to extensive damage to forests (Figure 16.7), lakes and marine life.

**Figure 16.7** This forest has been devastated by acid rain.

The amount of sulphur dioxide in the atmosphere has increased dramatically over recent years. There has always been some sulphur dioxide in the atmosphere, from natural processes such as volcanoes and rotting vegetation. Over Europe, however, around 80% of the sulphur dioxide in the atmosphere is formed from the combustion of fuels containing sulphur (Figure 16.8). After dissolving in rain to produce sulphurous acid, it further reacts with oxygen to form sulphuric acid.

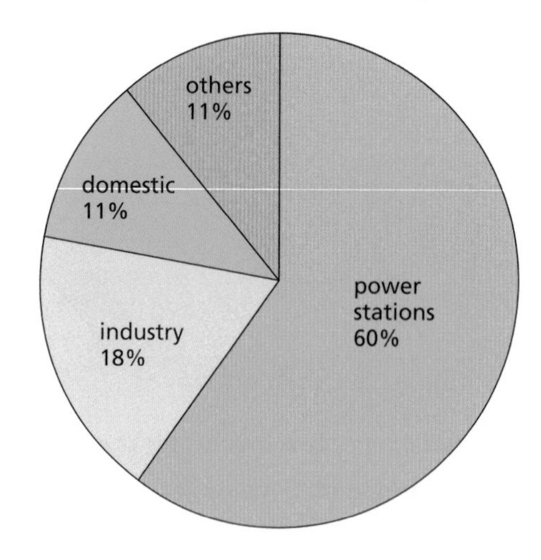


**Figure 16.8** Sources of sulphur dioxide.

### Questions

1 How could the amount of sulphur dioxide being produced by the above sources be reduced?

2 Devise an experiment which you could carry out in the school laboratory to determine the amount of sulphur in two different types of coal.

# *Industrial manufacture of sulphuric acid – the Contact process*

The major use of sulphur is in the production of sulphuric acid. This is probably the most important industrial chemical, and the quantity of it produced by a country has been linked with the economic stability of the country. Many millions of tonnes of sulphuric acid are produced in the UK each year. It is used mainly as the raw material for the production of many substances (Figure 16.9).

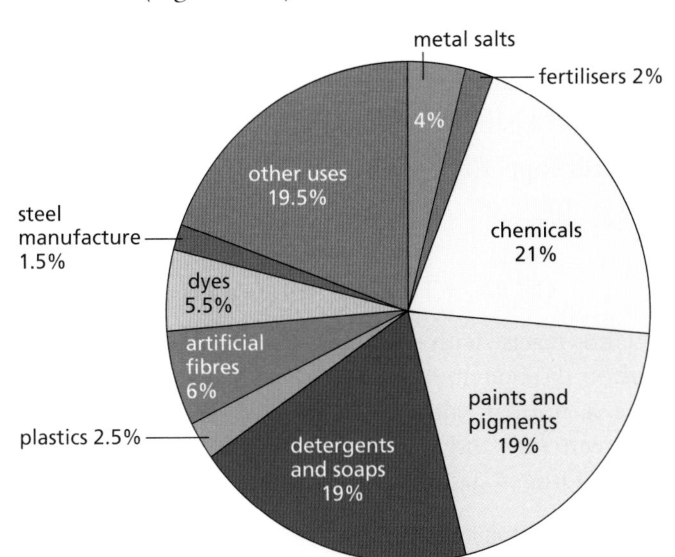

**Figure 16.9** Products made from sulphuric acid include fertilisers, paints and pigments.

The process by which sulphuric acid is produced is known as the **Contact process** (Figure 16.10).

**Figure 16.10** A Contact process plant used for making sulphuric acid

The process has the following stages.

- Sulphur dioxide is first produced, primarily by the reaction of sulphur with air.

$$\text{sulphur} + \text{oxygen} \rightarrow \text{sulphur dioxide}$$
$$S(s) + O_2(g) \rightarrow SO_2(g)$$

- Any dust and impurities are removed from the sulphur dioxide produced, as well as any unreacted oxygen. These 'clean' gases are heated to a temperature of approximately 450 °C and fed into a reaction vessel, where they are passed over a catalyst of vanadium(V) oxide ($V_2O_5$). This catalyses the reaction between sulphur dioxide and oxygen to produce sulphur trioxide (sulphur(VI) oxide, $SO_3$).

$$\text{sulphur dioxide} + \text{oxygen} \rightleftharpoons \text{sulphur trioxide}$$
$$2SO_2(g) + O_2(g) \rightleftharpoons 2SO_3(g)$$
$$\Delta H = -197\,\text{kJ mol}^{-1}$$

This reaction is reversible and so the ideas of Le Chatelier (Chapter 15, p. 220) can be used to increase the proportion of sulphur trioxide in the equilibrium mixture. The forward reaction is exothermic and so would be favoured by low temperatures. The temperature of 450 °C used is an optimum temperature which produces sufficient sulphur trioxide at an economical rate. Since the reaction from left to right is also accompanied by a decrease in the number of molecules of gas, it will be favoured by a high pressure. In reality, the process is run at atmospheric pressure. Under these conditions, about 96% of the sulphur dioxide and oxygen are converted into sulphur trioxide. The heat produced by this reaction is used to heat the incoming gases, thereby saving money.

- If this sulphur trioxide is added directly to water, sulphuric acid is produced. This reaction, however, is very violent and a thick mist is produced.

sulphur trioxide + water → sulphuric acid

$$SO_3(g) + H_2O(l) \rightarrow H_2SO_4(l)$$

This acid mist is very difficult to deal with and so a different route to sulphuric acid is employed. Instead, the sulphur trioxide is dissolved in concentrated sulphuric acid (98%) to give a substance called **oleum**.

sulphuric acid + sulphur trioxide → oleum

$$H_2SO_4(aq) + SO_3(g) \rightarrow H_2S_2O_7(l)$$

The oleum formed is then added to the correct amount of water to produce sulphuric acid of the required concentration.

oleum + water → sulphuric acid

$$H_2S_2O_7(l) + H_2O(l) \rightarrow 2H_2SO_4(l)$$

## Question

1 Produce a flow diagram to show the different processes which occur during the production of sulphuric acid by the Contact process. Write balanced chemical equations showing the processes which occur at the different stages, giving the essential raw materials and conditions used.

# *Properties of sulphuric acid*

## Dilute sulphuric acid

Dilute sulphuric acid is a typical strong **dibasic** acid. A dibasic acid is one with two replaceable hydrogen atoms which may produce two series of salts – normal and acid salts (Chapter 7, p. 99).

It will react with bases such as sodium hydroxide and copper(II) oxide to produce normal salts, called **sulphates**, and water.

- With sodium hydroxide:

sodium hydroxide + sulphuric acid → sodium sulphate + water

$$2NaOH(aq) + H_2SO_4(aq) \rightarrow Na_2SO_4(aq) + H_2O(l)$$

- With copper(II) oxide:

copper(II) oxide + sulphuric acid → copper(II) sulphate + water

$$CuO(s) + H_2SO_4(aq) \rightarrow CuSO_4(aq) + H_2O(l)$$

It also reacts with carbonates to give normal salts, carbon dioxide and water, and with reactive metals to give a normal salt and hydrogen gas. The reaction between zinc and sulphuric acid is often used to prepare hydrogen gas in the laboratory (Figure 16.11).

zinc + sulphuric acid → zinc sulphate + hydrogen

$$Zn(s) + H_2SO_4(aq) \rightarrow ZnSO_4(aq) + H_2(g)$$

The preparation of the acid salt with sodium hydroxide requires twice the volume of acid as that used in the preparation of the normal salt. Therefore, if 25 $cm^3$ of dilute sulphuric acid were required to form the normal salt from a given volume of alkali of the same concentration then 50 $cm^3$ of the same acid solution would be required to produce the acid salt, sodium hydrogensulphate, from the same volume of alkali.

sodium hydroxide + sulphuric acid → sodium hydrogensulphate + water

$$NaOH(aq) + H_2SO_4(aq) \rightarrow NaHSO_4(aq) + H_2O(l)$$

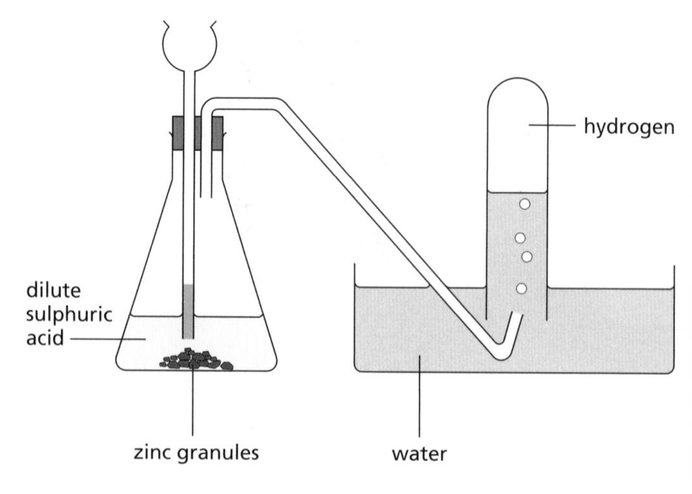


**Figure 16.11** The laboratory preparation of hydrogen gas.

## Sulphates

The salts of sulphuric acid, sulphates, can be identified by a simple test-tube reaction. To test for a sulphate, add a few drops of dilute hydrochloric acid to your unknown followed by a few drops of barium chloride. If a sulphate is present, a white precipitate of barium sulphate is produced.

$$\text{barium ions} + \text{sulphate ions} \rightarrow \text{barium sulphate}$$
$$Ba^{2+}(aq) + SO_4^{2-}(aq) \rightarrow BaSO_4(s)$$

Many sulphates have very important uses, as can be seen from Table 16.1.

**Table 16.1** Uses of some metal sulphates.

| Salt | Formula | Use |
|---|---|---|
| Ammonium sulphate | $(NH_4)_2SO_4$ | Fertiliser |
| Barium sulphate | $BaSO_4$ | 'Barium meal' used in diagnostic medical X-ray studies |
| Calcium sulphate | $CaSO_4.\frac{1}{2}H_2O$ | 'Plaster of Paris' used to set bones |
| Magnesium sulphate | $MgSO_4$ | In medicine it is used as a laxative |

## Concentrated sulphuric acid

Concentrated sulphuric acid is a powerful **dehydrating agent**. This means it will take water from a variety of substances. One such substance is cane sugar, or sucrose (Figure 16.12).

$$\text{sucrose (sugar)} \xrightarrow{\text{conc } H_2SO_4} \text{carbon} + \text{water}$$
$$C_{12}H_{22}O_{11}(s) \longrightarrow 12C(s) + 11H_2O(l)$$

Concentrated sulphuric acid will also take water from hydrated copper(II) sulphate crystals, leaving only anhydrous copper(II) sulphate. If a few drops of concentrated sulphuric acid are added to some blue hydrated copper(II) sulphate crystals, they slowly turn white as the water of crystallisation is removed by the acid. Eventually, only a white powder – anhydrous copper(II) sulphate – remains.

$$\text{hydrated copper(II) sulphate} \xrightarrow{\text{conc } H_2SO_4} \text{anhydrous copper(II) sulphate} + \text{water}$$
$$CuSO_4.5H_2O(s) \longrightarrow CuSO_4(s) + 5H_2O(l)$$

**Figure 16.12** The concentrated sulphuric acid has removed the elements of water from the sugar, leaving black carbon.

Concentrated sulphuric acid should be treated very carefully, because it will also remove water from flesh! It is a very corrosive substance and should always be handled with care.

Diluting concentrated sulphuric acid must be done with great care because of its affinity for water. The concentrated sulphuric acid should always be added to the water, not the other way around.

## Questions

1 If you were given an unlabelled bottle which was thought to be dilute sulphuric acid, how would you show that the solution contained sulphate ions ($SO_4^{2-}$(aq)), how would you show that it was an acid and how would you determine the concentration of the acid?

2 Write balanced chemical equations for the reactions between dilute sulphuric acid and:
 a zinc oxide
 b potassium carbonate
 c aluminium.

## *Manufacture of a soapless detergent*

A more recent use of sulphuric acid is in the production of soapless detergents. These are detergents that can be used more effectively than soap in hard water areas, are fairly cheap to make and are gradually replacing soaps (Chapter 14, p. 209).

The general process involves the reaction of a long, straight-chain alkene, such as dodecene, with benzene.

$$\text{benzene} + \text{dodecene} \rightarrow \text{dodecylbenzene}$$
$$C_6H_6(l) + CH_3(CH_2)_9CH{=}CH_2(l) \rightarrow C_6H_5(CH_2)_{11}CH_3(l)$$

The molecular formula of dodecylbenzene is $C_{18}H_{30}$. This compound is then reacted with concentrated sulphuric acid to give a compound that is known as a sulphonic acid.

$$\text{dodecylbenzene} + \text{sulphuric acid} \rightarrow \text{dodecylbenzene sulphonic acid} + \text{water}$$
$$C_{18}H_{30}(l) + H_2SO_4(l) \rightarrow C_{18}H_{29}SO_3H(aq) + H_2O(l)$$

Finally, this is reacted with the alkali sodium hydroxide, NaOH.

$$\text{dodecylbenzene sulphonic acid} + \text{sodium hydroxide} \rightarrow \text{sodium dodecylbenzene sulphonate} + \text{water}$$
$$C_{18}H_{29}SO_3H(aq) + NaOH(aq) \rightarrow C_{18}H_{29}SO_3^-Na^+(aq) + H_2O(l)$$

(soapless detergent)

Soapless detergents such as this are to be found in most washing powders.

## Checklist

**After studying Chapter 16 you should know and understand the following terms.**

- **Allotropes** Different structural forms of the same element having the same physical state. For example, carbon exists as the allotropes diamond, graphite and buckminsterfullerene, and sulphur as rhombic and monoclinic sulphur.
- **Contact process** The industrial manufacture of sulphuric acid using the raw materials sulphur and air.
- **Dibasic acid** An acid which contains two replaceable hydrogen atoms per molecule of the acid, for example sulphuric acid, $H_2SO_4$.
- **Frasch process** The process of obtaining sulphur from sulphur beds below the Earth's surface. Superheated water is pumped down a shaft to liquefy the sulphur, which is then brought back to the surface.
- **Soapless detergents** Substances which are more effective than soap at producing lathers, especially in hard water areas. They are large organic molecules, produced using sulphuric acid.
- **Sulphate** A salt of sulphuric acid formed by the reaction of the acid with carbonates, bases and some metals. It is possible to test for the presence of a sulphate by the addition of dilute hydrochloric acid and some barium chloride solution. A white precipitate of barium sulphate is formed if a sulphate is present.
- **Transition temperature** The temperature boundary at which one allotropic form of an element is converted into another allotropic form.

# Sulphur

# *Additional questions*

**1** What do you understand by the following terms?

**a** Dehydrating agent.
**b** Oxidising agent.
**c** Optimum temperature.
**d** Acid rain.

**2** Explain the following.

**a** Chemical plants that produce sulphuric acid are often located on the coast.
**b** Even though more sulphuric acid could be produced using high pressures, normal atmospheric pressure is used.
**c** Natural rubber cannot be used to produce car tyres but vulcanised rubber can.
**d** Sulphur dioxide gas is regarded as a pollutant.
**e** Coal-fired and oil-fired power stations produce sulphur dioxide. Some of them are being fitted with flue gas desulphurisation (FGD) units.

**3** A type of coal contains 0.5% of sulphur by mass.

**a** Write an equation for the formation of sulphur dioxide gas when this coal is burned.
**b** If 1500 tonnes of coal was burned, what mass of sulphur would it contain?
**c** What mass of sulphur dioxide gas would be formed if 1500 tonnes of coal were burned?
**d** What volume would this mass of sulphur dioxide gas occupy, measured at room temperature and pressure (rtp)? ($A_r$: O = 16; S = 32. One mole of a gas occupies 24 $dm^3$ at rtp.)

**4** Fossil fuels, such as oil, coal and natural gas, all contain some sulphur. When these fuels are burned they produce many different gases. Concern has grown in recent years about the effects of one of these gases, sulphur dioxide. When sulphur dioxide dissolves in rainwater it forms an acidic solution which has become known as acid rain.

Money has been made available to solve the problem of acid rain. Attempts are being made to clean gases being released from power stations and to look into ways in which the effects of acid rain can be reversed.

The table below and Figure 16.8 (p. 230) give some data about the emission of sulphur dioxide.

| | **Million tonnes per year** |
|---|---|
| USA | 26 |
| Russia/Ukraine | 18 |
| Germany | 7 |
| UK | 5 |
| Canada | 5 |
| France | 3 |
| Poland | 3 |
| Italy | 3 |
| Other countries | 30 |

**a** Using the figures in the table, produce a bar chart to show the amount of sulphur dioxide produced by each of the countries listed.
**b** What percentage of the world's sulphur dioxide is produced in:
(i) the UK?
(ii) North America?
**c** Using the information above, explain why countries such as the US, Russia and Germany are at the top of the list of sulphur dioxide producers.
**d** If the total amount of sulphur dioxide produced by the UK is 5 million tonnes per year, what amount is produced by:
(i) power stations?
(ii) domestic users?
(iii) industry?

**5** Study the following reaction scheme:

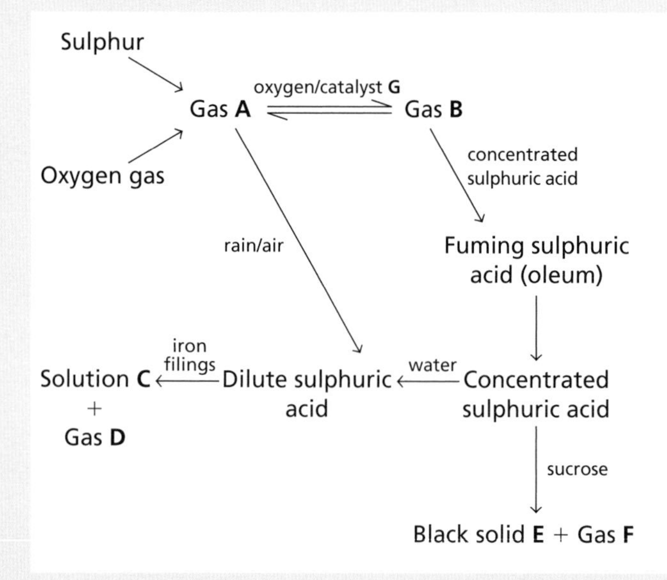


**a** Identify the substances **A** to **H** by giving their names and formulae.
**b** Write a balanced chemical equation for the formation of gas **B**.
**c** (i) Describe a chemical test, and give the positive result of it, to identify gas **D**.
(ii) Describe a chemical test, and give the positive result of it, to identify gas **F**.
**d** How would you obtain solid **C** from the solution **C**?
**e** Which pathway shows the formation of acid rain?
**f** In which way is the concentrated sulphuric acid acting in its reaction with sucrose?
**g** Where does the oxygen gas come from to form gas **A**?

6 In a neutralisation experiment, 25 $cm^3$ of dilute sulphuric acid was required to react completely with 40 $cm^3$ of a solution of 0.25 mol $dm^{-3}$ potassium hydroxide.
   a Write a balanced chemical equation for the reaction between dilute sulphuric acid and potassium hydroxide.
   b Calculate the number of moles of potassium hydroxide solution used in the reaction.
   c How many moles of dilute sulphuric acid would this number of moles of potassium hydroxide react with?
   d Calculate the concentration of the dilute sulphuric acid.
   e Which indicator could have been used to determine when neutralisation had just occurred?

7 Describe how you would prepare some crystals of hydrated copper(II) sulphate from copper(II) oxide and dilute sulphuric acid. Draw a diagram of the apparatus you would use and write a balanced chemical equation for the reaction.

8 When sulphur is extracted from sulphur beds below the Earth's surface, superheated water is pumped down a shaft into the beds to melt the solid sulphur.
   a What is meant by superheated water?
   b Why does the water have to be superheated? Why would boiling water not work?
   c (i) When the molten sulphur is pumped to the surface it solidifies. Which allotrope of sulphur forms first?
   (ii) What eventually happens to this allotrope as the temperature falls?
   Most sulphur obtained from these sulphur beds, in countries such as Poland and France, is exported.
   d In which form do you think the sulphur is loaded on to sulphur tankers? Explain your answer.
   e What hazards do you think are faced by people working in industries that use sulphur? Give reasons for your answer.

# 17 The planet Earth



## Volcanoes

**SLEEPER AWAKENS AFTER 100 YEARS**

This was the headline which announced the eruption of Mount St Helens in Washington State, USA, in May 1980. When it exploded into life it sent huge amounts of volcanic ash for over 1000 km and left a crater which was over 3 km wide.

More recently, Europe's most active volcano Mount Etna in Sicily erupted in July 2001 spewing out smoke, ash and lava, threatening nearby villages (Figure 17.1). The last major eruption had been in 1992.

**Figure 17.1** The eruption of Mount Etna in July 2001.

Volcanoes are formed when **magma**, the molten rock material containing dissolved gases and water beneath the Earth's crust, escapes to the surface through cracks (**fissures**) or holes (**vents**) in the crust (Figure 17.2a). The magma appears at the surface as **lava** (Figure 17.2b). Lava flow can engulf vast areas of land around the volcano.

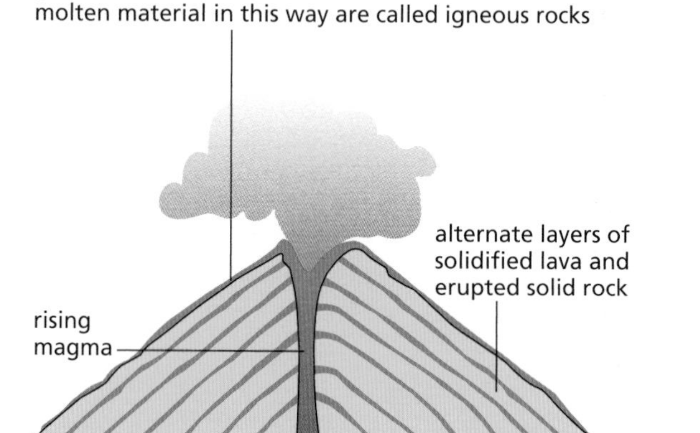



**a** A volcano.

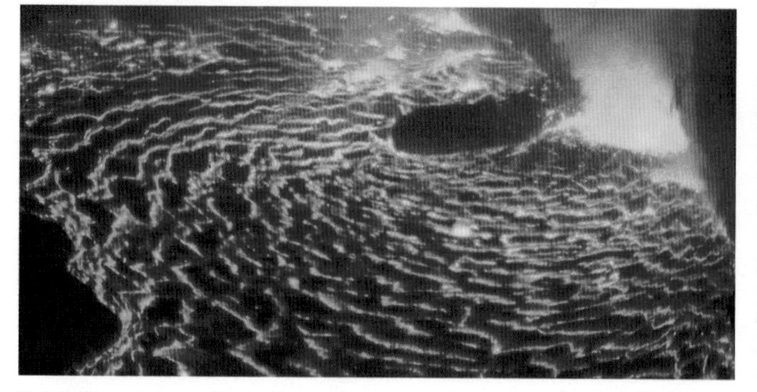

**b** Molten lava often flows from a volcano and engulfs land about it.

**Figure 17.2**

## Question

**1** Explain how a volcano is formed.

## The structure of the Earth

How do we know that the Earth has the structure shown in Figure 17.3? In their studies scientists have used sound waves as well as the waves sent out by earthquakes. From the information gathered they have concluded that the Earth consists of the following.

- Core – this is made up of very dense molten metal, which consists mainly of iron and nickel, under great pressure. It is approximately 6930 km in diameter.
- Mantle – this surrounds the core and is made up of cooler, less dense rock which contains a lot of iron-rich minerals. It is between 40 km and 2900 km below the Earth's surface.
- Crust – this is the thin, less dense, solid outer layer. The thickness of the crust under the oceans varies from 5 km to 10 km, while that under the continents varies from 6 km to 90 km.

The core has a temperature of about 4300 °C. This temperature drops as you go into the mantle and the temperature just below the crust is only about 1000 °C!

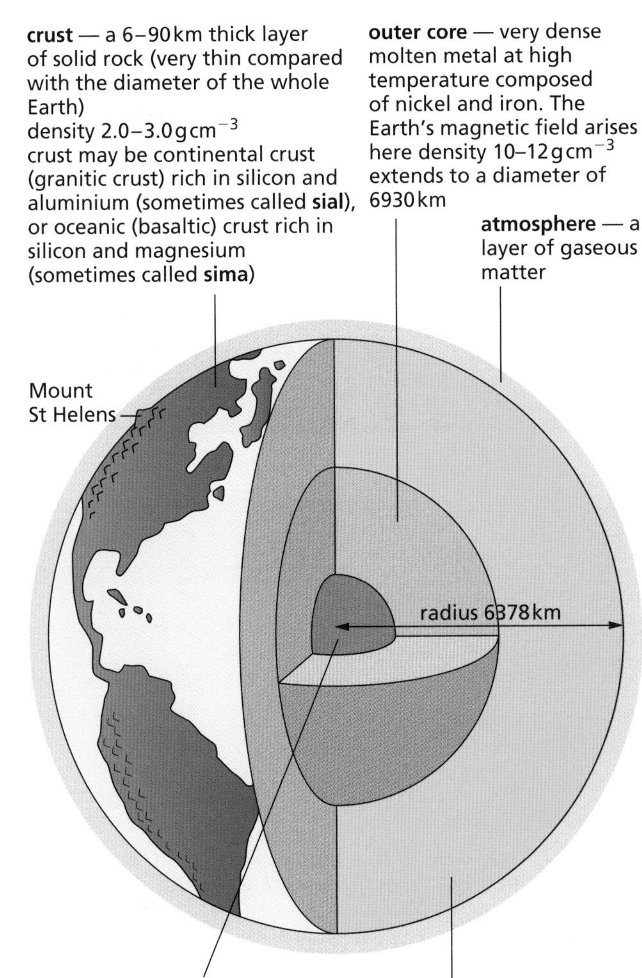


**Figure 17.3** The Earth.

These high temperatures are maintained mainly by:

- the inside of the Earth being insulated by the outer layers
- the radioactive isotopes of the elements potassium, thorium and uranium – the nuclei of these isotopes are unstable and break up, giving out large amounts of energy as they change into smaller nuclei.

### What is the crust made of?

There are many different rocks in the Earth's crust. These different rocks vary in the minerals they contain and in the shape and size of the mineral grains. Geologists have shown that there are three main groups of rocks – igneous, sedimentary and metamorphic.

**Igneous rocks** are formed when hot magma from the Earth's mantle or lower crust rises, cools and hardens. Igneous rocks are usually crystalline. There are two main types of igneous rock: intrusive and extrusive.

- Intrusive igneous rocks are formed by crystallisation of the magma underground. Granite is an example of this type of igneous rock (Figure 17.4).
- Extrusive igneous rocks are formed by crystallisation of the magma on the Earth's surface. Basalt is an example of this type of igneous rock (Figure 17.5).

**Figure 17.4** A granite erratic moved during glaciation.

**Figure 17.5** Vesicular lava, otherwise known as basalt.

**Sedimentary rocks** cover approximately 75% of the continents. These are formed when solid particles carried in seas or rivers are deposited. Sediment is also carried by wind and moving ice. Dissolved materials may later be extracted from water by plants or animals, or by evaporation to produce sediments. Layers of sediment can pile up for millions of years, and the sediment at the bottom of the pile experiences great pressure; the grains become cemented together, forming the sedimentary rock. Sedimentary rocks have definite layers, or **strata**, associated with them, and you can often see these layers running through the rocks (Figure 17.6a). There is a large variation in their hardness and grain size. Sedimentary rocks often contain fossils (Figure 17.6b).

**a** Limestone strata.

**b** Fossils are often found in limestone.

**Figure 17.6**

Limestone is a sedimentary rock which formed beneath the sea. Although it was formed beneath the sea, it is often found well above sea level due to the movement of the Earth's crust. This happens during the process of **uplift**. Uplift occurs mainly because of the large-scale lateral forces at work on the Earth's crust, resulting in its crumpling, for example at plate boundaries (Figure 17.12, p. 245).

Limestone is composed mainly of calcium carbonate, which effervesces when it comes into contact with a dilute acid (Chapter 8, p. 122). This property is often used to show the presence of limestone in a rock sample. Sandstone is also a sedimentary rock.

**Metamorphic rocks** are formed when rocks buried deep beneath the Earth's surface are altered by the action of great heat and pressure. Marble is a metamorphic rock and is formed by this type of action on limestone (Figure 17.7, top). Slate is another example of metamorphic rock (Figure 17.7, bottom), which is formed from mudstone.

**Figure 17.7** The Taj Mahal in India and roofing material are made from the metamorphic rocks marble and slate.

## Questions

**1** Describe the structure of the Earth.

**2** What is:
   **a** igneous rock?
   **b** sedimentary rock?
   **c** metamorphic rock?
   Give an example of each type of rock.

## Fossils

Fossils are the remains or impressions, in rocks, made by animals or plants when they die. When these organisms decay, they leave their impressions in the surrounding sediment. When the sediment becomes rock, impressions are left in the rock (Figure 17.6b). In some cases the organism decays and dissolves leaving a space in the rock. When this happens, certain minerals may seep into the space and take up the shape remaining, producing a cast of the original organism.

As stated earlier in this chapter, fossils are found in sedimentary rocks such as limestone. Sedimentary rocks are layered with the oldest layers (strata) found deeper underground than the younger rock. Geologists have been able to divide time into three **eras**, based on the type of fossil found in the different rock strata.

- Cenozoic era – this is the most recent era and covers the present to 65 million years ago.
- Mesozoic era – this era covers the time from 65 million years ago to 225 million years ago.
- Palaeozoic era – this era covers the time from 225 million years ago to 570 million years ago.

Figure 17.8 shows the way in which the eras have been divided into periods and the **periods** divided further into **epochs**.

### Questions

1 What is a fossil?

2 Use a diagram to help you to describe how fossils were formed.

| ERAS | PERIODS | EPOCHS | Present | SOME FOSSILS |
|---|---|---|---|---|
| CENOZOIC<br>The Earth's climate became much colder, resulting in several Ice Ages | QUATERNARY | Holocene | 0.01 | human skull, mammoth (tooth) |
| | | Pleistocene | 2 | |
| The age of the mammals as well as insects and flowering plants | TERTIARY | Pliocene | 7 | snail, bivalve shellfish |
| | | Miocene | 26 | |
| | | Oligocene | 38 | |
| Opening of the North Sea | | Eocene | 54 | |
| | | Paleocene | 65 | |
| MESOZOIC<br>The age of the dinosaurs | CRETACEOUS | | 136 | ammonite, lampshell, sea urchin |
| The great supercontinent of Pangaea began to break up, forming most of the continents as we know them | JURASSIC | | 190 | lampshell, ammonite, sea urchin, coral |
| | TRIASSIC | | 225 | bivalve shellfish, fish (tooth) |
| PALEOZOIC<br>Initially, most life was in the sea. Plants appeared on the land in the Silurian era, followed after a few million years by the first amphibians. Towards the end of this period the first reptiles appeared on land | PERMIAN | | 280 | bivalve shellfish, fish (tail), lampshell |
| | CARBONIFEROUS | | 355 | tree root, coral, amphibian (skull) |
| | DEVONIAN | | 395 | lampshell, fish |
| | SILURIAN | | 440 | coral, trilobite, lampshell, graptolite |
| | ORDOVICAN | | 500 | graptolite, lampshell, trilobite |
| | CAMBRIAN | | 570 | lampshell, trilobite, trilobite |

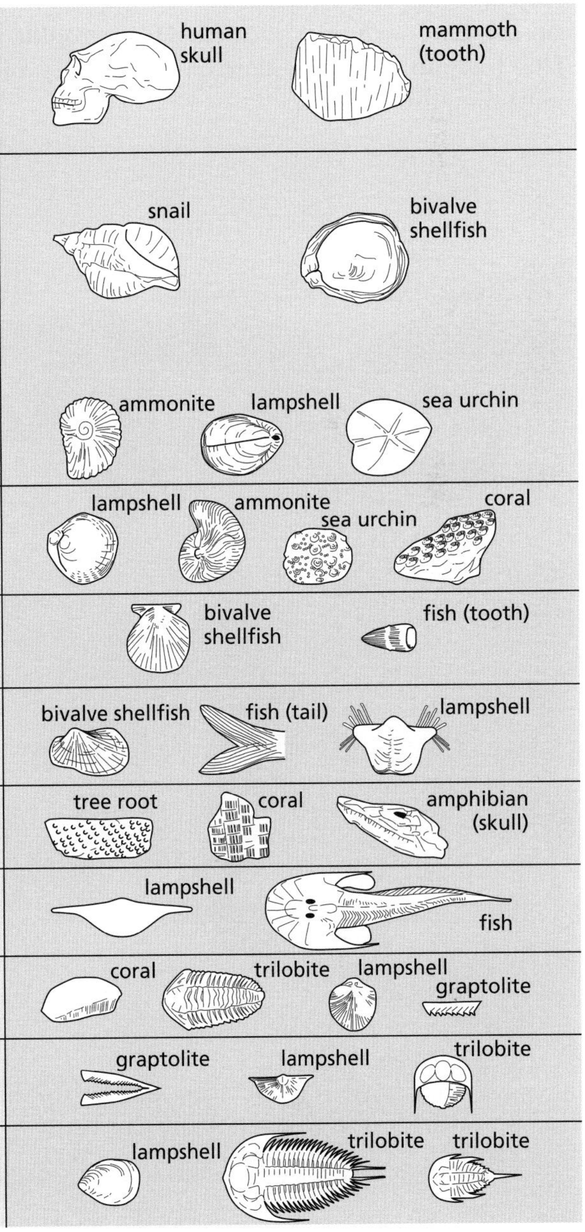

**Figure 17.8** Geological time (figures refer to millions of years before present).

## Weathering

Weathering is the actual breakdown of exposed rock on the Earth's surface. There are two main ways that rock can be broken down: by chemical means and by physical means.

### Chemical means

Rainwater contains dissolved carbon dioxide (Chapter 8, p. 118) as well as other gases such as sulphur dioxide (Chapter 16, p. 230) and nitrogen dioxide (Chapter 10, p. 156). The effect of these substances is to reduce the pH to quite a low value. This means that this now 'acid rain' can dissolve particular rocks, such as limestone, quite easily.

calcium carbonate + carbon dioxide + water → calcium hydrogencarbonate

$$CaCO_3(s) + CO_2(aq) + H_2O(l) \rightarrow Ca(HCO_3)_2(aq)$$

Minerals may also be oxidised. Oxygen from the air can combine with iron silicates to form iron(III) oxide. This leads to a brown stain on the surface of rocks containing this mineral.

Some minerals combine with water molecules and take them into the crystal structure. They become hydrated, which causes expansion, leading to stresses within the rock structure. This causes the rocks to break up. An example of this type of weathering takes place with haematite ($Fe_2O_3$). With water it forms limonite ($Fe_2O_3.H_2O$).

### Physical means

These include the actions of water and temperature. Rainwater enters cracks in rocks. When water freezes, its volume expands and it forces the rock apart. Stresses can also be built up in a rock formation by temperature changes. Minerals within the rock will expand and contract with changes in temperature at different rates. In the temperate areas of the world, such as the UK, where alternate freezing and thawing happens a lot, you find that the pieces of rock which break off fall down mountain sides, forming **scree** (Figure 17.9).

**Figure 17.9** This scree slope in Gorsdale Scar, Yorkshire, was formed by physical weathering.

## Erosion

Erosion involves the wearing away of rock and its transportation to another place. The photographs in Figure 17.10 show the four main ways by which erosion takes place.

Taking into account the four main methods of erosion, researchers have found that the erosion rate for the land area is between 8 cm and 9 cm of depth per 1000 years.

## Questions

1 What is the difference between weathering and erosion?

2 Describe, using a diagram, how rock from a mountainous region can get into the sea.

**a** This gorge was formed by the eroding action of the river.

**c** This stack was formed by wave action.

**b** Glaciers erode the mountain to which they are attached.

**d** Wind erosion caused these formations.

**Figure 17.10**

## Soil

The smaller pieces of rock produced by the different weathering processes are transported by the different methods shown in the photographs in Figure 17.10 and are deposited to cover the surface of the Earth.

Humus, which is decayed (or partly decayed) organic material from plants and animals, mixes with the different types of rock material. This mixture of humus and rock particles is called soil. It takes 400 years for 1 cm of soil to form.

The soil provides nutrients and minerals as well as water for the plants to grow. Plants remove nitrates (Chapter 15, p. 218) as well as elements such as sodium, potassium, magnesium and copper from the soil.

The pH of a soil depends on the type of rock from which the soil was formed as well as the amount of humus present. For example, soil formed in limestone areas tends to be alkaline, while soil formed in granite areas tends to be acidic. Also, the more humus present, the lower the pH of the soil.

There are different types of soil.

- **Loams**. These are the ideal soils for agriculture. This type of soil has sufficient clay (20%) to retain nutrients and hold moisture, sufficient sand (40%) to ensure the soil is well aerated and to prevent water-logging and silt (40%) which acts like an adhesive, holding the clay and sand together.
- **Sandy soils**. These are well drained and aerated. They are, therefore, easy to cultivate as they allow crop roots to penetrate them. However, because they lack humus they are vulnerable to drought. Also, they need large amounts of fertiliser as nutrients drain away quite quickly after rain.
- **Peaty soils**. These grow excellent crops but unfortunately the 'peaty' material can oxidise to carbon dioxide. It also gets blown away quite easily, which means that the level of this type of soil tends to fall by many metres a century.

## Questions

1 Explain how soil is produced from rock.

2 Describe experiments you could carry out on a soil sample to determine:
  a the moisture content of the soil
  b the amount of humus it contains.

## The rock cycle

The pattern of change we have been discussing, which takes place on the Earth's surface, is known as the **rock cycle**. This concept was first developed by James Hutton. Figure 17.11 shows the full rock cycle. In this cycle the rocks on the upland areas are weathered and the particles are carried away by erosion to form sediments which eventually become sedimentary rocks. These are then brought to the surface by the Earth's movements (uplifting) or they may be heated and compressed to form metamorphic rocks. If these metamorphic rocks are pushed deep below the surface, they will melt to form magma in the mantle. The magma may then rise upwards towards the surface, and igneous rocks are formed. So one type of rock may be recycled to form another type of rock.

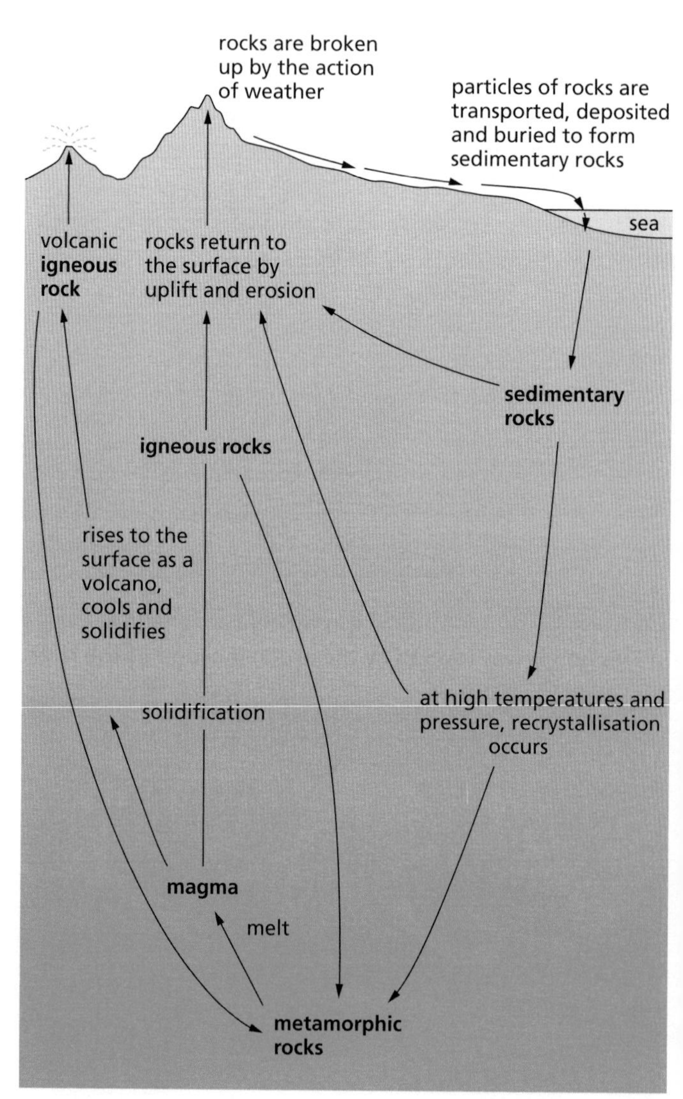


**Figure 17.11** The rock cycle.

## Question

1 Describe the different processes involved in the rock cycle.

## Plate tectonics

Evidence from geologists shows that the Earth's **lithosphere** is not a continuous structure but is divided into sections called **plates** (Figure 17.12). The lithosphere is the near-rigid outer shell of the Earth, made up of the crust and the outermost layer of the mantle. The majority of these plates have continents sitting on top of them. These plates are actually moving very slowly about the Earth's surface. The driving force behind this movement is thought to be convection currents within the mantle.

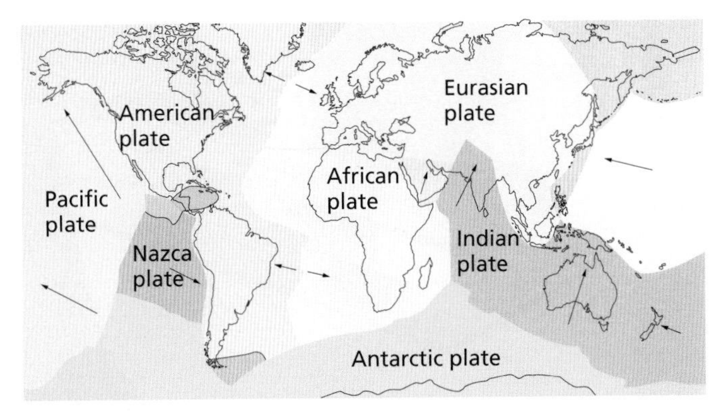


**Figure 17.12** The world's plate boundaries. The arrows show the directions in which the plates are moving.

The study of the way these plates behave is called **plate tectonics**. Geologists think that the continents were originally joined together into one giant continent, which they called Pangaea (Figure 17.13). It is thought that it has taken approximately 200 million years for the continents, as we know them today, to drift to the positions they are in now through their plate movement. Evidence for this has been obtained from fossils found in the US. There are great similarities in the fossils found in the US and in Europe, dating from 200 million years ago.

**Figure 17.13** The giant continent was called Pangaea. Much of the interior of this huge land mass was hot and dry.

**a** The 1994 earthquake in Los Angeles caused buildings and bridges to collapse.

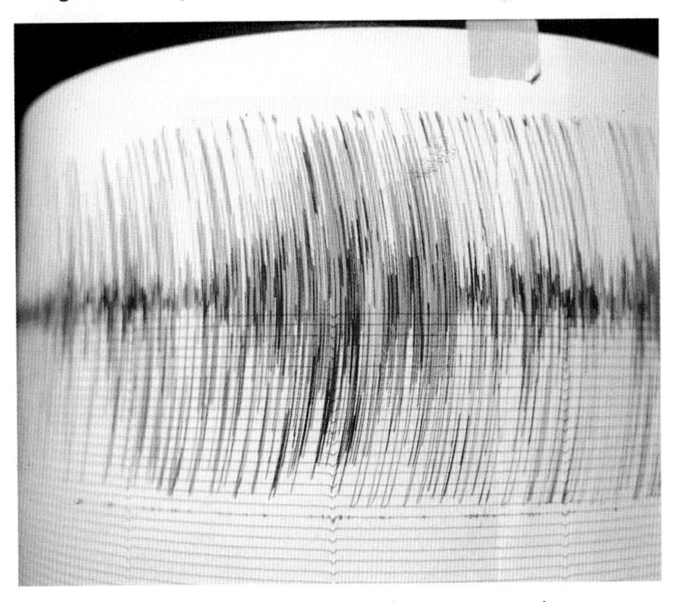

**b** The drum seismogram reading of the Los Angeles earthquake.

**Figure 17.14**

Where plates join, enormous forces are generated. This can, and does, create earthquakes (Figure 17.14) as well as volcanoes and mountains. Earthquakes mainly occur along fault zones:

- where plates are scraping past each other, for example California
- at margins where one plate is descending into the mantle, for example west coast of South America
- at constructive margins, where new rock is being formed from the mantle, for example the Mid-Atlantic Ridge.

a A seismometer site.

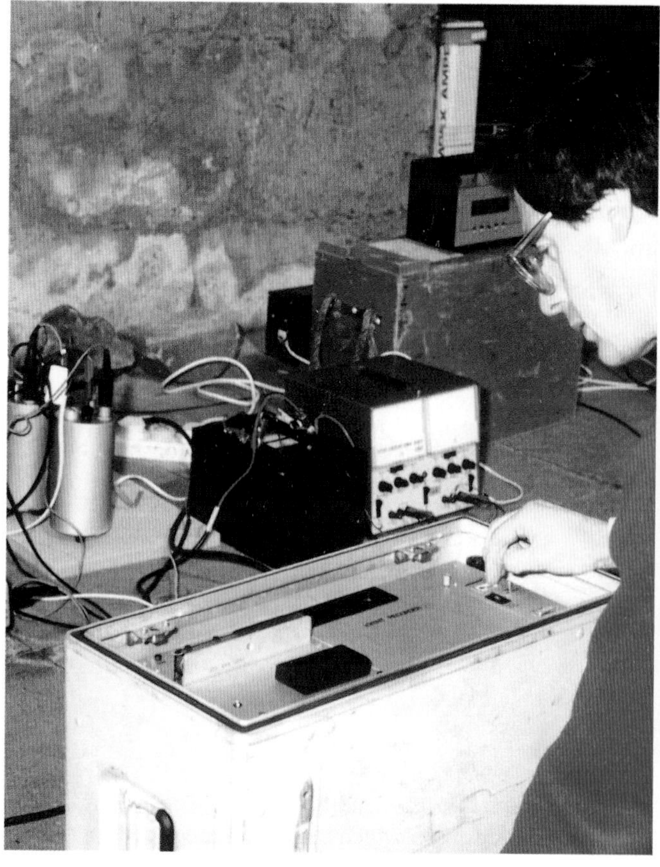

b Taking a reading from the seismometer.

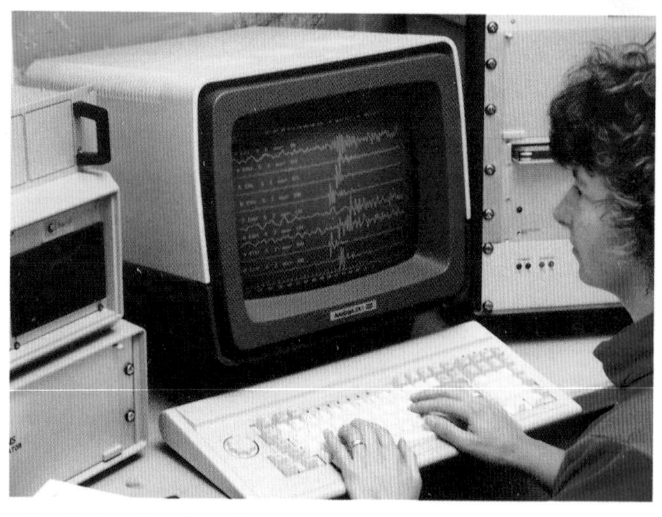

c Analysing the data from the seismometer.

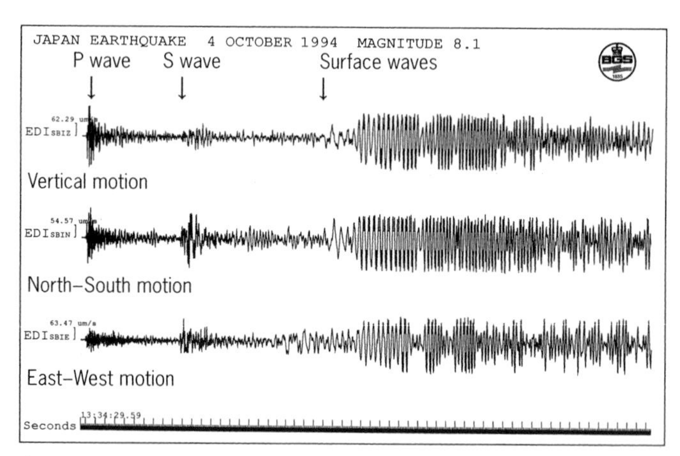


d A seismogram showing the three different waves.

**Figure 17.15**

Geologists monitor the earthquake activity around the Earth through hundreds of **seismic stations**. These stations are able to detect earthquake waves using an apparatus called a **seismometer** (Figure 17.15). Geologists hope to be able to predict earthquake activity through gathering information in this manner.

There are three different types of waves produced by earthquakes.

- **Primary (P) waves**. These are longitudinal waves, produced by pushing and pulling forces which cause the rock to shake backwards and forwards. This type of wave travels at speeds of up to 13 km per second and travels through both solids and liquids.
- **Secondary (S) waves**. These are transverse waves and cause the rock to shake at right angles to the direction of movement of the waves. This type of wave travels at speeds of up to 7 km per second and travels only through solids.
- **Surface (L) waves**. These waves have long wavelengths and are responsible for the largest land movements and, therefore, cause the most damage. This type of wave travels more slowly than either the P or the S wave.

The energy released by an earthquake is measured and described using the **Richter scale**. Each unit on this scale represents a ten-fold increase on the previous unit. In theory, the scale does not have a limit; however, earthquakes above 8 are rarely encountered (Table 17.1).

**Table 17.1** The Richter scale.

| Richter scale unit | Destruction level for a nearby earthquake |
|---|---|
| 2–3 | Hardly noticed |
| 3–4 | Slightly noticed |
| 4–5 | Minor |
| 5–6 | Damaging |
| 6–7 | Destructive |
| 7–8 | Major destruction |
| 8+ | Enormously destructive |

The January 1995 earthquake in Kobe, Japan, measured 7.2 on the Richter scale. Almost 5300 people died and approximately 27 000 were injured. In terms of devastation, around 110 000 buildings were damaged or destroyed. The cost of rebuilding has been estimated at £65 billion.

Another area of the world which suffers frequent small tremors and, occasionally, a larger earthquake, as in 1994, is around Los Angeles in the US. This area is prone to earthquakes because it lies on the San Andreas fault – the boundary between the Pacific plate and the American plate. The plates are moving almost parallel to one another, but in opposite directions (Figure 17.16).

**Figure 17.16** The displacement of these rows of trees shows how the plates are moving along the San Andreas fault.

The rough edges grind away as the plates slide across one another but occasionally particularly rough sections meet and 'lock' together. The pressure builds up until suddenly they slip across one another, causing the ground on each side to shudder violently and earthquake damage on the surface.

Along the western edge of South America you can see a boundary region between two plates that are moving in opposite directions. Continental crust will meet oceanic crust at the boundary between the two plates. When this happens, the thinner oceanic plate is pushed down underneath the thicker continental plate and melts.

Sediments scraped from the oceanic plate are pushed up to form fold mountains, such as the Andes. Close to the edge of the continental plate the crust is weakened. Magma forces its way through these areas of weaknesses, forming volcanic regions (Figure 17.17).

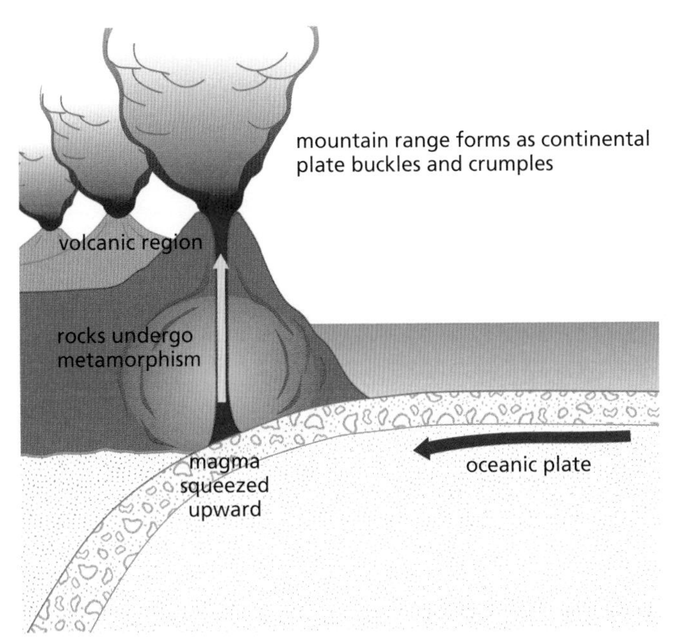


**Figure 17.17** Formation of mountain ranges such as the Andes.

Although it may be difficult to believe, there are mountain ranges under the sea that are much higher than those on land. They occur where two plates are moving apart. One such region lies in the middle of the Atlantic Ocean (see Figure 17.12, p. 245, and Figure 17.18, overleaf).

Along the line where the plates meet, lava pours through huge fissures or cracks. The magma cools quickly in the colder depths of the ocean, forming 'bubbles'. These are quickly burst by the pressure of the lava below, and fresh lava erupts on top of the previously erupted and now hardened layer. Layer upon layer of basalt is formed, forming a high ridge called the Mid-Atlantic Ridge. As more magma emerges, the older rock layers are pushed further apart – a process known as ocean-floor spreading.

On each side of ocean ridges the rocks show a clear pattern of magnetic stripes. These are caused by iron crystals in the magma which, as the magma hardens, line up in the direction of the magnetic North and South poles. The iron particles themselves are weakly magnetised, with their North and South poles aligned in the Earth's magnetic field.

At times during the Earth's history, the magnetic North and South poles have suddenly reversed. The reversal of the magnetically aligned particles in the stripes confirms the changes in the Earth's magnetic field.

The symmetrical nature of the magnetic stripes shows that the basalts on each side of the ridge were intruded into the ridge and became magnetised before the basalt was broken in two and the parts moved apart (Figure 17.19).

The meeting of two continental plates leads to the formation of high mountain ranges, for example the Himalayas. This is where the crust is at its thickest.

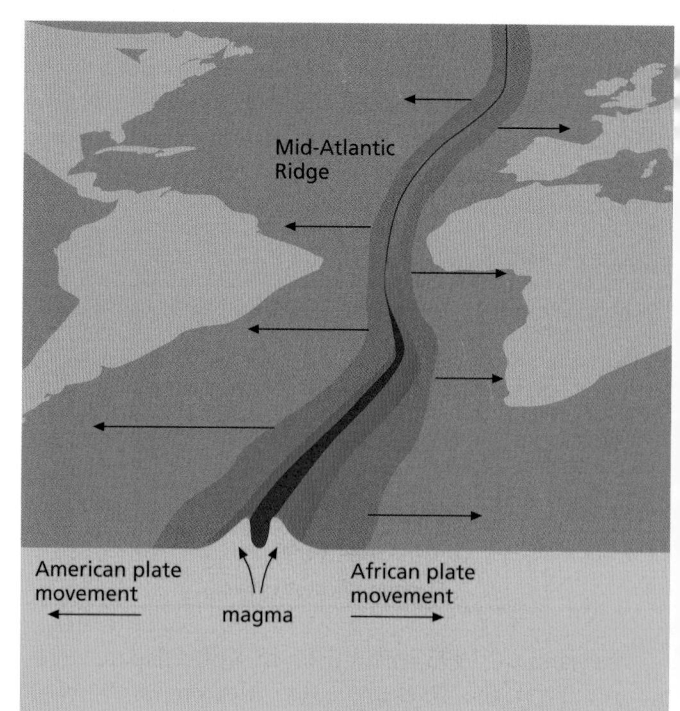


**Figure 17.18** The formation of the Mid-Atlantic Ridge.

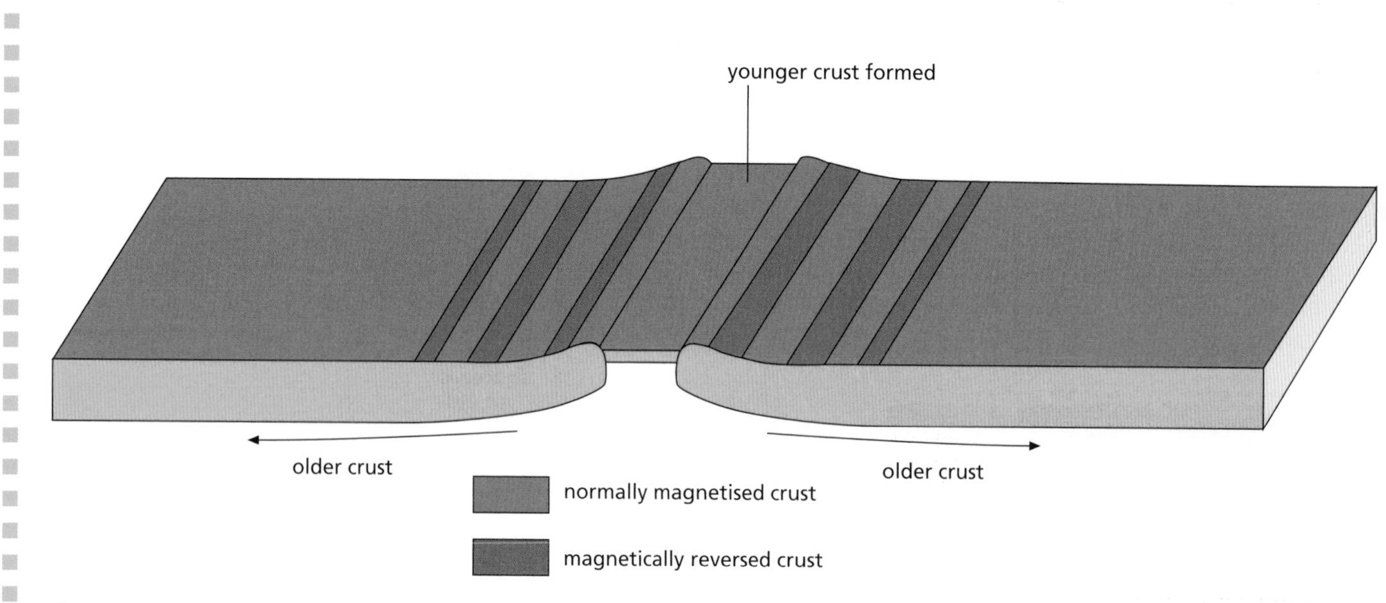


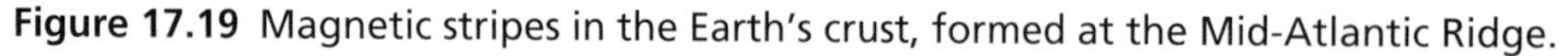


**Figure 17.19** Magnetic stripes in the Earth's crust, formed at the Mid-Atlantic Ridge.

## Questions

1. Explain why earthquakes usually happen at the boundaries between the world's tectonic plates.
2. Use your research skills to find the earthquake rating, on the Richter scale, of the Los Angeles earthquake of 1994.

# Checklist

**After studying Chapter 17 you should know and understand the following terms.**

- **Core** The central part of the Earth, composed of iron and nickel.
- **Crust** The outermost layer of the Earth, to an average depth of about 40 km.
- **Earthquake** The movement of the Earth's surface caused by plates scraping past each other on a 'fault', or at margins where one plate is descending into the mantle.
- **Erosion** The removal and transportation of material.
- **Fossils** Traces of prehistoric life which have been preserved by natural processes in rocks.
- **Geological time** The geological division of time into three eras: cenozoic, mesozoic and palaeozoic. Each era is subdivided into periods and these are further subdivided into epochs.
- **Igneous rocks** Rocks formed when magma cools and solidifies. Igneous rocks are usually crystalline. There are two main types of igneous rocks: intrusive (for example granite) and extrusive (for example basalt).
- **Lava** Molten rock material that surges from a volcanic vent or fissure.
- **Lithosphere** The near-rigid outer shell of the Earth, made up of the crust and the outermost layer of the mantle.
- **Magma** Molten rock which includes dissolved water and gases.
- **Mantle** The part of the Earth between the core and the crust, 40–2900 km below the Earth's surface.
- **Metamorphic rocks** These are formed when rocks buried deep beneath the Earth's surface are altered by the action of great heat and pressure. For example, marble is a metamorphic rock and is formed by this type of action on limestone.
- **Plate tectonics** The Earth's lithosphere is not a continuous structure but is divided into sections called plates. The majority of these plates have continents sitting on top of them. These plates are actually moving very slowly about the Earth's surface. The driving force behind this movement is thought to be convection currents within the mantle. Plate tectonics is the study of the way these plates behave.
- **Richter scale** The scale used to measure the energy released by an earthquake.
- **Rock cycle** The cycle of natural rock change in which rocks are uplifted, eroded, transported, deposited and possibly changed into another type of rock and then uplifted to start a new cycle.
- **Sedimentary rocks** These rocks cover approximately 75% of the continents. They are formed when solid particles carried in seas or rivers are deposited. Sediment is also carried by wind and moving ice. Dissolved materials may later be extracted from water by plants or animals, or by evaporation to produce sediments. Layers of sediment can pile up for millions of years, and the sediment at the bottom of the pile experiences great pressure, causing compaction. The grains become cemented together by minerals in solution passing through the layers of sediment. Sedimentary rocks, for example limestone, have definite layers or strata associated with them, and you can often see these layers running through the rocks.
- **Seismometer** An instrument used to monitor the magnitude of earthquake waves.
- **Soil** A mixture of mineral particles and organic matter or humus.
- **Volcano** A hole (vent) or crack (fissure) in the Earth's crust through which molten rock (magma) and hot gases escape to the surface during an eruption.
- **Weathering** The action of water and temperature on rock.

# The planet Earth

# *Additional questions*

1 If mountains are created where plates are joined, are the plates moving away from each other or towards each other? Explain your answer.

2 Dilute hydrochloric acid can be used to distinguish between pieces of limestone and granite.
   a Which of the two rock samples would give a reaction with the dilute acid?
   b Write a word and balanced chemical equation to represent the reaction taking place in **a**.
   c Name two other types of rock which would give a similar reaction with dilute hydrochloric acid.

3 The average size of the crystals in a sample of granite is larger than that found in a sample of basalt.
   a What types of rock are granite and basalt?
   b Explain the difference in the average size of the crystals found in the two samples.
   c Granite is an *intrusive* rock and basalt is an *extrusive* rock. Explain the meaning of the terms in italics.

4 The *core* of the Earth is maintained at a temperature of about 4300 °C. This temperature drops as you go into the *mantle* and the temperature just below the *crust* is only about 1000 °C! One of the reasons for these high temperatures is the *radioactive decay* of the *isotopes* of *elements* such as potassium, thorium and uranium.
   a Explain the meaning of the terms in italics.
   b Why does the radioactive decay of the isotopes mentioned help to maintain the high temperatures?

5 a With reference to the rock cycle (Figure 17.11), describe how magma undergoes a change to a metamorphic rock.
   b Describe the differences in the driving forces behind the water cycle and the rock cycle.

6 Plate tectonics is the study of the movement of the Earth's plates.
   a What are tectonic plates?
   b Why are these plates able to move?
   c Explain, as fully as you can, one way the movement of these plates can cause:
      (i) earthquakes to happen
      (ii) mountains to be formed.

7 A seismometer is an instrument used to monitor the magnitude of earthquake waves.
   a Describe how a seismometer works.
   b There are three types of waves produced by earthquakes.
      (i) Name the three types of waves.
      (ii) Which of these waves would be able to travel through the Earth's mantle? Explain your answer.
   c The waves produced are either *longitudinal* or *transverse*. Describe the meaning of these terms as applied to waves.

8 The diagram below shows a volcano.

Copy the diagram into your notes and label the following:
   a (i) a vent
      (ii) lava
      (iii) volcanic ash.
   b Also label the place where:
      (i) basalt is likely to be found
      (ii) weathering and erosion will take place.
   c Explain how a volcano is formed

9 a Fossils of sea creatures are found high up on Mount Everest.
      (i) What does this suggest about how the Himalayan mountain range was formed?
      (ii) Explain your answer.
   b What evidence could be gathered to show that the east coast of America and the west coast of Europe were once joined?

# GCSE exam questions

**1 a** Materials are classified as solid, liquid or gas according to their properties. For each state give **two** typical properties.
Solid **(2)**
Liquid **(2)**
Gas **(2)**

**b** The melting and boiling points of six substances are given below.

| Substance | Melting point/°C | Boiling point/°C |
|---|---|---|
| Nitrogen | −210 | −196 |
| Carbon disulphide | −112 | 46 |
| Ammonia | −78 | −34 |
| Bromine | −7 | 59 |
| Phosphorus | 44 | 280 |
| Mercury(II) chloride | 276 | 302 |

(Room temperature is taken as 20 °C.)

(i) Which **element** is a solid at room temperature? **(1)**
(ii) Which **compound** is a liquid at room temperature? **(1)**
(iii) Which **compound** is a gas at room temperature? **(1)**
(iv) Which **element** will condense when cooled to room temperature from 100 °C? **(1)**
(v) Which **compound** will freeze first on cooling from room temperature to a very low temperature? **(1)**
(vi) Which of the six substances is a liquid over the widest range of temperature? **(1)**
(vii) Draw diagrams to show how the particles are arranged in bromine and in ammonia at room temperature. **(4)**

**c** Using the ideas of the kinetic theory explain why
(i) a metal expands on heating **(3)**
(ii) a gas changes to a liquid on cooling **(3)**
(iii) a sample of water left in a dish at room temperature will evaporate over a period of time. **(3)**

**d** A student used the apparatus below to observe smoke particles.

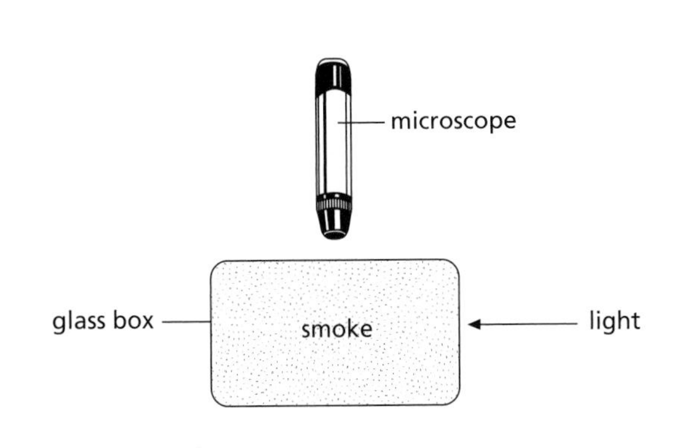


The student observed that the smoke particles moved randomly with a jerky, haphazard movement.
(i) Explain the student's observations. **(3)**
(ii) What is the name given to this type of movement? **(1)**

*(CCEA, GCSE, Paper 2, June 1999)*

**2** This question is about the changes of state which take place when a pure substance is heated.

Some of the pure substance is placed in a test-tube and is heated steadily. The temperature of the substance is measured at regular intervals. The results of this experiment are shown on the graph.

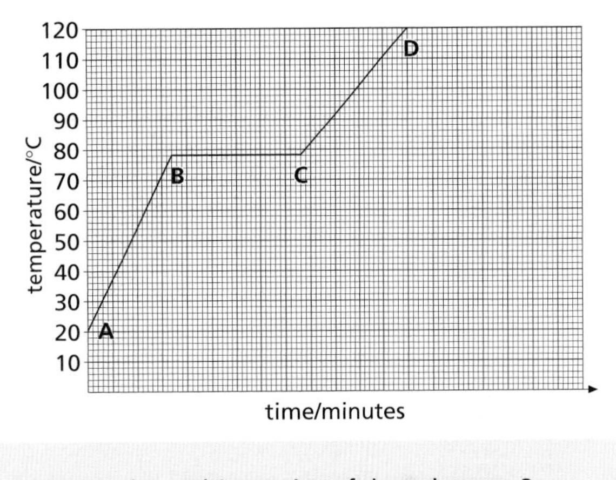


**a** What is the melting point of the substance? **(1)**
**b** On which part of the graph is the substance completely liquid? **(1)**
**c** Explain the shape of the graph.
Use your knowledge of energy and forces between particles in your answer. **(4)**

*(OCR, GCSE, Paper 2, June 1999)*

**3** A fridge has three tubes to show if the temperature gets too high. Each tube contains a solid that melts at a different temperature. The solid stays at the top of the tube. When it melts, it drops to the bottom.

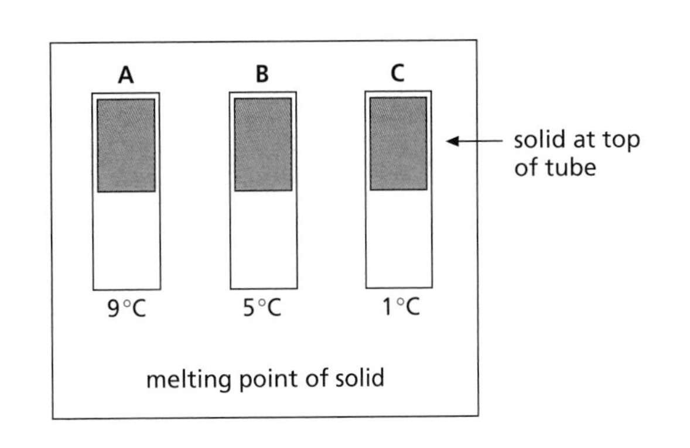

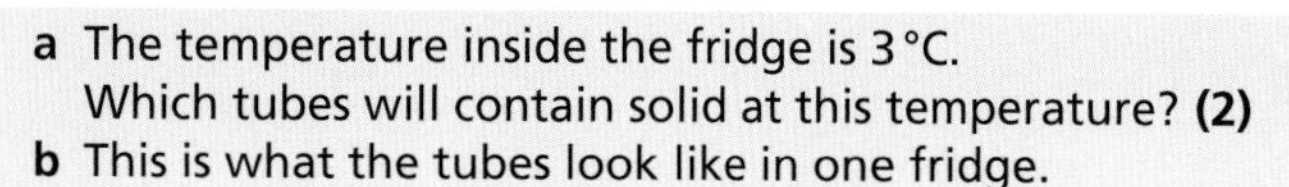

**a** The temperature inside the fridge is 3 °C.
Which tubes will contain solid at this temperature? **(2)**

**b** This is what the tubes look like in one fridge.

A B C

solid

liquid

9 °C 5 °C 1 °C

melting point of solid

Copy and complete these two sentences.
The temperature of the fridge is **above** ——°C.
The temperature of the fridge is **below** ——°C. **(2)**

The contents of the tubes are made of particles.

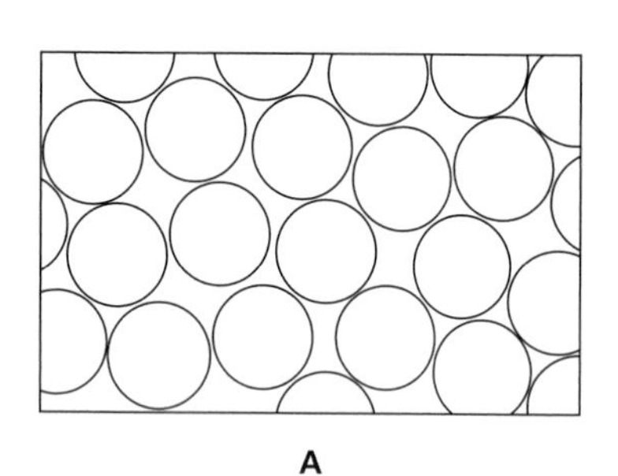

A

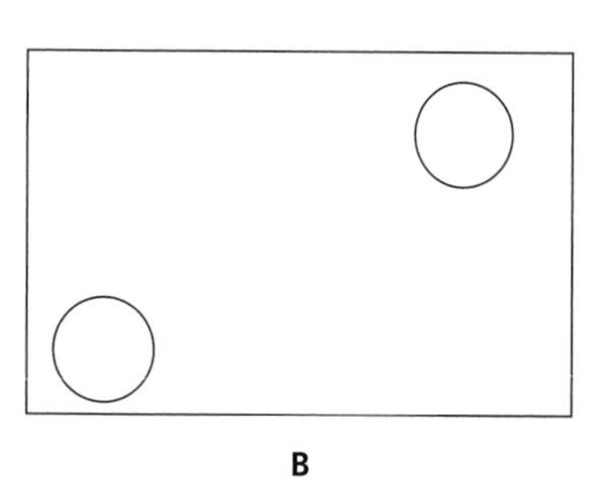

B

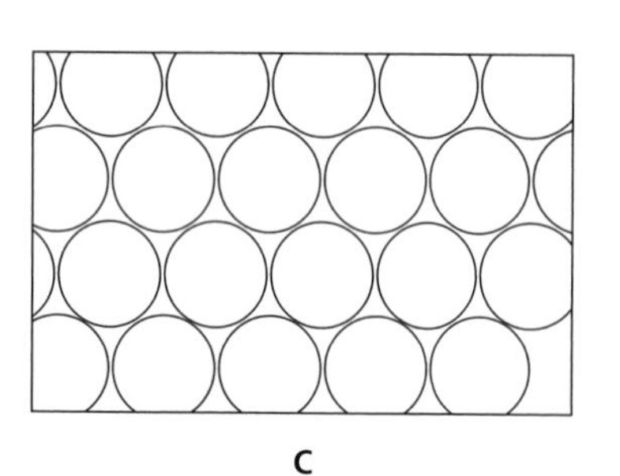

C

**c** Which box shows the particles in a solid? **(1)**

**d** Which box shows the particles in a liquid? **(1)**

**e** What happens to the particles when the solid melts? **(1)**

*(OCR, GCSE, Paper 1, June 2000)*

**4 a** The diagram below shows a burning candle.
Draw the arrangements of the particles at positions (i), (ii) and (iii). **(3)**

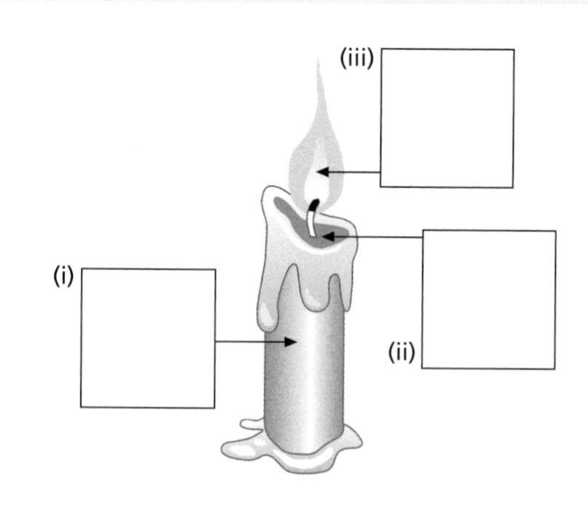


**b** It is possible to interchange the states of matter.
The following diagram shows these changes.

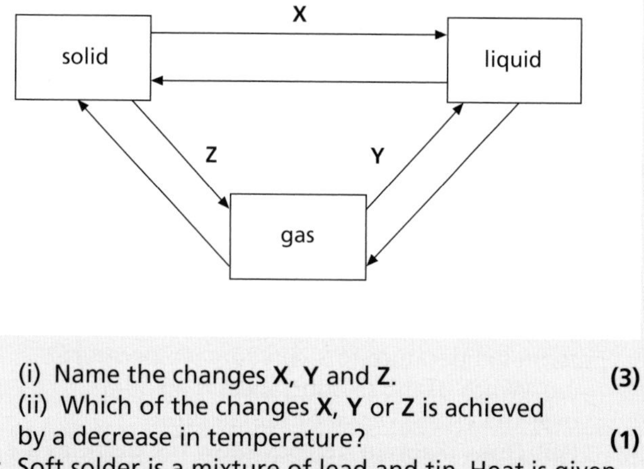


(i) Name the changes **X, Y** and **Z**. **(3)**
(ii) Which of the changes **X, Y** or **Z** is achieved by a decrease in temperature? **(1)**

**c** Soft solder is a mixture of lead and tin. Heat is given out when it changes from liquid to solid. Explain this in terms of particle theory. **(2)**

**d** Explain the essential difference between the **evaporation** of water and the **decomposition** of water. **(2)**

**e** The following apparatus was set up to investigate the movement of two gases.

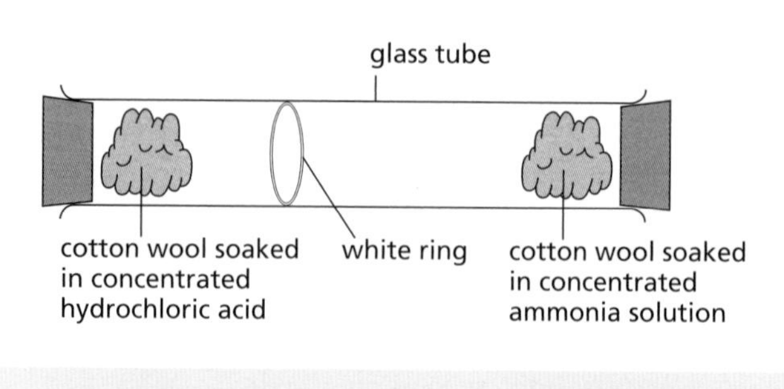


(i) What name is given to the movement of gases? **(1)**
(ii) Name the gas given off by concentrated hydrochloric acid. **(1)**
(iii) Explain fully why the white ring was formed closer to the concentrated hydrochloric acid end of the tube. **(2)**

**f** A gas syringe contains 80 $cm^3$ of gas at 280 K and 1 atmosphere pressure. What pressure would this same amount of gas exert if the volume was decreased to 40 $cm^3$ and the temperature increased to 350 K? **(4)**

*(CCEA, GCSE, Paper 2, June 2000)*

**5 a** Ice melts when it is heated. Explain clearly what happens to the water molecules as a piece of ice melts. **(4)**

**b** Solid carbon dioxide is sometimes known as **dry ice**. Under normal circumstances dry ice sublimes as it warms up. Explain clearly the difference between melting and subliming. **(2)**

**c** In an experiment, a student collected 100 $cm^3$ of carbon dioxide at 300 K and 1 atmosphere pressure. What volume would this amount of gas occupy if the conditions were changed to 450 K and 2 atmospheres pressure? **(4)**

*(CCEA, GCSE, Paper 2, June 1998)*

**6** John Dalton was a famous chemist who lived 200 years ago. He made a list of substances he thought were elements. He gave symbols to these elements. Here is a copy of his table.

ELEMENTS

| | W.t | | W.t |
|---|---|---|---|
| Hydrogen | 1 | Strontian | 46 |
| Azote | 5 | Barytes | 68 |
| Carbon | 5.4 | Iron | 50 |
| Oxygen | 7 | Zinc | 56 |
| Phosphorus | 9 | Copper | 56 |
| Sulphur | 13 | Lead | 90 |
| Magnesia | 20 | Silver | 190 |
| Lime | 24 | Gold | 190 |
| Soda | 28 | Platina | 190 |
| Potash | 42 | Mercury | 167 |

**a** We now know that some substances (such as hydrogen, carbon, oxygen and zinc) are elements. Write down the **names** of two other substances in his list that we now know are elements. **(2)**

**b** Here are three compounds shown using Dalton's symbols. Write down the names of the compounds. One has been done for you. **(2)**

Z

zinc oxide

(i) ____________

(ii) ____________

**c** Dalton stated that:

**1** All elements are made up of atoms.

**2** Atoms cannot be split up into simpler particles.

We now know that atoms contain smaller particles. Describe the structure of an atom such as carbon. **(4)**

*(OCR, GCSE, Paper 2, June 1999)*

**7 a** Elements can be divided into metals and non-metals according to their physical properties. Metals in general have a lustre, are solid at room temperature, malleable and ductile.

(i) Give **two** other physical properties of metals not mentioned above. **(2)**

(ii) What is the meaning of the following terms?

Lustre **(1)**

Malleable **(1)**

Ductile **(1)**

**b** The chemical properties of elements are also used to distinguish between metals and non-metals. Most metals react with acids.

(i) Give a word equation for the reaction of a metal with a dilute acid. **(2)**

(ii) Name **one** metal which reacts dangerously with a dilute acid. **(1)**

(iii) Name **one** metal which does not react with dilute acids. **(1)**

(iv) One metal which also reacts with alkalis is aluminium. Give a balanced symbol equation for the reaction of aluminium with sodium hydroxide solution. **(2)**

*(CCEA, GCSE, Paper 2, June 1999)*

**8 a** Name the following hazard symbols: **(2)**

(i)

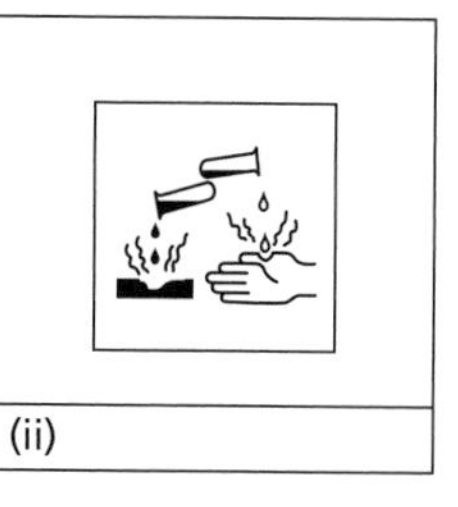

(ii)

**b** The following is a list of some classes of materials:

metals ceramics glass plastics fibres.

Choose from the above list the class to which the following materials A–E belong.

(i) A is flexible, easily melted, can be moulded and does not conduct electricity. **(1)**

(ii) B is hard, non-transparent, strong when compressed but weak when stretched. It is brittle, high melting and resistant to heat. **(1)**

(iii) C has the same properties as B but is transparent. **(1)**

(iv) D is strong, hard, can be bent and is a good conductor of both heat and electricity. **(1)**

(v) E is flexible and is formed from long, strong, hair-like strands. **(1)**

**c** Copper metal is often used in central heating systems. Give **two** properties of copper which makes it suitable for this purpose. **(2)**

*(CCEA, GCSE, Paper 2, June 1998)*

**9** This question is about the periodic table.

**a** Complete the following paragraph.

The greatest contributor to the development of the periodic table was the Russian scientist (i) ________ in 1869. He stated 'when elements are arranged in order of increasing (ii) ________, similar properties recur at intervals'. He recorded what he knew about each element on a separate card and then sorted the cards into 'piles of elements' with similar properties. His inspiration was to leave 'gaps' for (iii)________ elements. Nowadays the elements of the periodic table are arranged in order of increasing (iv) ________. The table is divided into rows and columns. The rows are called (v)________ and the columns are called (vi) ________. **(6)**

**b** The following diagram shows the positions of some elements in the periodic table. Answer the following questions using **only** those elements shown.

(i) Name the element in row 2 of column 3. **(1)**

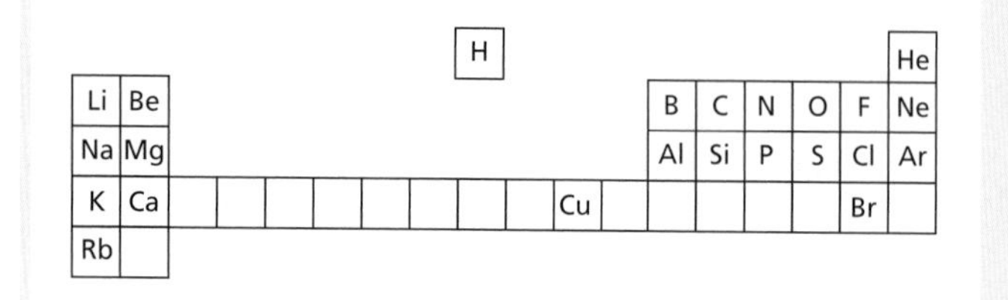


(ii) How many electrons would you expect in the outer shell of a bromide **ion**? **(1)**

(iii) Write down the symbol of the most reactive element in column 2. **(1)**

(iv) Write down the symbol of an element which has a filled outer shell of two electrons. **(1)**

(v) Which **two** elements would react together most vigorously? **(2)**

(vi) Write a balanced, symbol equation for the reaction in part (v). **(2)**

**c** Elements can be broadly classified as metals or non-metals. Name one element which is classified as a 'semi-metal' and briefly explain why it may be classified in this way. **(2)**

*(CCEA, GCSE, Paper 2, June 1998)*

**10** John Newlands attempted to classify the elements in 1866. He tried to arrange all the known elements in order of their atomic weights. The first 21 elements in Newlands' Table are shown below.

| | Column | | | | | | |
|---|---|---|---|---|---|---|---|
| | a | b | c | d | e | f | g |
| Symbol<br>Atomic weight | H<br>1 | Li<br>2 | Be<br>3 | B<br>4 | C<br>5 | N<br>6 | O<br>7 |
| Symbol<br>Atomic weight | F<br>8 | Na<br>9 | Mg<br>10 | Al<br>11 | Si<br>12 | P<br>13 | S<br>14 |
| Symbol<br>Atomic weight | Cl<br>15 | K<br>16 | Ca<br>17 | Cr<br>18 | Ti<br>19 | Mn<br>20 | Fe<br>21 |

Use a periodic table in a data book to help you answer these questions.

**a** In two of Newlands' columns, the elements match the first three elements in two groups of the modern periodic table.
Which two columns, **a** to **g**, are these? **(1)**

**b** (i) A group in the modern periodic table is completely missing from Newlands' Table.
What is the number of this group?
(ii) Suggest a reason why this group of elements is missing from Newlands' Table. **(1)**

**c** Give **one** difference between iron, Fe, and the other elements in column **g** of Newlands' Table. **(1)**

**d** Give the name of the block of elements in the modern periodic table which contains Cr, Ti, Mn and Fe. **(1)**

**e** Both Newlands and Mendeleev based their tables on atomic weights.
Explain why the modern periodic table is based on proton (atomic) numbers. **(2)**

**f** The atoms of elements in Group 1 of the modern periodic table increase in size going down the group.
Explain, in terms of electrons, how this increase in size affects the reactivity of these elements. **(3)**

*(AQA, GCSE, Paper 2372, June 2000)*

**11** Use a periodic table in a data book to help you answer these questions.

**a** A Russian chemist named Mendeleev produced a periodic table. His periodic table had the elements in order of increasing atomic mass. Find the elements **potassium** and **argon** in the periodic table.

(i) What problem is caused by using atomic mass to place these elements in order? **(1)**

(ii) Show how this problem is solved for potassium and argon in a modern periodic table. **(2)**

The table below gives information about some elements in the third period of the periodic table.

| Element | Symbol | Electron arrangement | Formulae of chlorides |
|---|---|---|---|
| Sodium | Na | | NaCl |
| Magnesium | Mg | 2,8,2 | |
| Aluminium | Al | | $AlCl_3$ |
| Silicon | Si | | $SiCl_4$ |
| Phosphorus | P | | $PCl_5$ |
| Sulphur | S | | $S_2Cl_2$ |
| Chlorine | Cl | | – |

**b** There is a pattern in the electron arrangements of atoms of elements in this period.

(i) Complete the missing electron arrangements in the table. **(2)**

(ii) What is the connection between electron arrangement and the position of the element in the periodic table? **(2)**

**c** There is a pattern in the formulae of chlorides in this period. Suggest the formula for magnesium chloride. **(1)**

**d** Sodium reacts with cold water.

(i) Write down the names of the **two** products of this reaction. **(1)**

(ii) Write a balanced equation for this reaction. **(2)**

**e** Potassium is in the same group of the periodic table as sodium.

(i) Write down the electron arrangement in a potassium atom. **(1)**

(ii) Explain why potassium reacts faster than sodium with cold water. **(2)**

*(OCR, GCSE, Paper 2, June 1999)*

**12** In some nuclear reactors sodium metal is used to transfer heat. The sodium is heated to about 500 °C by the reactor. The heat is transferred to water to produce steam.

**a** Use a data book to help you answer this question. Is sodium a solid, a liquid or a gas when it has been heated by the nuclear reactor? **(1)**

**b** One disadvantage of sodium is that it reacts with water.

What happens when a small piece of sodium reacts with water?

You should describe what you would see and state what substances are formed. **(4)**

**c** Use a data book to help you answer this question.

When sodium reacts with water it forms sodium ions. The diagrams represent the electron arrangements of some atoms and ions.

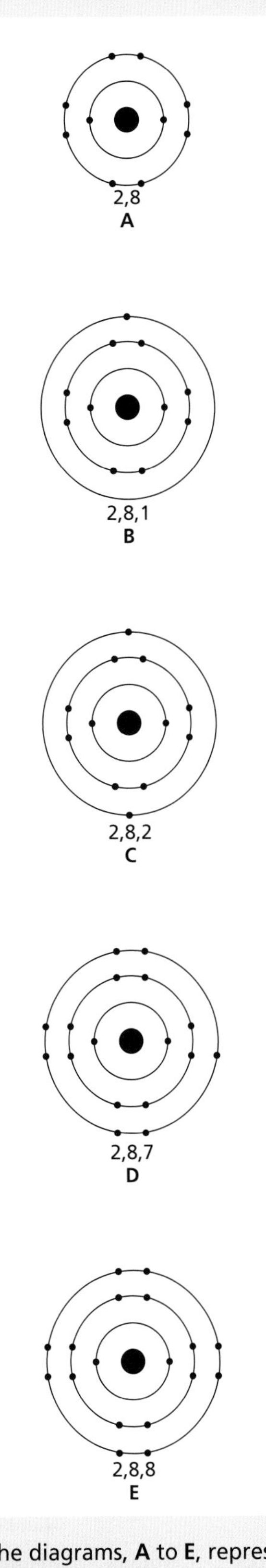


Which of the diagrams, **A** to **E**, represents the electron arrangement of each of the following?

(i) a sodium atom, Na

(ii) a sodium ion, $Na^+$. **(2)**

*(AQA, GCSE, Paper 2372, June 1999)*

**13 a** Use a periodic table in a data book to answer the following questions.

(i) What is the mass number of fluorine? **(1)**

(ii) How many protons are present in an atom of beryllium? **(1)**

(iii) How many neutrons are present in an atom of aluminium? **(1)**

(iv) Which element has the electronic structure 2, 8, 6? **(1)**

(v) Which Group 0 element has the largest atom? **(1)**

**b** Using **X** to represent an electron, copy and complete the following diagram to show the electronic structure for an atom of phosphorus. **(1)**

(*WJEC, GCSE, June 2000*)

**14** The elements in Mendeleev's periodic table were arranged in order of increasing atomic mass. Part of the modern periodic table is shown.

| | | | | | | | |
|---|---|---|---|---|---|---|---|
| | | H 1 | | | | | He 2 |
| Li 3 | Be 4 | | B 5 | C 6 | N 7 | O 8 | F 9 | Ne 10 |
| Na 11 | Mg 12 | | Al 13 | Si 14 | P 15 | S 16 | Cl 17 | Ar 18 |
| K 19 | Ca 20 | | | | | | | |

**a** Complete the sentence by writing out the missing words.

The modern periodic table is arranged in order of increasing __________________ . **(1)**

**b** (i) Name a metal in the same group as lithium. **(1)**

(ii) Name a non-metal in the same period as magnesium. **(1)**

**c** The table contains some information about **two** elements.

| Element | Symbol | Number of protons | Number of neutrons | Number of electrons |
|---|---|---|---|---|
| Fluorine | F | 9 | 10 | 9 |
| Chlorine | Cl | 17 | 18 | 17 |
| Chlorine | Cl | 17 | 20 | 17 |

(i) In terms of atomic structure, give **one** feature that both these elements have in common. **(1)**

(ii) There are two **isotopes** of chlorine shown in the table. Explain what **isotope** means. **(2)**

(iii) Explain, in terms of electron arrangement, why fluorine is more reactive than chlorine. **(2)**

**d** Sodium reacts with chlorine to form the compound sodium chloride.

$2Na + Cl_2 \rightarrow 2NaCl$

Describe, in terms of electron arrangement, the type of bonding in:

(i) a molecule of chlorine **(3)**

(ii) the compound sodium chloride. **(4)**

(*SEG, GCSE, Paper 6/4, Summer 2000*)

**15 a** Find nickel on a periodic table.

(i) What name is given to metals in this part of the periodic table? **(1)**

(ii) What is the symbol for nickel? **(1)**

(iii) What is the atomic number of nickel? **(1)**

(iv) What is meant by the atomic number of an element?

**A** Number of neutrons in an atom

**B** Number of protons in an atom

**C** Total number of protons and neutrons in an atom

Which is correct (**A, B** or **C**)? **(1)**

**b** Use words from the box to complete the sentence.

| **gas liquid solid** |
|---|

When nickel melts, it changes from a (i) ________ to a (ii) ________. **(1)**

**c** Nickel is made of atoms.

Copy the following diagram and draw circles in the box to show the arrangement of nickel atoms when it is a solid. One has been drawn for you.

(*Edexcel, GCSE, Paper 2F/1F, June 2000*)

**16 a** Sodium, atomic number 11, reacts with chlorine, atomic number 17, to form sodium chloride.
(i) Give the electronic structures of the two elements, sodium and chlorine. **(2)**
(ii) Explain, by means of a diagram or otherwise, the electronic changes that take place during the formation of sodium chloride. Include the charges on the ions. **(4)**
(iii) The table below shows some physical properties of sodium chloride.
Using the information in the table above, state the type of **structure** found in sodium chloride. **(1)**

| Melting point/°C | Boiling point/°C | Solubility in water |
|---|---|---|
| 801 | 1413 | Soluble |

**b** Chlorine gas, $Cl_2$, consists of molecules. By means of a diagram, show the bonding in a chlorine molecule. **(2)**
Name this type of bonding. **(1)**

*(WJEC, GCSE, June 2000)*

**17 a** Lithium, atomic number 3, reacts with oxygen, atomic number 8, to form lithium oxide.
(i) Give the electronic structures of the two elements, lithium and oxygen. **(2)**
(ii) Explain, by means of a diagram or otherwise, the electronic changes that take place during the formation of lithium oxide. **(4)**
(iii) Give the chemical formula for lithium oxide. **(1)**

**b** Oxygen, atomic number 8, forms a compound with hydrogen, atomic number 1, called water.
By means of a labelled diagram, show how the atoms of oxygen and hydrogen are bonded together in water. **(2)**

*(WJEC, GCSE, 0125/2, June 1999)*

**18** Uranium metal can be produced by reacting uranium hexafluoride with calcium.

$$UF_6 + 3Ca \rightarrow 3CaF_2 + U$$

**a** Describe how calcium and fluorine bond together to form calcium fluoride. The electron arrangement of each atom is shown. **(5)**

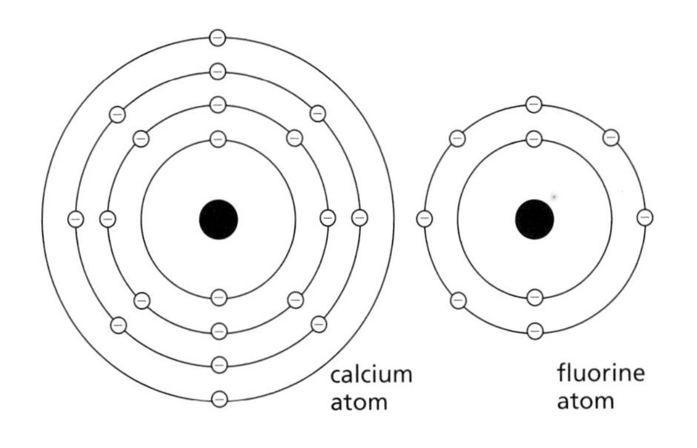


**b** Uranium has two main isotopes, $^{235}_{92}U$ and $^{238}_{92}U$.
Use these as examples to explain what is meant by the word isotope. **(4)**

*(SEG, GCSE, Paper 6/4 June 1999)*

**19 a** Complete the following passage:

Substances which are made up of atoms and cannot be simplified by chemical methods are called (i) _______. Atoms often combine with different types of atoms to form (ii) _______. Sometimes atoms will gain or lose electrons in chemical reaction to form charged particles called (iii) _______. **(3)**

**b** Copy and complete the following table relating to particles **A**, $\mathbf{B^+}$ and $\mathbf{C^{2-}}$. (The letters do not represent the symbols of the elements.) **(9)**

| Particle | Atomic number | Mass number | Number of protons | Number of neutrons | Number of electrons |
|---|---|---|---|---|---|
| A | 6 | | | 6 | |
| $B^+$ | | 39 | 19 | | |
| $C^{2-}$ | | 16 | | | 10 |

**c** (i) Draw a 'dot and cross' diagram to show how oxygen atoms become bonded together in oxygen gas. (Only the outer shell of electrons need be shown.) **(3)**
(ii) Name the type of bond in part **c** (i). **(1)**

**d** Sodium is a soft silvery white metal with melting point 98 °C.
Using a diagram, describe how the atoms are held together in a piece of sodium metal. **(3)**

**e** Substances may be classified in terms of their physical properties. Use the table below to answer the following questions:

| Substance | Melting point/°C | Boiling point/°C | Electrical conductivity | |
|---|---|---|---|---|
| | | | As solid | As liquid |
| A | 3720 | 4827 | Good | Poor |
| B | −95 | 69 | Poor | Poor |
| C | 327 | 1760 | Good | Good |
| D | 3550 | 4827 | Poor | Poor |
| E | 776 | 1500 | Poor | Good |

(i) Which substance could be sodium chloride? **(1)**
Explain your answer. **(2)**
(ii) Which substance consists of small covalent molecules? **(1)**
Explain your answer. **(2)**
(iii) Explain why substance **A** could **not** be diamond. **(2)**

*(CCEA, GCSE, Paper 2, June 1999)*

**20 a** Chlorine is a non-metallic element which has an **atomic number** of 17 and can exist as **isotopes**.
Explain what is meant by the terms in bold type.
(i) Atomic number **(1)**
(ii) Isotopes **(2)**

**b** Chlorine exists as diatomic molecules. Show clearly how atoms of chlorine combine to form chlorine molecules. (Outer electrons only need to be shown.) **(3)**

**c** Chlorine can form a range of compounds with both metals and non-metals.
(i) Explain carefully, using full electronic arrangements, how atoms of chlorine combine with atoms of calcium to form calcium chloride. (You may use a diagram to help you.) **(6)**
(ii) Name the type of bonding found in calcium chloride. **(1)**
(iii) Show clearly how atoms of chlorine combine with carbon to form tetrachloromethane $CCl_4$. (Outer electrons only need to be shown.) **(3)**
(iv) Name the type of bonding found in $CCl_4$. **(1)**

**d** The properties of compounds depend very closely on their bonding. In the following table give the **correct** word to show some of the expected properties of calcium chloride and tetrachloromethane. **(4)**

| Compound | Solubility in water | Relative melting point |
|---|---|---|
| Calcium chloride | (i) Soluble/Insoluble | (ii) Low/High |
| Tetrachloromethane | (iii) Soluble/ Insoluble | (iv) Low/High |

**e** The bonding in the **elements** calcium and carbon is very different. Describe clearly the bonding in
(i) calcium. **(3)**
(ii) carbon in the form of graphite. **(3)**
(iii) Both calcium and graphite can conduct electricity. State **two** properties of calcium which are different from graphite. **(2)**

*(CCEA, GCSE, Paper 1, June 2000)*

**21** This advertisement appeared in a Do-It-Yourself magazine.

Drill bits that power through almost anything

These amazing new drill bits penetrate virtually every material. Tipped with space-age Wolfram, they have diamond-like hardness – in fact, you'll probably never need to replace them.

**a** The drill bits have a 'diamond-like' hardness.

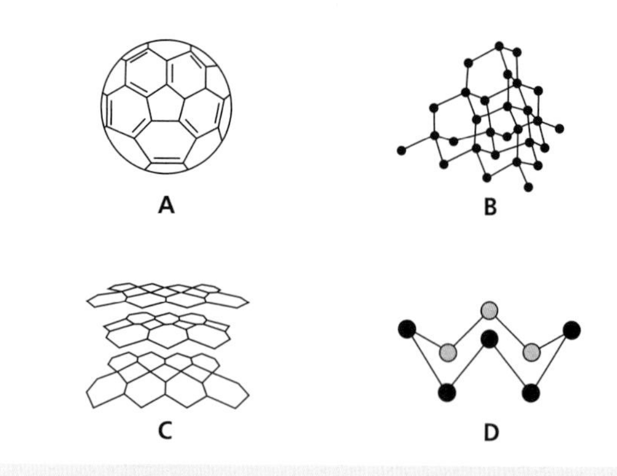


Which of these is a diamond structure? **(1)**

**b** Use ideas about chemical bonds to explain why diamond is so hard. **(4)**

**c** The drill tip has to be very hard, so it is made out of a different substance from the rest of the drill, the shank. The shank of the drill is made out of iron.

The advertisers have made the material of the drill tip sound new and exciting by calling it 'space-age Wolfram'. Wolfram is the old name for an element in the periodic table which is now called something else.
(i) What property **in the table** below makes Wolfram better than iron for use as a drill tip? Explain why. **(2)**
(ii) Wolfram and iron are in the same part of the periodic table. Using the atomic numbers of the elements, find Wolfram.
What is the usual name for Wolfram? **(1)**

**d** The tip of the drill becomes discoloured after it has been used. This is possibly because the Wolfram forms a layer of oxide, $WO_3$, on its surface.
Use the periodic table to help you predict the formula of
(i) Wolfram sulphide **(1)**
(ii) Wolfram chloride. **(1)**

**e** In a laboratory experiment Wolfram was converted into Wolfram oxide, $WO_3$.
(i) What mass of oxygen will react with 184 g of Wolfram? ($A_r$: W = 184; O = 16) **(2)**
(ii) A different oxide of Wolfram is made of 9.2 g of Wolfram and 1.6 g of oxygen.
What is the formula of this oxide?
Show your working. ($A_r$: W = 184; O = 16) **(3)**

*(OCR, GCSE, Paper 2, June 2000)*

| Name | Symbol | Atomic number | Melting point/°C | Boiling point/°C | Density/ $g\,cm^{-3}$ | Electrical conductivity |
|---|---|---|---|---|---|---|
| Iron | Fe | 26 | 1535 | 2750 | 7.9 | Good |
| Wolfram | W | 74 | 3410 | 5660 | 19.4 | Fair |

**22** *To obtain full marks, the steps in the calculations in this question must be shown.*

**a** Calcium nitrate decomposes on heating according to the equation

$$2Ca(NO_3)_2 \rightarrow 2CaO + 4NO_2 + O_2$$

(i) Calculate the mass of calcium nitrate required to be heated in order to produce 2.8 g of calcium oxide. (Relative atomic masses: Ca = 40, N = 14, O = 16.) **(4)**

(ii) Calculate the volume of oxygen produced in the same reaction, measured at room temperature and pressure.
[1 mole of gas at room temperature and pressure occupies a volume of 24 $dm^3$ (litres).] **(3)**

**b** (i) State Avogadro's law. **(3)**

(ii) The hydrocarbon ethane, $C_2H_6$, undergoes complete combustion according to the following equation.

$$2C_2H_6 + 7O_2 \rightarrow 4CO_2 + 6H_2O$$

What volume of oxygen is required to completely combust 5 $dm^3$ of ethane? **(2)**

(*CCEA, GCSE, Paper 1, June 2000*)

**23** *To obtain full marks the steps in the calculations in this question must be shown.*

**a** The salt, lead(II) nitrate, may be prepared in the laboratory by the reaction between lead(II) oxide and nitric acid. The equation for the reaction is

$$PbO + 2HNO_3 \rightarrow Pb(NO_3)_2 + H_2O$$

In an experiment to prepare lead(II) nitrate crystals, a student placed 50 $cm^3$ of nitric acid, concentration 0.5 mol $dm^{-3}$ (moles per litre), in a beaker and warmed the solution. Solid lead(II) oxide was added with stirring until no more solid dissolved.

(i) Calculate the number of moles of nitric acid present in the 50 $cm^3$ of nitric acid solution. **(2)**

(ii) Calculate the maximum mass of lead(II) nitrate which could be obtained from the above experiment. (Relative atomic masses: N = 14, O = 16, Pb = 207.) **(4)**

**b** When solid lead(II) nitrate is heated it decomposes according to the equation

$$2Pb(NO_3)_2(s) \rightarrow 2PbO(s) + 4NO_2(s) + O_2(g)$$

In an experiment a student heated 3.31 g of lead(II) nitrate until no further change took place.

(i) Calculate the mass of lead(II) oxide formed. (Relative atomic masses: N = 14, O = 16, Pb = 207.) **(4)**

(ii) Calculate the volume of nitrogen dioxide gas, measured at room temperature and pressure, formed.
[1 mole of gas at room temperature and pressure occupies a volume of 24 $dm^3$ (litres).] **(3)**

**c** (i) State Avogadro's law. **(3)**

(ii) The equation for the catalytic oxidation of ammonia is

$$4NH_3(g) + 5O_2(g) \rightarrow 4NO(g) + 6H_2O(g)$$

What volume of oxygen is needed to produce 10 $dm^3$ of nitrogen monoxide? **(2)**

(*CCEA, GCSE, Paper 2, June 1998*)

**24** Copper oxide reacts with hydrogen. The hydrogen combines with the oxygen and copper is left.

In an experiment, some copper oxide is put into a porcelain boat inside a tube. Hydrogen gas is passed over the heated copper oxide until all the copper oxide has changed into copper. The apparatus used is shown in the diagram.

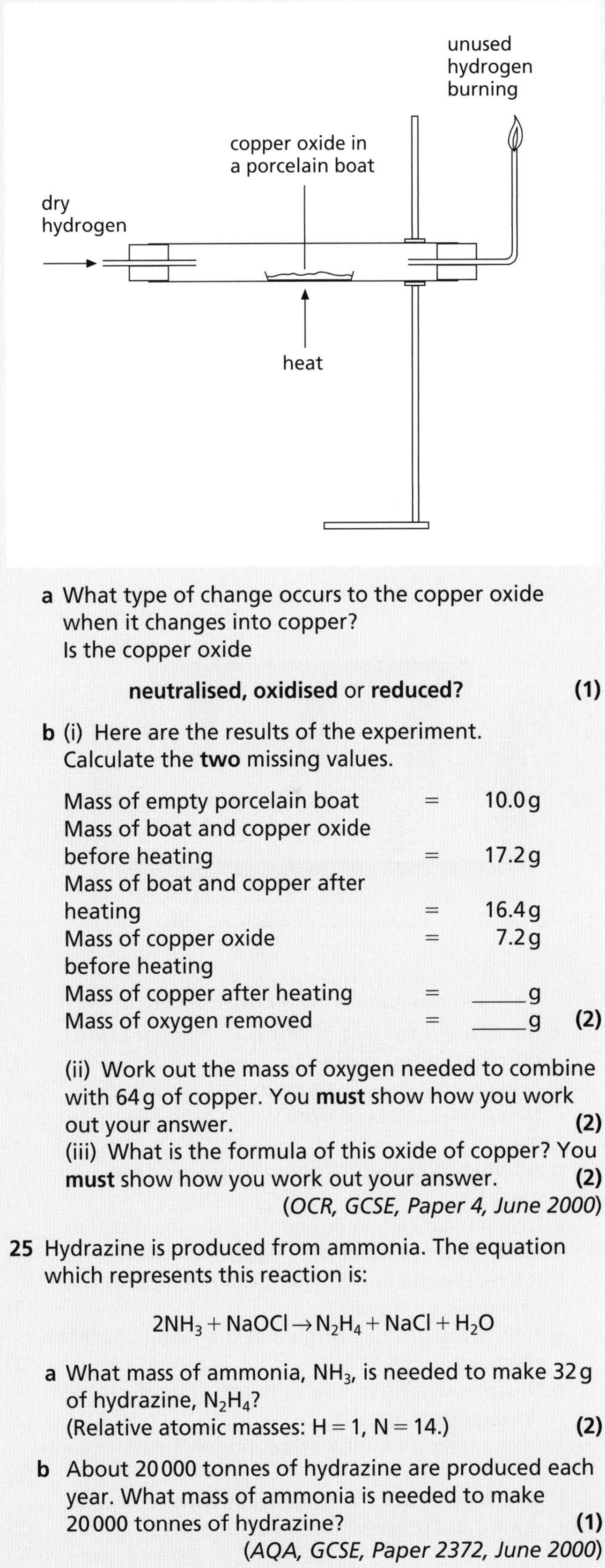


**a** What type of change occurs to the copper oxide when it changes into copper?
Is the copper oxide

**neutralised, oxidised** or **reduced?** **(1)**

**b** (i) Here are the results of the experiment. Calculate the **two** missing values.

| | | |
|---|---|---|
| Mass of empty porcelain boat | = | 10.0 g |
| Mass of boat and copper oxide before heating | = | 17.2 g |
| Mass of boat and copper after heating | = | 16.4 g |
| Mass of copper oxide before heating | = | 7.2 g |
| Mass of copper after heating | = | ____ g |
| Mass of oxygen removed | = | ____ g |

**(2)**

(ii) Work out the mass of oxygen needed to combine with 64 g of copper. You **must** show how you work out your answer. **(2)**

(iii) What is the formula of this oxide of copper? You **must** show how you work out your answer. **(2)**

(*OCR, GCSE, Paper 4, June 2000*)

**25** Hydrazine is produced from ammonia. The equation which represents this reaction is:

$$2NH_3 + NaOCl \rightarrow N_2H_4 + NaCl + H_2O$$

**a** What mass of ammonia, $NH_3$, is needed to make 32 g of hydrazine, $N_2H_4$?
(Relative atomic masses: H = 1, N = 14.) **(2)**

**b** About 20 000 tonnes of hydrazine are produced each year. What mass of ammonia is needed to make 20 000 tonnes of hydrazine? **(1)**

(*AQA, GCSE, Paper 2372, June 2000*)

**26 a** What is the name given to the block of elements in the middle of the periodic table which includes vanadium? **(1)**

**b** Some of the properties of vanadium are shown in this list.

- It has a high melting point.
- It is a solid at room temperature.
- It is a conductor of electricity.
- It is a good conductor of heat.
- It forms coloured compounds.
- It forms crystalline compounds.
- It forms compounds that are catalysts.

Select two properties, from the list above, which are not typical of a Group 1 metal. **(2)**

**c** One compound of vanadium is vanadium oxide. A sample of vanadium oxide contained 10.2 g of vanadium and 8.0 g of oxygen. Calculate the formula of this vanadium oxide. You must show **all** your working to gain full marks.
(Relative atomic masses: V = 51, O = 16.) **(3)**

*(AQA, GCSE, Paper 2372, June 1999)*

**27** Aluminium metal is extracted from pure aluminium oxide by electrolysis using the cell shown below.

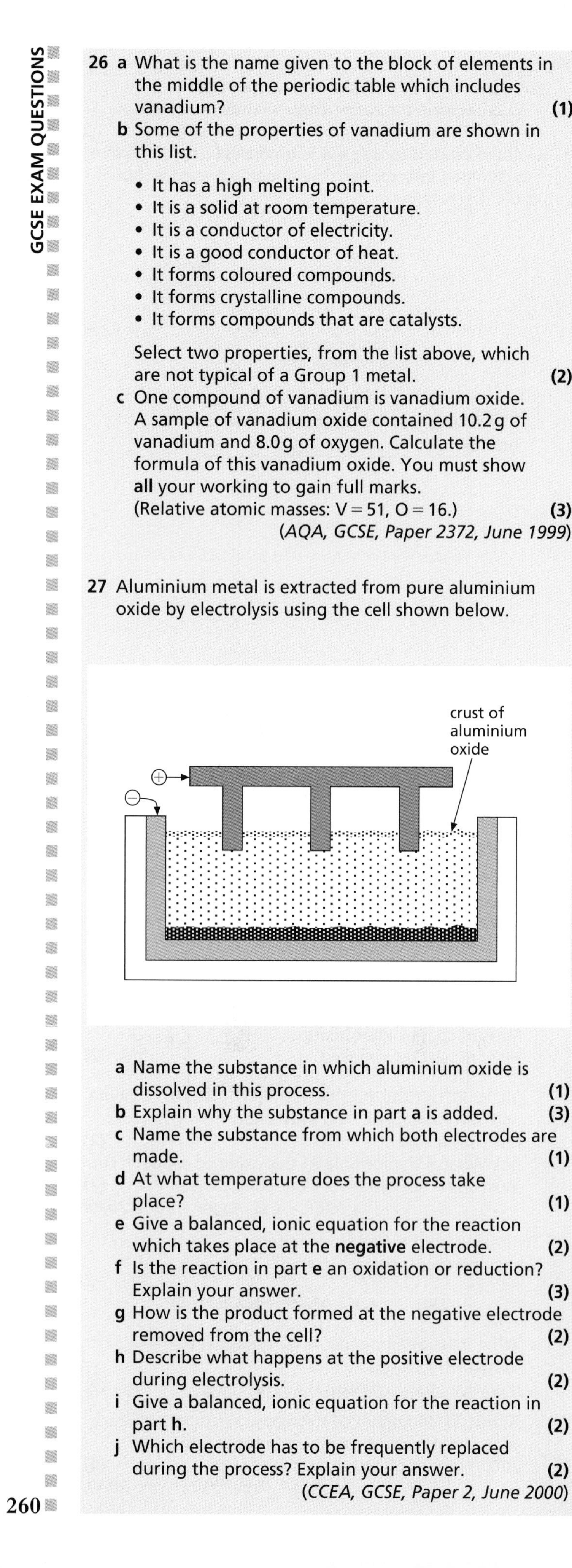


**a** Name the substance in which aluminium oxide is dissolved in this process. **(1)**

**b** Explain why the substance in part **a** is added. **(3)**

**c** Name the substance from which both electrodes are made. **(1)**

**d** At what temperature does the process take place? **(1)**

**e** Give a balanced, ionic equation for the reaction which takes place at the **negative** electrode. **(2)**

**f** Is the reaction in part **e** an oxidation or reduction? Explain your answer. **(3)**

**g** How is the product formed at the negative electrode removed from the cell? **(2)**

**h** Describe what happens at the positive electrode during electrolysis. **(2)**

**i** Give a balanced, ionic equation for the reaction in part **h**. **(2)**

**j** Which electrode has to be frequently replaced during the process? Explain your answer. **(2)**

*(CCEA, GCSE, Paper 2, June 2000)*

**28** The industrial **electrolysis** of an aqueous solution of sodium chloride is shown.

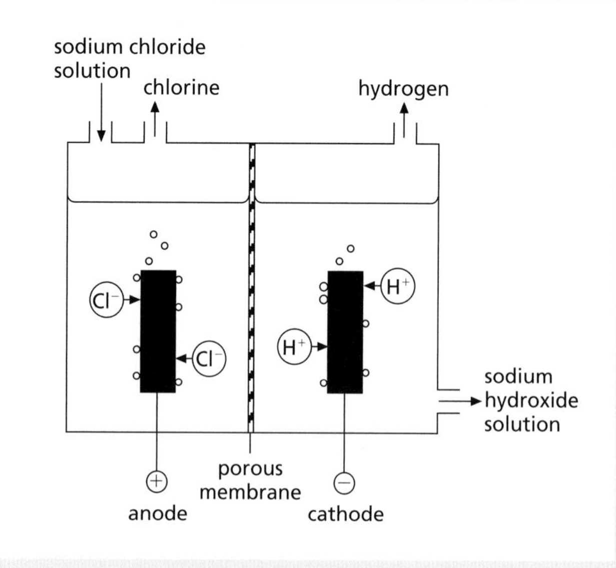


**a** What is meant by **electrolysis**? **(2)**

**b** Explain how hydrogen and sodium hydroxide solution are produced. **(5)**

*(SEG, GCSE, Paper 5, Summer 2000)*

**29** Electrolysis of acidified water can be carried out in this apparatus.
Hydrogen and oxygen are formed during the electrolysis.

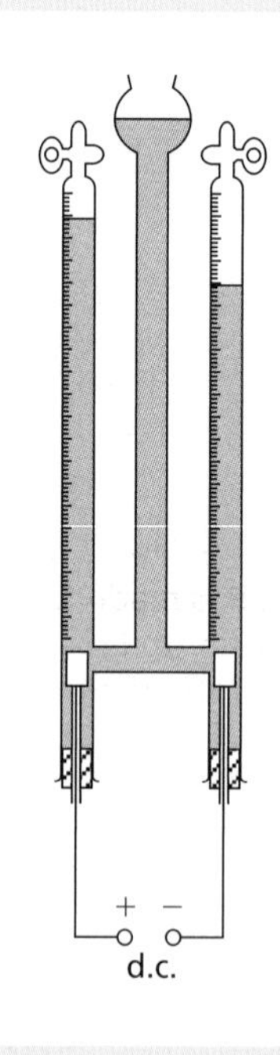


**a** The water, $H_2O$, is acidified with dilute sulphuric acid, $H_2SO_4$. The ions present in the acidified water are $\mathbf{H^+}$, $\mathbf{OH^-}$ and $\mathbf{SO_4^{2-}}$. Copy and finish the table by writing the symbols of the ions which move to each electrode. **(2)**

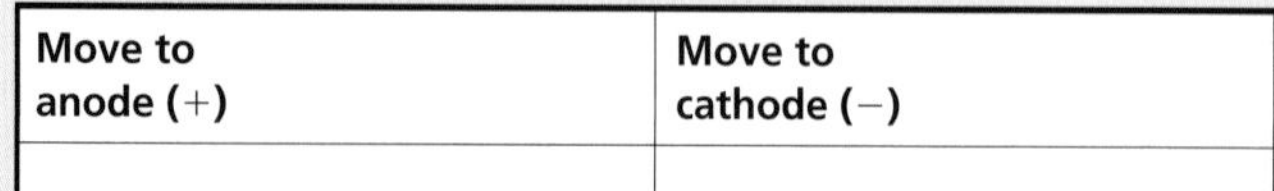

| Move to anode (+) | Move to cathode (−) |
|---|---|
| | |

**b** The electrolysis was carried out for ten minutes.
The hydrogen and oxygen were then removed.
The experiment was started again.
The volumes of hydrogen and oxygen collecting were measured every two minutes.
The current was not changed.
The table shows how much hydrogen was formed.

| Time/min | 0 | 2 | 4 | 6 | 8 | 10 |
|---|---|---|---|---|---|---|
| Volume of hydrogen/$cm^3$ | 0 | 4 | 8 | 12 | 17 | 20 |

(i) Plot these results on a grid.
Draw the best fit line for this graph. **(3)**
(ii) Draw another straight line on the grid to show the volume of oxygen collected during the experiment. **(1)**
(iii) If the electrolysis is repeated using fresh acidified water, less oxygen is collected.
Suggest why. **(1)**

**c** Copy and complete this ionic equation for the reaction taking place at the anode.

______ $OH^- \rightarrow$ ______ $H_2O + O_2 + 4$ ______ **(3)**

**d** The reaction at the cathode is represented by this equation.

$$2H^+ + 2e^- \rightarrow H_2$$

In an electrolysis 20 $cm^3$ of hydrogen are released, measured at room temperature and pressure.
(i) Calculate the number of moles in this 20 $cm^3$ of hydrogen gas.
(1 mole of gas occupies 24 000 $cm^3$ at room temperature and pressure.) **(1)**
(ii) Show, by calculation, that the quantity of electricity required to produce this amount of hydrogen gas is 160 coulombs.
(One faraday is 96 000 coulombs.) **(2)**
(iii) What current would be needed to produce this amount of hydrogen gas in 5 minutes? You must show how you work out your answer. **(2)**
*(OCR, GCSE, Paper 4, June 2000)*

**30** A small leaf is to be made into a piece of jewellery by plating it with silver. The leaf is first covered in graphite paste. Then it is made the cathode (negative electrode) in a circuit set up for electrolysis.
Here is a diagram of the circuit.

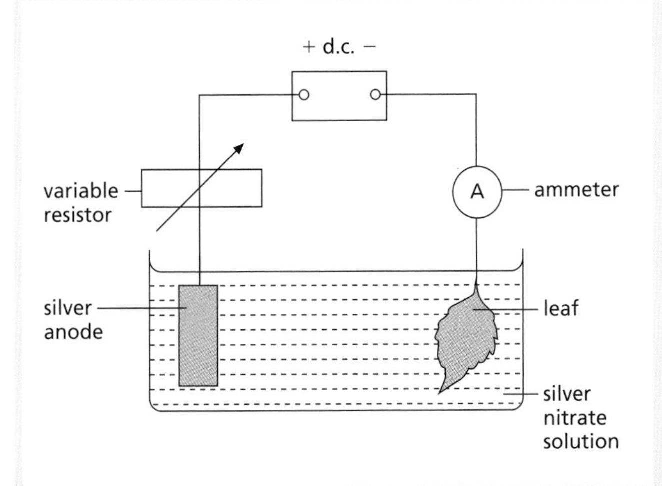


**a** Why is the leaf coated with graphite paste? **(1)**
**b** In carrying out this electrolysis, a constant current is used. How is the current kept constant in this circuit? **(1)**
**c** Which of the following conditions is likely to give the best coating?

**A** a large current for a long time
**B** a large current for a short time
**C** a small current for a long time
**D** a small current for a short time

**d** Direct current (d.c.) is used in this process. Why is alternating current (a.c.) **not** used? **(1)**
**e** A constant current of 0.1 A passes through the electrolysis cell.
0.108 g of silver is deposited on the leaf.
The reaction at the cathode is shown by the equation.

$$Ag^+(aq) + e^- \rightarrow Ag(s)$$

(i) How many moles of electrons are needed to deposit 1 mole of silver? **(1)**
(ii) How many **coulombs** are needed to deposit 0.108 g of silver?
You **must** show how you work out your answer.
(1 mole of electrons = 96 000 coulombs; relative atomic mass: Ag = 108.) **(2)**
(iii) For how long must a constant current of 0.1 A pass through the cell to deposit 0.108 g of silver? You **must** show how you work out your answer. **(2)**
*(OCR, GCSE, Paper 4, June 2000)*

**31 a** The diagram shows a method for obtaining pure copper from impure copper.

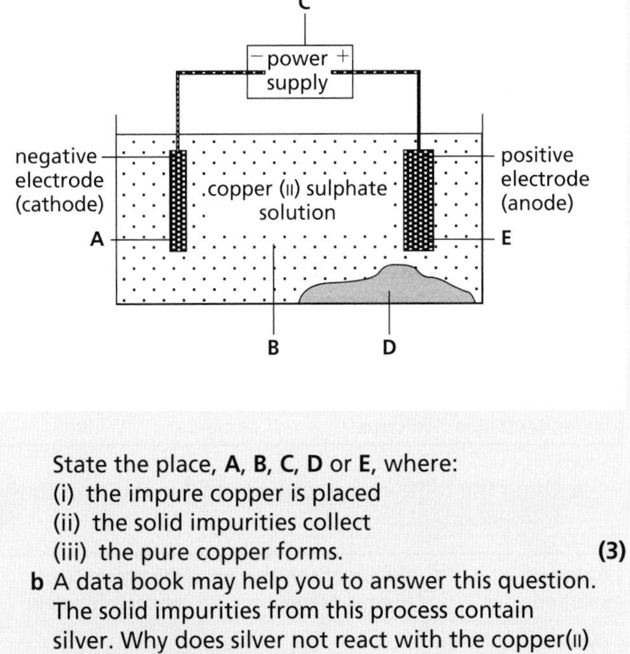


State the place, **A**, **B**, **C**, **D** or **E**, where:
(i) the impure copper is placed
(ii) the solid impurities collect
(iii) the pure copper forms. **(3)**

**b** A data book may help you to answer this question.
The solid impurities from this process contain silver. Why does silver not react with the copper(II) sulphate solution? **(1)**
**c** One of the reactions which takes place in this process is represented by the equation.

$$Cu(s) - 2e^- \rightarrow Cu^{2+}(aq)$$

Why is this reaction described as oxidation? **(1)**

**d** Copper and silver conduct electricity. Explain, in terms of particles, why they are good conductors of electricity. **(3)**

**e** Silver is changed into silver chloride in two stages.
(i) In Stage 1 silver is reacted with nitric acid to make silver nitrate, water and nitrogen oxide (NO).
Copy and balance the symbol equation for this reaction. **(1)**

$$3Ag + 4HNO_3 \rightarrow \_\_\_\_\_\_ AgNO_3 + \_\_\_\_\_\_ H_2O + NO$$

(ii) Copy and complete the word equation for Stage 2. **(1)**

silver nitrate + ______ → silver chloride + sodium nitrate

**f** Photographic film can be made by coating paper with silver chloride. Explain what happens to the silver chloride when the film is exposed to light. **(2)**

*(AQA, GCSE, Paper 2372, June 2000)*

**32 a** The relative molecular mass ($M_r$) of sodium hydroxide, NaOH, is 40. If 8.0 g of sodium hydroxide are present in 1000 cm³ (1 dm³) of aqueous solution
(i) What is the concentration of this solution in mol $dm^{-3}$? **(1)**
(ii) How many moles would be present in 25 cm³ of the sodium hydroxide solution? **(2)**

**b** Ethanoic acid, $CH_3COOH$, and sodium hydroxide react in a 1 : 1 ratio. When ethanoic acid was neutralised by sodium hydroxide solution, it was found that 25 cm³ of the sodium hydroxide solution required 20 cm³ of ethanoic acid solution.
(i) How many moles of ethanoic acid are present in 20 cm³ of ethanoic acid solution? **(1)**
(ii) How many moles would be present in 1 dm³ of ethanoic acid solution? **(2)**

**c** If the relative molecular mass ($M_r$) of ethanoic acid is 60, calculate the number of **grams** of ethanoic acid present in 1 dm³ of the solution. **(2)**

*(WJEC, GCSE, June 2000)*

**33** Methyl orange and phenolphthalein are two useful indicators in the laboratory. They change colour when they are mixed with acids or alkalis.

| Indicator | Colour in acid | Colour in neutral solution | Colour in alkali |
|---|---|---|---|
| Methyl orange | Pink | Yellow | Yellow |
| Phenolphthalein | Colourless | Colourless | Red |

**a** Using the table above, copy and fill in the gaps in the following table. **(4)**

| Aqueous solution | Colour of solution in methyl orange | the presence of phenolphthalein |
|---|---|---|
| Hydrogen chloride | | |
| Ammonia | | |
| Sodium chloride | | |
| Sulphur dioxide | | |

**b** A solution is made up by mixing 25 cm³ of NaOH(aq) with 15 cm³ of $H_2SO_4$(aq). Three drops of methyl orange indicator are added. The NaOH(aq) and $H_2SO_4$(aq) were of equal concentration before addition. What colour would you expect for the resultant solution? **(1)**

**c** Would the indicators above be suitable to decide if a solution of bromine in water is acidic, neutral or alkaline? Explain your answer. **(3)**

**d** Copper nitrate, a blue crystalline solid, may be obtained by reacting copper carbonate with dilute nitric acid and crystallising the resulting solution. The following method is often followed:

50 cm³ of nitric acid, concentration 1 mol/dm³ (moles per litre) are placed in a 250 cm³ beaker and copper carbonate is added with stirring until no further reaction takes place. The mixture is filtered into an evaporating basin and the solution reduced to about $\frac{1}{3}$ in volume by evaporation. The liquid is left to cool until blue crystals have formed. The crystals are filtered off, washed with a little cold water and dried between filter papers. Finally the crystals are weighed.
(i) Write a balanced, symbol equation for the reaction. **(2)**
(ii) How would you know when the reaction had finished? **(1)**
(iii) What was the purpose of the first filtration? **(2)**
(iv) Why was the final filtrate **not** evaporated to dryness? **(1)**
(v) In an experiment to make copper nitrate a student obtained 5.50 g of crystals. Calculations show that 6.05 g of crystals could have been obtained from the amounts of the starting materials which were used. Suggest **two** reasons why the student did not obtain the full 6.05 g of crystals. **(2)**

*(CCEA, GCSE, Paper 2, June 2000)*

**34 a** Five solutions, **A–E**, were tested with universal indicator solution, to find their pH. The results are shown below.

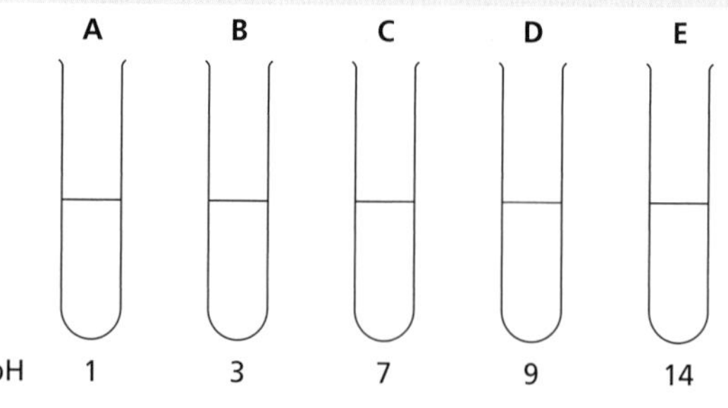


(i) The five solutions were known to be ammonia solution, potassium hydroxide, sodium chloride, sulphuric acid and vinegar. Identify each of the solutions **A–E**. **(5)**
(ii) Give a balanced symbol equation for the reaction which occurs when solutions **A** and **D** are mixed. **(2)**

**b** Complete the following passage by inserting the correct words/formulae in the spaces.
Acids are compounds which dissolve in water producing (i) _______ ions which have formula (ii) _______.
Alkalis are compounds which dissolve in water producing (iii) _______ ions which have formula (iv) _______.
The ionic equation for the reaction which takes place on mixing acids and alkalis is (v) _______. **(6)**

**c** Describe in detail how you would prepare a **pure dry** sample of sodium chloride crystals in the laboratory starting with solutions of sodium hydroxide and hydrochloric acid. **(8)**

**d** (i) Describe a test which you would use to confirm the presence of sodium ions in a sample of the salt prepared in part **c**. **(3)**
(ii) Describe a test which you would use to confirm the presence of chloride ions in a sample of the salt prepared in part **c**. **(4)**

*(CCEA, GCSE, Paper 2, June 1999)*

**35 a** Aluminium hydroxide reacts with both acids and alkalis. Give balanced symbol equations for the following reactions of aluminium hydroxide
(i) with dilute hydrochloric acid **(2)**
(ii) with sodium hydroxide solution. **(2)**

**b** Sodium hydroxide solution is added in excess to a green solution of a metal salt. A green precipitate is observed which slowly turns reddish-brown on leaving exposed to air.
(i) Identify the metal ion present in the original green solution. **(1)**
(ii) Explain why the green precipitate changed colour. **(3)**

*(CCEA, GCSE, Paper 2, June 1999)*

**36 a** The following table shows the colours of universal indicator at different pH values.

| Colour | Red | Orange | Yellow | Green | Blue | Navy blue | Purple |
|---|---|---|---|---|---|---|---|
| pH | 0–2 | 3–4 | 5–6 | 7 | 8–9 | 10–12 | 13–14 |

(i) Sodium carbonate solution turns universal indicator navy blue. Give the pH range of this solution. **(1)**
(ii) The pH of hydrochloric acid solution is 1. Give the colour universal indicator would turn in hydrochloric acid solution. **(1)**

**b** (i) Name the colourless solution and colourless gas formed when sodium carbonate reacts with dilute hydrochloric acid. **(2)**
(ii) Describe a test you would carry out to identify the gas formed. **(1)**
(iii) What would you **see** if the gas was present? **(1)**

*(WJEC, GCSE, June 2000)*

**37 a** The solubility of a solid in water changes as the temperature changes. What do you understand by the term 'solubility'? **(4)**

**b** A student investigates the solubility of two different solids at various temperatures. The table that follows shows the results obtained.

| Temperature/°C | Solubility of solid/g/100 g water | |
|---|---|---|
| | Sodium chloride | Potassium chlorate |
| 0 | 33.0 | 3.5 |
| 20 | 33.5 | 7.5 |
| 40 | 34.0 | 14.0 |
| 60 | 34.5 | 24.0 |
| 80 | 35.0 | 37.5 |

(i) Plot the results for the two different solids on graph paper. **(8)**
(ii) Compare the effect of increasing the temperature on the solubility of **each** solid. **(2)**
(iii) At which temperature are the solubilities of the two solids the same? **(1)**
(iv) What is the solubility of the two solids at the temperature in part (iii)? **(1)**
(v) If a saturated solution of potassium chlorate containing 50 g of water is cooled from 70 °C to 30 °C, what mass of the solid would crystallise? **(4)**
(vi) If a solution is made by dissolving 15 g of each solid in the same 100 g of water at 70 °C and the solution is cooled to 20 °C, crystals are observed. Which solid crystallises? Explain your answer with reference to both solids. **(4)**

*(CCEA, GCSE, Paper 1, June 2000)*

**38** *To obtain full marks the steps in the calculations must be shown.*
A series of experiments was carried out on a given solution of sulphuric acid.

**a** In the first of these experiments a titration was carried out to determine the concentration of the acid. 20.0 $cm^3$ of the acid required 32.0 $cm^3$ of 0.15 moles per litre ($mol/dm^3$) sodium hydroxide solution.
(i) Calculate the number of moles of sodium hydroxide used in the titration. **(2)**
The equation of the titration reaction is

$$2NaOH + H_2SO_4 \rightarrow Na_2SO_4 + 2H_2O$$

(ii) Calculate the number of moles of $H_2SO_4$ titrated. **(3)**
(iii) Calculate the concentration of the sulphuric acid used in moles per litre ($mol/dm^3$). **(2)**

**b** In a second experiment, a group of pupils proceeded to make nickel sulphate crystals by adding an excess of nickel carbonate to 50.0 $cm^3$ of the **same** sulphuric acid. The equation for the reaction is:

$$H_2SO_4 + NiCO_3 \rightarrow NiSO_4 + H_2O + CO_2$$

(i) Calculate the number of moles of sulphuric acid used. **(2)**
(ii) What is the **minimum** mass of $NiCO_3$ which must be used to react with all the sulphuric acid?

(Ni = 59, C = 12, O = 16) **(4)**

(iii) Nickel sulphate crystallises as $NiSO_4.7H_2O$. What is the **maximum** mass of crystals which could be obtained from this preparation.

(Ni = 59, S = 32, O = 16, H = 1) **(4)**

(iv) At the end of the experiment, one student collected 1.20 g of crystals. Suggest **one** reason why the mass collected was less than that calculated in part **b** (iii) above. **(1)**

c A possible way to make ammonium sulphate is to react ammonia gas with sulphuric acid.

If 25.0 $cm^3$ of the sulphuric acid from part **a** were used, what volume, in $cm^3$, of ammonia gas would be needed to react exactly with the acid.
(1 mole of any gas occupies 24 $dm^3$ at room temperature.)
The equation for the reaction is

$$H_2SO_4 + 2NH_3 \rightarrow (NH_4)_2SO_4$$ **(6)**

*(CCEA, GCSE, Paper 1, June 1999)*

**39** A student carried out a titration to find the concentration of a solution of sulphuric acid. 25.0 $cm^3$ of the sulphuric acid solution was neutralised exactly by 34.0 $cm^3$ of a potassium hydroxide solution of concentration 2.0 $mol/dm^3$. The equation for the reaction is:

$$2KOH(aq) + H_2SO_4(aq) \rightarrow K_2SO_4(aq) + 2H_2O(l)$$

a Describe the experimental procedure for the titration carried out by the student. **(4)**

b Calculate the number of moles of potassium hydroxide used. **(2)**

c Calculate the concentration of the sulphuric acid in $mol/dm^3$. **(3)**

*(SEG, GCSE, Paper 5, Summer 2000)*

**40** Limestone is a useful mineral. Every day, large amounts of limestone are heated in limekilns to produce lime. Lime is used in the manufacture of iron, cement and glass and for neutralising acidic soils.

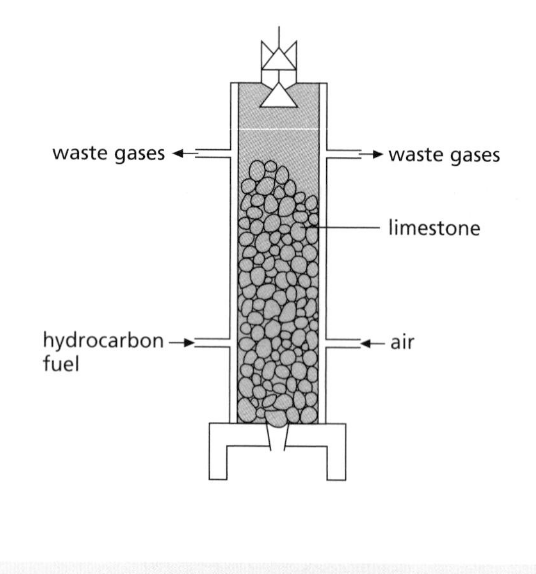

$$CaCO_3 \rightleftharpoons CaO + CO_2$$

a (i) The decomposition of limestone is a **reversible** reaction. Explain what this means. **(2)**

(ii) Calculate the mass of lime, CaO, that would be produced from 250 tonnes of limestone, $CaCO_3$. (Relative atomic masses: C 12; O 16; Ca 40.) **(3)**

b Large amounts of carbon dioxide are produced when using limekilns, both from burning hydrocarbons to provide heat and from the decomposition of the limestone.

Give **two** ways by which carbon dioxide is removed naturally from the atmosphere. **(2)**

c Limestone is added to the blast furnace during the extraction of iron. The lime formed reacts as shown by the equation.

$$CaO + SiO_2 \rightarrow CaSiO_3$$

Describe and explain the importance of this reaction in the blast furnace. **(3)**

*(SEG, GCSE, Paper 6/4, Summer 2000)*

**41** a (i) Cement is an important building material. Describe how cement is made. **(3)**

(ii) When cement is mixed with water, sand and crushed rock, a slow chemical reaction takes place which produces another important building material. Name this important building material. **(1)**

b Mortar is used to hold bricks and stonework in position. It is a much older building material than cement and has been used since Roman times. Mortar is made by mixing calcium hydroxide (slaked lime), sand and water into a paste. Mortar hardens over many years on standing in air. This is due to two processes.

- The water evaporates.
- The slaked lime reacts **slowly** with carbon dioxide in the air to form calcium carbonate.

$$Ca(OH)_2(s) + CO_2(g) \rightarrow CaCO_3(s) + H_2O(l)$$

(i) Suggest why the calcium carbonate forms very slowly. **(1)**

(ii) Use your answer to part **b** (i) and a data book to help you answer this question.

An archaeologist found two pieces of mortar. One piece was very old and from a Roman villa built about AD 300. The other piece was from a modern cottage built in 1995.
Describe and give the result of an experiment the archaeologist could do which would prove that one of the pieces of mortar was much older than the other piece. **(3)**

(iii) The outer layer of mortar slowly changes. The calcium carbonate reacts with carbon dioxide and water in the air to form calcium hydrogencarbonate.

$$CaCO_3(s) + H_2O(l) + CO_2(g) \rightarrow Ca(HCO_3)_2(aq)$$

Rainwater removes the calcium hydrogencarbonate from between the bricks. The gaps then need to be filled in (pointed) with new mortar or cement.
Use the equation to suggest why calcium hydrogencarbonate is easily removed by rainwater. **(1)**

*(AQA, GCSE, Paper 2372, June 2000)*

**42** The diagram shows three calcium compounds.

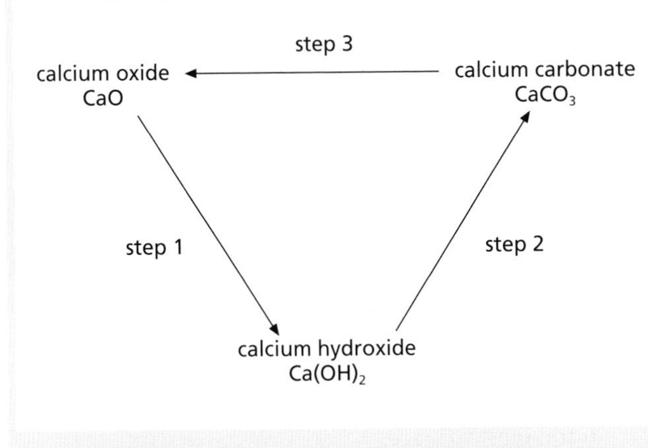


**a** (i) Which compound occurs widely in nature? **(1)**
(ii) Which compound is present in solution in limewater? **(1)**
(iii) What state symbol is used to show limewater is a solution? **(1)**
(iv) Which compound is formed on boiling temporary hard water? **(1)**
(v) What is added to calcium oxide to carry out step 1? **(1)**
(vi) How can step 3 be carried out? **(1)**
(vii) Which step occurs during the test for carbon dioxide? **(1)**

**b** Two different white powders are thought to be calcium carbonate and calcium hydroxide.
(i) Describe a test to prove that both powders are calcium compounds. **(2)**
(ii) Describe a test to find out which one of the powders is calcium carbonate. **(2)**

*(Edexcel, GCSE, Paper 2F, June 2000)*

**43** Analysis of a water supply produced the following data.

| Analysis | | |
|---|---|---|
| **Ion** | | **Concentration in mg per dm$^3$** |
| Calcium | ($Ca^{2+}$) | 104.0 |
| Magnesium | ($Mg^{2+}$) | 1.4 |
| Sodium | ($Na^{+}$) | 8.0 |
| Potassium | ($K^{+}$) | 1.0 |
| Iron | ($Fe^{3+}$) | 0.02 |
| Hydrogencarbonate | ($HCO_3^{-}$) | 293.0 |
| Chloride | ($Cl^{-}$) | 15.0 |
| Sulphate | ($SO_4^{2-}$) | 12.0 |
| Nitrate | ($NO_3^{-}$) | 5.0 |
| Fluoride | ($F^{-}$) | 0.1 |

**a** (i) Which **two** elements in this water supply are in Group 2 of the periodic table? **(1)**
(ii) Write down the name of a transition metal which is present in this water supply. **(1)**
(iii) Write down the names of **two** ions in this water supply which can combine together to form a compound of formula $XY_2$ where X is a metal. **(2)**

**b** Sana tests for the ions found in this water supply. She concentrates the ions by evaporating some of the water before doing the tests. Finish the table. There are **four** gaps.

| Test | What is seen | Ion present | |
|---|---|---|---|
| Add dilute hydrochloric acid.<br><br>Test the gas given off with limewater. | It fizzes, and the gas given off turns the limewater (i) ______. | $HCO_3^{-}$ | **(1)** |
| Add dilute nitric acid and silver nitrate solution. | A white precipitate is formed. | (ii) ______ | **(1)** |
| Add dilute nitric acid and barium chloride solution. | A (iii) ______ precipitate is formed. | $SO_4^{2-}$ | **(2)** |
| Add sodium hydroxide solution. | A brown precipitate is formed. | (iv) ______ | **(2)** |

**c** A hospital process has to let 30 dm$^3$ of water flow through pipes in 24 hours. If more than 120 mg per hour of calcium flows through the pipes, the process becomes damaged.
Decide whether this water supply is suitable for use in the process. You **must** show how you work out your answer. **(2)**

**d** **All** the hardness in this water has to be removed before the water is used in a steam-iron.
Which **two** of these methods could be used successfully? **(2)**

**A** Boil the water, cool it and then filter it.
**B** Distil the water.
**C** Trickle the water down a zeolite ion-exchange column.
**D** Treat the water with calcium hydroxide then filter it.

**e** Glyn always makes tea using this water supply. He boils the water in an electric kettle.
Describe and explain the problem caused by the frequent use of this water supply in the kettle. **(3)**

*(OCR, GCSE, Paper 4, June 2000)*

**44** A sample of tap water contains the following dissolved salts:

**calcium hydrogencarbonate**
**calcium sulphate**
**magnesium sulphate**
**potassium chloride**
**sodium chloride**

**a** (i) Hard water can be formed when water is in contact with the rock, gypsum. Which salt in the list is present in gypsum? **(1)**
(ii) Name **two** other salts in the list which make water hard. **(2)**

**b** Some methods of treating water are given below.

**A** adding chlorine
**B** adding a fluoride (fluoridation)
**C** adding sodium carbonate
**D** boiling the water

Which **one** of the methods (**A**, **B**, **C** or **D**) removes:
(i) temporary hardness but **not** permanent hardness? **(1)**
(ii) both temporary hardness **and** permanent hardness? **(1)**

**c** Ten drops of soap solution are shaken with a sample of hard tap water. The mixture turns cloudy but it does not form a lather. On shaking with ten more drops of soap solution, a lather is formed.
(i) What causes the cloudiness when soap solution is first mixed with this tap water? **(1)**
(ii) Predict what you would **see** when distilled water is shaken with ten drops of soap solution. Explain your answer. **(2)**

**d** Calcium hydrogencarbonate is formed when water and carbon dioxide are in contact with limestone. This reaction removes carbon dioxide gas from the atmosphere.
(i) Write the word equation for this reaction. **(1)**
(ii) Plants are also able to remove carbon dioxide from the atmosphere, forming glucose and oxygen gas in the process.
Give the name of this process and state **one** essential condition needed for it to take place. **(2)**

**e** If the percentage of carbon dioxide in the Earth's atmosphere increases, the average temperature of the atmosphere may also increase. What is the name given to this effect? **(1)**

(*Edexcel, GCSE, Paper 2F, June 2000*)

**45** The equations show three displacement reactions involving metals and solutions of metal nitrates.

$$Cu + 2AgNO_3 \rightarrow Cu(NO_3)_2 + 2Ag$$
$$Pb + Cu(NO_3)_2 \rightarrow Pb(NO_3)_2 + Cu$$
$$Zn + Pb(NO_3)_2 \rightarrow Zn(NO_3)_2 + Pb$$

**a** (i) Use this information to find the order of reactivity of the four metals, with the most reactive first. **(1)**
(ii) Calculate the mass of copper needed to displace 5.0 g of silver from silver nitrate solution.

$$Cu + 2AgNO_3 \rightarrow Cu(NO_3)_2 + 2Ag$$

(Relative atomic masses: Ag = 108, Cu = 63.5.) **(3)**
(iii) Write an equation, including state symbols, for the reaction between lead and aqueous silver nitrate. **(3)**

**b** Zinc and bromine undergo displacement reactions as shown by the equations below. Zinc is oxidised and bromine is reduced during these reactions.

$$Zn(s) + 2H^+(aq) \rightarrow Zn^{2+}(aq) + H_2(g)$$
$$Br_2(aq) + 2I^-(aq) \rightarrow 2Br^-(aq) + I_2(aq)$$

(i) Name one compound containing $H^+$ ions and another containing $I^-$ ions, which would be suitable for these reactions. **(2)**
(ii) Explain why zinc is said to be oxidised in its reaction. **(1)**
(iii) Complete the half equation to show the oxidising action of bromine.
______ $I^- \rightarrow I_2 +$ ______ **(2)**

(*Edexcel, GCSE, Paper 3H, June 1999*)

**46** The use of most metals depends on their reactivity.

**a** Reactivity of metals can be compared by using displacement reactions. The reactions of four metals **R**, **S**, **T** and **U** with their salt solutions are shown. (These letters are not the chemical symbols for the metals.)

| Metal salt solution | Metal | | | |
|---|---|---|---|---|
| | R | S | T | U |
| R | | ✗ | ✗ | ✓ |
| S | ✓ | | ✗ | ✓ |
| T | ✓ | ✓ | | ✓ |
| U | ✗ | ✗ | ✗ | |

✓ = reaction ✗ = no reaction

(i) Use the information to arrange the metals **R**, **S**, **T** and **U** in order of reactivity, with the most reactive first. **(2)**
(ii) Metal **R** was zinc and metal **T** was copper. State what causes the colour changes that you see when zinc reacts with copper sulphate solution. **(3)**

(*SEG, GCSE, Paper 6/4, June 1999*)

**47** By observing the reactions of metals with water and dilute sulphuric acid it is possible to put metals in order of their reactivity.

**a** **A**, **B**, **C** and **D** represent four metals.

| Metal | Reaction with water | Reaction with dilute sulphuric acid |
|---|---|---|
| A | No reaction | Reacts slowly at first |
| B | No reaction | No reaction |
| C | Little or no reaction | Reacts quickly |
| D | Vigorous reaction | Violent – dangerous reaction |

(i) Put metals **A**, **B**, **C** and **D** in order of their reactivity with the most reactive first. **(2)**
(ii) The metals used were copper, magnesium, sodium and zinc. Use the information in the table to identify which of these metals was **A**, **B**, **C** and **D**. **(2)**

**b** A student tried to make some magnesium sulphate. Excess magnesium was added to dilute sulphuric acid. During this reaction fizzing was observed due to the production of a gas.

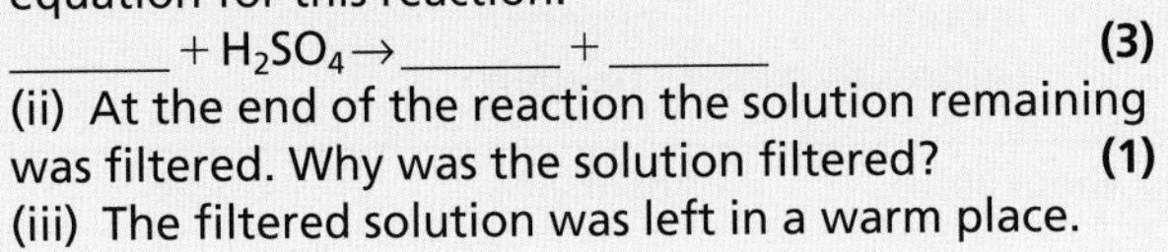


(i) Copy, complete and balance the chemical equation for this reaction.

_______ + $H_2SO_4 \rightarrow$ _______ + _______ **(3)**

(ii) At the end of the reaction the solution remaining was filtered. Why was the solution filtered? **(1)**

(iii) The filtered solution was left in a warm place.

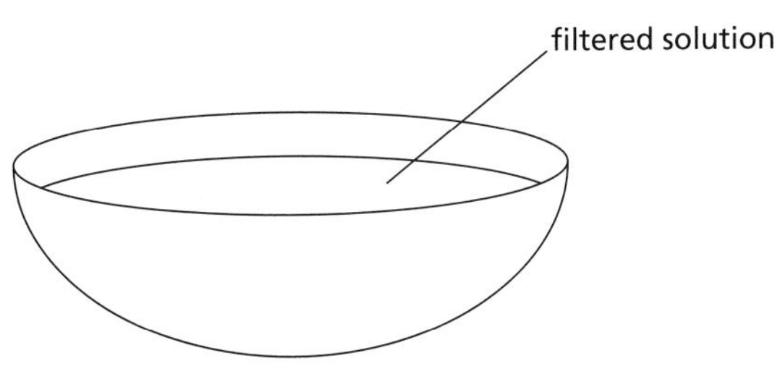


Explain why the filtered solution was left in a warm place. **(2)**

*(SEG, GCSE, Paper 6/4, Summer 2000)*

**48** Most of the cans used for drinks are made from aluminium.

**a** (i) Aluminium is an element. What does this mean? **(1)**

(ii) Metals are malleable and this makes them suitable to make drinks cans. What does malleable mean? **(1)**

**b** The arrangement of electrons in an aluminium atom is:

aluminium atom

(i) On the diagram label the nucleus of the aluminium atom. **(1)**

(ii) How many protons does an aluminium atom have? **(1)**

(iii) To which group of the periodic table does aluminium belong? **(1)**

(iv) To which period of the periodic table does aluminium belong? **(1)**

**c** The reaction between aluminium and iron oxide is used to join lengths of railway track. It is called the thermit reaction.

$$Fe_2O_3(s) + 2Al(s) \rightarrow Al_2O_3(s) + 2Fe(l)$$

(i) Why does aluminium react with iron oxide? **(1)**

(ii) What does the (l) after Fe in the chemical equation mean? **(1)**

(iii) Suggest why the thermit reaction can be used to join lengths of railway track. **(2)**

*(SEG, GCSE, Paper 6/4, June 1999)*

**49 a** Describe how steel is manufactured using molten iron obtained from the blast furnace.

Your answer should include:

- the types of reaction occurring;
- the details of the conditions used;
- energy changes involved. **(5)**

**b** Suggest **two** factors which influence the location of plants associated with the manufacture of steel. **(2)**

**c** Give **two** reasons why it is important to recycle steel. **(2)**

*(AQA, GCSE, Paper 2372, June 2000)*

**50 a** Iron is extracted from iron(III) oxide in a blast furnace. One of the main reactions in the furnace is

$$Fe_2O_3 + 3CO \rightarrow 2Fe + 3CO_2$$

(Relative atomic mass: $A_r(O) = 16$; $A_r(Fe) = 56$.)

(i) Calculate the relative molecular mass ($M_r$) of iron(III) oxide, $Fe_2O_3$. **(2)**

(ii) Use the given equation for the extraction of iron to calculate how many tonnes of metal could be obtained from 80 tonnes of iron(III) oxide. **(3)**

**b** 14.4 g of another oxide of iron was found to contain 11.2 g of iron.

Calculate the simplest formula for this oxide of iron.

**Show your working.**

(Relative atomic mass: $A_r(O) = 16$; $A_r(Fe) = 56$.) **(3)**

**c** Iron(II) and iron(III) salts in solution form different coloured precipitates with sodium hydroxide solution.

State the **colour** of the precipitate formed with

(i) iron(II) salts

(ii) iron(III) salts. **(2)**

*(WJEC, GCSE, 0125/2, June 1999)*

**51 a** Three test-tubes were set up as shown to investigate the rusting of iron. After one week rusting was observed in tube **C** but not in tubes **A** and **B**.

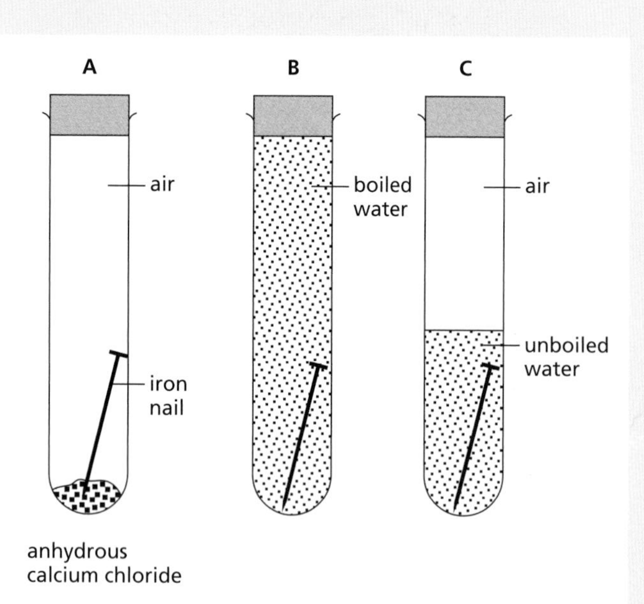


(i) Why is anhydrous calcium chloride placed in tube **A**? **(1)**
(ii) Why is the water in tube **B** boiled? **(2)**
(iii) What would you expect to happen to the nail in tube **B** if the stopper was removed and the tube left for a further week? **(1)**
(iv) From the results of the above experiments state the conditions necessary for the rusting of iron. **(2)**

**b** Rusting may be prevented using a variety of methods. Which method is **normally** used to protect the following from rusting?
(i) School railings
(ii) Garden tools
(iii) Bicycle handlebars
(iv) Scissor blades **(4)**

**c** Underground pipes made from steel are protected from rusting by attaching bars of magnesium to the pipes.
(i) Give the name of this type of protection. **(1)**
(ii) Explain why magnesium is used. **(2)**
(iii) Name **two** other metals which could be used to protect the iron instead of magnesium. **(2)**

**d** Many foods are packed in 'tin' cans which are made from steel coated with a fine layer of tin.
(i) Give **two** reasons why the cans are coated with tin. **(2)**
(ii) The tin is deposited on the steel in a process called _______. **(1)**

**e** Ordinary steel exhaust systems rust much faster than any other part of a car. Explain this observation. **(3)**

**f** What is the chemical name for rust? **(2)**

**g** Is rusting of iron an oxidation or reduction process? Explain your answer in terms of electrons. **(3)**

*(CCEA, GCSE, Paper 1, June 1999)*

**52 a** The transition elements have the typical properties of metals.
(i) Transition metals are good conductors of electricity. Explain why. **(2)**
(ii) Transition metals have high melting points. Explain why. **(2)**

**b** Iron is a transition metal. Iron, because of its strength, is used in the building of ships' hulls. The main problem of using iron is that it rusts unless protected.
(i) Explain how painting stops an iron hull from rusting. **(2)**
(ii) Explain how attaching zinc blocks to an iron hull stops it from rusting. **(2)**

*(SEG, GCSE, Paper 6/4, June 1999)*

**53** This question is about the composition of the atmosphere.

**a** Copy and finish the table to show the composition of the atmosphere.
Use **all** the names from the list.

**argon** **nitrogen**
**carbon dioxide** **oxygen** **(3)**

| About 80% | About 20% | Traces |
|---|---|---|
| | | |

**b** Millions of years ago the atmosphere of the Earth had a different composition.
(i) How have the amounts of carbon dioxide and oxygen changed since then? **(1)**
(ii) Suggest how the evolution of plants has helped to produce the present composition of the atmosphere. **(3)**

**c** The composition of the atmosphere is also affected by the weathering of vast amounts of igneous rocks containing iron compounds. Some of the iron compounds are changed into iron(II) carbonate. This is readily oxidised, on contact with air, to form iron(III) oxide. The equation for this reaction can be represented by

$$4FeCO_3(s) + O_2(g) \rightarrow 2Fe_2O_3(s) + 4CO_2(g)$$

How does this process affect the composition of the atmosphere? **(2)**

*(OCR, GCSE, Paper 2, June 1999)*

**54** For 200 million years the proportions of the different gases in the atmosphere have been much the same as they are today.

Over the past 150 years the amount of carbon dioxide in the atmosphere has increased from 0.03% to 0.04%.

**a** Describe how carbon dioxide is released into the atmosphere
(i) by human and industrial activity **(2)**
(ii) from carbonate rocks by geological activity. **(2)**

**b** Explain how the seas and oceans can decrease the amount of carbon dioxide in the atmosphere. **(3)**

**c** (i) Give **one** reason why the amount of carbon dioxide in the atmosphere is increasing gradually. **(1)**
(ii) Give **one** effect that increasing levels of carbon dioxide in the atmosphere may have on the environment. **(1)**

*(AQA, GCSE, Paper 2372, June 1999)*

**55 a** The main components of the **original** Earth's atmosphere were carbon dioxide and water. The pie chart below shows the approximate composition of dry air in the atmosphere **today**.

(i) State the **source** of the original Earth's atmosphere. **(1)**

noble gases

oxygen

carbon dioxide

nitrogen

(ii) Name the gas in the pie chart which is entirely biological in origin. **(1)**

(iii) There has been a drastic reduction in the amount of water vapour in the air over geological time. Explain how this decrease occurred. **(2)**

**b** The Earth's atmosphere is surrounded by a layer of ozone. State the importance of the ozone layer to our health. **(1)**

**c** Give **one** use to which oxygen is put. **(1)**

**d** The carbon cycle helps to maintain atmospheric composition. Name **two** processes which have the opposite effect to photosynthesis in the carbon cycle. **(2)**

*(WJEC, GCSE, June 2000)*

**56** This graph shows how we think the amounts of gases in our atmosphere have changed over millions of years.

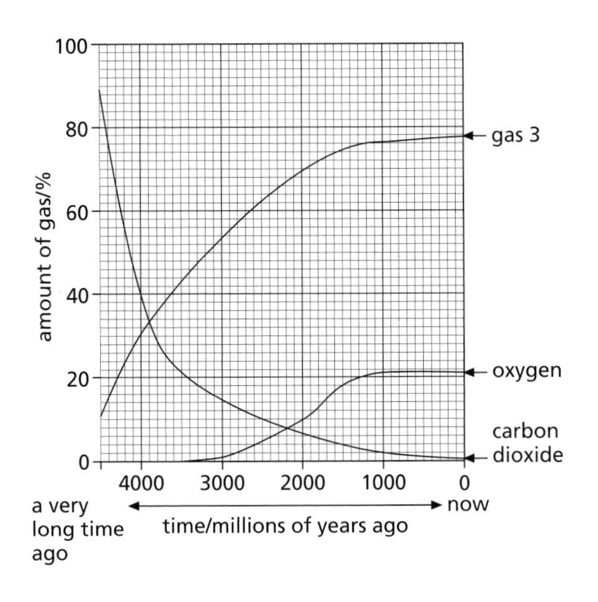


**a** The first plants appeared on Earth 3000 million years ago.

(i) What happened to the amount of oxygen after the plants appeared? **(1)**

(ii) After animals appeared on Earth the amount of oxygen stayed the same. When did land animals appear? **(1)**

**b** Describe how the amount of carbon dioxide has changed. **(2)**

**c** Scientists think the amount of carbon dioxide in the atmosphere has been increasing for the last 200 years. What environmental problem could this cause? **(1)**

**d** There is another gas on the graph, labelled 'gas 3'.

(i) How much of 'gas 3' is in our atmosphere today? **(1)**

(ii) What is the name of this gas? **(1)**

*(OCR, GCSE, Paper 1, June 2000)*

**57 a** Water is essential for life. Two of the stages involved in the treatment of public water supplies are the use of filter beds and chlorination.

State the purpose of

(i) filter beds **(1)**

(ii) chlorination. **(1)**

**b** Describe how you would show that a colourless liquid is **pure** water. Include in your answer the result you would expect. **(2)**

**c** (i) A small amount of soap solution was added to samples of soft water and hard water and the mixtures were shaken. How would you identify the hard water? **(1)**

(ii) What is dissolved in water which makes it hard? **(1)**

(iii) Give **one** advantage of hard water. **(1)**

(iv) Give **one** method of softening hard water other than by boiling. **(1)**

*(WJEC, GCSE, June 2000)*

**58** Sue studied the reaction between calcium carbonate and hydrochloric acid. The equation for this reaction is

$$CaCO_3(s) + 2HCl(aq) \rightarrow CaCl_2(aq) + H_2O(l) + CO_2(g)$$

Sue carried out three experiments to study the effect of the surface area of the calcium carbonate. She used calcium carbonate in the form of small lumps or medium lumps or large lumps. In each experiment she used the same mass of calcium carbonate. She measured the volume of gas collected in a gas syringe at intervals.

**a** Write down **two** other things she should keep the same in each experiment. **(2)**

**b** In each experiment all the calcium carbonate had reacted within five minutes. Sue's graph line for medium lumps is shown on the grid.

Copy the graph and sketch graph lines for small lumps and large lumps. Label your lines **S** for small lumps and **L** for large lumps. **(3)**

volume of gas/$cm^3$

M

0 1 2 3 4 5 6 7 8

time/minutes

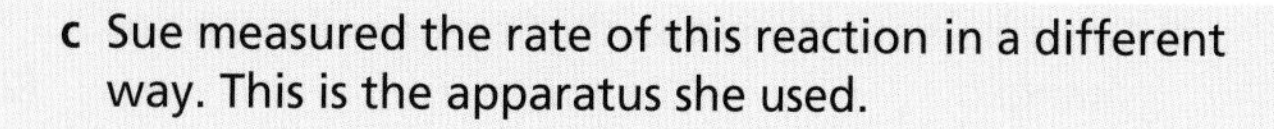

**c** Sue measured the rate of this reaction in a different way. This is the apparatus she used.

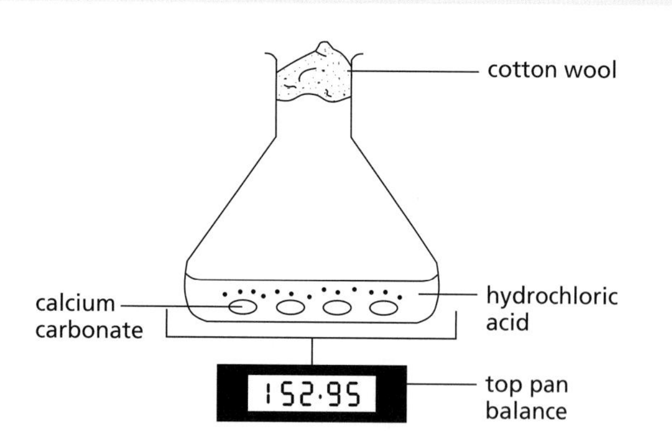


(i) How could she use measurements of mass to follow the rate of reaction? **(2)**
(ii) Why did she put cotton wool in the neck of the flask? **(1)**

*(OCR, GCSE, Paper 2, June 1999)*

**59** This item appeared in the Wolverhampton *Express & Star* on October 31st, 1997.

> **Fumes scare at factory**
>
> Workers were forced to flee a factory after a chemical alert.
>
> The building was evacuated when a toxic gas filled the factory.
>
> It happened when nitric acid spilled on to the floor and mixed with magnesium metal powder.

Read the passage and answer the questions that follow.

**a** Explain, in terms of particles, how the toxic gas was able to fill the factory quickly. **(2)**

**b** The reaction of nitric acid with magnesium metal powder is more dangerous than if the acid had fallen on to the same mass of magnesium bars. Explain why. **(1)**

**c** Water was sprayed on to the magnesium and nitric acid to slow down the reaction. Explain, in terms of particles, why the reaction would slow down. **(2)**

**d** (i) Copy and balance the equation for the reaction between magnesium and nitric acid:

$Mg + \_\_\_\_\_ HNO_3 \rightarrow Mg(NO_3)_2 + \_\_\_\_\_ H_2O + 2NO_2$ **(1)**

(ii) The toxic gas was nitrogen dioxide, $NO_2$. Calculate the mass of nitrogen dioxide produced when 96 g of magnesium reacts completely with nitric acid.
(Relative atomic masses: N = 14, O = 16, Mg = 24.) **(3)**

*(AQA, GCSE, Paper 2372, June 1999)*

**60** A student studied the effect of temperature on the rate of reaction between hydrochloric acid and sodium thiosulphate.

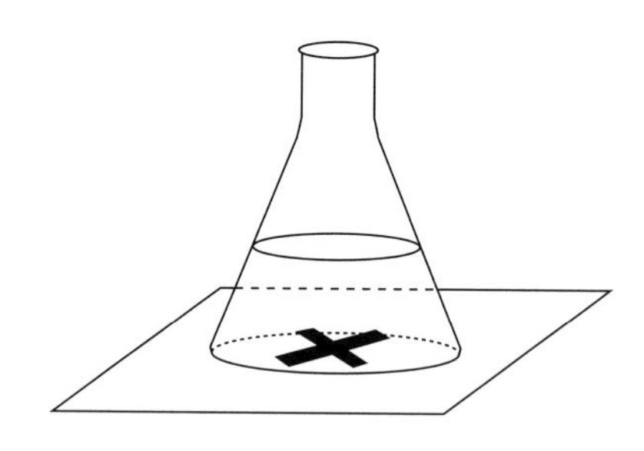

- The student mixed 50 $cm^3$ of a sodium thiosulphate solution and 5 $cm^3$ of hydrochloric acid in a flask.
- The flask was placed over a cross.
- The student timed how long after mixing the cross could no longer be seen.

**a** (i) Balance the chemical equation for this reaction.

$Na_2S_2O_3(aq) + HCl(aq) \rightarrow NaCl(aq) + H_2O(l) + SO_2(g) + S(s)$ **(1)**

(ii) What causes the cross to be seen no longer? **(1)**

**b** A graph of the results is shown.

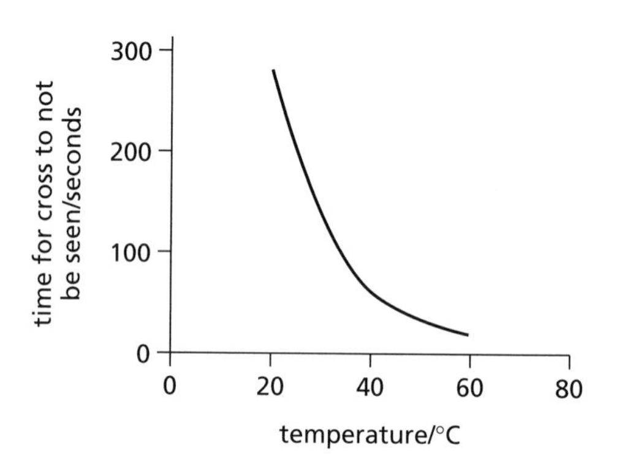


(i) What effect does temperature have on the rate of this reaction? **(1)**
(ii) Explain why temperature has this effect on the rate of reaction. **(2)**

*(SEG, GCSE, Paper 6/4, Summer 2000)*

**61** Magnesium ribbon was reacted with **excess** dilute hydrochloric acid to produce hydrogen gas. The volume of hydrogen produced in the reaction was measured using a gas syringe. The experiment was carried out at a temperature of 20 °C.

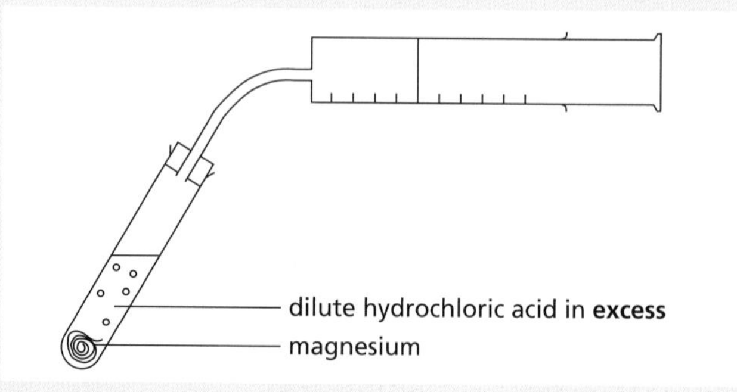

The table below shows the results obtained in the experiment.
**One of the results is unreliable.**

| Time/ seconds | 0 | 10 | 20 | 30 | 40 | 50 | 60 | 70 |
|---|---|---|---|---|---|---|---|---|
| Volume of gas/$cm^3$ | 0 | 50 | 80 | 100 | 102 | 118 | 120 | 120 |

**a** Plot the data on graph paper and draw a smooth curve through the reliable points. **(3)**

**b** Sketch carefully on the grid the graph that would have been obtained if
(i) the same experiment was repeated at 40 °C. Label this graph **B**. **(2)**
(ii) the magnesium ribbon had been replaced by a piece half as long and reacted with the same acid, which is still in **excess**. Label this graph **C**. **(2)**

**c** State and explain, using **particle theory**, what would happen to the rate if a **higher** concentration of acid than in the original experiment was used. **(2)**

*(WJEC, GCSE, 0125/2, June 1999)*

**62** Hydrogen peroxide, $H_2O_2$, is often used as a bleach. It decomposes forming water and oxygen.

**a** (i) Write the balanced chemical equation for the decomposition of hydrogen peroxide. **(3)**
(ii) Give a test for oxygen, and the result for that test. **(2)**

**b** The rate of decomposition of hydrogen peroxide at room temperature is very slow. Manganese oxide is a catalyst which can be used to speed up the decomposition. Copy and complete the sentence.
A catalyst is a substance which speeds up a chemical reaction. At the end of the reaction, the catalyst is _______. **(1)**

**c** Two experiments were carried out to test if the amount of manganese oxide, $MnO_2$, affected the rate at which the hydrogen peroxide decomposed.
(i) Copy and complete the diagram to show how you could measure the volume of oxygen formed during the decomposition. **(2)**

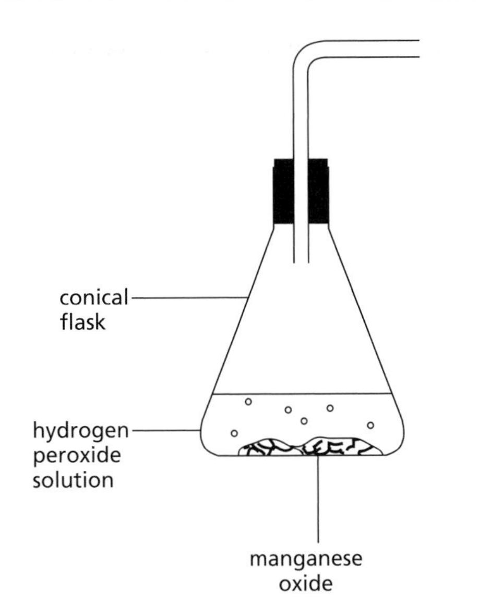


(ii) The results are shown in the table.

| Time (min) | 0 | 0.5 | 1 | 1.5 | 2 | 2.5 | 3 | 3.5 |
|---|---|---|---|---|---|---|---|---|
| Volume of gas, using 0.25 g $MnO_2$ ($cm^3$) | 0 | 29 | 55 | 77 | 98 | 116 | 132 | 144 |
| Volume of gas, using 2.5 g $MnO_2$ ($cm^3$) | 0 | 45 | 84 | 118 | 145 | 162 | 174 | 182 |

Copy the grid and draw a graph of these results. The graph for 0.25 g $MnO_2$ has been drawn for you.

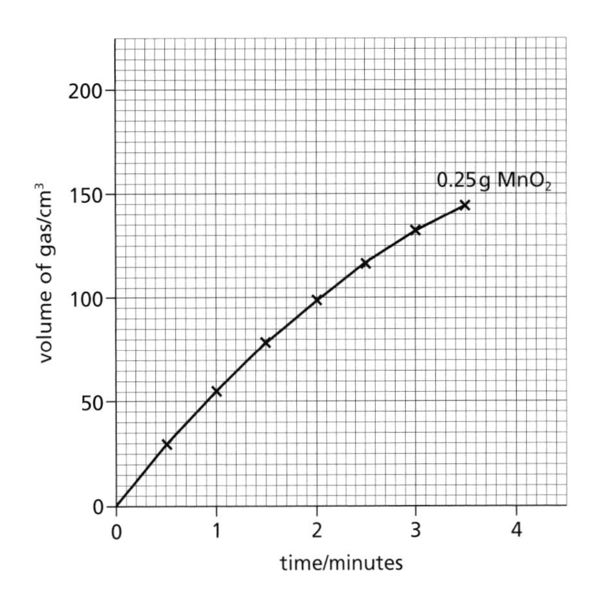


(iii) Explain why the slopes of the graphs become less steep during the reaction. **(2)**
(iv) The same volume and concentration of hydrogen peroxide solution was used for both experiments. What **two** other factors must be kept the same to make it a fair test? **(2)**

*(SEG, GCSE, Paper 6/4, June 1999)*

**63** This question is about catalase, an **enzyme** in vegetables.
Catalase acts as a catalyst for the splitting up of hydrogen peroxide.

hydrogen peroxide → water + oxygen

**a** Sam does an experiment at 20 °C. She uses 25 $cm^3$ of hydrogen peroxide solution and 1 $cm^3$ of catalase solution. She measures the volume of gas given off each minute for five minutes. The table shows her results.

| Time/min | 0 | 1 | 2 | 3 | 4 | 5 |
|---|---|---|---|---|---|---|
| Volume of gas/$cm^3$ | 0 | 25 | 40 | 53 | 50 | 50 |

(i) On a piece of graph paper, label and choose the best scale for the vertical axis. **(1)**
(ii) Plot the points on the grid. **(1)**
(iii) Finish the graph by drawing the curve of best fit. **(1)**
(iv) Sam does the experiment again but this time at 30 °C. Draw, on the same grid, the graph she would expect to get. **(2)**

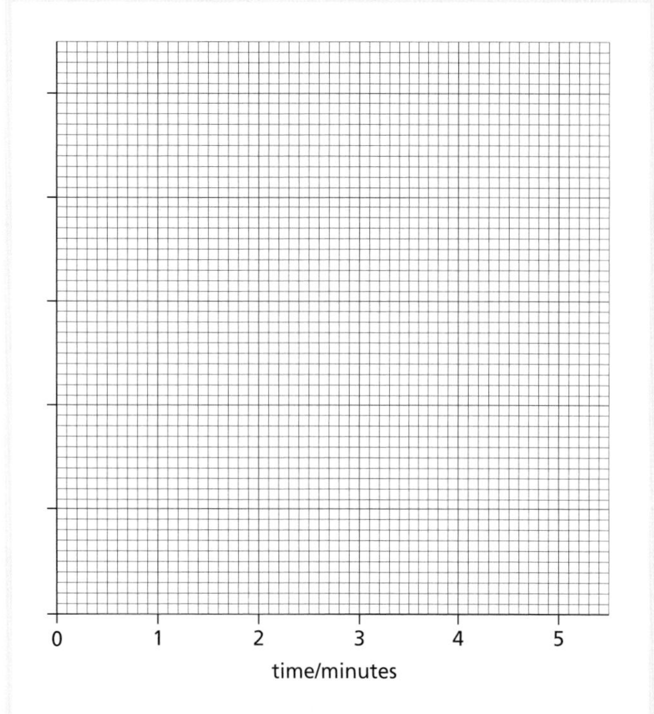


**b** The strength of a hydrogen peroxide solution is sometimes given as 'volume strength'. A 10-volume solution produces 10 $cm^3$ of oxygen for each 1 $cm^3$ of solution used.

What is the 'volume strength' of the hydrogen peroxide Sam used? **(1)**

*(OCR, GCSE, Paper 2, June 1999)*

**64 a** Petrol is a fossil fuel and so its supply is limited. Alternative fuels will be needed as it runs out. The table shows data from 1998 for petrol and some alternative fuels.

| Fuel | Cost of 100 g/ pence | Energy per 100 g/kJ | Energy per penny/kJ |
|---|---|---|---|
| Petrol | 6.8 | 4800 | 706 |
| Diesel oil | 6.4 | 4700 | 734 |
| Ethanol | 8.5 | 2900 | 341 |
| Hydrogen | 20.0 | 14 300 | 715 |
| Vegetable oil | 9.0 | 3800 | 422 |

(i) Use the data in the table to explain why diesel oil seems to be a good alternative to petrol. **(1)**
(ii) From your knowledge of fuels, give **one** disadvantage of using diesel oil as a replacement fuel for petrol. **(1)**
(iii) From the table, hydrogen seems to be a good alternative to petrol. Suggest **one** advantage and **two** disadvantages of using hydrogen as a fuel for cars. **(3)**

**b** Petrol is a mixture of hydrocarbons. There are several compounds in petrol which have the molecular formula $C_8H_{18}$.

Name the homologous series to which these compounds belong. **(1)**

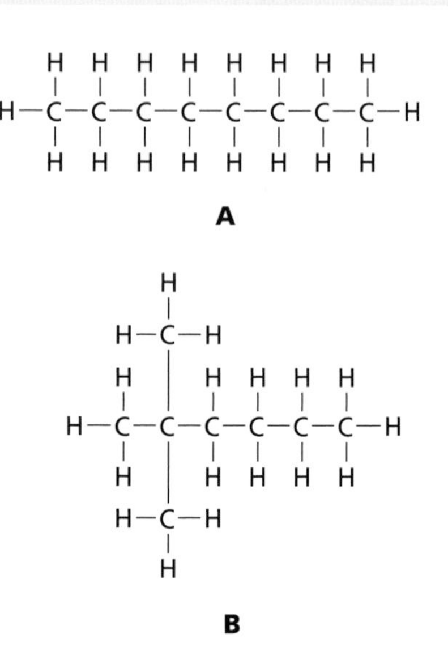


**c** The structures of two compounds with the formula $C_8H_{18}$ are shown below.
(i) Compound **A** has a higher boiling point than Compound **B**. Explain why. **(2)**
(ii) Draw the structure of **two other** compounds with the same formula, $C_8H_{18}$. **(2)**
(iii) What name is given to compounds with the same molecular formula, but with different structures? **(1)**

**d** Compound **A**, $C_8H_{18}$, can be cracked to produce pentane and propene.

Copy and complete the symbol equation which represents this reaction.

$C_8H_{18} \rightarrow C_5H_{12} + \_\_\_\_\_\_$ **(1)**
pentane propene

**e** Propene can be polymerised.
(i) Name the polymer formed. **(1)**
(ii) Explain the meaning of the term polymerisation. **(2)**

*(AQA, GCSE, Paper 2372, June 2000)*

**65 a** Crude oil (petroleum) is a mixture of compounds called hydrocarbons. Crude oil is heated and then gradually cooled as it passes up through a fractionating column.

The diagram below shows the levels at which some fractions condense.

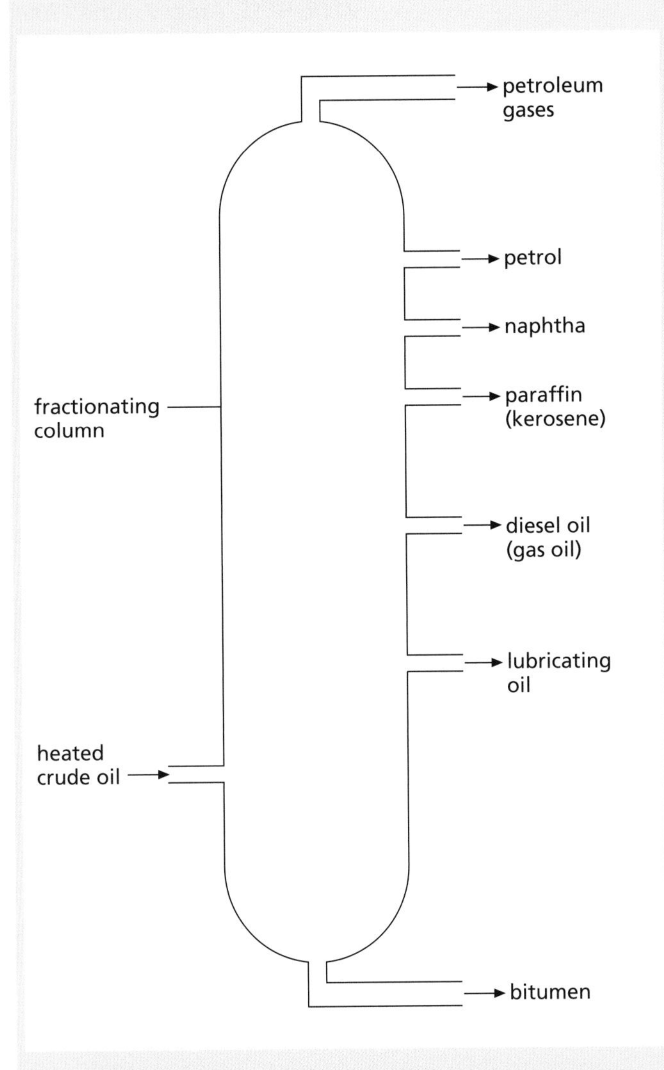


(i) Give the names of the **two** elements found in hydrocarbons. **(1)**
(ii) Give the name of the process which is used to separate crude oil into fractions. **(1)**
(iii) State the **physical** property of the hydrocarbons on which the method of separation depends. **(1)**
(iv) The petrol fraction is used to supply fuel for motor cars. Name **one** other fraction used as a fuel. **(1)**
(v) Give **one** reason why British oil refineries are usually situated near the coast. **(1)**

**b** Hydrocarbons are used as fuels in everyday life.
(i) Name the **two** products formed during the **complete** combustion of a hydrocarbon fuel. **(2)**
(ii) Explain the danger of the **incomplete** combustion of hydrocarbon fuels. **(2)**

*(WJEC, GCSE, 0125/2, June 1999)*

**66** Three fuels are:
**petrol**
**kerosene**
**diesel**

**a** These fuels are all obtained from the same raw material.
(i) Name this raw material. **(1)**
(ii) Name the process used to separate these fuels from the raw material. **(2)**
(iii) Which **two** of these fuels are used most often in car engines? **(1)**

**b** Each of these fuels contains a mixture of hydrocarbons.
The table shows the formula and boiling point of one hydrocarbon present in each of the fuels.

| Fuel | Formula of one hydrocarbon present in the fuel | Boiling point of hydrocarbon/ °C |
|---|---|---|
| Petrol | $C_8H_{18}$ | 126 |
| Kerosene | $C_{11}H_{24}$ | 196 |
| Diesel | $C_{17}H_{36}$ | 303 |

Use information from the table to answer these questions.
(i) Name the **two** elements present in these hydrocarbons. **(2)**
(ii) Give the formula of the hydrocarbon with the lowest boiling point. **(1)**
(iii) Give the formula of the hydrocarbon with the smallest molecules. **(1)**
(iv) Copy and complete the sentence using a word or phrase from the box.

| **increases** | **decreases** | **stays the same** |
|---|---|---|

As the size of the hydrocarbon molecule increases, the boiling point ________. **(1)**

**c** When the hydrocarbon $C_8H_{18}$ burns, this reaction takes place.

$$2C_8H_{18}(g) + 25O_2(g) \rightarrow 16CO_2(g) + 18H_2O(g)$$

(i) Name the element which reacts with the hydrocarbon. **(1)**
(ii) What is meant by $H_2O(g)$? **(2)**
(iii) During this reaction heat is released. What word describes a reaction which gives out heat? **(1)**

*(Edexcel, Paper 2F/1F, June 2000)*

**67** Natural gas and crude oil are important resources.
**a** Describe how these resources were **formed** in the geological past. **(2)**
**b** Once formed, crude oil becomes trapped in rock such as sandstone. Briefly explain why crude oil becomes trapped in sandstone.
Use a labelled diagram in your answer, if you wish to do so. **(3)**

*(WJEC, GCSE, June 2000)*

**68** The table gives the names and formulae of four **alkanes** in crude oil.

| Name | Formula |
|---|---|
| Methane | $CH_4$ |
| Hexane | $C_6H_{14}$ |
| Decane | $C_{10}H_{22}$ |
| Hexadecane | $C_{16}H_{34}$ |

**a** Copy and complete the formula of the alkane containing 8 carbon atoms.

**C H** **(1)**

**b** Cracking of hexane can produce two hydrocarbons each containing two carbon atoms. One compound has a molecular formula $C_2H_6$ and the other $C_2H_4$.

(i) Write a balanced equation for this cracking reaction.

$C_6H_{14} \rightarrow$ ______ + ______ **(2)**

(ii) Suggest **two** conditions used for this cracking reaction. **(2)**

(iii) Explain why the cracking of long chain alkanes from crude oil is important commercially. **(3)**

(iv) The graphical (displayed) formulae of the two hydrocarbons are

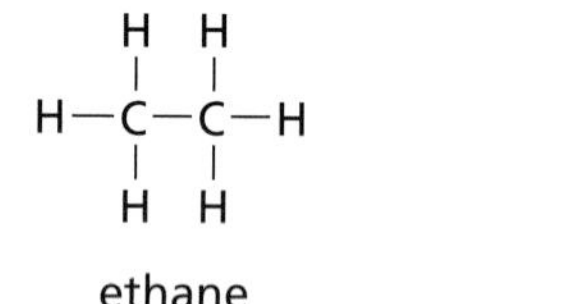


How can the two compounds be distinguished by a simple chemical test? **(3)**

**c** Chlorethene is produced from a hydrocarbon. The formula for chloroethene is

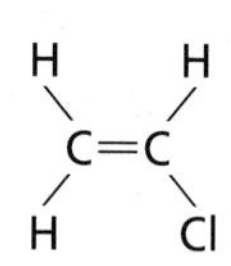

Draw a graphical (displayed) formula for poly(chloroethene). **(2)**

**d** This list includes some of the products formed when methane burns.

**carbon**
**carbon dioxide**
**carbon monoxide**
**hydrogen**
**water**

(i) Write down the names of the **two** products formed by the **complete** combustion of methane. **(2)**

(ii) Some water heaters use methane as a fuel. These water heaters need to be checked and serviced regularly. Some people die from breathing the fumes from heaters which have not been serviced. Explain how these fumes are produced and why they are dangerous. **(3)**

*(OCR, GCSE, Paper 2, June 1999)*

**69** The molecular formulae of two hydrocarbons **M** and **N** are given.

**M** = $C_4H_{10}$
**N** = $C_4H_8$

**a** **M** reacts with chlorine to form $C_4H_9Cl$.

(i) Write a balanced chemical equation for the reaction of chlorine with **M**. **(2)**

(ii) Name this type of reaction. **(1)**

**b** A displayed structural formula for **N** is:

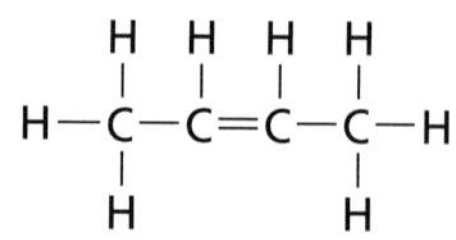

Draw a displayed structural formula of a compound which is an isomer of **N**. **(1)**

**c** Complete the boxes to show the displayed structural formula for each of the products formed.

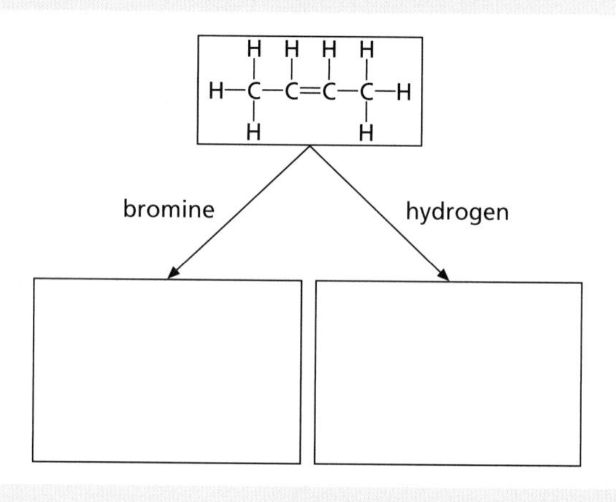


*(SEG, GCSE, Paper 5, Summer 2000)*

**70** The gas used as a fuel for heating in most homes is methane, $CH_4$.

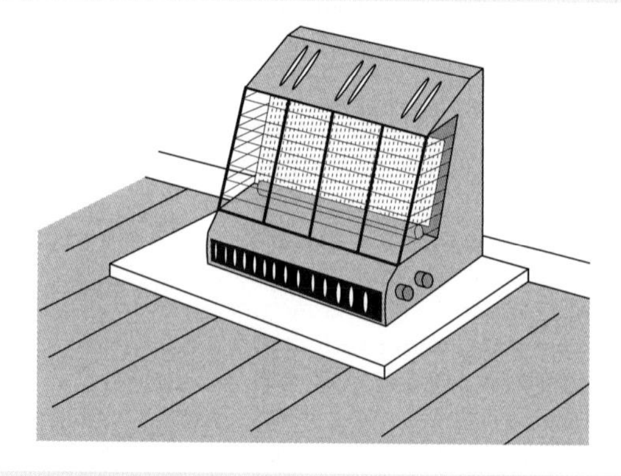

**a** It is very important to have a good air supply when methane burns. Explain why. **(2)**

**b** The word equation when methane burns in a good air supply is:

methane + oxygen → carbon dioxide + water

(i) Copy and balance the chemical equation for this reaction.

$\_\_\_\_\_ CH_4(g) + \_\_\_\_\_ O_2(g) \rightarrow \_\_\_\_\_ CO_2(g) + \_\_\_\_\_ H_2O(g)$ **(1)**

(ii) Why is this reaction called an exothermic reaction? **(1)**

**c** The experiment shown was used to test the gases formed when methane burns in a good air supply.

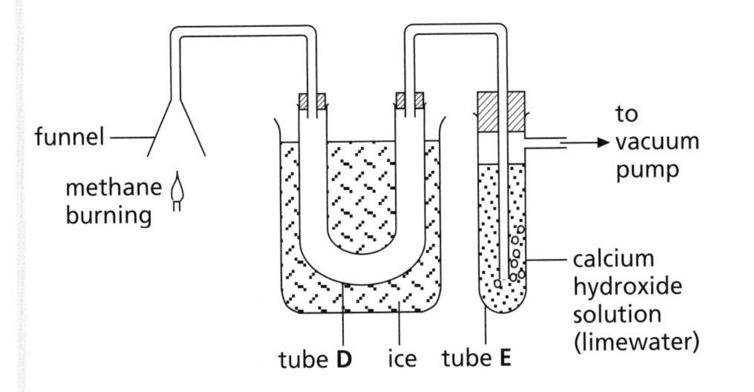


(i) Explain why the water formed collects in tube **D**. **(2)**
(ii) Give a chemical test for water. **(2)**
(iii) The reaction that happens in tube **E** is:

$$Ca(OH)_2(aq) + CO_2(g) \rightarrow CaCO_3(s) + H_2O(l)$$

Describe and explain the change you would see in tube **E**. **(2)**

*(SEG, GCSE, Paper 6/4, June 1999)*

**71** Some power stations use fossil fuels which are burned and form the pollutant gas, sulphur dioxide. This sulphur dioxide can be removed by a process called Flue Gas Desulphurisation (FGD process).

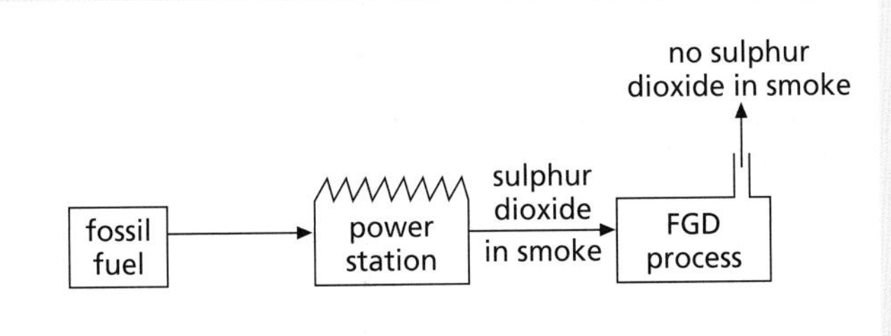


**a** Name a fossil fuel. **(1)**

**b** Why is it important to prevent sulphur dioxide going into the air? Explain your answer. **(2)**

**c** Limestone can be used in the FGD process. Limestone reacts with the sulphur dioxide to form calcium sulphite.
(i) What is the chemical name for limestone? **(1)**
(ii) Calculate the relative formula mass of calcium sulphite, $CaSO_3$.
(Relative atomic masses: O 16; S 32; Ca 40.) **(2)**

**d** A solution of magnesium oxide in water can also be used to remove the sulphur dioxide. The reaction is:

$$MgO + SO_2 \xrightarrow{\text{water}} MgSO_3$$

Magnesium sulphite, $MgSO_3$, can be changed into magnesium oxide and sulphur dioxide.

$$MgSO_3 \rightarrow MgO + SO_2$$

(i) Give a pH value for a solution of magnesium oxide in water. **(1)**
(ii) Suggest why sulphur dioxide reacts with magnesium oxide. **(2)**
(iii) Suggest how magnesium sulphite can be changed into magnesium oxide and sulphur dioxide. **(1)**
(iv) Give **two** uses of sulphur dioxide. **(2)**

*(SEG, GCSE, Paper 5, Summer 2000)*

**72** Cars in Brazil use ethanol as a fuel instead of petrol (octane). The ethanol is produced by the fermentation of sugar solution from sugar cane.

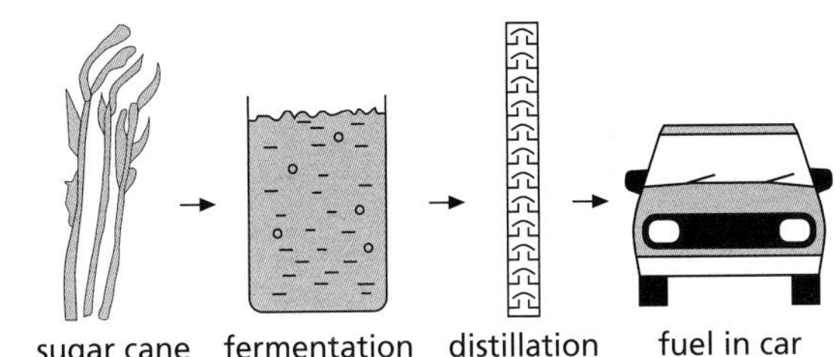


**a** What must be added to sugar solution to make it ferment? **(1)**

**b** What is the most suitable temperature for a fermentation?
0 °C 10 °C 30 °C 70 °C 100 °C **(1)**

**c** (i) What compounds are formed by the complete combustion of ethanol? **(2)**
(ii) Why are these compounds **not** harmful to the environment? **(1)**

**d** Suggest why pollution from cars is less when using ethanol instead of petrol. **(1)**

**e** Give **one** reason why ethanol is **not** used as a fuel for cars in Britain. **(1)**

**f** Some information about octane and ethanol is shown.

| Property | Octane | Ethanol |
|---|---|---|
| Melting point/°C | −57 | −113 |
| Boiling point/°C | 125 | 78.5 |
| Density/g/cm$^3$ | 0.70 | 0.79 |
| Heat produced/ kJ/mol | 5512 | 1367 |

Explain a similarity between octane and ethanol that allows ethanol to be used as a fuel in cars. **(2)**

*(SEG, GCSE, Paper 5, Summer 2000)*

**73** The table gives the energy required to break some bonds. The number of bonds broken is the same in each case.

| Bond | Energy required (kJ) |
|---|---|
| C—H | 435 |
| Cl—Cl | 243 |
| H—Cl | 432 |
| C—Cl | 346 |

The energy required to break a bond is the same as the energy given out when the bond forms.

In the presence of sunlight, methane will react with chlorine as in the equation.

```
   H            Cl           Cl            H
   |            |            |             |
H—C—H    +     Cl    →    H—C—H    +      Cl
   |                         |
   H                         H
```

Calculate the energy transfer in this reaction. You **must** show your working. **(3)**

*(OCR, GCSE, Paper 2, June 1999)*

**74 a** The equation shows how hydrazine reacts with oxygen.

```
      H
      |                                  O
H—N—N—H   +   O=O   →   N≡N  +  2 (  /  \  )
   |                                H    H
   H
```

Use the bond energy values given in the table to calculate the overall energy change for this reaction. **(3)**

| Bond | Bond energy/kJ per mole of bonds |
|---|---|
| N—N | 158 |
| N—H | 389 |
| O=O | 497 |
| N≡N | 945 |
| O—H | 464 |

**b** Suggest a reason by hydrazine has been used as a rocket fuel. **(1)**

*(AQA, GCSE, Paper 2372, June 2000)*

**75 a** Ethene can undergo polymerisation to form poly(ethene), commonly called polythene.

(i) Copy and complete and balance the equation below to show the polymerisation of ethene. **(2)**

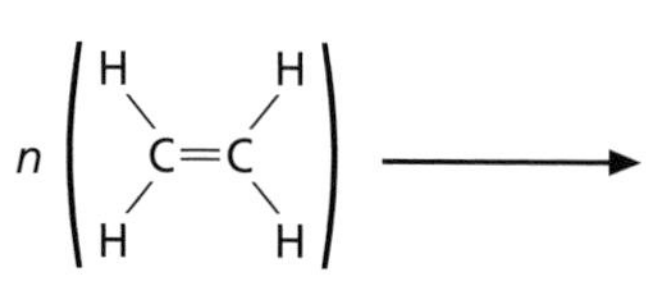

(ii) Give the name for this type of polymerisation. **(1)**

**b** When ethene reacts with hydrogen, ethane is formed.

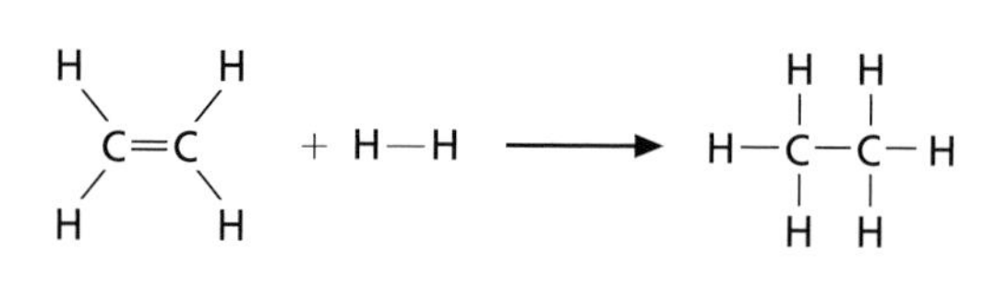

The relative amounts of energy needed to break the bonds in the above reaction are shown in the table.

| Bond | Amount of energy needed to break the bond/kJ |
|---|---|
| C=C | 612 |
| H—H | 436 |
| C—H | 413 |
| C—C | 347 |

*NOTE: The amount of energy needed to make a bond is equal and opposite to that needed to break the bond.*

(i) Using the bond energy values in the table calculate the relative energy

I needed to break **all** the bonds in the reactants, **(2)**

II evolved when the bonds in the product are formed. **(2)**

(ii) Using your answers to parts I and II explain why the relative overall energy change is exothermic. **(1)**

*(WJEC, GCSE, 0125/2, June 1999)*

**76 a** Ethene is a more useful compound than ethane. One use of ethene is to make the polymer polythene. Write an equation to show how ethene is converted into polythene. **(2)**

**b** Another important polymer is PVC. Give **two** uses of this polymer and state one property of PVC which makes it suitable for this use. **(4)**

**c** Ethene may also be converted into ethanol by direct combination with water using a catalyst. Write a balanced, symbol equation for this reaction. **(2)**

*(CCEA, GCSE, Paper 2, June 2000)*

**77 a** (i) Polyvinyl chloride (PVC) is a very important and useful plastic. By using an equation show how vinyl chloride molecules link together to form part of a PVC molecule. **(3)**

(ii) Give **two** uses of polyvinyl chloride and state the property of polyvinyl chloride which makes it suitable for the use given. **(4)**

**b** It has been proposed to construct a factory for the production of plastics on a derelict site close to Belfast.

Give **one** advantage and **one** disadvantage of siting the factory in this area. **(2)**

*(CCEA, GCSE, Paper 1, June 1999)*

**78** Home-made wine can be made by putting yeast into grape juice. Here is a diagram of the apparatus which can be used.

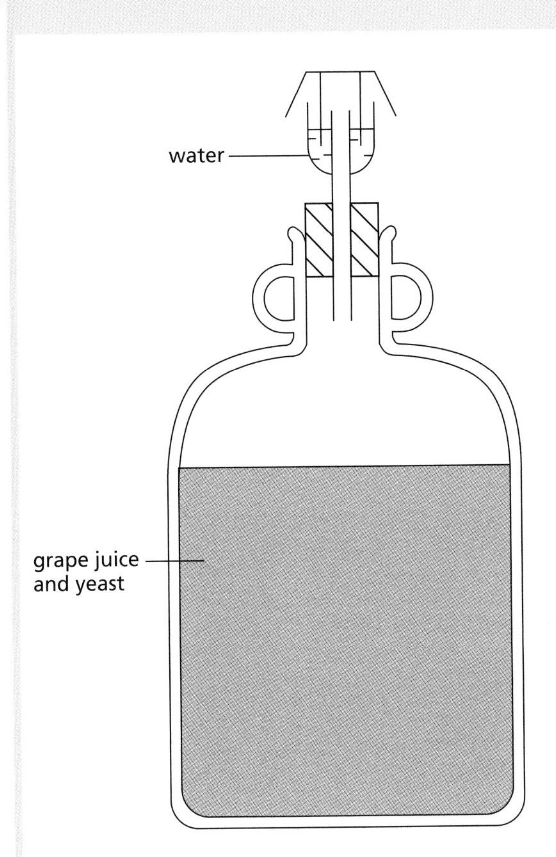


**a** (i) Write down the name of this process. **(1)**
(ii) Write down the name of the gas produced in this process. **(1)**
**b** Explain the job of the yeast in this process. **(2)**
*(OCR, GCSE, Paper 4, June 2000)*

**79** Glucose can be fermented with yeast to produce ethanol and carbon dioxide. The equation which represents this reaction is shown below.

$$C_6H_{12}O_6(aq) \rightarrow 2C_2H_5OH(aq) + 2CO_2(g)$$

**a** (i) Yeast is needed for the fermentation. Why is yeast not written in the equation? **(1)**
(ii) Suggest why temperatures above 45 °C are not used for fermentation. **(1)**
**b** Use the equation above to help you with the following calculations.
0.10 mole of glucose was completely fermented.
Calculate:
(i) the number of moles of ethanol produced. **(1)**
(ii) the mass in grams of ethanol produced.
(Relative atomic masses: H = 1, C = 12, O = 16.) **(2)**
(iii) the volume, at room temperature and pressure, of carbon dioxide gas produced.
(1 mole of any gas occupies 24 litres at room temperature and pressure.) **(1)**
**c** Most of the pure ethanol made in the United Kingdom is made from ethene. Ethene is reacted with steam in the presence of a catalyst. The equation which represents the reaction is:

$$C_2H_4 + H_2O \rightarrow C_2H_5OH$$

Explain why **this** process is preferred to fermentation by the chemical industry. Your answer should include:

- information about the raw materials
- the type of process
- the rate of reaction
- the quality of the product. **(4)**

**d** Ethanol can be converted to ethanoic acid, $CH_3COOH$.
(i) What type of reaction takes place when ethanol is converted to ethanoic acid? **(1)**
(ii) Ethanol and ethanoic acid can react in the presence of an acid catalyst to produce ethyl ethanoate. What type of substance is ethyl ethanoate? **(1)**
**e** Ethanol and ethanoic acid are both colourless liquids at room temperature.
Describe what you would **see** if sodium carbonate solution was added to each liquid in separate test-tubes. **(2)**
*(AQA, GCSE, Paper 2372, June 2000)*

**80** Ethanol can be made from sugar solution by the process of fermentation. Fermentation is the method used to make alcoholic drinks.
**a** State why distillation can be used to separate the ethanol from the fermented mixture. **(1)**
**b** (i) Give **one** health problem associated with alcoholic abuse. **(2)**
(ii) Give **one** social problem associated with alcoholic abuse. **(2)**
**c** Ethanol in wine is easily changed to ethanoic acid (vinegar).
(i) What causes this change? **(1)**
(ii) Give the name for the chemical process taking place. **(1)**
*(WJEC, GCSE, June 2000)*

**81** The flow diagram shows two ways of making ethanol, and some reactions of ethanol.

**a** Give the names of
(i) gas **A**
(ii) process **B**
(iii) catalyst **C**
(iv) metal **D**
(v) type of reaction **E**
(vi) catalyst **F** **(6)**

**b** Fermentation takes place at a temperature of about 35 °C.
(i) What happens to this reaction if the temperature is raised to 70 °C?
(ii) Give a reason for your answer. **(2)**

**c** Describe the advantages and disadvantages of each of the two processes for making ethanol. **(4)**

(*AQA, GCSE, Paper 2372, June 1999*)

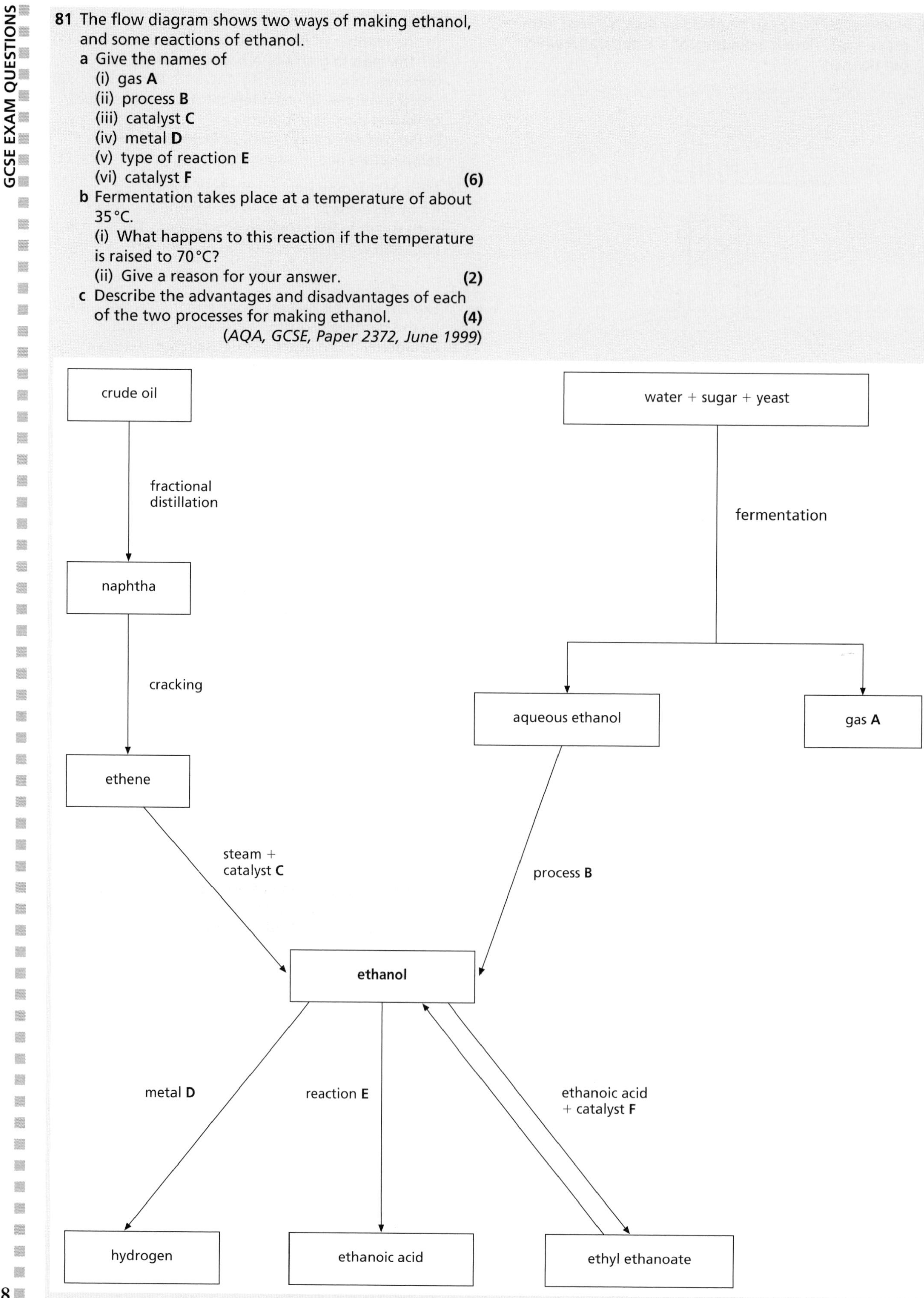

**82** Ethanoic acid, $CH_3COOH$, forms a **weak acid** when added to water. Some reactions of ethanoic acid are shown.

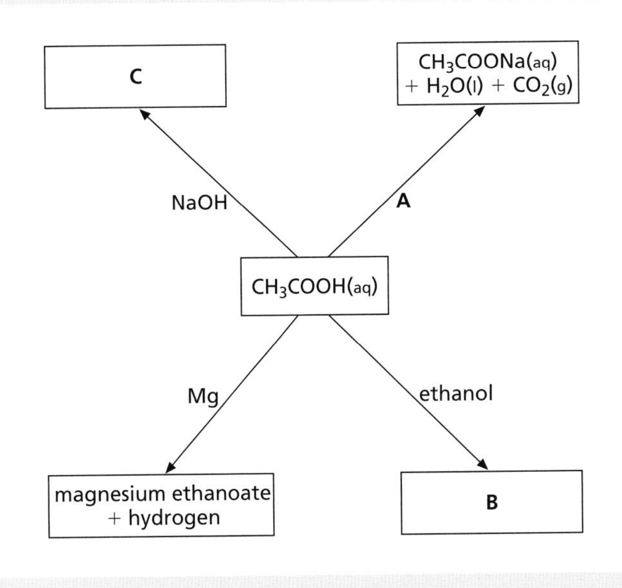


**a** Explain what is meant by a **weak acid**. **(2)**
**b** Name the substance **A** that is added to ethanoic acid. **(1)**
**c** Substance **B** is formed when ethanoic acid reacts with ethanol. What type of substance is **B**? **(1)**
**d** Draw a displayed structural formula for salt **C**. **(1)**
**e** Write a balanced chemical equation for the reaction between magnesium and ethanoic acid. **(2)**

*(SEG, GCSE, Paper 5, Summer 2000)*

**83** This question is about ammonia and fertilisers. Ammonia is made from nitrogen and hydrogen in the Haber process. The equation is

$$N_2 + 3H_2 \rightleftharpoons 2NH_3$$

**a** Write down the name of the catalyst in this process. **(1)**
**b** The diagram shows the energy change when ammonia is made.

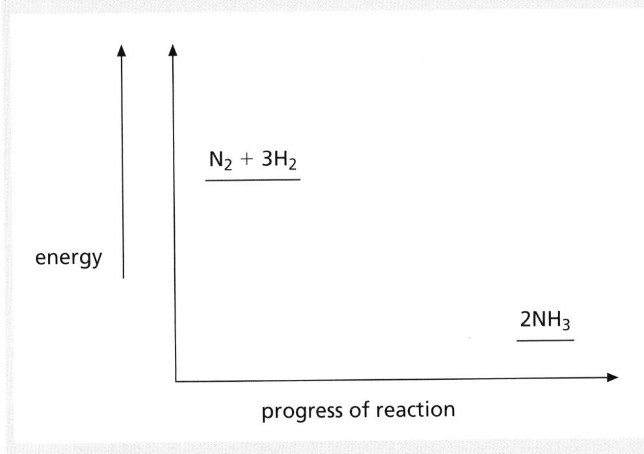


What can you conclude from this energy level diagram? **(1)**

**c** **Ammonium nitrate** is a fertiliser made from ammonia and nitric acid. Nitric acid is made from ammonia in three stages.

Stage 1

ammonia + oxygen $\xrightarrow{\text{platinum}}$ nitrogen monoxide + steam

Stage 2

nitrogen monoxide + oxygen $\longrightarrow$ nitrogen dioxide

Stage 3

nitrogen dioxide + water + oxygen $\longrightarrow$ nitric acid

(i) Ammonia is a raw material in this process. What are the other two raw materials? **(2)**
(ii) In Stage 1, the platinum has to be heated to 900 °C to start the reaction. Then the temperature of the catalyst stays at 900 °C without the need for further heating. What does this tell you about the first stage? **(1)**

*(OCR, GCSE, Paper 2, June 1999)*

**84** Read the following poem about the manufacture of ammonia, $NH_3$, and answer the questions that follow.

**The Haber Process**
*by Martin Perry*

Dry nitrogen and hydrogen gas.
Over a finely divided iron catalyst are passed.
The gases in ratio one to three,
At a pressure of 200 and a temperature of
450 degrees C.

Ammonia gas is produced,
From its choking smell this is deduced.
From the ammonia is made ammonium sulphate,
And also the fertiliser, ammonium nitrate.

(Taken from Chemistry Poems, *Education Today*, Volume 38 No. 2)

**a** This word equation represents the reaction between nitrogen and hydrogen.

nitrogen + hydrogen $\rightleftharpoons$ ammonia

(i) Write the balanced symbol equation for this reaction. **(2)**
(ii) The symbol '$\rightleftharpoons$' means that the reaction is reversible. Explain what is meant by a reversible reaction. **(1)**
(iii) The reaction producing ammonia is exothermic. In industry, the conditions used are a pressure of 200 atmospheres and a temperature of 450 °C.
Explain fully why **both** of these conditions are chosen. **(5)**

**b** (i) Calculate the relative formula mass ($M_r$) of ammonia, $NH_3$.
(Relative atomic masses: H = 1, N = 14.)
(ii) Calculate the percentage of nitrogen in ammonia. **(3)**

**c** (i) Ammonium nitrate is a fertiliser. What is a fertiliser? **(1)**
(ii) Ammonium nitrate is formed when ammonia reacts with nitric acid. Nitric acid is manufactured from ammonia. Describe how nitric acid is made from ammonia.
Your answer should include:

- the types of reaction which occur
- the conditions used
- the energy changes involved. **(4)**

*(AQA, GCSE, Paper 2372, June 2000)*

**85** Hydrazine, $N_2H_4$, is a useful chemical. These are some of its properties:

- colourless liquid
- boiling point, melting point and density are similar to water
- reacts with water to form an alkaline solution
- good at reducing other substances
- produces a lot of energy when reacted with oxygen.

**a** Describe a pH test and its results to show that a liquid is hydrazine and not pure water. **(2)**

**b** When aqueous hydrazine reacts with hydrochloric acid, the salt called hydrazinium chloride is produced.
Copy and complete the name of the salt produced when aqueous hydrazine reacts with nitric acid. **(1)**

**c** Hydrazine is added to the water supplied to industrial boilers. It removes the oxygen dissolved in the water and so helps prevent corrosion of the boiler and pipes.
What property of hydrazine given in the list at the start of the question allows it to remove the oxygen from the water? **(1)**

**d** The structural formula of a hydrazine molecule is shown below.

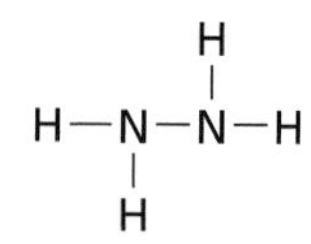


Use a data book to help you to answer this question.
Copy and complete the diagram below to show how the outer energy level (shell) electrons are arranged in a hydrazine molecule. Show the electrons as dots and crosses. **(2)**

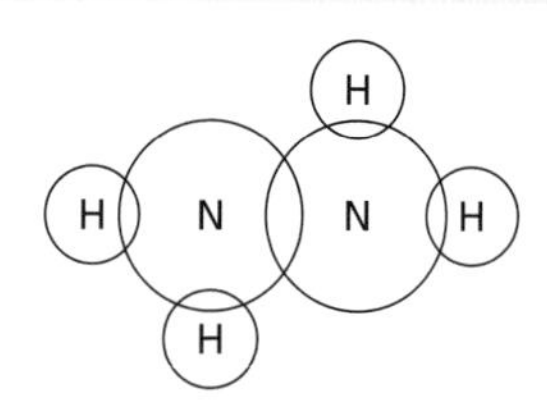


**e** Explain why hydrazine has a low boiling point. **(2)**

*(AQA, GCSE, Paper 2372, June 2000)*

**86** **a** The Haber process is used to make ammonia, $NH_3$.
The table shows the percentage yield of ammonia at different temperatures and pressures.

| Pressure/ atmospheres | Percentage (%) yield of ammonia at 350 °C | Percentage (%) yield of ammonia at 500 °C |
|---|---|---|
| 50 | 25 | 5 |
| 100 | 37 | 9 |
| 200 | 52 | 15 |
| 300 | 63 | 20 |
| 400 | 70 | 23 |
| 500 | 74 | 25 |

(i) Use the data in the table to draw two graphs on a grid. Draw one graph for a temperature of 350 °C and the second graph for a temperature of 500 °C. Label each graph with its temperature. **(4)**
(ii) Use your graphs to find the conditions needed to give a yield of 30% ammonia. **(1)**
(iii) On the grid, sketch the graph you would expect for a temperature of 450 °C. **(1)**

**b** This equation represents the reaction in which ammonia is formed.

$$N_2(g) + 3H_2(g) \rightleftharpoons 2NH_3(g) + \text{heat}$$

(i) What does the symbol $\rightleftharpoons$ in this equation tell you about the equation? **(1)**
(ii) Use your graphs and your knowledge of the Haber process to explain why a temperature of 450 °C and a pressure of 200 atmospheres are used in industry. **(5)**

**c** Ammonium nitrate is one type of artificial fertiliser.
(i) Calculate the relative formula mass of ammonium nitrate, $NH_4NO_3$.
(Relative atomic masses: H = 1, N = 14, O = 16.) **(1)**
(ii) Use your answer to part **c** (i) to help you calculate the percentage by mass of nitrogen present in ammonium nitrate, $NH_4NO_3$. **(2)**

*(AQA, GCSE, Paper 2372, June 1999)*

**87** Ammonia ($NH_3$) is manufactured from hydrogen and nitrogen in the Haber process.

**a** (i) Write a balanced equation for the formation of ammonia in the Haber process. **(2)**
(ii) Draw a dot and cross diagram to show the bonding in a molecule of ammonia. **(2)**
(iii) Explain, in terms of the bonds broken and formed, why the formation of ammonia from nitrogen and hydrogen is exothermic. **(3)**

**b** The manufacture of methanol from carbon monoxide and hydrogen requires similar conditions to those used in the Haber process.
The equation for the manufacture of methanol is

$$CO(g) + 2H_2(g) \rightleftharpoons CH_3OH(g)$$

This reaction is exothermic. The reaction conditions are a pressure of 200 atm and a temperature of 400 °C.
(i) State **one** advantage of using a pressure higher than 200 atm. Explain your answer. **(3)**
(ii) State **one** disadvantage of using a pressure higher than 200 atm. **(1)**
(iii) State **one** advantage of using a temperature lower than 400 °C. Explain your answer. **(3)**
(iv) State **one** disadvantage of using a temperature lower than 400 °C. Explain your answer. **(2)**

*(Edexcel, GCSE, Paper 3H, June 1999)*

**88** **a** Ammonia is manufactured by the Haber process.

$$N_2 + 3H_2 \rightleftharpoons 2NH_3$$

The graph that follows shows the yield of ammonia at different temperature and pressure conditions.
Use the graph to answer parts (i) and (ii).
(i) State what happens to the yield of ammonia as the
I temperature increases **(1)**
II pressure increases. **(1)**

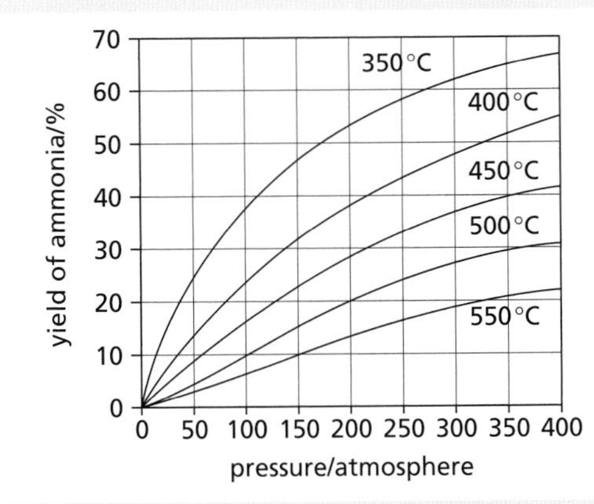


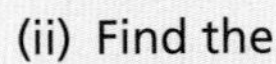
(ii) Find the
I pressure needed to obtain 40% yield of ammonia at 450 °C **(1)**
II temperature needed to obtain 20% yield of ammonia at a pressure of 200 atmospheres. **(1)**

**b** State how the **rate** of ammonia production is increased, apart from changing the temperature and pressure conditions. **(1)**

**c** Ammonia is used to make the nitrogenous fertiliser, ammonium nitrate, $NH_4NO_3$. Calculate the relative molecular mass ($M_r$) of ammonium nitrate. **(2)**
(Relative atomic masses: $A_r(H) = 1$; $A_r(N) = 14$; $A_r(O) = 16$.)

*(WJEC, GCSE, June 2000)*

**89 a** Fertiliser tablets contain compounds of ammonia. Ammonia can be made in the laboratory.
Part of the equation for making ammonia is shown below.
This reaction is reversible.
Copy and complete the equation by putting the correct symbol in the box.

$N_2 + 3H_2$ ☐ $2NH_3$ **(1)**

**b** The diagram below shows the apparatus used to make ammonia.

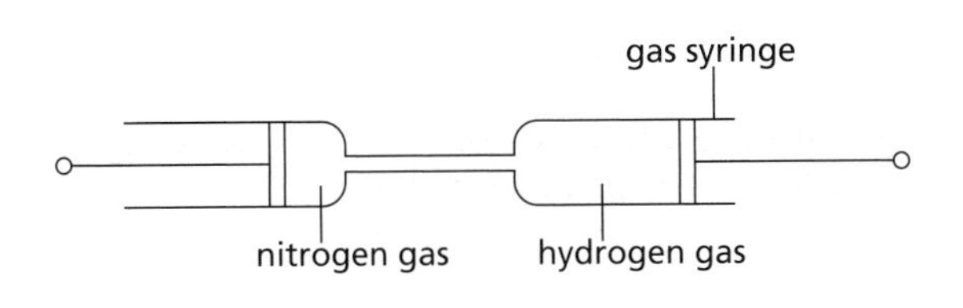


The syringes can be pushed backwards and forwards to mix the gases. The experiment uses a hot iron catalyst. Put a cross on the diagram to show where the catalyst should go. **(1)**

**c** In industry, catalysts are used because they make the process cheaper to run. However, catalysts are very expensive to buy. Explain why it is cost effective to use catalysts in the long term. **(2)**

**d** Three unlabelled test-tubes of gas contain hydrogen, ammonia and nitrogen. You need to identify the three gases. Describe the tests you would do and the results you would expect. **(3)**

*(OCR, GCSE, Paper 2, June 2000)*

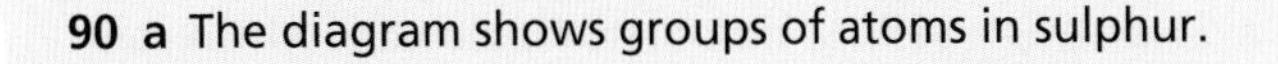
**90 a** The diagram shows groups of atoms in sulphur.

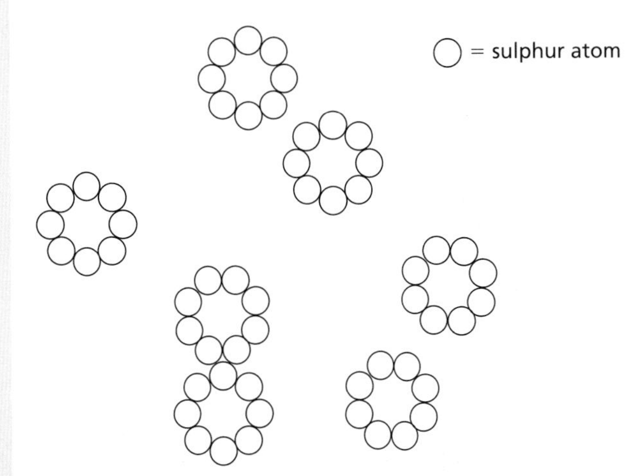


(i) What is the name given to these groups of atoms? **(1)**
(ii) Give the correct formula for the structure in part **a**. **(1)**

**b** Sulphur melts at 119 °C. Rhombic and monoclinic sulphur are **allotropes** which form from molten sulphur.

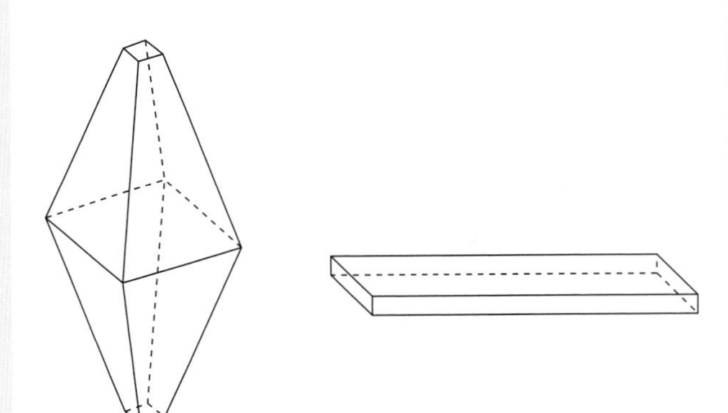


Some information about the allotropes is in the table.
(i) What is meant by **allotropes**? **(2)**

| Allotrope | Relative density |
|---|---|
| Rhombic<br>Stable below 96 °C | 2.07 |
| Monoclinic<br>Stable above 96 °C | 1.96 |

(ii) Suggest why the rhombic allotrope has the higher relative density. **(2)**
(iii) Explain the changes which happen when molten sulphur is slowly cooled to room temperature. **(3)**

*(SEG, GCSE, Paper 5, Summer 2000)*

**91 a** Sulphur is a yellow, non-metallic element which can exist as **allotropes**. Define what is meant by the term allotrope and name the allotropes of sulphur.
(i) Definition of allotrope **(2)**
(ii) The allotropes of sulphur are _________. **(3)**

**b** The most important chemical which can be made from sulphur is sulphuric acid. Describe how sulphur is converted into sulphuric acid stating relevant conditions of temperature, pressure and catalyst as necessary. Write balanced, symbol equations where appropriate. **(12)**

**c** Concentrated sulphuric acid shows reactions which are different from those of dilute sulphuric acid. Describe what you would **observe** when concentrated sulphuric acid is added to
(i) sugar **(3)**
(ii) sodium chloride. **(2)**

*(CCEA, GCSE, Paper 2, June 2000)*

**92 a** Sulphuric acid is a very important chemical. Over two million tonnes of the acid are produced in the United Kingdom each year. It has many uses including the production of fertilisers.
(i) Give **three** other uses of sulphuric acid. **(3)**
(ii) Describe the various stages in the manufacture of sulphuric acid by the Contact process. Give the names of the raw materials required, the conditions involved and balanced, symbol equations for the reactions. **(12)**

**b** Sulphuric acid is also important in the school laboratory. As a concentrated acid it reacts with copper sulphate crystals and with sodium chloride.
(i) Describe what is observed when concentrated sulphuric acid is added to copper sulphate crystals. **(3)**
(ii) In the reaction in part **b** (i), what has the concentrated sulphuric acid acted as? **(1)**
(iii) Give the balanced, symbol equation for the reaction of concentrated sulphuric acid with sodium chloride. **(2)**

**c** When concentrated sulphuric acid is diluted it acts as a typical acid.
(i) Give the precautions you would take in diluting concentrated sulphuric acid. **(3)**
(ii) Describe what would be observed when a small amount of copper oxide is added to dilute sulphuric acid and the mixture warmed. **(2)**
(iii) Give a balanced, symbol equation for the reaction in part **c** (ii). **(2)**
(iv) Describe what would be observed when sodium carbonate solution is mixed with dilute sulphuric acid. **(2)**
(v) Give a balanced, symbol equation for the reaction in part **c** (iv). **(2)**

*(CCEA, GCSE, Paper 2, June 1998)*

**93 a** Sulphuric acid is produced in the United Kingdom from sulphur. The three main reactions for the production of sulphuric acid are represented by the equations below.

$$S + O_2 \rightarrow SO_2$$
$$2SO_2 + O_2 \rightleftharpoons 2SO_3$$
$$SO_3 + H_2O \rightarrow H_2SO_4$$

(i) Which reaction uses a catalyst? **(1)**
(ii) Great care is taken to ensure that gases do not escape from the plant. Explain why. **(2)**
(iii) Suggest a reason why sulphuric acid plants are often located near a port. **(1)**

**b** Copper sulphate crystals can be used to show that a sample of sulphuric acid is concentrated. Describe what colour change you would **see** and why the colour changes. **(2)**

**c** (i) A student diluted some concentrated sulphuric acid with water. The student thought the dilute acid was weak. The teacher said that it was still a strong acid.
Why is the acid described as **strong**? **(1)**
(ii) The teacher gave the student two solutions. One was a strong acid and the other was a weak acid. The solutions were of the same concentration.
Describe a test the student could do to show which solution was the strong acid and which was the weak acid. Give the results of the test with both solutions. **(3)**

*(AQA, GCSE, Paper 2372, June 1999)*

**94 a** Describe and explain the three main stages in the Contact process for the manufacture of 98% sulphuric acid (concentrated sulphuric acid).
For each stage give the equation for the reaction and details of the reaction conditions.
Stage 1 Production of sulphur dioxide
Stage 2 Conversion of sulphur dioxide to sulphur trioxide
Stage 3 Conversion of sulphur trioxide to 98% sulphuric acid **(8)**

**b** The quantity of sulphuric acid manufactured each year is said to be a good indicator of the success of the industrial economy of the United Kingdom.
Explain, with examples, why the amount of sulphuric acid being manufactured is a good measure of the industrial performance of a nation. **(3)**

*(Edexcel, GCSE, Paper 4H, June 1999)*

**95** The three main steps in the manufacture of sulphuric acid are:

| | | | | | |
|---|---|---|---|---|---|
| Step 1 | sulphur | + | oxygen | $\rightarrow$ | sulphur dioxide |
| Step 2 | sulphur dioxide | + | oxygen | $\rightleftharpoons$ | sulphur trioxide |
| Step 3 | sulphur trioxide | + | water | $\rightarrow$ | sulphuric acid |

**a** Give the name of the raw material which supplies the oxygen in Steps 1 and 2. **(1)**

**b** What is meant by $\rightleftharpoons$ in Step 2? **(1)**

**c** The box below shows conditions used in some industrial processes.
Select the conditions used to make sulphur trioxide in Step 2. **(3)**
Temperature
Pressure
Catalyst

**d** The addition of sulphur trioxide to water is too dangerous to carry out in practice. State how sulphur trioxide is converted into sulphuric acid in the industrial process. **(2)**

**atmospheric pressure   200–300 atmospheres pressure**
**400–500 °C   room temperature   iron   vanadium oxide**

**e** State **one** large scale use of sulphuric acid. **(1)**

*(WJEC, GCSE, 0125/2, June 1999)*

**96** The diagram represents the layers of the Earth.

**a** (i) Give the name of layer **X** **(1)**

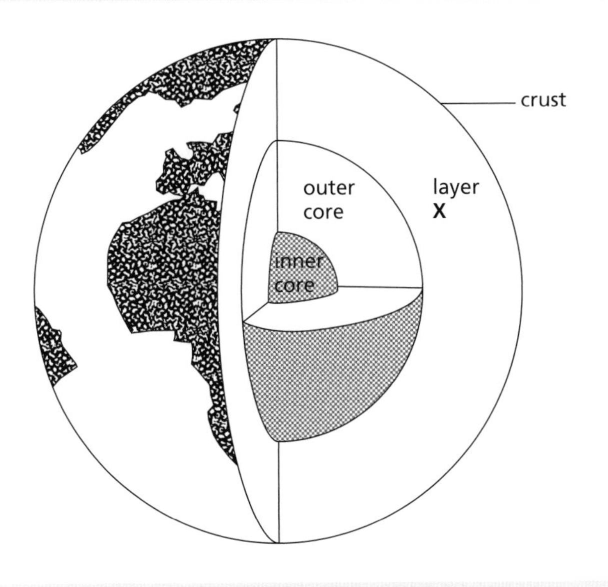


(ii) The inner core is smaller and nearer to the centre of the Earth than the outer core. Give **one** other difference between the inner and outer parts of the core. **(1)**

**b** A student was shown two igneous rocks **A** and **B**. Rock **A** had large crystals. Rock **B** had small crystals.

(i) Describe how igneous rocks are formed. **(2)**

(ii) Explain fully why the crystals in Rock **A** are larger than the crystals in Rock **B**. **(1)**

**c** Explain why most igneous rocks form near the boundaries of tectonic plates. **(2)**

*(AQA, GCSE, Paper 2372, June 2000)*

**97** This question is about different types of rocks.

**a** The lists show the names of some rocks and descriptions of them.

Link each name to the correct description. Each name must be joined to a different description. One has been done for you.

| name of rock | description of rock |
|---|---|
| basalt | grains arranged in layers of dark and light bands |
| conglomerate | crystals smaller than 0.5 mm, mainly dark in colour |
| gneiss | small stones bound together by cementing material |
| granite | hard, brittle, grey rock that splits into sheets |
| slate | crystals bigger than 0.5 mm, mainly light in colour |

**(3)**

**b** Write down the name of an example of each of the following rock types.

Choose your answers from this list.

**basalt conglomerate gneiss granite slate**

(i) a sedimentary rock **(1)**
(ii) a metamorphic rock **(1)**
(iii) an extrusive igneous rock **(1)**
(iv) an intrusive igneous rock **(1)**

*(OCR, GCSE, Paper 2, June 1999)*

**98 a** The diagrams show two rocks, **Q** and **R**. These rocks do **not** fizz with acid.

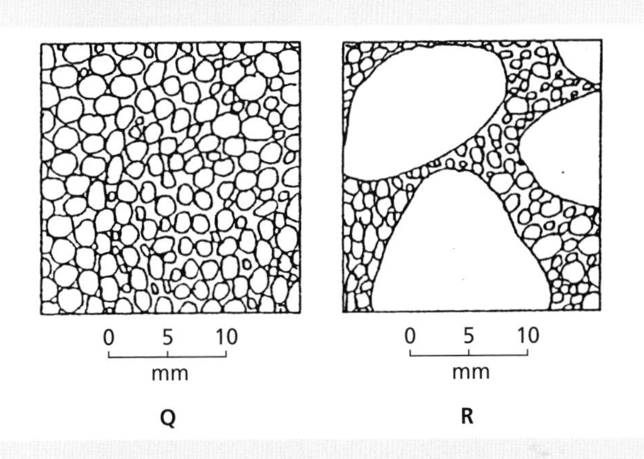


Use information from a data book to suggest the correct name for Rock **Q** and Rock **R**. **(2)**

**b** The diagram below shows a rock **P**, which was found within a mountain. The rock was formed deep in the Earth's crust.

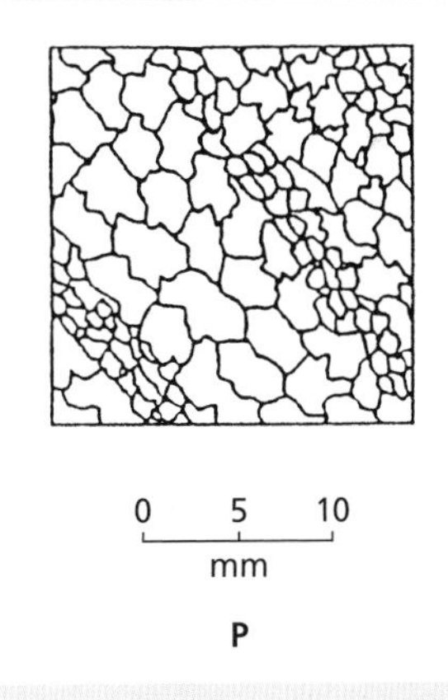


Is rock **P** igneous or metamorphic?

Use the diagram to give a reason for your answer. **(1)**

**c** Major features of the Earth's crust include:

- the present-day continents and oceans
- mountain ranges on land
- deep trenches and volcanic ridges under the oceans.

An early theory suggested these features were the result of the crust shrinking as the Earth cooled down. The modern theory involves the movement of tectonic plates.

Describe the evidence which led scientists to reject the earlier theory in favour of the modern one. **(5)**

*(AQA, GCSE, Paper 2372, June 1999)*

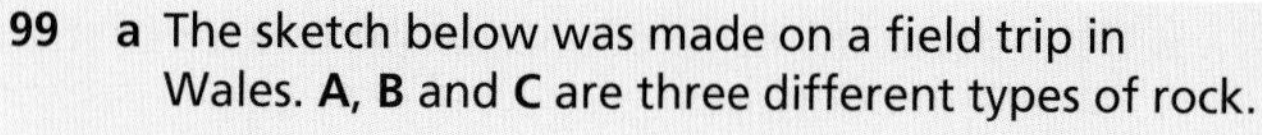

**99** **a** The sketch below was made on a field trip in Wales. **A**, **B** and **C** are three different types of rock.

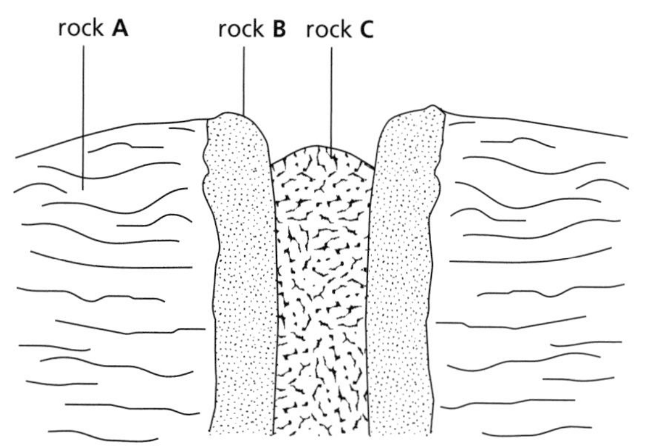


The diagrams below show what the three rock types look like under a microscope.

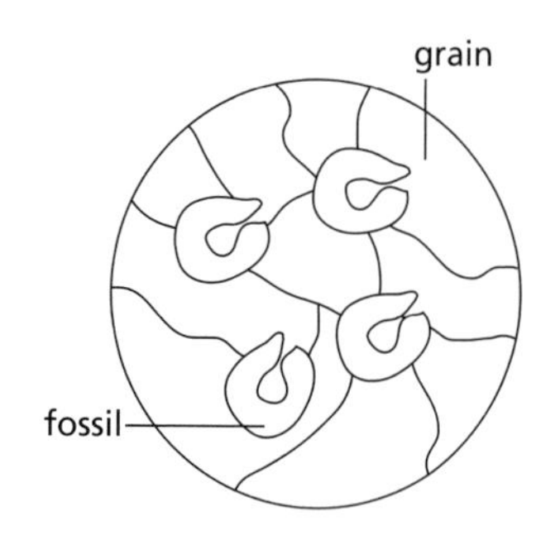


**A**

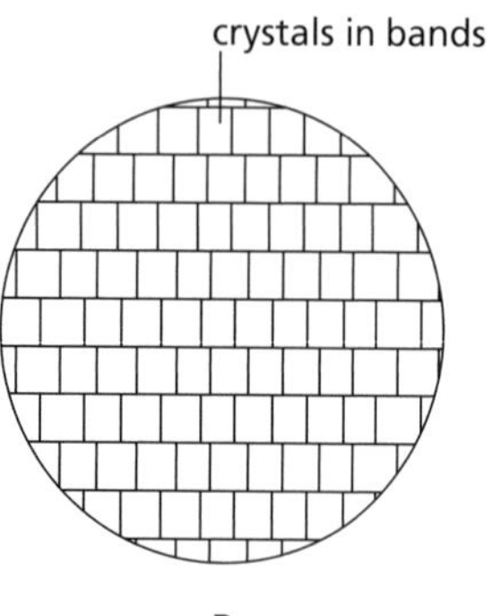

**B**

**C**

(i) Using the words in the box below complete a copy of the table that follows:

**granite** **igneous** **limestone** **marble** **metamorphic** **sedimentary**

| Rock | Type of rock | Name of rock |
|---|---|---|
| A | | |
| B | | |
| C | | |

(ii) Place the rocks, **A**, **B** and **C** in order of their age, with youngest first. **(1)**

**b** Describe how the following rock types are formed.

(i) Sedimentary **(2)**

(ii) Metamorphic **(2)**

(*WJEC, GCSE, 0125/2, June 1999*)

**100** The Earth's crust is cracked into a number of large pieces which are moving very slowly.

**a** Give the name for these large pieces. **(1)**

**b** State the result of these large pieces slowly moving

(i) apart **(1)**

(ii) towards each other. **(1)**

(*WJEC, GCSE, 0125/2, June 1999*)

**101** Two samples of rock from different parts of a volcanic island had the same chemical composition but different crystal sizes.

**a** Name this type of rock and describe how it was formed, accounting for the difference between the samples. **(4)**

**b** Analysis of another rock showed that it contained an oxide of tin in which 3.57 g of tin was combined with 0.96 g of oxygen.

i) Calculate the empirical formula of the tin oxide present in the rock.
(Relative atomic masses: O = 16, Sn = 119.) **(3)**

(ii) The melting point of the tin oxide was found to be over 1000 °C. Explain why tin oxide has a high melting point and suggest the type of structure it has. **(2)**

(*Edexcel, GCSE, Paper 3H, June 1999*)

**102** Write a **brief** account of **three** of the following. Relevant chemical equations and/or diagrams *may* be included in your answer where appropriate.

**a** Explain how addition polymers can be made from alkenes.

Discuss whether the advantages of plastics in everyday life outweigh the disadvantages. **(5)**

**b** Describe and explain the purification of copper. How do the uses of copper in everyday life relate to its properties? **(5)**

**c** Describe how the appearance of rocks is used as evidence to explain their formation and hence to classify rock type. **(5)**

**d** Outline the manufacture of sulphuric acid. **(5)**

(*WJEC, GCSE, June 2000*)

**103** The diagram below shows one of the plates under the Pacific Ocean. It is always moving, very slowly, towards and under the South American land mass.

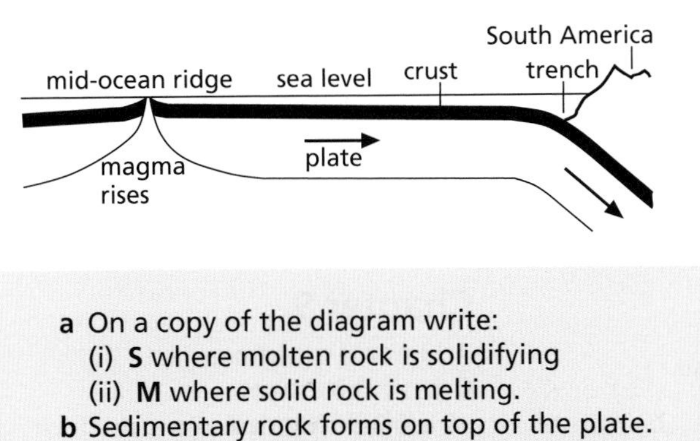


**a** On a copy of the diagram write:
(i) **S** where molten rock is solidifying **(1)**
(ii) **M** where solid rock is melting. **(1)**

**b** Sedimentary rock forms on top of the plate.
Describe how sedimentary rock is formed. **(3)**

*(Edexcel, GCSE, Paper 2F/1F, June 2000)*

**104** Write a **brief** account of **three** of the following. Relevant chemical equations and/or diagrams *may* be included in your answer where appropriate.

**a** Outline the industrial extraction of aluminium.
State and explain the major factors that determine the siting of a new aluminium extraction plant. **(5)**

**b** Outline the similarities and differences between alkanes and alkenes.
State and explain the importance of alkanes and alkenes in everyday life. **(5)**

**c** Substances can be classified as simple molecular or giant ionic. Give an example for each type of structure and show how the physical properties are related to each structure. **(5)**

**d** State and explain how ethanol can be obtained from sugar.
Discuss the dangers of alcohol abuse. **(5)**

*(WJEC, GCSE, 0125/2, June 1999)*

# Answers to numerical questions

## Chapter 1

### Gas laws
3 54.27 $cm^3$
4 23.81 $cm^3$
5 14.35 $cm^3$

### Additional questions
8 214.6 °C

## Chapter 5

### Calculating moles
1 a 0.10 mole
b 0.16 mole
c 2 moles
2 a 3.20 g
b 160 g
c 5.75 g
3 a 0.10 mole
b 1 mole
c 10 moles
4 a 162 g
b 8.50 g
c 55.83 g
5 a 0.08 mole
b 10 moles
c 0.0008 mole
6 a 7.20 $dm^3$
b 2.40 $dm^3$
c 48 $dm^3$
7 a 2 mol $dm^{-3}$
b 0.20 mol $dm^{-3}$
8 a 7.98 g
b 40.40 g

### Calculating formulae
1 CaO
2 Empirical formula = $CH_3$
Molecular formula = $C_2H_6$

### Moles and chemical equations
1 32 g
2 320 g
3 127 g
4 24 $dm^3$
5 500 $cm^3$
6 0.27 mol $dm^{-3}$

### Additional questions
1 a (i) 71 g
(ii) 72 g
b (i) 74 g
(ii) 8.5 g
2 a 300 $dm^3$
b 3.60 $dm^3$
3 a 0.0015 mole
b 6 moles
4 $SiO_2$
5 a $CHO_2$
b $C_2H_2O_4$
6 d 448 tonnes
7 a VuCl
8 c 0.002 mole
d 48 $cm^3$
e 40 $cm^3$

## Chapter 6

### Calculations in electrolysis
2 a 2 faradays
b 1 faraday
c 3 faradays
d 2 faradays
3 a 193 000 coulombs
b 96 500 coulombs
c 193 000 coulombs
4 0.2 faradays

### Additional questions
4 27 971 s (466.18 min)
5 216 $dm^3$
6 0.15 faradays

## Chapter 7

### Crystal hydrates
1 a 36.07%
b 62.94%
c 36.29%

### Solubility of salts in water
1 a 24.2 g per 100 g of water
b 18.5 g per 100 g of water
c 32.4 g per 100 g of water
3 a 56.5 g per 100 g of water
b 73.8 g per 100 g of water
c 102.3 g per 100 g of water
4 a 17.3 g (± 0.50 g)
b 45.8 g (± 0.50 g)
c 28.5 g (± 0.50 g)
5 6.25 g (± 0.50 g)

### Titration
1 0.19 mol $dm^{-3}$
2 0.18 mol $dm^{-3}$

### Additional questions
4 c 2.33 mol $dm^{-3}$
6 b (i) 34.0 g per 100 g of water
41.0 g per 100 g of water
65.0 g per 100 g of water
90.0 g per 100 g of water
(ii) 31.0 g (± 0.50 g)
(iii) 7 g (± 0.50 g)
8 e (i) 51.22%
(ii) 0.07 mole of water
0.01 mole of magnesium sulphate
(iii) $MgSO_4.7H_2O$

## Chapter 8

### Additional questions
5 e (iii) 2.40 $dm^3$
8 a 110 tonnes

## Chapter 9

### Additional questions
6 c (i) 4.8 tonnes
(ii) 9.2 tonnes
8 c 1590 tonnes

## Chapter 10

### Additional questions
1 b (i) 42.9 $cm^3$
(ii) 21.45%
5 c 28 720 s (or 478.67 min)
d (ii) 103 400 tonnes
6 a 184 000 tonnes
b $3.09 \times 10^{12}$ $dm^3$
7 a $4.4 \times 10^{10}$ $dm^3$

## Chapter 11

### Surface area
2 c 26 $cm^3$ (± 0.5 $cm^3$)
d 1 min 51 s (± 3 s)

### Additional questions
2 f 46 s (± 1 s)
g 43 s (± 1 s)
4 g 12 min
6 c 0.309 g
7 e (i) 6.16 g
(ii) 3.36 $dm^3$
(iii) 13.44 $dm^3$

## Chapter 12

### Alkanes
1 a 98.5 °C
b 126 °C

### Other uses of alkanes
2 a 1 mole $CH_4$ : 1 mole $Cl_2$
b 1 mole $CH_4$ : 4 moles $Cl_2$

### Alkenes
a 54 °C

### Additional questions
7 a CH
b $C_6H_6$

## Chapter 13

### Chemical energy

1 a $-1161\ kJ\ mol^{-1}$
  d ethanol 25 kJ
    heptane 49.5 kJ
2 a 364 kJ
  b 3640 kJ
  c 182 kJ
3 a 114 kJ
  b 14.25 kJ
  c 114 kJ

### Additional questions

1 b (i) $-1461\ kJ\ mol^{-1}$
4 b (iii) $-114$ kJ
  c (iii) 143.75 kJ
5 a 0.50 g
  b 10 080 J
  c 10.08 kJ
  d 20.16 kJ
  e 46 g
  f 927.36 kJ
6 b 100 g
  c 2100 J (2.10 kJ)
  d 0.05 mol
  e 42 000 J (42 kJ)
8 d (i) 110 g
    (ii) 1110 kJ
    (iii) 555 kJ
    (iv) 185 000 kJ

## Chapter 14

### Biotechnology

2 a whisky/brandy: $400\ cm^3$
  b 35%
    $350\ cm^3$
  c 16 litres ($dm^3$)

### Additional questions

5 d (i) 363 kJ
    2904 kJ
    90.75 kJ
    (ii) $12\ dm^3$

## Chapter 15

### Artificial fertilisers

1 Ammonium nitrate: 35.00%
  Ammonium phosphate: 28.19%
  Ammonium sulphate: 21.21%
  Urea: 46.67%

### Additional questions

4 a Sodium nitrate: 16.47%
    Potassium nitrate: 13.86%
  e 39.38 kJ
5 d (i) 32%
    (ii) 19%
  e 25%

6 a 10.25 10.1 10.1 10.15
  b $10.12\ cm^3$
  c (ii) KOH: 1 mole
    $HNO_3$: 1 mole
  d (i) 0.0038 mole
    (ii) 0.0038 mole
  e $0.37\ mol\ dm^{-3}$

## Chapter 16

### Additional questions

3 b 7.5 tonnes
  c 15 tonnes
  d $5.6 \times 10^6\ dm^3$
4 b (i) 5%
    (ii) 31%
  d (i) $3 \times 10^6$ tonnes
    (ii) 550 000 tonnes
    (iii) 900 000 tonnes
6 b 0.01 mole
  c 0.005 mole
  d $0.2\ mol\ dm^{-3}$ (or 0.2 M)

# Periodic table

| Period | Group 1 | Group 2 | | | | | | | |
|---|---|---|---|---|---|---|---|---|---|
| 1 | | | | | $^{1}_{1}\text{H}$ Hydrogen | | | | |
| 2 | $^{7}_{3}\text{Li}$ Lithium | $^{9}_{4}\text{Be}$ Beryllium | | | | | | | |
| 3 | $^{23}_{11}\text{Na}$ Sodium | $^{24}_{12}\text{Mg}$ Magnesium | | | | | | | |
| 4 | $^{39}_{19}\text{K}$ Potassium | $^{40}_{20}\text{Ca}$ Calcium | $^{45}_{21}\text{Sc}$ Scandium | $^{48}_{22}\text{Ti}$ Titanium | $^{51}_{23}\text{V}$ Vanadium | $^{52}_{24}\text{Cr}$ Chromium | $^{55}_{25}\text{Mn}$ Manganese | $^{56}_{26}\text{Fe}$ Iron | $^{59}_{27}\text{Co}$ Cobalt |
| 5 | $^{85}_{37}\text{Rb}$ Rubidium | $^{88}_{38}\text{Sr}$ Strontium | $^{89}_{39}\text{Y}$ Yttrium | $^{91}_{40}\text{Zr}$ Zirconium | $^{93}_{41}\text{Nb}$ Niobium | $^{96}_{42}\text{Mo}$ Molybdenum | $^{99}_{43}\text{Tc}$ Technetium | $^{101}_{44}\text{Ru}$ Ruthenium | $^{103}_{45}\text{Rh}$ Rhodium |
| 6 | $^{133}_{55}\text{Cs}$ Caesium | $^{137}_{56}\text{Ba}$ Barium | | $^{178.5}_{72}\text{Hf}$ Hafnium | $^{181}_{73}\text{Ta}$ Tantalum | $^{184}_{74}\text{W}$ Tungsten | $^{186}_{75}\text{Re}$ Rhenium | $^{190}_{76}\text{Os}$ Osmium | $^{192}_{77}\text{Ir}$ Iridium |
| 7 | $^{223}_{87}\text{Fr}$ Francium | $^{226}_{88}\text{Ra}$ Radium | | $^{261}_{104}\text{Rf}$ Rutherfordium | $^{262}_{105}\text{Db}$ Dubnium | $^{263}_{106}\text{Sg}$ Seaborgium | $^{262}_{107}\text{Bh}$ Bohrium | $^{269}_{108}\text{Hs}$ Hassium | $^{268}_{109}\text{Mt}$ Meitnerium |

| | | | | | |
|---|---|---|---|---|---|
| $^{139}_{57}\text{La}$ Lanthanum | $^{140}_{58}\text{Ce}$ Cerium | $^{141}_{59}\text{Pr}$ Praseodymium | $^{144}_{60}\text{Nd}$ Neodymium | $^{147}_{61}\text{Pm}$ Promethium | $^{150}_{62}\text{Sm}$ Samarium |
| $^{227}_{89}\text{Ac}$ Actinium | $^{232}_{90}\text{Th}$ Thorium | $^{231}_{91}\text{Pa}$ Protactinium | $^{238}_{92}\text{U}$ Uranium | $^{237}_{93}\text{Np}$ Neptunium | $^{244}_{94}\text{Pu}$ Plutonium |

**Key**

- reactive metals
- transition metals
- poor metals
- metalloids
- non-metals
- noble gases

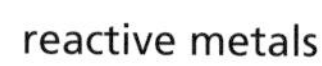

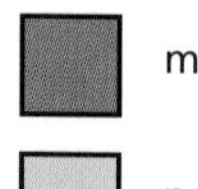

| | | | 3 | 4 | 5 | 6 | 7 | 0 |
|---|---|---|---|---|---|---|---|---|
| | | | | | | | | $^{4}_{2}$He Helium |
| | | | $^{11}_{5}$B Boron | $^{12}_{6}$C Carbon | $^{14}_{7}$N Nitrogen | $^{16}_{8}$O Oxygen | $^{19}_{9}$F Fluorine | $^{20}_{10}$Ne Neon |
| | | | $^{27}_{13}$Al Aluminium | $^{28}_{14}$Si Silicon | $^{31}_{15}$P Phosphorus | $^{32}_{16}$S Sulphur | $^{35.5}_{17}$Cl Chlorine | $^{40}_{18}$Ar Argon |
| $^{59}_{28}$Ni Nickel | $^{63.5}_{29}$Cu Copper | $^{65}_{30}$Zn Zinc | $^{70}_{31}$Ga Gallium | $^{73}_{32}$Ge Germanium | $^{75}_{33}$As Arsenic | $^{79}_{34}$Se Selenium | $^{80}_{35}$Br Bromine | $^{84}_{36}$Kr Krypton |
| $^{106}_{46}$Pd Palladium | $^{108}_{47}$Ag Silver | $^{112}_{48}$Cd Cadmium | $^{115}_{49}$In Indium | $^{119}_{50}$Sn Tin | $^{122}_{51}$Sb Antimony | $^{128}_{52}$Te Tellurium | $^{127}_{53}$I Iodine | $^{131}_{54}$Xe Xenon |
| $^{195}_{78}$Pt Platinum | $^{197}_{79}$Au Gold | $^{201}_{80}$Hg Mercury | $^{204}_{81}$Tl Thallium | $^{207}_{82}$Pb Lead | $^{209}_{83}$Bi Bismuth | $^{209}_{84}$Po Polonium | $^{210}_{85}$At Astatine | $^{222}_{86}$Rn Radon |
| $^{269}_{110}$Uun Ununnilium | $^{272}_{111}$Uuu Unununium | $^{277}_{112}$Uub Ununbium | | $^{285}_{114}$Uuq Ununquadium | | $^{289}_{116}$Uuh Ununhexium | | $^{293}_{118}$Uno Ununoctium |

| | | | | | | | | |
|---|---|---|---|---|---|---|---|---|
| $^{152}_{63}$Eu Europium | $^{157}_{64}$Gd Gadolinium | $^{159}_{65}$Tb Terbium | $^{162}_{66}$Dy Dysprosium | $^{165}_{67}$Ho Holmium | $^{167}_{68}$Er Erbium | $^{169}_{69}$Tm Thulium | $^{173}_{70}$Yb Ytterbium | $^{175}_{71}$Lu Lutetium |
| $^{243}_{95}$Am Americium | $^{247}_{96}$Cm Curium | $^{247}_{97}$Bk Berkelium | $^{251}_{98}$Cf Californium | $^{252}_{99}$Es Einsteinium | $^{257}_{100}$Fm Fermium | $^{258}_{101}$Md Mendelevium | $^{259}_{102}$No Nobelium | $^{260}_{103}$Lr Lawrencium |

# Acknowledgements

The authors would like to thank Irene, Katharine, Michael and Barbara for their never-ending patience and encouragement throughout the production of this textbook. In addition, great thanks is given to Mr Dennis Richards, Headteacher, St Aidan's Church of England High School, Harrogate, for his help and support.

## Examination questions

Exam questions have been reproduced with kind permission from the following examination boards:

AQA (Assessment and Qualifications Alliance)
AEB (Associated Examining Board)
NEAB (Northern Examining and Assessment Board)
SEG (Southern Examining Board)
CCEA (Northern Ireland Council for the Curriculum, Examinations and Assessment)
Edexcel
OCR (Oxford, Cambridge and RSA Examinations)
WJEC (Welsh Joint Education Committee)

## Source acknowledgements

**pp.58, 59, 60, 61, 175, 178, 179, 197, 198, 199, 203**
The molecular models shown were made using the Molymod® system available from Molymod® Molecular Models, Spiring Enterprises Limited, Billingshurst, West Sussex RH14 9NF England

## Photo credits

Thanks are due to the following for permission to reproduce copyright photographs:

**Cover** Peter Aprahamian/Science Photo Library; **p.1** *l* John Townson/Creation, *tr* Colorsport, *br* Powerstock/Zefa; **p.2** *l* John Townson/Creation, *tr* Andrew Lambert, *br* Science Photo Library/David Scharf; **p.3** *tl* Science Photo Library/James King-Holmes, *bl* Science Photo Library/Andrew Syred; **p.4** *tr* Science Photo Library/Michael W Davidson, *br* John Townson/Creation; **p.5** Redferns; **p.6** *all* Andrew Lambert; **p.7** *both* John Townson/Creation; **p.9** *t* Mary Evans Picture Library, *b* Powerstock/Zefa; **p.13** *l* Science Photo Library, *r* Mary Evans Picture Library; **p.14** *tr & cr* Ace Photo Library, *br* John Townson/Creation; **p.15** *tl* Science Photo Library/Ron Sutherland, *cl* Bridgeman Art Library/Christie's Images, *bl* John Townson/Creation; **p.16** Andrew Lambert; **p.17** Andrew Lambert; **p.18** Science Photo Library/Simon Fraser/ Searle Pharmaceuticals; **p.19** *l* Science Photo Library/Geoff Lane/CSIRO, *tr* Powerstock/Zefa, *cr & br* Andrew Lambert; **p.21** *both* Andrew Lambert; **p.22** *tl & bl* Science Photo Library, *tr* Andrew Lambert, *br* Bruce Coleman; **p.23** *l* Andrew Lambert, *tr* Zefa, *br* Still Pictures/Al Grillo; **p.24** *both* Andrew Lambert; **p.25** *tl* Rex Features, *bl* BOC Gases, *tr* Science Photo Library/Alex Bartel, *br* Andrew Lambert; **p.26** *l* Andrew Lambert, *tr* Science Photo Library/J C Revy, *br* Bruce Coleman; **p.27** *l* Science Photo Library/Geoff Tompkinson, *r (all)* John Townson/Creation; **p.28** *tl & tr* John Townson/Creation, *cr & br* Andrew Lambert; **p.29** Science Photo Library/Dr Jeremy Burgess, *tr* Science Photo Library/Eye of Science, *br* The Laser Centre, Banbury; **p.30** *l* Science Photo Library/Manfred Kage, *r* Biophoto Associates; **p.34** Science Photo Library/Mehau Kulyk; **p.39** Science Photo Library; **p.41** *tl & cr* John Townson/Creation, *bl* Powerstock/Zefa, *tr* Debby Moxon, *br* Science Photo Library/Mehau Kulyk; **p.42** Andrew Lambert; **p.43** *tl & bl* Andrew Lambert, *r* Science Photo Library/Charles D Winters; **p.44** Andrew Lambert; **p.45** *l* Andrew Lambert, *r* John Townson/Creation; **p. 46** *all* Andrew Lambert; **p.47** *tl, bl & tr* John Townson/Creation, *br* Trip Photo Library; **p.48** *both* Science Photo Library; **p.49** *tl, tr, cr & br* John Townson/ Creation, *bl* Rex Features; **p.50** *tl* Andrew Lambert, *bl* John Townson/Creation; **p.53** *tl & bl* John Townson/Creation, *tr (both)* Andrew Lambert, *br* John Townson/Creation; **p.56** Science Photo Library/Science Source; **p.58** Andrew Lambert; **p.59** Andrew Lambert; **p.60** *both* Andrew Lambert; **p.61** Andrew Lambert; **p.62** *tl* Science Photo Library/E R Degginger, *bl* Image Bank; **p.63** Science Photo Library/Philippe Plailly; **p.64** *l* Robert Harding, *r (top to bottom)* Science Photo Library/Richard Megna/Fundamental, Allsport, Science Photo Library/Sheila Terry, Science Photo Library/Manfred Kage; **p.65** RIBA British Architectural Library; **p.66** *l* John Townson/Creation, *r* Science Photo Library/NASA; **p.70** *l* Science Photo Library/Rosenfeld Images Ltd, *tr & br* Andrew Lambert; **p.71** Andrew Lambert; **p.76** Alan Murray Tilcon Ltd; **p.80** *l & br* John Townson/Creation, *tr* Robert Harding; **p.81** Colorific; **p.82** Rex Features; **p.83** *tl* Powerstock/Zefa, *bl* Rex Features, *r* Paul Brierley; **p.84** Zefa Pictures; **p.85** *l* Kind permission of ICI Chemicals & Polymers, *r* Andrew Lambert; **p.88** Andrew Lambert; **p.90** Science Photo Library/Adam Hart Davis; **p.91** *tl & bl* Andrew Lambert, *r* Mary Evans Picture Library; **p.95** *both* John Townson/Creation; **p.96** *tl, bl & tr* Andrew Lambert, *br* Science Photo Library/Jean Loup Charmet; **p.97** *l* Science Photo Library/Sheila Terry, *r* Andrew Lambert; **p.98** Robert Harding; **p.100** *tl & bl* John Townson/Creation, *r* Andrew Lambert; **p.101** *both* Andrew Lambert; **p.102** *all* Andrew Lambert; **p.103** *tl & bl* John Townson/Creation, *r* Andrew Lambert; **p.104** *both* Andrew Lambert; **p.105** *both* Andrew Lambert; **p.112** Science Photo Library/Philippe Plailly; **p.113** *tl & r* John Townson/Creation, *bl* Science Photo Library/Martin Land; **p.114** *l* Roger Scruton, *tr* Rex Features, *br* Robert Harding; **p.115** Alan Murray, Tilcon Ltd; **p.116** *all* Andrew Lambert; **p.117** *l (left)* Science Photo Library/Roberto de Gugliemo, *l (right)* GeoScience Features, *r* Science Photo Library/Maximillian Stock Ltd; **p.118** *l* Andrew Lambert, *tr* John Townson/Creation, *br* Science Photo Library/Sheila Terry; **p.119** *l* Powerstock/Zefa, *r* Andrew Lambert; **p.120** Andrew Lambert; **p.122** Andrew Lambert; **p.123** Andrew Lambert; **p.127** *l* Andrew Lambert, *tr* Powerstock/Zefa, *br* Bruce Coleman; **p.128** Andrew Lambert; **p.129** *tr* Morso, *cr* Anthony Blake Photo Library, *br* British Airways; **p.130** *tr* Andrew Lambert, *br* Robert Harding; **p.131** *all* Andrew Lambert; **p.132** *all* GeoScience Features; **p.133** Corus; **p.134** *tr* Science Photo Library/Dr Jeremy Burgess, *br* Science Photo Library/Martin Bond; **p.135** *all* John Townson/Creation; **p.136** *both* Bruce Coleman; **p.137** *tl* Science Photo Library/Hank Morgan, *bl* Zefa; **p.138** *tl* Scotland in Focus, *cl, bl, tr & br* John Townson/Creation; **p.139** *l* Paul Brierley, *r* Planet Earth/K Scholey; **p.140** *l* John Townson/Creation, *r* Rex Features; **p.141** Science Photo Library/Maximillian Stock Ltd;

**p.142** *l* Robert Harding, *r* Science Photo Library/Manfred Kage; **p.146** *l* Zefa, *tr* Science Photo Library/Kent Wood, *br* Science Photo Library/O I Miller; **p.147** *tr* Powerstock/Zefa, *br* John Townson/Creation; **p.149** Andrew Lambert; **p.150** *l* BOC Gases, *r* Science Photo Library/Penny Tweedie; **p.151** *tl, bl & tr* John Townson/Creation, *cr* Science Photo Library/Alexander Tsiaras, *br* Science Photo Library/Vaughan Fleming; **p.153** *tl* Colorsport, *bl* John Townson/Creation; **p.155** *tl* Planet Earth/Sean Avery, *bl* Powerstock/Zefa; **p.157** Planet Earth/Chris Howes; **p.161** *tl* Holt Studio, *bl* Roger Scruton, *tr & br* Zefa; **p.162** *all* Andrew Lambert; **p.163** *both* Andrew Lambert; **p.166** Science Photo Library/Astrid & Hanns-Frieder Michler; **p.167** *l* Science Photo Library/Alfred Pasieka, *tr* John Townson/Creation, *br* National Medical Slide Bank; **p.168** *l* Science Photo Library/Rosenfeld Images Ltd, *r* Anthony Blake Photo Library; **p.172** *tl, bl, tr, cr & br* John Townson/Creation; **p.173** *tl* Andrew Lambert, *cl, bl, cr & br* John Townson/Creation, *tr* Science Photo Library; **p.174** Science Photo Library/Stevie Grand; **p.175** *all* Andrew Lambert; **p.176** *tr* British Gas, *br* Andrew Lambert; **p.177** *both* John Townson/Creation; **p.178** *l* Robert Harding Picture Library; *r (all)* Andrew Lambert; **p.179** Andrew Lambert; **p.180** *all* Andrew Lambert; **p.184** *tr* GeoScience Features, *br* Zefa; **p.185** *l* Science Photo Library/Richard Folwell, *r* Science Photo Library/Martin Bond; **p.186** Science Photo Library /Martin Bond; **p.187** *l* Science Photo Library/Novosti, *r* Science Photo Library/John Mead; **p.188** Andrew Lambert; **p.189** *tl* Science Photo Library/Philippe Plailly, *bl* Bruce Coleman, *r* Science Photo Library/David Hall; **p.192** Andrew Lambert; **p.193** Science Photo Library/NASA; **p.197** *both* Andrew Lambert; **p.198** *both* Andrew Lambert; **p.199** *all* Andrew Lambert; **p.200** *tr* Science Photo Library/James Holmes/Zedcor, *br (both)* John Townson/Creation; **p.201** *tr* Ecoscene, *br* Andrew Lambert; **p.202** Andrew Lambert; **p.203** *all* Andrew Lambert; **p.204** *tl* Science Photo Library/Biophoto Associates, *bl* Science Photo Library, *r (both)* Andrew Lambert; **p.205** *tl* Rex Features, *bl* Bruce Coleman, *r* Rex Features; **p.206** *tl* Science Photo Library, *bl* Rex Features/The Times, *r* Andrew Lambert; **p.207** *all* Andrew Lambert; **p.208** *tl* Bruce Coleman, *bl* John Townson/Creation; **p.209** *both* John Townson/Creation; **p.211** *all* Andrew Lambert; **p.212** *tl* Science Photo Library/Richardson, *bl* Science Photo Library/James King Holmes; **p.213** Rex Features; **p.222** Andrew Lambert; **p.223** Science Photo Library/Malcom Fielding, Johnson Matthey plc; **p.224** Andrew Lambert; **p.228** *tl* GeoScience Features, *bl* Andrew Lambert; **p.230** Still Pictures; **p.231** *tl & bl* Andrew Lambert, *r* Kind Permission of ICI Chemicals & Polymers; **p.233** *all* Andrew Lambert; **p.238** *l* Popperfoto/Reuters, *r* Bruce Coleman; **p.239** *tr* Ecoscene, *br* GeoScience Features; **p.240** *tl* GeoScience Features, *bl & tr* Zefa, *br* Rex Features; **p.242** Andrew Lambert; **p.243** *tl, tr & br* Planet Earth, *bl* John Townson/Creation; **p.245** *both* Rex Features; **p.246** *all* By permission of Director of British Geological Surveys; **p.247** California Institute of Technology; **p.253** Science Museum/Science & Society Picture Library.

*t* = top, *b* =bottom, *l* = left, *r* = right, *c* = centre.

Every effort has been made to contact copyright holders, and the publishers apologise for any omissions which they will be pleased to rectify at the earliest opportunity.

# Index